AF894936

Ecological Studies

Analysis and Synthesis

Volume 228

Series editors
Martyn M. Caldwell
Logan, Utah, USA

Sandra Díaz
Cordoba, Argentina

Gerhard Heldmaier
Marburg, Germany

Robert B. Jackson
Durham, North Carolina, USA

Otto L. Lange
Würzburg, Germany

Delphis F. Levia
Newark, Delaware, USA

Harold A. Mooney
Stanford, California, USA

Ernst-Detlef Schulze
Jena, Germany

Ulrich Sommer
Kiel, Germany

Ecological Studies is Springer's premier book series treating all aspects of ecology. These volumes, either authored or edited collections, appear several times each year. They are intended to analyze and synthesize our understanding of natural and managed ecosystems and their constituent organisms and resources at different scales from the biosphere to communities, populations, individual organisms and molecular interactions. Many volumes constitute case studies illustrating and synthesizing ecological principles for an intended audience of scientists, students, environmental managers and policy experts. Recent volumes address biodiversity, global change, landscape ecology, air pollution, ecosystem analysis, microbial ecology, ecophysiology and molecular ecology.

More information about this series at http://www.springer.com/series/86

Ramesh D. Gulati • Egor S. Zadereev •
Andrei G. Degermendzhi
Editors

Ecology of Meromictic Lakes

Springer

Editors
Ramesh D. Gulati
Netherlands Institute of Ecology (NIOO)
Wageningen, The Netherlands

Egor S. Zadereev
SB RAS
Institute of Biophysics
Krasnoyarsk, Russia

Andrei G. Degermendzhi
SB RAS
Institute of Biophysics
Krasnoyarsk, Russia

ISSN 0070-8356 ISSN 2196-971X (electronic)
Ecological Studies
ISBN 978-3-319-49141-7 ISBN 978-3-319-49143-1 (eBook)
DOI 10.1007/978-3-319-49143-1

Library of Congress Control Number: 2016963751

© Springer International Publishing AG 2017
This work is subject to copyright. All rights are reserved by the Publisher, whether the whole or part of the material is concerned, specifically the rights of translation, reprinting, reuse of illustrations, recitation, broadcasting, reproduction on microfilms or in any other physical way, and transmission or information storage and retrieval, electronic adaptation, computer software, or by similar or dissimilar methodology now known or hereafter developed.
The use of general descriptive names, registered names, trademarks, service marks, etc. in this publication does not imply, even in the absence of a specific statement, that such names are exempt from the relevant protective laws and regulations and therefore free for general use.
The publisher, the authors and the editors are safe to assume that the advice and information in this book are believed to be true and accurate at the date of publication. Neither the publisher nor the authors or the editors give a warranty, express or implied, with respect to the material contained herein or for any errors or omissions that may have been made. The publisher remains neutral with regard to jurisdictional claims in published maps and institutional affiliations.

Printed on acid-free paper

This Springer imprint is published by Springer Nature
The registered company is Springer International Publishing AG
The registered company address is: Gewerbestrasse 11, 6330 Cham, Switzerland

Preface

This book is a not only a state-of-the-art book on Ecology of Meromictic Lakes but is also perhaps the first detailed published record focused exclusively on such lakes. Geller et al. (2011) had aptly remarked in the preface of a Springer book dealing with Acid Pit Lakes that most of these pit lakes appear to become meromictic during the course of their development. I (RDG) as one of the three editors of this book was very happy to get an invitation from Walter Geller late in 2012 to review their detailed book on the Acid Pit Lakes in the SIL newsletter. (This SIL newsletter is both a scientific and social forum for the limnologists and in fact all those interested in limnology the world over.) While reviewing the book on Acid Pit Lakes by Geller et al., it had struck me that meromictic lakes were an under-represented subject area of the lake limnology. My feelings were substantiated when I was invited by my Russian colleagues (Prof. Andrei Degermendzhi and Dr. Egor Zadereev, my co-editors of this book) to join them as a co-editor of this planned Springer book, the first one on the meromictic lakes. The most important task, also the most difficult one of choosing the book structure, was the format and chapter divisions and chapter authors. Both these aspects were taken care of by my Russian editorial counterpart (EZ and AD). Next, we decided to have the book to comprise three parts with 13 chapters.

The introduction (Chap. 1) is followed by Part I. Part I is a more general part including three concise chapters, one each on physical features, chemical features and the biology of meromictic lakes. These four chapters provide state-of-the-art information based on recent scientific literature on meromictic lakes. Strangely, we detected a striking bias in the expanse of scientific literature from Europe and North America with relatively the most literature on one hand and the remaining five continents with much less or no information.

Part II of the book has eight chapters, comprising scientific accounts of more than a dozen meromictic lakes, which are included here as case studies. The information ranges from one lake per chapter to several lakes in some other chapters. Chapter 12 that deals with lakes of a wide range of area, volume and magnitude in Mexico departs in style and contents from the preceding chapters.

This chapter is more of a critical review of the presence/absence of such lakes, with some information about meromictic lakes in the Mexico State, Central America. Unfortunately, information on the existence of meromictic lakes in countries on the main continent of South America is altogether lacking, let alone some data on the physico-chemical features or biology of such lakes. Much of the information in Part II is based partly on the published scientific literature and partly on the unpublished results. The authors of these Chaps. (5–12) are well known, having wide international experience on the limnology of meromictic lakes.

It was not an easy task to select lakes where the meromixis is supported by different processes and to cover various aspects of meromixis in the case-study chapters. We could not follow a geographical approach so as to cover all the continents. On the one hand, we tried to select meromictic lakes which have been comprehensively studied and, on the other hand, to include those lakes where research is ongoing. For example, we found in the literature that there were active studies in progress on microbiology of some meromictic lakes in Australia some years ago, but these lakes are being not so actively pursued anymore. At the first round of book preparation, we contacted a wider community of meromictic lake scientists. However, the final selection was also determined by the speed of response by the potential book authors and their desire to actively participate. Studies on meromictic lakes are intensively developed in several cases. For example, for Ace lake in Antarctica, recent, detailed publication on food-web structure appeared (Laybourn-Parry and Bell 2014) after we had already started with the preparation of the book. So, we realise that there are other important studies and well-studied meromictic lakes that are not included in the book as case study chapters. Nevertheless, we do refer to some such studies, e.g., on Lake Mahoney in Canada and Ace Lake in Antarctica in Chap. 4 on the biological aspects of meromictic lakes.

We feel that there are also some aspects of meromixis that probably should get more attention, but the researches on those topics are either ongoing or not well developed. Specifically, the hydrophysical and biogeochemical model predictions and simulations of regime shift from holomictic to meromictic conditions and vice versa (such shifts are described for Mono Lake in Chap. 11 and for Lake Rogoznica in Chap. 6) are extremely important from both the theoretical and practical limnology vewpoints. We discuss briefly in Chap. 9 that mathematical modelling is important for such predictions and enumerate some general considerations about the methodology of such simulations. This topic would require a separate chapter, but probably the researches on different lakes are not equally deep enough. We hope that this aspect will be studied more intensively in the near future, and the theoretical research on the stability of meromixis will improve our knowledge further on the mixing regime shifts in lakes.

Last of all, Part III of the book that deals with some General Conclusions (Chap. 13 by RDG and EZ) is derived partly from the summary conclusions provided by the authors of the foregoing 12 chapters to the editors.

Considering that it was our maiden attempt to edit a book on meromictic lakes, it was not always feasible to stick to the deadlines. Therefore, the preparation of the book was delayed by some months. We also greatly appreciate the patience of the Springer staff (Andrea Schlitzberger and colleagues) dealing with the administrative part of the book preparation.

Last but not least, as Editors we are highly grateful to both the chapter authors (47 co-authors from 12 countries) and the reviewers for their cooperation and help.

Wageningen, The Netherlands — Ramesh D. Gulati
Krasnoyarsk, Russia — Egor S. Zadereev
Krasnoyarsk, Russia — Andrei G. Degermendzhi

References

Geller W, Schultze M, Kleinmann R, Wolkersdorfer C (eds) (2012) Acidic pit lakes: the legacy of coal and metal surface mines. Springer (Reviewed by R.D. Gulati (2013), SILnewsletter, Vol 62, pp 23–26)

Laybourn-Parry J, Bell EM (2014) Ace Lake: three decades of research on a meromictic, Antarctic Lake. Polar Biol 37:1685–1699

Contents

Chapter 1
Introduction: Meromictic Lakes, Their Terminology and Geographic Distribution

Egor S. Zadereev, Bertram Boehrer, and Ramesh D. Gulati

Meromixis has never really been main focus of the limnological research. However, many workers have felt the need during the last decades, and some recent scientific advances demonstrate the need to understand these meromictic lake ecosystems and their unique problems more thoroughly than hitherto. We cite here half a dozen instances, which support our contention to pay more scientific attention to what is already known of meromictic lakes:

- The disaster in 1986 at Lake Nyos, and Lake Monoun in 1984, both in Cameroon (Africa), where large amounts of CO_2 gushed out under pressure from deeper waters and escaped forcefully from the lake surface killing about 1700 people in the lake vicinity (Kling et al. 1987). We need understanding and theory to predict and minimise probability or consequences of such catastrophic events.
- Opencast mining now in progress leaves depressions, which get filled with water. Many of these lakes tend to become meromictic, i.e. they do not circulate completely. To assess the water quality in such lakes, we need to understand the mechanism of meromixis (Böhrer et al. 1998; Schultze et al. 2016). We know that some lakes have been turned meromictic on purpose to improve surface water quality (e.g. Island Copper Mine, Canada: Fisher and Lawrence 2006).
- It is planned to exploit the rich methane deposits in deeper waters of meromictic Lake Kivu bordering the Democratic Republic of the Congo and Rwanda in Africa for power production without letting the surface waters to deteriorate. To

E.S. Zadereev (✉)
Institute of Biophysics SB RAS, Akademgorodok, Krasnoyarsk, Russia
e-mail: egor@ibp.ru

B. Boehrer
Helmholtz Centre for Environmental Research (UFZ), Magdeburg, Germany

R.D. Gulati
Netherlands Institute of Ecology (NIOO), Wageningen, The Netherlands

© Springer International Publishing AG 2017

R.D. Gulati et al. (eds.), *Ecology of Meromictic Lakes*, Ecological Studies 228,
DOI 10.1007/978-3-319-49143-1_1

minimize the environmental impact, transport processes for gases must be understood.

- Climate change might impact the stability of stratification and circulation patterns in many lakes and hence the lake food webs and ecosystems (Danis et al. 2004). In some cases, anthropogenic influences turn holomictic lakes to meromictic ones (e.g. Mono Lake: Melack and Jellison 1998; see also in Chap. 11). If a lake shifts to another mixing regime, the food web and biological communities are stressed. To predict and understand the response of ecosystem to altered mixing regime, comparative studies of meromictic and non-meromictic lakes are needed.
- Lake sediments from meromictic lakes represent some of the best climate records, which range from a few hundred years to more than ten thousand years. Moreover, the paleolimnological studies tell us a lot about the past climate and thus improve our understanding about meromixis.
- The different chemical settings in meromictic lakes support diverse microbial processes concerning carbon fixation and oxidation. A very obvious example is the anoxygenic photosynthesis by phototrophic sulphur bacteria. Examples that are more specific are the anaerobic oxidation of ammonia (Anammox) by obligately anaerobic chemolithoautotrophs or primary production by photoferrotrophic bacteria. Studies on these bacteria can provide us with a more all-inclusive understanding of different primary production processes and have high biotechnological value.

1.1 Terminology

We follow Hutchinson's definition of meromixis: "A lake in which [a chemically different] water remains partly or wholly unmixed with the main water mass at the circulation periods is said to be *meromictic*. Findenegg (1935), who introduced these terms, called the deep layer of a meromictic lakes the *monimolimnion*, and Hutchinson (1937) introduced the term *mixolimnion* for the part above the monimolimnion in which free circulation periodically can occur. The boundary layer between the mixolimnion and the monimolimnion is known as the *chemocline*" (Hutchinson 1957, for details see Chap. 2). With only some modifications, this definition still holds after 60 years.

It was necessary to add *a chemically different* at the beginning of the definition, as Hutchinson (1957) took for granted that lakes without full circulation would develop pronounced chemical gradients: he thus defined the transition between mixolimnion and monimolimnion as "the chemocline" and considered the monimolimnion as stagnant layer. In 1957, it was not known about the lakes that are permanently stratified but do not develop a pronounced chemical gradient. This comprises deep lakes with thermobaric stratification (stable density stratification which results from the temperature dependence of the compressibility of water; see Boehrer et al. 2014 for details) and lakes that generate deep water by partial deep-water renewal (e.g. Issyk-Kül: Peeters et al. 2003). Referring to chemical gradients in the definition also clearly distinguishes between meromictic lakes and oligomictic lakes (which experience a complete overturn once every few years—without developing a pronounced chemical gradient, e.g. Lago Maggiore: Ambrosetti and Barbanti 1999). Strictly speaking, here we depart from defining the term *meromictic* as opposite of the *holomictic* (see Hall and Northcote 2012).

If intervals of no overturn result in intermittent chemical deterioration of the deep water, it is prudent to call it a meromictic period or phase (e.g. Melack and Jellison 1998 and Chap. 11). Therefore, we removed the word "perennially" from the definition. This also confirms that we know that monimolimnia can vary in age (e.g. Kaden et al. 2010) and that its water and solutes are exchanged with the overlying mixolimnion. Consequently, monimolimnetic water can be younger or newer in existence than the meromixis in the lake (e.g. Vollmer et al. 2002). We have also eliminated the word "stagnant", as we know that there is also motion in the deep-water layers determined by the pressure differences with the surface waters and drag (fluid resistance) between layers. Monimolimnia receive inflows from the groundwater (crenogenic meromixis—see below), but also monimolimnetic water can creep into the groundwater space. Beyond this, also macroscopic vertical circulation within monimolimnia has been documented, e.g. by double diffusive convection—a form of convection driven by two different density gradients, which have different rates of diffusion (Newman 1976; Imboden and Wüest 1995), which also has been confirmed for temperate lakes (von Rohden et al. 2010) and which can comprise the entire monimolimnion in one convection layer without breaking or even terminating its meromictic nature (Boehrer et al. 2009).

Protected against direct gas exchange with the atmosphere, oxygen-depleting processes overcome oxygen production in the very low-light or dark conditions of monimolimnion. Thus, the redox potential becomes more negative in the monimolimnion, which is usually anoxic. This creates a special setting for chemical gradients in meromictic lakes, which can be colonized by diverse micro-zooplankton and microbial communities (see Chap. 4).

The classification of meromictic lakes by Hutchinson (1957) is based on processes that initiate the density stratification. This classification differentiates ectogenic, crenogenic and biogenic meromixis, i.e. meromixis caused by external inflows, entry of groundwater and degradation of organic material, respectively. Later, Walker and Likens (1975) refined Hutchinson's classification into two groups of ectogenic and endogenic meromictic lakes: ectogenic meromictic lakes remain permanently stratified due to inflow of different solute concentration, inflow of high turbidity or inflow of groundwater of higher salinity. Endogenic meromictic lakes are formed either by degradation of organic material in the deep water or by salt exclusion when ice is formed during winter. Hakala (2004) recognized meromixis that results from (1) flow/precipitation of saline water over freshwater or the other way round; (2) superficial diffuse nutrient load or turbidity currents, or both, from the catchment; (3) subsurface inflow of groundwater; and (4) inadequate mixing due to the lake morphology and surrounding topography. Boehrer and

Schultze (2008) listed all processes that are said to sustain meromixis (see also Chap. 2) and reasoned that as soon as meromictic conditions are established, more processes can be activated that contribute to keeping the stratification stable. Many lakes hence do not belong to just one class. Boehrer and Schultze (2008) conclude that all sustaining processes must be considered collectively. As long as these processes can balance mixing and diffusive effects that erode the stratification, the lake will remain meromictic. In Chap. 2, all the processes that have been documented to form and sustain meromixis are divided into two groups: (1) those which operate purely mechanically and (2) those in which geochemistry of lake waters is a controlling factor. It is important to recall that organisms also regulate many geochemical processes.

Meromixis traditionally has been regarded as a permanent setting. In some lakes, e.g. in Lake Salsvatn (Norway) (Bøyum 1973) and Powell Lake (Canada) (Sanderson et al. 1986), deep water was sealed off since the end of the last ice age. However, recent scientific contributions have shown a more dynamic picture of meromixis. Monimolimnia get eroded by mixing effects (e.g. Wallendorfer See and Rassnitzer: Boehrer et al. 2014); also seasonal variation in chemocline location has been quantitatively tracked (e.g. von Rohden et al. 2009; Zadereev et al. 2014; Nixdorf and Boehrer 2015). Holomictic lakes can turn meromictic due to climate change, and hence a new hydrologic situation can occur as for Lake Van, Turkey (Kaden et al. 2010), or due to the combined effect of climate and anthropogenic impacts (Lake Ikeda, Japan: Boehrer et al. 2008). On the other hand, such phases can also terminate (e.g. Dead Sea: Nissenbaum 1969; Imboden and Wüest 1995; Lake Mono: Melack and Jellison 1998). Hakala (2004) even regarded meromixis as a typical phase of lake formation on the Aland islands in the Baltic. Even if stratification is sustained continuously, exchange between mixolimnion and monimolimnion can occur by mixing across the chemocline, by seasonal progression and recession of the monimolimnion and by partial deep-water renewal (see Chap. 2). Hence, the age of monimolimnetic water does not necessarily have to comply with duration of the meromixis (e.g. Vollmer et al. 2002).

Finally, classifying a lake as meromictic has to be done carefully. It is clear that regular sampling during different seasons should confirm the presence or absence of mixing. Moreover, to classify a lake as meromictic, based on mixing regime (e.g. as dimictic or holomictic) is here not our aim per se. The lake once classified as meromictic can switch to another annual circulation pattern. Also, the age of monimolimnetic waters can be much younger than the duration of meromixis. What is relevant is to understand the current circulation pattern in the lake (e.g. meromixis during a specific time period, annual mixing, etc.) and based on regular monitoring either reaffirm it or refute it.

1.2 Worldwide Distribution of Meromictic Lakes

Hall and Northcote (2012) refer to the global distribution of meromictic lakes. They recognize 177 meromictic lakes on the Globe, located in nine geographical regions on five continents, with South America excluded. It is not surprising that most meromictic lakes are reported from North America (79 lakes) and Europe (28 lakes), since limnological studies are much more intensive and well recorded in these two continents. Interestingly, a recent review of 365 lakes with hypoxia (some of them are meromictic) by Jenny et al. (2016) appears to be also geographically biased: most lakes with hypoxia are also found in Europe and North America.

Presumably, the number of meromictic lakes is much higher if we also carefully investigate lake meromixis in other parts of the world. Aptly, although Hall and Northcote (2012) found information for only nine meromictic lakes in Asia, we assume that many more meromictic lakes exist in the arid and huge mountainous regions of Asia. For example, recently Dr. Rogozin (pers. com.) who regularly sampled Lake Uchum (South Siberia, Russia), which is located close to meromictic lakes Shira and Shunet in Siberia (see Chap. 5), found that Lake Uchum was not mixed at least during the 1-year field study. Rogozin believes that this lake, based on its size, salinity, morphology, geography and the laminated sediment, is meromictic. Hall and Northcote (2012) found no evidence from the literature on the presence of meromictic lakes in South America. In South America, however, we discovered the existence of meromictic pit lake at the Kori Kollo Mine, Bolivia (see Chap. 9). Also, literature search reveals references to a small (15 ha) and deep (45 m) meromictic lake Laguna El Ocho (3250 m a.s.l.) situated about 80 km southeast of Santiago, Chile, in South America (von Gunten et al. 2009).

Anderson et al. (1985) documented about 160 lakes in North America: they are all meromictic and (or) have a laminated sediment. Such a sediment is generally considered as typical for meromictic lakes. Indeed, meromictic lakes are not mixed down to the bottom. Also, they lack benthic organisms in their monimolimnion or in their anoxic bottom sediment. Obviously, the laminated sediment is formed not necessarily in meromictic lakes but also if conditions in the lake in and above the sediment do not lead to bioturbation (Scharf et al. 2010; Wendt-Potthoff et al. 2016). Some lakes in the list of Anderson et al. (1985) appear obscure, especially those reported as "seasonally meromictic" or if "they overturn occasionally". Nevertheless, the authors list some 100 lakes that are known to be, or are likely to be, meromictic.

In this book, we discuss either in detail or refer to published studies on 83 meromictic lakes located in five of the six continents (Fig. 1.1). It is interesting to contemplate about the global distribution of meromictic lakes and the potential mechanisms that support meromixis and geographical areas or climate regions where these mechanisms are conceivable. There is a resemblance between the map on the distribution of meromictic lakes mentioned in this book (Fig. 1.1) and the map of the worldwide prevalence of saline lakes (Williams 2002). Some meromixis sustaining processes are linked to the salt content, e.g. the salt exclusion during ice formation (see Chap. 2). Such a mechanism supports meromixis in

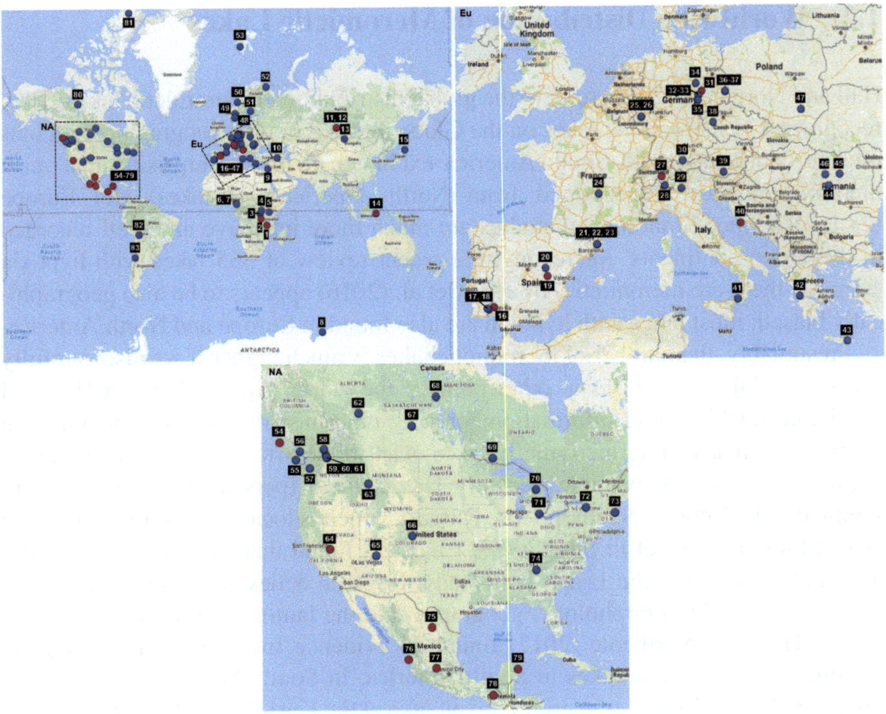

Fig. 1.1 Global distribution of meromictic lakes mentioned in this book on ecology of meromictic lakes. *Red dots*—lakes described in Special Study chapters (Part II of this book). *Blue dots*—lakes mentioned in different chapters of this volume. This map is available online at https://www.google.com/maps/d/embed?mid=16dUc71xcIaoLb0WdF1YG7QmQNyM. The list of continents and lakes: **Africa:** 1—Lake Malawi (Chap. 10), 2—Lake Tanganyika (Chap. 10), 3—Lake Kivu, 4—Lake Nyahirya, 5—Lake Sonachi, 6—Lake Nyos, 7—Lake Monoun; **Antarctica:** 8—Ace Lake; **Asia:** 9—Dead Sea (was meromictic before 1979), 10—Lake Van, 11—Lake Shira (Chap. 5), 12—Lake Shunet (Chap. 5), 13—Oigon Lake, 14—Lake Matano (Chap. 10), 15—Lake Harutori; **Europe (EU):** 16—Cueva de la Mora (Chap. 9), 17—Nuestra Señora del Carmen Mine Pit, 18—Guadiana Pit Lake, 19—Laguna de La Cruz (Chap. 8), 20—El Tobar Lake, 21—Lake Banyoles, 22—Lake Ciso, 23—Lake Vilar, 24—Lake Pavin, 25—Schalkenmehrener Maar, 26—Weinfelder Maar, 27—Lake Cadagno (Chap. 7), 28—Lake Lugano, 29—Lake Idro, 30—Lake Alatsee, 31—Lake Goitsche (Chap. 9), 32—Rassnitzer See, 33—Wallendorfer See, 34—Felsensee, 35—Lake Vollert-Süd, 36—Lake Waldsee, 37—Lake Moritzteich, 38—Hromnice Lake, 39—Längsee, 40—Rogoznica Lake (Chap. 6), 41—Lake Faro, 42—Aitoliko Lake, 43—Salty Lake, 44—Fara Fund Lake, 45—Ursu Lake, 46—Ocnei Lake, 47—Piaseczno reservoir, 48—Rørholtfjorden, 49—Nordbytjernet, 50—Salsvatn, 51—Lake Alinen Mustajarvi, 52—Lake Mogilnoe, 53—Kongressvatn; **North America (NA):** 54—Island Copper Mine Pit Lake (Chap. 9), 55—Nitinat Lake, 56—Sakinaw Lake, 57—Hall Lake, 58—Brenda Mine, 59—Lake Mahoney, 60—Blue Lake, 61—Hot Lake, 62—Roi Lake, 63—Berkeley Pit, 64—Mono Lake (Chap. 11), 65—Powell Lake, 66—Soda Lake, 67—Waldsea Lake, 68—Camp Lake, 69—Caland pit lake, 70—Hemlock Lake, 71—Third Sister Lake, 72—Fayetteville Green Lake, 73—Lower Mystic Lake, 74—South Mine pit lake, 75—Cuatro Ciénegas (Chap. 12), 76—Isabela Lake (Chap. 12), 77—Rincón de Parangueo (Chap. 12), 78—Dos Lagos (Chap. 12), 79—Nohoch Hol (Chap. 12), 80—Zone 2 Pit Lake, 81—Lake A; **South America:** 82—Kori Kollo Mine, 83—Laguna El Ocho

Siberian, e.g. lakes Shira and Shunet (Chap. 5), and in Canadian meromictic lakes (e.g. Lake Mahoney: Northcote and Hall 1990). Thus, areas where the saline mero mictic lakes are supported by similar mechanisms are in dry regions where evaporation exceeds precipitation, with severe cold climate as in Siberia or parts of Canada.

Saline lakes in general are quite susceptible to the effects of climate change, as both lake volume and salinity and its gradients depend on the hydrologic condition. For example, Aral Sea has significantly declined in volume during the last 60 years or so. This water-level decline has resulted in a dramatic salinity increase. The Western Large Aral Sea has now a very strong halocline between 6 m and 8 m depth that reminds us about the situation typical of meromictic salt lakes (Izhitskiy et al. 2016). Thus, we expect meromictic lakes to occur in regions with saline lakes after a drop of water level (drought, water diversions, any other natural or anthropogenic factors that accelerate the water-level drop in saline lake) or in dry territories with saline lakes after the superficial freshwater runoff due to ice melt and accidental heavy rains, if there is flow of freshwater.

Recent predictions reveal that the effect of climate change will cause salinization of many lakes in Mediterranean climate zone (Jeppesen et al. 2015). Lewis et al. (2015) demonstrated that the subarctic lakes in Alaska are shrinking with the substantial increases of nutrients and other ions which lead them to become eutrophic or saline. Another causal factor for salinization of many inland lakes is water from salt applied to roads in winter. Kelting et al. (2012) demonstrated that the concentration of sodium and chloride ions in lakes within the watershed of US national roads is positively correlated with the density of roads and the extent of salts applied in wintertime. Sibert et al. (2015) recently described the effect of road salt on the chemical stratification of an urban lake as *cultural meromixis*. Thus, many factors, including both natural and anthropogenic, can augment the salinity of many freshwater lakes and under favourable circumstances turn such lakes to become meromictic.

Some small meromictic lakes are located in karst areas where deep vertical holes are filled with water. When such holes are located close to the sea, the saline water is either trapped in the holes or can infiltrate the system through the ground. In both cases, meromixis can be established and sustained long. The maps of global distribution of carbonate rocks, which usually overlap with karst areas, reveal that these rocks are widespread over the world (Ford and Williams 2007) and some meromictic lakes mentioned in this book are located exactly in the karst areas, e.g. Lake Rogoznica, Chap. 6; Laguna de La Cruz, Chap. 8; and Mexican meromictic lakes, Chap. 12. Probably many small karst holes filled with water have not yet been investigated but appear to be meromictic.

Many meromictic lakes are, indeed, man-made lakes established in former surface mines deep enough to intersect the local water table, so-called pit lakes (see Chap. 9). Usually such lakes have a steep lake basin that prevents mixing in the water column. The geochemical variability of water within the pit lake catchment (i.e. total dissolved solids, chemical composition, pH) also plays an important role in stratification. The number of such pit lakes is rising worldwide because mine pits are abandoned after exploitation and flooded and new open pit mines are excavated (Castendyk and Eary 2009). The geological setting for different minerals varies. Hence the chemical characteristics of mine pit lakes also depend on the materials mined (rock quarries, gravel, kaolinite, sulphur, lignite, coal and metal ore mines). In general, mine pit lakes are more prone to show meromixis than natural lakes (Schultze et al. 2016).

Global distribution of meromictic lakes can be partly explained by the global and regional conditions that favour the development of meromixis. However, several natural and anthropogenic factors can affect the territorial patterns of meromixis. It is, therefore, difficult to map the presence of all meromictic lakes because some lakes will lose meromixis, while others may turn from holomictic to meromictic. Regular monitoring is the key to track the condition of many lakes and classify their current mixing regimes.

1.3 Introduction to This Volume

Currently, there are no books devoted exclusively to meromictic lakes. The information on meromixis in limnology textbooks is often short and fragmentary (e.g. Kalff 2003) and the account on meromixis in the thematic chapter in the Encyclopedia of Inland Waters (Stewart et al. 2009) does not appear to be sufficient in many respects. However, there has been great progress regarding our understanding of meromixis more recently that we cover in our book. This includes the mechanisms of stabilization of meromixis, details of the biogeochemical and ecological processes involved in meromixis. Data and measurements that are more accurate are available and span over time periods that allow a better description of processes controlling meromixis in lakes. Now, limnology can provide quantitative data on issues connected with meromixis. With this book, we comply with the need for a comprehensive collection of data and studies on meromixis.

In this book we refer to 83 meromictic lakes located in five continents excluding Australia (Fig. 1.1). There are too many meromictic lakes and the variability among them is too large to include all in this book. We select a few prominent cases that are also instructive for our general understanding. In Part I of this book, we discuss information on salient features of meromictic lakes, focusing on physical (see Chap. 2), chemical (see Chap. 3) and biological (see Chap. 4) properties of these lakes. The Case Study chapters in Part II (Chaps. 5–12) are each dedicated to one particular meromictic lake or a group of such lakes. There are chapters on meromictic karst holes and pit lakes and large meromictic lakes located in hot tropical region and, and contrasting these, small and saline meromictic lakes in the cold and generally dry Siberian climate. Chapter 13 highlights in brief the Conclusions based on the summaries of the preceding 12 chapters. Hence, we hope to provide guidance to the reader to collate detailed information of single lakes. We hope this book will provide a practical guide to meromictic lakes for limnologists.

Several study aspects covered in this volume differ from those in the existing literature on meromictic lakes mainly in:

(a) Detailed studies on different meromictic lakes with mechanisms that cause and sustain meromixis in such lakes
(b) Reviews of information on the food web structure and on organisms adapted to meromictic conditions, especially microorganisms forming the microbial food web and how this food web is connected with macrobial or the major food web
(c) Description of meromictic lakes as paleolimnological archives due to preserved layers of intact sedimenting material in these lakes during meromictic conditions and the presence of specific paleolimnological markers (e.g. markers of purple sulphur bacteria, or the so-called PSB)
(d) Updated information on bacterial diversity in the meromictic lakes supported by molecular biology techniques and information on microbial processes, including those not adequately covered in earlier literature, such as anaerobic methane oxidation, anaerobic ammonium oxidation and sulphate reduction at low pH.

It is, however, not possible to answer all questions on ecology of meromictic lakes. There are some uncertainties regarding terminology of meromixis; moreover, we believe that many meromictic lakes are still undiscovered. Most probably, the discovery of new meromictic lakes in extreme habitats or with specific thermal, chemical and biological regimes will bring new knowledge on the functioning of biological communities. Rather than the final answer, we consider this volume as a relatively comprehensive update reference that can be used in the future studies on meromictic lakes.

Acknowledgement We thank Martin Schultze for his comments on this chapter. EZ was partially supported by the Council on grants from the President of the Russian Federation for support of leading scientific schools (grant NSh-9249.2016.5).

References

Ambrosetti W, Barbanti L (1999) Deep water warming in lakes: an indicator of climate change. J Limnol 58:1–9

Anderson RY, Dean WE, Bradbury P et al (1985) Meromictic lakes and varved lake sediments in North America. U.S. geological survey bulletin 1607. United States Government Printing Office, Washington, DC

Boehrer B, Dietz S, von Rohden C et al (2009) Double-diffusive deep water circulation in an iron-meromictic lake. Geochem Geophys Geosyst 10:Q06006. doi:10.1029/2009GC002389

Boehrer B, Fukuyama R, Chikita K (2008) Stratification of very deep, thermally stratified lakes. Geophys Res Lett 35:L16405. doi:10.1029/2008GL034519

Boehrer B, Kiwel U, Rahn K et al (2014) Chemocline erosion and its conservation by freshwater introductionto meromictic salt lakes. Limnologica 44:81–89

Boehrer B, Schultze M (2008) Stratification of lakes. Rev Geophys 46:RG2005. doi:10.1029/2006RG000210

Böhrer B, Heidenreich H, Schimmele M et al (1998) Numerical prognosis for salinity profiles of future lakes in the open cast mine Merseburg-Ost. Int J Salt Lake Res 7:235–260

Bøyum A (1973) Salsvatn, a lake with old sea water. Schweiz Zeitschr Hydrol 35:262–277

Castendyk DN, Eary LE (2009) The nature and global distribution of pit lakes. In: Castendyk DN, Eary LE (eds) Mine Pit Lakes: characteristics, predictive modeling, and sustainability. Society for Mining, Metallurgy, and Exploration, Littleton, CO, pp 1–11

Danis P-A, von Grafenstein U, Masson-Delmotte V et al (2004) Vulnerability of two European lakes in response to future climatic changes. Geophys Res Lett 31:L21507
Findenegg I (1935) Limnologische Untersuchungen im Kärntener Seengebiete. Ein Beitrag zu Kenntnis des Stoffhaushaltes in Alpenseen. Int Rev Gesamten Hydrobiol Hydrogr 32:369–423
Fisher TSR, Lawrence GA (2006) Treatment of acid rock drainage in a meromictic pit lake. J Environ Eng 132:515–526
Ford DC, Williams PW (2007) Karst hydrology and geomorphology. Wiley, London
Hakala A (2004) Meromixis as a part of lake evolution—observations and a revised classification of true meromictic lakes in Finland. Boreal Environ Res 9:37–53
Hall KJ, Northcote TG (2012) Meromictic lakes. In Bengtsson L, Herschy RW, Fairbridge RW (eds) Encyclopedia of lakes and reservoirs. Springer, Dordrecht, pp 519–524
Hutchinson GE (1937) A contribution to the limnology of arid regions: primarily founded on observations made in the Lahontan basin. Trans Connecticut Acad Arts Sci 33:47–132
Hutchinson GE (1957) A treatise on limnology, Geography, physics and chemistry, vol 1. John Wiley, New York
Imboden DM, Wüest A (1995) Mixing mechanisms in lakes. In: Lerman A, Imboden DM, Gat JR (eds) Physics and chemistry of lakes. Springer, Berlin, pp 83–138
Izhitskiy AS, Zavialov PO, Sapozhnikov PV et al (2016) Present state of the Aral Sea: diverging physical and biological characteristics of the residual basins. Sci Rep 6:23906. doi:10.1038/srep23906
Jenny J-P, Francus P, Normandeau A et al (2016) Global spread of hypoxia in freshwater ecosystems during the last three centuries is caused by rising local human pressure. Glob Chang Biol 22:1481–1489
Jeppesen E, Brucet S, Naselli-Flores L et al (2015) Ecological impacts of global warming and water abstraction on lakes and reservoirs due to changes in water level and related changes in salinity. Hydrobiologia 750:201–227
Kaden H, Peeters F, Lorke A et al (2010) Impact of lake level change on deep-water renewal and oxic conditions in deep saline Lake Van, Turkey. Water Resour Res 46:W11508. doi:10.1029/2009WR008555
Kalff J (2003) Limnology. Prentice Hall, Upper Saddle River, NJ
Kelting DL, Laxson CL, Yerger EC (2012) Regional analysis of the effect of paved roads on sodium and chloride in lakes. Water Res 46:2749–2758
Kling GW, Clark MA, Compton HR et al (1987) The 1986 Lake Nyos gas disaster in Cameroon, West Africa. Science 236:169–175
Lewis TL, Lindberg MS, Schmutz JA et al (2015) Pronounced chemical response of Subarctic lakes to climate-driven losses in surface area. Glob Chang Biol 21:1140–1152
Melack JM, Jellison R (1998) Limnological conditions in Mono Lake: contrasting monomixis and meromixis in the 1990s. Hydrobiologia 384:21–39
Newman FC (1976) Temperature steps in Lake Kivu: a bottom heated saline lake. J Phys Oceanogr 6:157–163
Nissenbaum A (1969) Studies in the geochemistry of the Jordan River-Dead Sea system. Dissertation, University of California
Nixdorf E, Boehrer B (2015) Quantitative analysis of biogeochemically controlled density stratification in an iron-meromictic lake. Hydrol Earth Syst Sci 19:4505–4515
Northcote TG, Hall KJ (1990) Vernal microstratification patterns in a meromictic saline lake: their causes and biological significance. Hydrobiologia 197:105–114
Peeters F, Finger D, Hofer M et al (2003) Deep-water renewal in Lake Issyk-Kul driven by differential cooling. Limnol Oceanogr 48(4):1419–1431
Sanderson B, Perry K, Pedersen T (1986) Vertical diffusion in meromictic Powell Lake, British Columbia. J Geophys Res 91(C6):7647–7655
Scharf BW, Boehrer B, Buettner O et al (2010) Local variability of sedimentation rate in Lake Arendsee, Germany. Limnologica 40:97–101

Schultze M, Castendyk D, Wendt-Potthoff K, Sánchez-España J, Boehrer B (2016) On the relevance of meromixis in pit lakes—an update. In: Proceedings IMWA 2016, Leipzig/Germany, pp 227–235
Sibert RJ, Koretsky CM, Wyman DA (2015) Cultural meromixis: effects of road salt on the chemical stratification of an urban kettle lake. Chem Geol 395:126–137
Stewart KM, Walker KF, Likens GE (2009) Meromictic lakes. In: Likens GE (ed) Encyclopedia of inland waters. Academic Press, Oxford, pp 589–602
Vollmer MK, Weiss RF, Bootsma HA (2002) Ventilation of Lake Malawi/Nyasa. In: Odada EO, Olago DO (eds) The East African great lakes: limnology, paleolimnology and biodiversity. Kluwer, Dordrecht, pp 209–233
von Gunten L, Grosjean M, Beer J et al (2009) Age modeling of young non-varved lake sediments: methods and limits. Examples from two lakes in Central Chile. J Paleolimnol 42:401–412
von Rohden C, Ilmberger J, Boehrer B (2009) Assessing groundwater coupling and vertical exchange in a meromictic mining lake with an SF6-tracer experiment. J Hydrol 372:102–108
von Rohden C, Boehrer B, Ilmberger J (2010) Evidence for double diffusion in temperate meromictic lakes. Hydrol Earth Syst Sci 14:667–674
Walker KF, Likens GE (1975) Meromixis and a reconsidered typology of lake circulation patterns. Verh Int Verein Limnol 19:442–458
Wendt-Potthoff K, Scharf BW, Boehrer B, Völkner C, Tittel J, Merkt J (2016) Varve formation in the acidic (pH 2.7) pit lake 111 (Lusatia,Germany). In: Proceedings IMWA 2016, Leipzig/Germany, pp 244–250
Williams WD (2002) Environmental threats to salt lakes and the likely status of inland saline ecosystems in 2025. Environ Conserv 29:154–167
Zadereev ES, Tolomeev AP, Drobotov AV et al (2014) Impact of weather variability on spatial and seasonal dynamics of dissolved and suspended nutrients in water column of meromictic Lake Shira. Contemp Prob Ecol 7:384–396

Part I
Special Features of Meromictic Lakes

Chapter 2
Physical Features of Meromictic Lakes: Stratification and Circulation

Bertram Boehrer, Christoph von Rohden, and Martin Schultze

2.1 General Features

In meromictic lakes, a chemically different deepwater layer "the monimolimnion" remains perennially as a consequence of insufficient mixing with the overlying water body "the mixolimnion" (e.g., Findenegg 1933, 1935, 1937; Hutchinson 1937, 1957; Boehrer and Schultze 2008).

The transition between *mixolimnion* above and *monimolimnion* below is called the *chemocline*, as many chemical conditions change over a short vertical distance (Hutchinson 1957). Usually higher concentrations of solutes in the monimolimnion and stable density stratification are sustained throughout the annual cycle. In most cases, a high density gradient restricts the vertical exchange of water parcels and hence the turbulent transport of dissolved substances as well as heat. As a consequence, strong chemical gradients are conserved, and a fine zonation in this depth range can establish. Scientists looking at different features of the chemocline (gradients of electrical conductivity, oxygen, organisms, etc.) refer to slightly different depth ranges within this zone when speaking of "chemocline" (see Fig. 2.1).

The mixolimnion shows stratification and circulation patterns as in holomictic lakes, i.e., lakes without a monimolimnion, with the usual vertical subdivision into epilimnion (upper layer) and hypolimnion (lower layer). At the end of the thermal stratification period, deep recirculation mixes both layers. A lake (or the

B. Boehrer (✉) • M. Schultze
Department Lake Research, UFZ Helmholtz Centre for Environmental Research, Magdeburg, Germany
e-mail: bertram.boehrer@ufz.de

C. von Rohden
Institute For Environmental Physics, University of Heidelberg, Heidelberg, Germany

Lindenberg Meteorological Observatory, German Weather Service, Lindenberg, Germany

© Springer International Publishing AG 2017
R.D. Gulati et al. (eds.), *Ecology of Meromictic Lakes*, Ecological Studies 228,
DOI 10.1007/978-3-319-49143-1_2

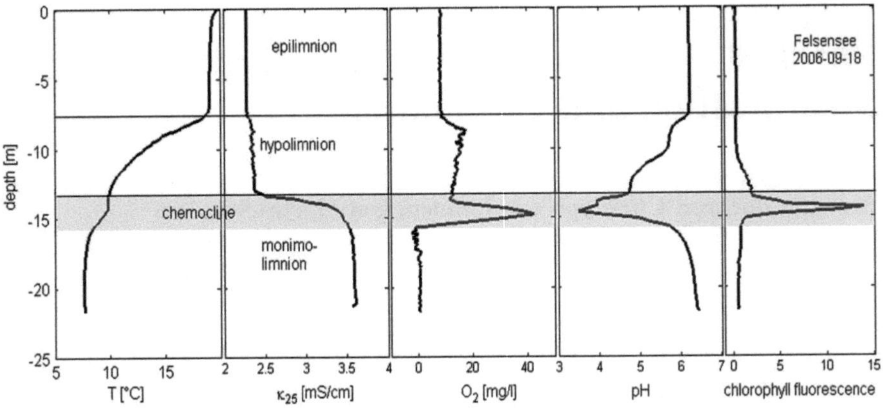

Fig. 2.1 Profiles of temperature, electrical conductance, oxygen concentration (numerically corrected for sensor response time), pH and fluorescence of the chlorophyll against depth in Felsensee (near Magdeburg, Germany). The *upper horizontal line* indicates the interface between epilimnion and hypolimnion, while the *lower horizontal line* marks the interface between hypolimnion and chemocline

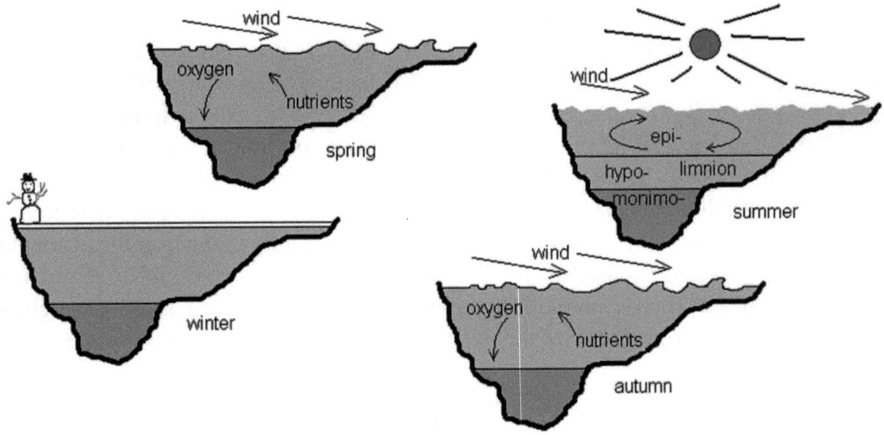

Fig. 2.2 Sketch of stratification and recirculation in a meromictic lake over an annual cycle

mixolimnion of a meromictic lake) can be monomictic (one recirculation period during the cold season) or dimictic (with a deep recirculation during spring and autumn—see Wetzel 2001—over an annual cycle depending on the climate zone; Fig. 2.2).

Similar to deeper holomictic lakes, a water layer at the surface heats up in spring as a consequence of increased solar irradiation and contact to a warming atmosphere. As a consequence, warmer water is formed, which floats on colder, denser water. While the upper layer, the epilimnion, is exposed to gas exchange and energy transfer with the atmosphere all year round, the hypolimnion below is shielded from direct impact during summer. Usually the hypolimnion remains density stratified over the summer stratification period. Hence, vertical exchange of water parcels requires energy, which is available only in a limited amount. As a consequence, vertical transport of solutes and heat is small during stratification periods.

Over the summer stratification in the mixolimnion, little is happening about the depth of the chemocline. Groundwater may enter the monimolimnion and contribute to its volume, and hence the chemocline slowly rises (von Rohden et al. 2010). In addition, diffusion and turbulent transport of solutes from the monimolimnion can raise the density gradient locally and lead to oxygen demand in the lowest zone of the hypolimnion. Affected water may change its properties from hypolimnetic to monimolimnetic and eventually become a part of the monimolimnion. An ice cover can add another period of quiet conditions (winter stagnation). In latitudes where ice cover extends for a long period, this can result in meromixis (e.g., lakes Shira and Shunet, Chap. 6), as the subsequent circulation in spring time is short or is completely missing. The ice cover together with the quick warming up due to high concentrations of colored humic substances (Eloranta 1999), basin shape, and an almost complete protection from wind probably cause high frequency of meromixis in the small, humus-rich forest lakes in Finland (Merilainen 1970; Salonen et al. 1984).

Later in the year, cooling at the lake surface removes the protecting thermal density stratification, and convection forces water motion down to the chemocline. Turbulent kinetic energy is used to shave off parts of the chemocline. As a consequence, the chemocline moves downward and gradients get sharper. The upper part of the monimolimnion gets included in the mixolimnion and so do the water and chemicals contained in it. This is of particular interest for nutrients available at higher concentrations in the monimolimnion. The volume of water introduced into the mixolimnion depends on weather conditions during the recirculation periods and may, therefore, greatly vary from year to year. The deeper monimolimnion remains unchanged as long as its density is greater than that of the mixolimnion and it is sufficient to withstand the erosion by advected and locally produced turbulent kinetic energy.

Typically meromictic lakes show increasing electrical conductance with depth because of increasing concentrations of ionic solutes with depth. In general, higher concentrations of solutes increase the density of water (see below paragraphs on density). Due to mixing at least once each year, differences between epilimnion and hypolimnion are small. However, gradients between mixolimnion and monimolimnion can be enormous. Gradients are known to range from salinity under freshwater conditions to salt concentrations exceeding beyond salinity in ocean water (Wallendorfer See, Germany; Boehrer et al. 2014) or exceeding 100 g/L as in hypersaline meromictic lakes (e.g., Hot Lake, USA; Zachara et al. 2016; see also Chap. 3).

Temperatures at the lake surface are determined by weather conditions, while in the hypolimnion, one usually finds temperatures prevailing during the last cold period (e.g., Boehrer et al. 2000). In lakes located in colder climates, hypolimnion temperatures are close to the temperature of maximum density under normal conditions (i.e., 4 °C for freshwaters but slightly lower at high salinities). Interestingly the monimolimnion is locked between hypolimnion above and ground/groundwater below. While its upper boundary is defined by temperatures during the cold period of the year, the lower boundary is warmer due to groundwater reflecting more an annual average temperature and geothermal heat flux. Usually a monimolimnion has a temperature gradient that reduces the density gradient imposed by dissolved substances.

Both, the permanent density stratification in the deep water and the seasonal circulation of the mixolimnion, determine the distribution of solutes, as we show for oxygen as one of the key solutes for organisms. At the water surface, the atmosphere implies the boundary condition for oxygen. As a consequence, the epilimnion usually shows oxygen concentrations close to the equilibrium with the atmosphere, i.e., close to 100 % saturation (by definition) at the respective temperature. The hypolimnion can have higher concentrations of oxygen than the epilimnion due to higher solubility of oxygen at lower temperatures during the last mixing period or due to photosynthesis, if the light penetrates to deeper layers. However, over the stratification period, oxygen in the hypolimnion is subject to depletion until the next circulation period when the oxygen levels are recharged. Oxygen concentrations drop to zero across the chemocline. The extent of availability of oxygen sets the conditions for the fate of other solutes. As the contact line between water masses of different chemistry, the chemocline is a zone of active chemical transformations (see Fig. 2.3). Consequently, organisms have to cope with the chemical conditions in this gradient zone. Some organisms manage to profit from unusual chemical conditions (see Chap. 4).

2.2 Processes Forming Meromixis

In general, water density is higher in the monimolimnion than in the mixolimnion so as to retain the monimolimnion throughout the year. To balance the gradual reduction of this density difference by diffusion and mixing, a process is required that transports substances into the higher concentration waters of the monimolimnion.

Hutchinson (1957) classified meromictic lakes according to the major processes that transport solutes. He could identify three classes, (1) ectogenic, (2) crenogenic, and (3) biogenic meromixis, i.e., meromixis caused by (1) external inflows, (2) groundwater, or (3) degradation of organic material, respectively. On a broader base of examples, Walker and Likens (1975) refined Hutchinson's classification into groups when lakes remain permanently stratified due to (1) inflows of different salinity, (2) inflows of high turbidity, or (3) inflowing groundwater. These three groups comprise ectogenically meromictic lakes (class A), while endogenic meromixis (class B) is formed either by (4) degradation of organic material in the deep water or (5) by salt exclusion when ice is formed during winter.

Fig. 2.3 Water sample from about 10 m depth from Moritzteich (south of Berlin, Germany), showing the transition from colorless water of mixolimnion (*upper* part of the sampling tube) to red water of monimolimnion (from Boehrer 2013). The turbidity in the chemocline is the result of oxidation and precipitation of iron

Since these cited works on classification of meromictic lakes, many meromictic lakes have been scientifically investigated and reported. Microbiologically controlled chemical reactions, e.g., iron meromixis (Hongve 2002), have been understood more in detail, and their contribution to the density of monimolimnetic waters can be evaluated (e.g., Dietz et al. 2012; Nixdorf and Boehrer 2015). Boehrer and Schultze (2008) refer to the importance of evaluating all processes that are known from scientific literature for sustaining meromixis, since several processes may be acting simultaneously. Here we explain the important processes that can sustain meromixis and support each of these processes by the clearest representations we know of in the environment.

To provide a better overview, we list all the processes that have been documented to form and sustain meromixis, before we go into the details and mention the representative examples form the literature. We mainly see a distinction into two groups: those which operate purely mechanically (where we include salt exclusion at ice formation) and those at which geochemistry of lake waters takes the control. Though we treat the geochemistry as a set of chemical reactions, many of these processes are mediated by organisms. For details, we refer to the subsequent sections and chapters of this book.

1. Purely mechanical:
   (a) Salty inflows into lakes:
   From external sources
   From groundwater
   (b) Freshwater onto salty lakes:
   From external sources
   Salt exclusion at ice formation
   (c) Partial deepwater renewal:
   Evaporation in side bays
   Cooling in side bays
   Salty intrusions from ice
2. Involving geochemistry:
   (a) Decomposition of organic material
   (b) Iron meromixis
   (c) Temperature-dependent solubility of mirabilite
   (d) Calcite precipitation

The local distribution of these mixing and stratifying processes is shown in a schematic display of a lake (Fig. 2.4).

Salty inflows that find their way to the deep layers of a lake can form a permanent stratification, no matter whether these inflows enter at the surface of lakes (Lake Nitinat, Canada: Ozretich 1975) or as groundwater (see Fig. 2.4; Kongressvatn, Norway: Bøyum and Kjensmo 1970; Wallendorfer See and Rassnitzer See, Germany: Böhrer et al. 1998; Heidenreich et al. 1999; Waldsee, Germany: von Rohden et al. 2009, 2010; see also Chaps. 6 and 12 for more

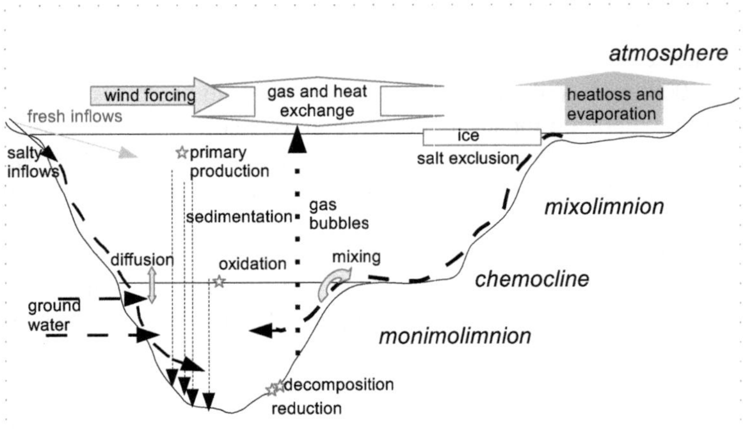

Fig. 2.4 Sketch of processes involved in sustaining meromixis

examples). Sibert et al. (2015) reported meromixis caused by salty wastes entering the lakes and formed by salt used for de-icing of roads; Scharf and Oehms (1992) also reported the same for meromixis in Lake Schalkenmehrener Maar (Germany). High-salinity water can also be produced in the lake itself by enhanced evaporation in a shallow side bay, as in the Dead Sea in the period before 1979 (Nissenbaum 1969). Because only a small portion of the monimolimnion is replaced, the monimolimnion chemistry (e.g., oxygen concentration) is not greatly affected. Salt exclusion during ice formation can also act as a source of saline water (Antarctic lakes: Kerry et al. 1977; Gibson 1999).

In addition, density stratification in the mixolimnion is reinforced when less saline ice melts in spring and forms a water layer of reduced density near the surface, which needs to be removed before turbulent kinetic energy can effectively erode the density stratification at the upper edge of the monimolimnion (e.g., Lake Shira, Russia, Chap. 5). Snowmelt in the vicinity of meromictic lakes and freshwater runoff from snowmelt add to the effect of melting lake ice (e.g., Hammer 1994; Zachara et al. 2016). During wet season, high runoff may also support meromixis by bridging the density stratification into the thermal stratification period especially in regions where precipitation occurs almost exclusively seasonally (e.g., Santofimia et al. 2012). They also reported a full overturn when the precipitation came too late in the year to accomplish the bridging. High precipitation and consequently higher freshwater inflows have stopped some lakes from circulating for several years (e.g., Lake Mono, USA, see Chap. 11: Jellison and Melack 1993; Jellison et al. 1998; Caspian Sea: Peeters et al. 2000; Lake Van, Turkey: Kaden et al. 2010). Similarly, meromictic lakes were formed where fjords have been disconnected from the ocean by land rising after glaciers had receded at the end of the ice age, e.g., Rørholtfjorden, Norway (Strøm 1957); Powell Lake, Canada (Williams et al. 1961); and lakes in coastal regions in the northern Baltic Sea (Lindholm 1996). Some of these lakes maintained their stratification for several thousand years. Seawater was capped with freshwater to create a meromictic lake in the former Island Copper Mine (Canada; see Chap. 9). Controlled freshwater input in lakes Wallendorf and Rassnitz (west of Leipzig, Germany) was used to quantify the effect of shielding the deepwater stratification by a freshwater introduction (Boehrer et al. 2014).

In principle, density-driven flows as described above can also be caused by temperature differences. However, density difference due to temperature is limited to 2 kg/m or 3 kg/m, while density differences due to solute concentration can be one to two orders of magnitude higher. In addition, molecular diffusion of heat is faster (factor ~100) than diffusion of solutes, and hence a density stratification due to temperature is more affected by diffusion. Thus, meromictic lakes due to temperature-driven flows are rare. Similarly, Dead Sea (before 1979), where evaporation created saltier and hence denser water in the south basin, is a good example of creation of meromixis: also surface cooling during southern winter at the south end of Lake Malawi/Nyasa (Malawi, Tanzania, Mozambique) produces sufficiently dense water to intrude the monimolimnion (Vollmer et al. 2002a). Lakes with partial deepwater renewal are called meromictic, only if a sufficient portion of the monimolimnion is replaced to sustain the stratification, but not enough to make dissolved oxygen to be detectable in the monimolimnion.

In addition to water currents, solute precipitation can effectively transport matter through the water column. If salts of high sodium and sulfate concentration are excluded from the ice formation, mirabilite ($Na_2SO_4 \cdot 10H_2O$) may precipitate. Hammer (1994) reports such precipitation below the ice at temperature close to 0 °C for Waldsea Lake (Canada). The settling mirabilite removes solutes—and hence their density contribution—from the cold surface water, but it gets redissolved in the monimolimnion due to the higher temperatures (5–7 °C; Hammer and Haynes 1978) and the strong dependence of mirabilite solubility on temperature (Marion and Farren 1999; see also Chap. 3). The density contribution is added to the monimolimnetic waters. The picture gets slightly more complex when biogenic calcite formation accompanies photosynthesis. With increasing photosynthetic activity, pH increases such that carbonate will form from dissolved bicarbonate in the epilimnion. Eventually the solubility product of calcite is exceeded, and calcite precipitates and sinks to deeper layers. High CO_2 partial pressure in the monimolimnion, e.g., from decomposed organic material, allows a redissolution of calcite (Lake La Cruz, Spain: Rodrigo et al. 2001; see also Chap. 8). Also in this case, the calcite removes the density contribution of solutes from the mixolimnetic waters and adds it to monimolimnetic waters.

Waldsea Lake (Canada) and Lake La Cruz (Spain) demonstrate that reactivity of solutes can inhibit the deep recirculation, but only, if this process transports solutes up the gradients, which means from low concentrations in the mixolimnion to high concentrations in the monimolimnion, effectively enough to compensate for the diffusive and turbulent transport down-gradient. This is facilitated by limited solubility of a substance: particulate solids are formed and precipitate and sink down to the monimolimnion. If conditions in the monimolimnion are favorable, these precipitates can redissolve at least in part. To keep this chemical cycling of matter going, an energy source must be accessible. In nearly all cases, energy is provided by organic material, which has gained its energy from photosynthesis, and feeds it into the chemical cycle on being decomposed. Organisms usually have a slightly higher density than water, and hence they settle to the lake bottom still alive or after they die. Organic material can be oxidized by microorganism, releasing CO_2, which contributes effectively to water density. The oxidizing agent in this process is important, as the net effect of released products must yield a considerable increase in density. Both oxygen (classic biogenic meromixis) and iron (iron meromixis) are good oxidizing agents to raise density sufficiently in the monimolimnion. Also other oxidizing agents (nitrate or sulfate) may be present, but they have not yet been demonstrated to be the primary factor for building the density gradient needed for meromixis.

Iron is present in many mine lakes but also in natural lakes originating from the soil and rocks in the catchments. However, if dissolved oxygen is present in the mixolimnion and if in the upper part of the chemocline, iron gets oxidized to ferric iron Fe(III), it precipitates as hydroxide in water at pH above 3.5 (e.g., Stumm and Morgan 1996). On the contrary, in a monimolimnion where no dissolved oxygen is

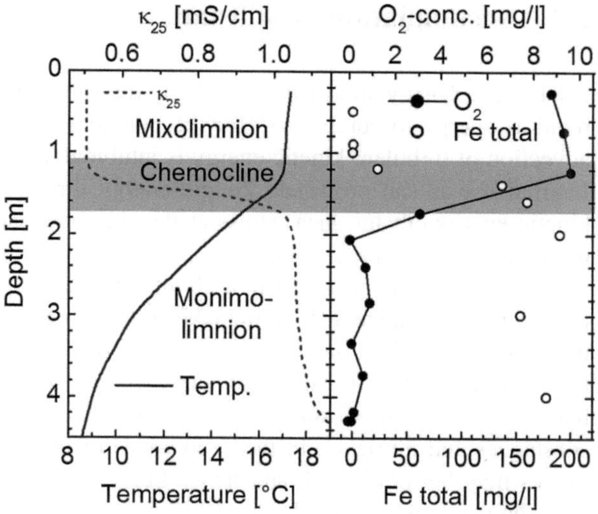

Fig. 2.5 Profiles of temperature and electrical conductance (*left panel*) and total iron and dissolved oxygen (*right panel*) from Lake Waldsee near Döbern, Germany (taken from Boehrer et al. 2009)

available but enough organic material (Fig. 2.5), ferric iron can be used as oxidizing agent and remain dissolved in the reduced form as ferrous iron Fe(II) (e.g., Campbell and Torgerson 1980). Consequently, both iron and the produced CO_2, as well as corresponding bicarbonate, contribute to density of monimolimnetic waters, while in the mixolimnion, iron concentrations and CO_2 concentrations are very low because of low solubility and the escape to the atmosphere (Kjensmo 1967, 1988; Campbell and Torgerson 1980; Hongve 1997, 2002). In Lake Waldsee (Germany), both iron and the carbonate system contribute about the same amount to the density difference between mixolimnion and monimolimnion (Dietz et al. 2012). Manganese can play a similar role as iron, but meromixis dominated by manganese is rare (Nordbytjernet, Norway; Hongve 1997, 2002). A detailed presentation of the biogeochemical processes mentioned above is given in Chap. 3. Also suspended material adds density to lake water. Casamitjana and Roget (1993) claim that the Lake Banyoles, Spain, is kept meromictic by continuously suspended particles. Similarly, Frey (1955) claimed that a turbidity inflow into Längsee (Austria) initiated meromixis in the lake.

Finally, there are lakes that do not fully turn over during the annual cycle: at any time, a profile will show stable density stratification. In these lakes, permanent stratification can be accomplished by thermobaric effects (Lake Shikotsu, Japan: Yoshimura 1935, 1936; Boehrer et al. 2008; Crater Lake, USA: Crawford and Collier 2007; Lake Baikal, Russia: Carmack and Weiss 1991) or by partial deep-water renewal (Issyk Kul, Kyrgyzstan: Vollmer et al. 2002b; Peeters et al. 2003; Lake Baikal, Russia: Weiss et al. 1991; Kondenev 2001). However, such lakes are usually not called meromictic as their deep waters are not permanently anoxic. In addition, the exchange in the monimolimnion is so fast that no considerable accumulation of chemicals is possible in the deep water, and, thus, no pronounced chemical differences are found through the water column.

2.3 Accumulation of Substances

If lakes are density stratified as is mostly the case, vertical transport of substances requires energy to compensate for the potential energy of the system. However, advection of turbulent kinetic energy is inhibited, as long as stratification persists. Internal waves can propagate kinetic energy through density stratification. This kinetic energy creates shear stress at the side boundaries which finally results in mixing. However, the amount of supplied energy is limited.

The complete overturn, which effectively transports solutes from the deepest point quasi-instantaneously to the lake surface, is missing in meromictic lakes. As demonstrated for iron meromixis above, we find an accumulation of substances in the monimolimnion. These substances include iron, which can precipitate as iron hydroxides, and also substances that are flocculated during this process, like organic substances and nutrients. In the deep zones of lakes, the binding of the coprecipitates may be terminated, and hence high concentrations of DOC and nutrients (e.g., phosphorus) can be the result. Primary production and connected uptake of nutrients below the chemocline are limited by low irradiance. Solutes can be stored in monimolimnia at high concentration. In general, concentration of dissolved ionic substances is limited by the solubility product of cations and anions. If this limit is exceeded, precipitation removes the excess portion. In summary, downward transport, incomplete mixing, and limited biogeochemical uptake promote the accumulation of dissolved substances in monimolimnia.

As the continuous supply of oxygen to the monimolimnion is suppressed, we can find substances in chemically reduced form (ferrous iron, ammonium, sulfide, dissolved organic matter (DOM)/DOC, etc.) at concentrations that are not encountered in holomictic lakes or only exceptionally in the mixolimnion. Methane is stable in the chemical setting of many monimolimnia. Reduced sulfur may be found as hydrogen sulfide, if no suitable cation is present to precipitate it from the water column. Hydrogen sulfide is poisonous to many organisms; if released from the water, its unpleasant odor is a nuisance also at distance from the shoreline and can restrict leisure activities on a lake and in its immediate surroundings (e.g., Lower Mystic Lake, USA: Ludlam and Duval 2001). Table 3.1 provides examples for high monimolimnetic concentrations of reduced chemical species (Sect. 3.1). Chapters 5–12 add further examples.

In addition, dissolved gases may accumulate in the monimolimnion (e.g., Aeschbach-Hertig et al. 1999). Monimolimnia are sheltered from direct exchange with the atmosphere. Hence, gases need to be transported upward by diffusive transport through the chemocline and the mixolimnion before they are exchanged with the atmosphere. In addition, the hydrostatic pressure in the deep layers of lakes facilitates the accumulation of gases, beyond atmospheric pressure. According to its concentration c_i, any dissolved volatile substance i produces a gas pressure (Henry law):

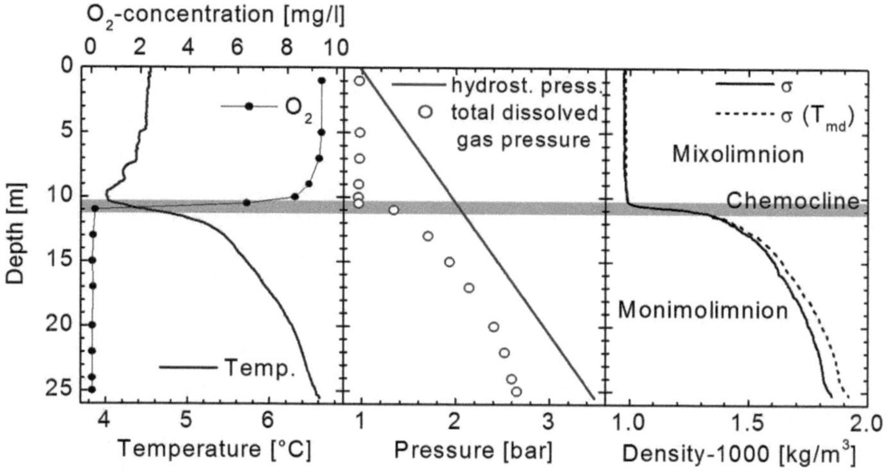

Fig. 2.6 Measured profiles of temperature, oxygen concentration (*left panel*) and gas pressure (symbols in *central panel*) in Lake Vollert-Süd (south of Leipzig, Germany). For comparison, absolute hydrostatic pressure (*central panel*) and density based on temperature and electrical conductivity measurements are included (*solid line*, *right panel*) and density at given electrical conductivity at a temperature of constant 4 °C (*dashed line*, *right panel*)

$$p_i = c_i/H_i \tag{2.1}$$

where H_i is a specific temperature-dependent constant (Henry coefficient) (e.g., Sander 1999).

If a strong gas source is available, gas accumulation may continue up to a critical point where the sum of gas pressures equals absolute pressure (see Fig. 2.6; atmospheric pressure p_a plus hydrostatic pressure p_h):

$$\sum_i p_i = p_a + p_h \tag{2.2}$$

From this point, ebullition occurs.

Because of its high solubility (large Henry coefficient), CO_2 can reach concentrations of several liters per liter of water, as reported for Lake Nyos (Cameroon; Kling 1988). In this lake, some trigger mechanism released a large amount of the stored CO_2 gas and asphyxiated 1700 human beings in the lake surroundings (Kling et al. 1987). At least three more lakes are known where gas accumulation is a matter of concern: Lake Monoun, Cameroon (Rouwet et al. 2015); Kabuno Bay of Lake Kivu, Rwanda/Congo (Schmid et al. 2004a); and Guadiana pit lake in Herrerias Mine, Spain (Sánchez-España et al. 2014). Controlled degassing can remove the risks related to gas accumulations (e.g., Halbwachs et al. 2004; Boehrer et al. 2016). Also methane (e.g., Black Sea, McGinnis et al. 2006; Lake Matano, Indonesia, Chap. 10) and hydrogen sulfide are known to accumulate in monimolimnia. In Lake Kivu, methane concentrations are high enough for commercial exploitation (Tietze et al. 1980; Wüest et al. 2012).

2.4 Density Approaches

In meromictic lakes, dissolved substances play an important role for sustaining the stratification. Each solute contributes its share in density. If the composition of solutes is known, density can be calculated accurately from coefficients of partial molal volumes listed in the scientific literature of physical chemistry. Boehrer et al. (2010) included the partial molal volumes of limnologically important solutes in a convenient algorithm RHOMV, i.e., density ρ from partial molal volumes.

In addition, electrical conductivity is used as a quantitative bulk measure for the amount of dissolved substances. Electrical conductivity is easy to measure, e.g., by automatic probes, at high resolution and high accuracy. In conclusion, electrical conductivity has been used to calculate the density contribution of solutes to natural waters. In general, oceanographic formulae (UNESCO, Fofonoff and Millard 1983, International Organization for Standardization (ISO) 1985, TEOS-10) do not reflect the effect of solutes accurately for limnic waters (also Chen and Millero 1986), especially as meromictic lakes often show either high concentrations or unusual composition of solutes at least from an oceanographic point of view (e.g., tabled saline lakes in books by Hammer 1986 and Kalff 2002). To improve density calculations, Moreira et al. (2016) provided a numerical approach to evaluate a correlation between electrical conductivity and density for any chemical lake water composition. A simple formula is proposed for adding the density contribution of solutes to the density of pure water ρ_W from temperature T and electrical conductivity κ_{25} out of CTD profiles. Only two lake-specific coefficients λ_0 and λ_1 need to be determined:

$$\rho = \rho_W(T) + \kappa_{25}(\lambda_0 + \lambda_1 \cdot (T - 25\ ^\circ\mathrm{C})) \tag{2.3}$$

Both RHOMV and Moreira's density calculator can be used or downloaded from the Internet (Fig. 2.7). Calculations show that solute contributions to density are underestimated by typically between 20 % and 100 % with oceanographic approaches. Better accuracy can be achieved by measuring density of lake water and correlating the results with electrical conductivity and temperature (e.g., Jellison et al. 1999; Karakas et al. 2003). Sufficiently precise methods for in situ measurements of density in lakes are not available (Gräfe et al. 2002).

Dietz et al. (2012) showed that density differences can easily be calculated from "specific density fraction" as long as concentrations are in the freshwater range (<3 g/L). The authors have shown that in iron-meromictic Waldsee (Germany), the density difference between mixolimnion and monimolimnion is caused by higher concentration of iron and carbon species (CO_2, bicarbonate, DOC) in roughly equal parts.

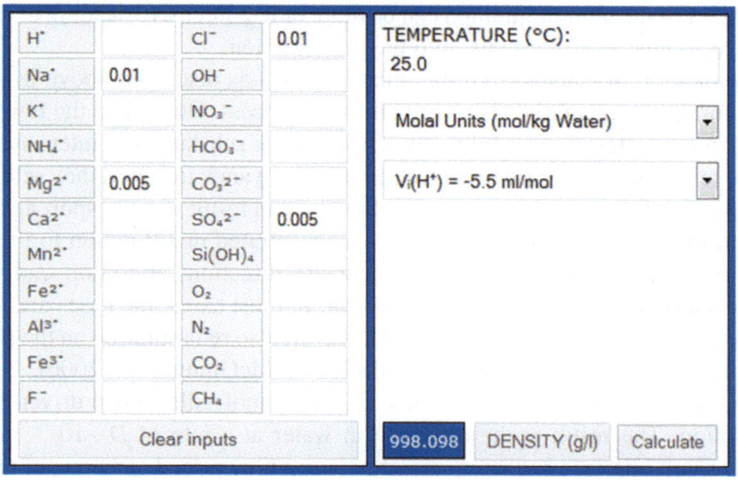

Fig. 2.7 The input mask for RHOMV on www.ufz.de/webax for easy calculation of density for given salt composition

2.5 Transport Under Conditions of Permanent Stratification

The stratification in meromictic lakes is often looked upon as very static phenomenon. Expressions like "permanently stratified" give the impression of an eternally sealed-off monimolimnion. This is not correct: we have already mentioned about discernible exchange processes between mixolimnion and monimolimnion and through the chemocline. The sharp interface would be smoothened by diffusion, but other transport processes cut the diffuse ends off and carry material away from (or to) the chemocline.

Meromixis sustained by salty inflows (surface or groundwater) is a result of flushing the monimolimnion. For example, renewal time for the Waldsee (Germany) monimolimnion is around one year (von Rohden et al. 2009). The volume of the monimolimnion would increase, if there were no volume losses due to monimolimnion erosion through turbulence and subsequent inclusion of the water into the mixolimnion. This is also true for lakes with partial renewal of the monimolimnion from sources within the lake (e.g., saltwater formation in south basin of Dead Sea before 1979: Nissenbaum 1969, Lerman et al. 1995; cold waters at south end of Lake Malawi/Nyasa: Vollmer et al. 2002a).

These erosive effects have been documented in Lake Mono (Jellison and Melack 1993, Jellison et al. 1998) and quantitatively measured in Waldsee (Germany; von Rohden et al. 2009) and Wallendorfer See and Rassnitzer See (Germany; Boehrer et al. 2014). In most cases, erosion of the monimolimnion takes place during intense recirculation periods, i.e., in winter. In Waldsee (Germany), this effect happens during nocturnal cooling during summer. Depending on weather conditions, the shaving off of the monimolimnion can be more or less intensive, leading to irregular inputs of solutes, e.g., nutrients, into the mixolimnion.

Precipitation of solutes from the mixolimnion—as discussed above in the processes forming meromixis—is a very effective process. Other than the transport of solutes, precipitates transport almost exclusively the substances of interest and only little water. As most precipitates have negative buoyancy in water, they settle to the lake bed. A redissolution in the monimolimnion can result in a transport against the concentration gradient, i.e., from the low-concentrated mixolimnion to the highly concentrated monimolimnion. In the opposite direction, ebullition can remove volatile substances from the monimolimnion. In most cases, the gases are released into the atmosphere at the lake surface. However, also redissolution on the path into mixolimnetic waters is possible (e.g., Black Sea: McGinnis et al. 2006).

Even under weak water movements, dissolved molecules move driven by thermal motion. The result is the diffusion in water at a rate of $D \sim 10^{-9}$ m^2/s. An initially sharp interface becomes smoother over a time period t at a rate of the order of magnitude of

$$\delta \sim (Dt)^{0.5} \tag{2.4}$$

Thus, an originally perfectly sharp interface will smoothen to an interface of close to 1 m in about a year. This is very slow and diffusive transports are less effective than turbulent transport processes.

Several mechanisms create instabilities in the open waters (e.g., Imboden and Wüest 1995), but for small- and middle-sized lakes, mixing in the deep water is dominated by processes acting close to the side walls (e.g., Goudsmit et al. 1997). Flows over rough terrain create current shear and hence turbulence, and together with the retention of water in the interspaces of coarser sediment, contribute the bulk of the transport of dissolved substances. High-density gradients result in small vertical excursions of water movements and hence small vertical transport. Under the assumption that at any depth z, the same amount of energy E is transferred into mixing at an efficiency of γ, we expect a relation for the vertical transport coefficients K_z of the form

$$K_z(z) = \gamma \cdot E/N^2 \quad \text{where} \quad N^2(z) = -\frac{g}{\rho}\frac{\partial \rho}{\partial z} \tag{2.5}$$

Brunt-Väisälä frequency (squared) represents the density ρ gradient along the vertical axis z against earth acceleration g. Transport coefficients measured by spiking the deep water with the artificial tracer SF_6 revealed the strong dependence of transport on density stratification (von Rohden and Ilmberger 2001). At levels of highest stratification in lakes Wallendorf and Rassnitz (Germany), vertical transport was close to molecular levels for both heat and solutes. In a number of fjord lakes, seawater was trapped in the deep layers for thousands of years (e.g., in Salsvatn (Norway) for ca. 3000 years, Bøyum 1973; in Rørholtfjorden (Norway) ca. 6000 years, Strøm 1963; and in Powell Lake (Canada) 9000–11,000 years, Williams et al. 1961). Vertical mixing was so low that even today part of this seawater can be found in the deepest areas of the lakes. In conclusion, the effective turbulent mixing was only little higher than molecular diffusion. In fact in Powell Lake (Canada), it is claimed that the different distribution of ions stems from the differences in molecular diffusivity of these substances (Sanderson et al. 1986).

In comparison with solutes, diffusive transport of heat is two orders of magnitude faster. Consequently, deep waters stratified by dissolved substances can show double diffusion, if temperature gradients inflict a reduction of the density gradient (see also Brandt and Fernando 1996; Boehrer 2012). In initially stable zones with disappearing turbulence, the greatly higher diffusivity of heat than that of solutes can impose locally unstable conditions, which form thin convection cells (decimeters) of large horizontal extension (kilometers) (Newman 1976; Schmid et al. 2004b; von Rohden et al. 2010), separated by even thinner layers of strong density gradients (Fig. 2.8). Vertical transport of solutes as a consequence is controlled by the local production of convection and turbulence as a result of double-diffusive convection and can, therefore, be considerably more efficient than molecular diffusivity (e.g., Schmid et al. 2004b). In Waldsee, the monimolimnetic overturn was clearly driven by double-diffusive effects modified by chemical reactivity of solutes (Boehrer et al. 2009).

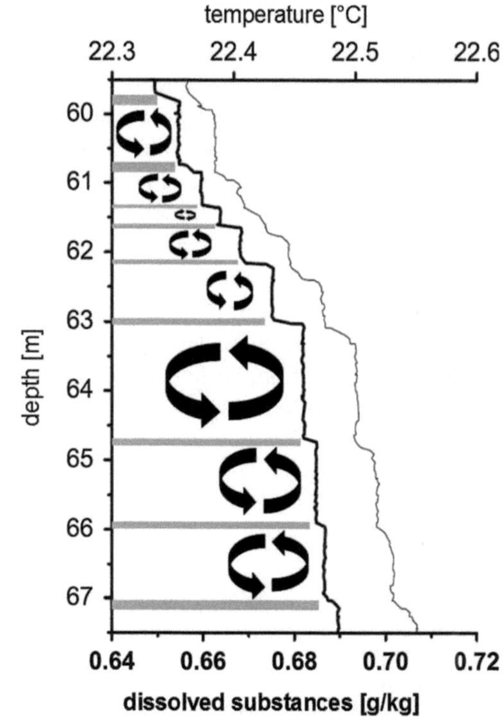

Fig. 2.8 Double-diffusive convection cells creating a staircase of the concentration of dissolved substances (*fine line*) and temperature (*thick line*) in a profile of Lake Nyos, Cameroon (modified from Schmid et al. 2004b). Convection cells, several decimeters thick, alternate with layers of no convection (in *gray*), where heat is transported diffusively (from Boehrer 2012)

Because transport through density gradients at chemoclines is slow, it has created interest in utilizing deep waters for the disposal or the confinement of undesirable substances (organics from lignite processing: Stottmeister et al. 1998; heavy metals: Fisher 2002; Fisher and Lawrence 2006; see also Sects. 9.4 and 9.5) in monimolimnia. Notably, the sealing is not perfect and transport still happens (Stevens and Lawrence 1997). Also heat can be trapped in monimolimnia of so-called solar ponds, to store energy or to culture organisms requiring higher water temperatures in colder climates (Weinberger 1964; Kirkland et al. 1983; Bozkurt et al. 2015). Since aspects of management are particularly relevant for meromictic pit lakes, such issues are discussed in Sect. 9.5, and an example is provided in Sect. 9.4 of Chap. 9.

2.6 Concluding Remarks

Meromictic waterbodies are subdivided into two chemically different water layers. While the upper layer (epilimnion or mixolimnion) is subject to exchange with the atmosphere, the deep water (monimolimnion) is isolated and hence does not receive any oxygen from the atmosphere. As this situation is generally maintained for long periods, other dissolved substances accumulate in the monimolimnion in a chemically reduced form.

The chemocline usually represents a sharp transition within a few decimeters (sometimes meters) between the mixolimnion and the monimolimnion. The higher concentration of solutes is usually responsible for the greater stability of the density-induced stratification. Consequently, the chemocline also represents a density gradient, i.e., pycnocline. Settling particles of a density between mixolimnion water and monimolimnion water can find their level of neutral buoyancy within the chemocline. Such particles may be found floating within the chemocline.

In the chemocline, substances from mixolimnion and monimolimnion get into contact and can react with each other, even supplying energy, e.g., dissolved iron from monimolimnion and dissolved oxygen from the mixolimnion. Such zones can be identified by their turbidity through visual inspection (Fig. 2.3). Although most of the reactions are mediated by organisms, the exothermal nature of reactions facilitates the reaction, without involvement of an additional energy source, e.g., light for photosynthesis.

References

Aeschbach-Hertig W, Hofer M, Kipfer R, Imboden DM, Wieler R (1999) Accumulation of mantle gases in a permanently stratified volcanic lake (Lac Pavin, France). Geochim Cosmochim Acta 63:3357–3372

Boehrer B (2012) Double-diffusive convection in lakes. In: Bentsson L, Herschy RW, Fairbridge RW (eds) Encyclopedia of lakes and reservoirs. Springer, Dordrecht, pp 223–224

Boehrer B (2013) Physical properties of acidic pit lakes. In: Geller W, Schultze M, Kleinmann R, Wolkersdorfer C (eds) Acidic pit lakes. Springer, Heidelberg, pp 23–42

Boehrer B, Schultze M (2008) Stratification of lakes. Rev Geophys 46:RG2005. doi:10.1029/2006RG000210

Boehrer B, Matzinger A, Schimmele M (2000) Similarities and differences in the annual temperature cycles of East German mining lakes. Limnologica 30:271–279

Boehrer B, Fukuyama R, Chikita K (2008) Stratification of very deep, thermally stratified lakes. Geophys Res Lett 35:L16405. doi:10.1029/2008GL034519

Boehrer B, Dietz S, von Rohden C, Kiwel U, Jöhnk KD, Naujoks S, Ilmberger J, Lessmann D (2009) Double-diffusive deep water circulation in an iron-meromictic lake. Geochem Geophys Geosyst 10:Q06006. doi:10.1029/2009GC002389

Boehrer B, Herzsprung P, Schultze M, Millero FJ (2010) Calculating density of water in geochemical lake stratification models. Limnol Oceanogr Methods 8:567–574. doi:10.4319/lom.2010.8.567

Boehrer B, Kiwel U, Rahn K, Schultze M (2014) Chemocline erosion and its conservation by freshwater introduction to meromictic salt lakes. Limnologica 44:81–89

Boehrer B, Yusta I, Magin K, Sanchez-España J (2016) Quantifying, assessing and removing the extreme gas load from meromictic Guadiana pit lake, Southwest Spain. Sci Total Environ 563–564:468–477

Böhrer B, Heidenreich H, Schimmele M, Schultze M (1998) Numerical prognosis for salinity profiles of future lakes in the open cast mine Merseburg-Ost. Int J Salt Lake Res 7:235–260

Bøyum A (1973) Salsvatn, a lake with old sea water. Schweiz Zeitschr Hydrol 35:262–277

Bøyum A, Kjensmo J (1970) Kongressvatn. A crenogenic meromictic lake at Western Spitsbergen. Arch Hydrobiol 67:542–552

Bozkurt I, Mantar S, Karakilcik M (2015) A new performance model to determine energy storage efficiencies of a solar pond. Heat Mass Transf 51:39–48

Brandt A, Fernando HJS (eds) (1996) Double diffusive convection. Geophys Monogr Ser, vol 94. AGU, Washington, DC, p 334

Campbell P, Torgersen T (1980) Maintenance of iron meromixis by iron redeposition in a rapidly flushed monimolimnion. Can J Fish Aquat Sci 37:1303–1313

Carmack EC, Weiss RF (1991) Convection in Lake Baikal: an example of thermobaric instability. In: Chu PC, Gascard, JC (eds) Deep convection and deep water formation in the Oceans. Elsevier, Amsterdam, pp 215–228,

Casamitjana X, Roget E (1993) Resuspension of sediment by focused groundwater in Lake Banyoles. Limnol Oceanogr 38:643–656

Chen C-TA, Millero FJ (1986) Precise thermodynamic properties for natural waters covering only the limnological range. Limnol Oceanogr 31:657–662

Crawford GB, Collier RW (2007) Longterm observations of hypolimnetic mixing in Crater Lake, Oregon. Hydrobiologia 574:47–68

Dietz S, Lessmann D, Boehrer B (2012) Contribution of solutes to density stratification in a meromictic lake (Waldsee/Germany). Mine Water Environ 31:129–137

Eloranta P (1999) Light penetration and thermal stratification in lakes. In: Keskitalo J, Eloranta P (eds) Limnology of humic waters. Backhuys Publishers, Leiden, pp 72–74

Findenegg I (1933) Alpenseen ohne Vollzirkulation. Int Rev Gesamten Hydrobiol Hydrogr 28: 295–311

Findenegg I (1935) Limnologische Untersuchungen im Kärntener Seengebiete. Ein Beitrag zu Kenntnis des Stoffhaushaltes in Alpenseen. Int Rev Gesamten Hydrobiol Hydrogr 32:369–423

Findenegg I (1937) Holomiktische und meromiktische Seen. Int Rev Gesamten Hydrobiol Hydrogr 35:586–610

Fisher TSR (2002) Limnology of the meromictic Island Copper Mine pit lake. Dissertation, University of British Columbia

Fisher TSR, Lawrence GA (2006) Treatment of acid rock drainage in a meromictic pit lake. J Environ Eng 132:515–526

Fofonoff NP, Millard RC Jr (1983) Algorithms for commutation of fundamental properties of seawater. UNESCO Tech Pap Mar Sci 44

Frey DG (1955) Längsee: a history of meromixis. Mem Ist Ital Idrobiol Suppl 8:141–164

Gibson JAE (1999) The meromictic lakes and stratified marine basins of the Vestfold Hills, East Antarctica. Antarct Sci 11:175–192

Goudsmit G-H, Peeters F, Gloor M, Wüest A (1997) Boundary versus internal diapycnal mixing in stratified natural waters. J Geophys Res 102(C13):27903–27914

Gräfe H, Boehrer B, Hoppe N, Müller SC, Hauptmann P (2002) Ultrasonic in situ measurements of density, adiabatic compressibility, and stability frequency. Limnol Oceanogr 47:1255–1260

Halbwachs M, Sabroux J-C, Grangeon J, Kayser G, Tochon-Danguy J-C, Felix A, Beard J-C, Villevielle A, Voter G, Richon P, Wuest A, Hell J (2004) Degassing the "Killer Lakes" Nyos and Monoun, Cameroon. Eos Trans AGU 85:281–285

Hammer UT (1986) Saline lake ecosystems of the world. Dr W Junk Publishers, Dordrecht

Hammer UT (1994) Life and times of five Saskatchewan saline meromictic lakes. Int Rev Gesamten Hydrobiol 79:235–248

Hammer UT, Haynes RC (1978) The saline lakes of Saskatche-wan. II. Locale, hydrography and other physical aspects. Int Rev Gesamten Hydrobiol 63:179–203

Heidenreich H, Boehrer B, Kater R, Hennig G (1999) Gekoppelte Modellierung geohydraulischer und limnophysikalischer Vorgänge in Tagebaurestseen und ihrer Umgebung. Grundwasser 4: 49–54

Hongve D (1997) Cycling of iron, manganese, and phosphate in a meromictic lake. Limnol Oceanogr 42:635–647

Hongve D (2002) Seasonal mixing and genesis of endogenic meromixis in small lakes in Southeast Norway. Nord Hydrol 33:189–206

Hutchinson GE (1937) A contribution to the limnology of arid regions: primarily founded on observations made in the Lahontan basin. Trans Connecticut Acad Arts Sci 33:47–132

Hutchinson GE (1957) A treatise on limnology, Geography, physics and chemistry, vol 1. Wiley, New York

Imboden DM, Wüest A (1995) Mixing mechanisms in lakes. In: Lerman A, Imboden DM, Gat JR (eds) Physics and chemistry of lakes. Springer, Berlin, pp 83–138

International Organization for Standardization (ISO) (1985) Water quality: determination of electrical conductivity, Standard 7888. ISO, Geneva

Jellison R, Melack JM (1993) Meromixis in hypersaline Mono Lake, California. 1. Vertical mixing and density stratification during the onset, persistence, and breakdown of meromixis. Limnol Oceanogr 38:1008–1019

Jellison R, Romero J, Melack JM (1998) The onset of meromixis during restoration of Mono Lake, California: unintended consequences of reducing water diversions. Limnol Oceanogr 43: 706–711

Jellison R, MacIntyre S, Millero FJ (1999) Density and conductivity properties of NA-CO_3-Cl-SO_4 brine from Mono Lake, California, USA. Int J Salt Lake Res 8:41–53

Kaden H, Peeters F, Lorke A, Kipfer R, Tomonaga Y, Karabiyikoglu M (2010) Impact of lake level change on deep-water renewal and oxic conditions in deep saline Lake Van, Turkey. Water Resour Res 46:W11508. doi:10.1029/2009WR008555

Kalff J (2002) Limnology, 2nd edn. Prentice Hall, Upper Saddle River, NJ

Karakas G, Brookland I, Boehrer B (2003) Physical characteristics of acidic Mining Lake 111. Aquat Sci 65:297–307

Kerry KR, Grace DR, Williams R, Burton HR (1977) Studies on some saline lakes of the Vestfold Hills, Antarctica. In: Llano GAS (ed) Adaptations within antarctic ecosystems. Gulf, Houston, Texas, pp 839–858

Kirkland DW, Platt Bradbury J, Dean WE (1983) The heliothermic lake—a direct method of collecting and storing solar energy. Arch Hydrobiol Suppl 65:1–60

Kjensmo J (1967) The development and some main features of "ironmeromictic" soft water lakes. Arch Hydrobiol 32:137–312

Kjensmo J (1988) Post-glacial sediments and the stagnation history of the iron-meromictic Lake Skjennungen, Eastern Norway. Arch Hydrobiol 113:481–499

Kling GW (1988) Comparative transparency, depth of mixing and stability of stratification in lakes of Cameroon, West Africa. Limnol Oceanogr 33:27–40

Kling GW, Clark MA, Compton HR, Devine JD, Evans WC, Humphrey AM, Koenigsberg EJ, Lockwood JP, Tuttle ML, Wagner GN (1987) The 1986 Lake Nyos gas disaster in Cameroon, West Africa. Science 236:169–175

Kodenev GG (2001) Deep-water renewal in Lake Baikal. Geol Geofiz 42:1127–1136

Lerman A, Imboden DM, Gat JR (eds) (1995) Physics and chemistry of lakes. 2nd edn, Springer, Berlin

Lindholm T (1996) Periodic anoxia in an emerging coastal lake basin in SW Finland. Hydrobiologia 325:223–230

Ludlam SD, Duval B (2001) Natural and management induced reduction in monimolimnetic volume and stability in a coastal, meromictic lake. Lake Reserv Manag 17:71–81

Marion GM, Farren RE (1999) Mineral solubilities in the Na-K-Mg-Ca-Cl-SO_4-H_2O system: a re-evaluation of the sulfate chemistry in the Spencer-Møller-Weare model. Geochim Cosmochim Acta 63:1305–1318

McGinnis DF, Greinert J, Artemov Y, Beaubien SE, Wüest A (2006) Fate of rising methane bubbles in stratified waters: how much methane reaches the atmosphere? J Geophys Res 111: C09007. doi:10.1029/2005JC003183

Merilainen J (1970) On the limnology of the meromictic Lake Valkiajarvi, in the Finnish lake district. Ann Bot Fenn 7:29–51

Moreira S, Schultze M, Rahn K, Boehrer B (2016) A practical approach to lake water density from electrical conductivity and temperature. Hydrol Earth Syst Sci Discuss 20:2975–2986. doi:10.5194/hess-2016-36

Newman FC (1976) Temperature steps in Lake Kivu: a bottom heated saline lake. J Phys Oceanogr 6:157–163

Nissenbaum A (1969) Studies in the geochemistry of the Jordan River-Dead Sea system. Dissertation, University of California

Nixdorf E, Boehrer B (2015) Quantitative analysis of biogeochemically controlled density stratification in an iron-meromictic lake. Hydrol Earth Syst Sci 19:4505–4515

Ozretich RJ (1975) Mechanisms for deep water renewal in Lake Nitinat, a permanently anoxic fjord. Estuar Coast Mar Sci 3:189–200

Peeters F, Kipfer R, Achermann D, Hofer M, Aeschbach-Hertig W, Beyerle U, Imboden DM, Rozanski K, Fröhlich K (2000) Analysis of deep-water exchange in the Caspian Sea based on environmental tracers. Deep-Sea Res I Oceanogr Res Pap 47:621–654

Peeters F, Finger D, Hofer M, Brennwald M, Livingstone DM, Kipfer R (2003) Deep-water renewal in Lake Issyk-Kul driven by differential cooling. Limnol Oceanogr 48:1419–1431

Rodrigo MA, Miracle MR, Vicente E (2001) The meromictic Lake La Cruz (Central Spain), patterns of stratification. Aquat Sci 63:406–416

Rouwet D, Christenson B, Tassi F, Vandemeulenbrouck J (eds) (2015) Volcanic Lakes. Springer, Heidelberg

Salonen K, Arvola L, Rask M (1984) Autumnal and vernal circulation of small forest lakes in Southern Finland. Verh Int Ver Limnol 22:103–107

Sánchez-España J, Boehrer B, Yusta I (2014) Extreme carbon dioxide concentrations in acidic pit lakes provoked by water/rock interaction. Environ Sci Technol 48:4273–4281

Sander R (1999) Compilation of Henry's law constants for inorganic and organic species of potential importance in environmental chemistry. http://www.mpch-mainz.mpg.de/~sander/res/henry.html. Accessed 15 Jan 2016

Sanderson B, Perry K, Pedersen T (1986) Vertical diffusion in meromictic Powell Lake, British Columbia. J Geophys Res 91(C6):7647–7655

Santofimia E, López-Pamo E, Reyes J (2012) Changes in stratification and iron redox cycle of an acidic pit lake in relation with climatic factors and physical processes. J Geochem Explor 116–117:40–50

Scharf BW, Oehms M (1992) Physical and chemical characteristics. Arch Hydrobiol Beih 38: 63–83
Schmid M, Lorke A, Dinkel C, Tanyileke G, Wüest A (2004a) Double-diffusive convection in Lake Nyos, Cameroon. Deep-Sea Res I 51:1097–1111
Schmid M, Tietze K, Halbwachs M, Lorke A, McGinnis D, Wüest A (2004b) How hazardous is the gas accumulation in Lake Kivu? arguments for a risk assessment in light of the Nyiragongo Volcano eruption of 2002. Acta Vulcanol 14(15):115–122
Sibert RJ, Koretsky CM, Wyman DA (2015) Cultural meromixis: effects of road salt on the chemical stratification of an urban kettle lake. Chem Geol 395:126–137
Stevens CL, Lawrence GA (1997) The effect of subaqueous disposal of tailings in standing waters. J Hydraul Res 35:147–159
Stottmeister U, Weißbrodt E, Becker PM, Pörschmann J, Kopinke FD, Martius GM, Wießner A, Kennedy C (1998) Analysis, behaviour and fate of a lignite pyrolysis wastewater deposit. In: Contaminated soil 98. Thomas Telford, London, pp 113–121
Strøm KM (1957) A lake with trapped sea-water? Nature 180:982–983
Strøm KM (1963) Trapped sea water. New Sci 274:384–386
Stumm W, Morgan JJ (1996) Aquatic chemistry, 3rd edn. Wiley, New York
TEOS-10: IOC, SCOR, and IAPSO (2010) The international thermodynamic equation of seawater–2010: calculation and use of thermodynamic properties. Comm Manuals Guide. http://www.TEOS-10.org. Accessed 15 Jan 2016
Tietze K, Geyh M, Müller H, Schröder L (1980) The genesis of methane in Lake Kivu. Geol Rundsch 69:452–472
Vollmer MK, Weiss RF, Bootsma HA (2002a) Ventilation of lake Malawi/Nyasa. In: Odada EO, Olago DO (eds) The East African great lakes: limnology, paleolimnology and biodiversity. Kluwer, Dordrecht, pp 209–233
Vollmer MK, Weiss RF, Williams RT, Falkner KK, Qiu X, Ralph EA, Romanovsky VV (2002b) Physical and chemical properties of the waters of saline lakes and their importance for deep-water renewal: lake Issyk-Kul, Kyrgyzstan. Geochim Cosmochim Acta 66:4235–4246
von Rohden C, Ilmberger J (2001) Tracer experiment with sulfurhexafluoride to quantify the vertical transport in a meromictic pit lake. Aquat Sci 63:417–431
von Rohden C, Ilmberger J, Boehrer B (2009) Assessing groundwater coupling and vertical exchange in a meromictic mining lake with an SF6-tracer experiment. J Hydrol 372:102–108
von Rohden C, Boehrer B, Ilmberger J (2010) Evidence for double diffusion in temperate meromictic lakes. Hydrol Earth Syst Sci 14:667–674
Walker KF, Likens GE (1975) Meromixis and a reconsidered typology of lake circulation patterns. Verh Int Ver Limnol 19:442–458
Weinberger H (1964) The physics of the solar pond. Sol Energy 8:45–56
Weiss RF, Carmack EC, Koropalov VM (1991) Deepwater renewal and biological production in Lake Baikal. Nature 349:665–669
Wetzel RG (2001) Limnology: lake and river ecosystems, 3rd edn. Academic, San Diego
Williams PM, Mathews WH, Pckard GL (1961) A lake in British Columbia containing old sea water. Nature 191:830–832
Wüest A, Jarc L, Bürgmann H, Pasche N, Schmid M (2012) Methane formation and future extraction in Lake Kivu. In: Descy J-P, Darchambeau F, Schmid M (eds) Lake Kivu—limnology and biogeochemistry of a tropical great lake. Springer, Dordrecht, pp 165–180
Yoshimura S (1935) A contribution to the knowledge of deep water temperatures of Japanese lakes. Part 1. Summer temperature. Jpn J Astron Geophys 13:61–120
Yoshimura S (1936) Deep water temperatures of lakes of Japan in winter. Sea Air 15:195–208
Zachara JM, Moran JJ, Resch CT, Lindemann SR, Felmy AR, Bowden ME, Cory AB, Fredrickson JK (2016) Geo- and biogeochemical processes in a heliothermal hypersaline lake. Geochim Cosmochim Acta 181:144–163. doi:10.1016/j.gca.2016.02.001

Chapter 3
Chemical Setting and Biogeochemical Reactions in Meromictic Lakes

Martin Schultze, Bertram Boehrer, Katrin Wendt-Potthoff, Sergei Katsev, and Erik T. Brown

3.1 Introduction

The chemical composition of meromictic lake waters varies widely, as it does in lakes in general. Concentrations of total dissolved substances (TDS) range from very low (<20 mg L^{-1}) to very high (>300 g L^{-1}), i.e. saturated with respect to particular salts. The pH varies from acidic (<3) to alkaline (>10), and the redox conditions range from well oxygenated and dominated by high concentrations of dissolved ferric iron (Eh about 600 mV in iron-rich acidic pit lakes) to strongly reduced (Eh < -100 mV). While the above-mentioned ranges of TDS and pH apply to both mixolimnia and monimolimnia, redox conditions are typically oxic for mixolimnia (except hypolimnia in some meromictic lakes during thermal stratification) and permanently anoxic for monimolimnia. Concentrations of reduced chemical species, e.g. ferrous iron, hydrogen sulphide and ammonia, vary over a wide range in monimolimnia depending on local conditions and the development of the lakes in the past (see Sect. 3.2).

There are three main special features of meromictic lakes related to biogeochemistry. Firstly, the different chemical compositions of mixolimnion and

M. Schultze (✉) • B. Boehrer (✉) • K. Wendt-Potthoff
Department Lake Research, UFZ Helmholtz-Centre for Environmental Research, Magdeburg, Germany
e-mail: martin.schultze@ufz.de; bertram.boehrer@ufz.de

S. Katsev
Large Lakes Observatory, University of Minnesota Duluth, Duluth, MN 55812, USA

Department of Physics, University of Minnesota Duluth, Duluth, MN 55812, USA

E.T. Brown
Large Lakes Observatory, University of Minnesota Duluth, Duluth, MN 55812, USA

Department of Earth & Environmental Sciences, University of Minnesota Duluth, Duluth, MN 55812, USA

© Springer International Publishing AG 2017

R.D. Gulati et al. (eds.), *Ecology of Meromictic Lakes*, Ecological Studies 228,
DOI 10.1007/978-3-319-49143-1_3

monimolimnion are essential for the stability of the permanent stratification in the meromictic lakes (e.g. Findenegg 1935; Hutchinson 1957; Boehrer and Schultze 2008). Secondly, a number of processes that occur in the water column of meromictic lakes are known to occur in sediments, but such processes usually do not occur in the water column of holomictic lakes. Such processes are in particular anoxygenic photosynthesis (e.g. Meyer et al. 2011; Crowe et al. 2008a), formation of pyrite (e.g. Perry and Pedersen 1993) and anaerobic ammonia oxidation (Schubert et al. 2006; Hamersley et al. 2009; Wenk et al. 2014). Thirdly, the permanently anoxic conditions above the monimolimnetic sediments of meromictic lakes provide better conditions for the conservation of settling organic material and prevent disturbance by bioturbation (e.g. Sanger 1988; Meyers and Ishiwatari 1995; Wanchun et al. 2001; Corella et al. 2011).

Based on these special conditions, we divide this chapter into three sections in addition to this introduction: in Sect. 3.2, we present ten selected examples to illustrate the variety of chemical conditions in meromictic lakes. We refer also to appropriate case studies presented in Chaps. 5–12. Section 3.3 is dedicated to biogeochemical processes that have the potential for creating and sustaining meromixis and that occur in the water column of meromictic lakes but usually not in the water column of holomictic lakes. Special biogeochemical aspects of monimolimnetic sediments in meromictic lakes are presented in Sect. 3.4 from a palaeolimnologic point of view.

3.2 Chemical Setting in Meromictic Lakes

Table 3.1 provides examples that illustrate a variety of chemical compositions in meromictic lakes. The first six examples are arranged in order of increasing TDS levels in the mixolimnion. They are followed by an example of trapped seawater in the monimolimnion, and finally one alkaline and two acidic lakes. Chapters 5–12 add further examples for chemical conditions that greatly differ.

The first example in Table 3.1 is Hall Lake (WA, USA), a small freshwater lake with very low concentrations of dissolved substances (Culver 1977) in both mixolimnion and monimolimnion (TDS 64 mg L^{-1} and 117 mg L^{-1}, respectively), as the catchment is poor in soluble minerals. Balistrieri et al. (1994) reported only slightly higher concentrations, particularly in the mixolimnion after a period of holomixis and re-establishment of meromixis (TDS mixolimnion 110 mg L^{-1} and monimolimnion 237 mg L^{-1}). Bicarbonate, carbon dioxide, dissolved iron and dissolved silica were key contributors to the density gradients.

TDS concentrations of Lake Malawi (Malawi, Tanzania, Mozambique; Chap. 10) are comparable with those of Hall Lake. Lake Tanganyika (Democratic Republic of the Congo, Burundi, Tanzania, Zambia; Chap. 10) and Lake Matano (Indonesia; Chap. 10) have slightly higher concentrations of dissolved substances. Lake Matano has relatively high concentrations of ferrous iron (ca. 8 mg L^{-1}) and

Table 3.1 Chemical conditions and morphometric characteristics in selected meromictic lakes

	Hall Lake	Nordbytjernet	Längsee	Fayetteville Green Lake	Waldsea Lake	Hot Lake	Powell Lake	Big Soda Lake	Moritzteich	Camp Lake
Country	USA	Norway	Austria	USA	Canada	USA	Canada	USA	Germany	Canada
pH	7.1	7.4	8.5	7.9	8.1	8.27	5.7	9.68	3.2	3.9
	6.6	***7.5***	***7.1***	***7.2***	***7.2***	***6.85***	***6.8***	***9.71***	***6.9***	***5.6***
Cl^-	5.0	5.1	12.7	34.1	2,700	2,040	29.0	6,900	19	dna
	4.0	***4.2***	***12.2***	***66.0***	***8,500***	***2,800***	***9,308***	***27,600***	***27***	***dna***
SO_4^{2-}	13	29.5	16.1	1,119	8,250	85,700	4.9	5,810	351	70
	3.7	***5.6***	***1.4***	***1,344***	***30,400***	***356,000***	***1,903***	***6,700***	***598***	***14,500***
HCO_3^-	18.3	71	212	200	279	671	1.6	4,060	0	0
	52.7	***350***	***372***	***433***	***850***	***2,360***	***4,991***	***24,100***	***756***	***2.5***
CO_2	1.8	10	1.9	dna	dna	dna	dna	dna	35.2	dna
	13.6	***27***	***1.1***	***dna***	***dna***	***dna***	***dna***	***dna***	***735***	***dna***
Na	4.8	3.3	5.6	17.0	2,400	8,350	1.6	8,130	12.4	2.4
	4.3	***3.3***	***5.4***	***29.9***	***8,000***	***37,000***	***4,991***	***26,800***	***16.1***	***33***
K	1.9	1.6	3.1	3.0	160	1,750	0.2	318	6	2.9
	2.4	***2.3***	***3.8***	***6.6***	***560***	***3,660***	***157***	***1,190***	***8.9***	***29***
Mg	2.7	3.0	27.1	71.9	1,650	18,400	0.26	148	17.1	4.8
	3.6	***3.5***	***28.4***	***80.2***	***6,000***	***71,900***	***673***	***5.7***	***28.1***	***439***
Ca	7.5	30.2	36.4	420	280	328	0.76	5.0	73	11.3
	9.2	***40.8***	***72.6***	***561***	***104***	***242***	***276***	***0.8***	***118***	***314***
SiO_2	8.9	dna	0.64	2.8	dna	<25	2.55	4.0	37.6	dna
	14	***32.1***	***16.0***	***3.4***	***dna***	***<31***	***16.4***	***208***	***51.8***	***dna***
S^{2-}	dna	dna	dna	0	dna	dna	0	<0.5	dna	<0.001
	dna	***dna***	***dna***	***38.4***	***dna***	***dna***	***162***	***386***	***dna***	***0.030***
Fe	0.01	0.22	<0.02	0.01	dna	5.5	0.005	0.02	21	9.03
	4.34	***34.2***	***0.048***	***0.04***	***dna***	***<2***	***0.003***	***0.07***	***720***	***9,370***

(continued)

Table 3.1 (continued)

	Hall Lake	Nordbytjernet	Längsee	Fayetteville Green Lake	Waldsea Lake	Hot Lake	Powell Lake	Big Soda Lake	Moritzteich	Camp Lake
Mn	dna	0.5	<0.002	0.038	dna	<0.001	0.001	0.002	0.69	0.097
	dna	***51.5***	***0.307***	***0.461***	***dna***	***0.78***	***0.041***	***0.004***	***1.57***	***18***
NO_3^-	0.186	1.15	0.741	1.80	dna	dna	0.180	0.058	0.36	0
	0.062	***0***	***0.013***	***0.006***	***dna***	***dna***	***0***	***<0.03***	***0***	***0***
NH_4^+	0	0.45	0.050	0.10	dna	dna	0	<0.1	3.21	<0.1
	4.27	***11.83***	***10.4***	***1.80***	***dna***	***dna***	***64.6***	***43.4***	***13.5***	***5.18***
TP	0.029	<0.01	0.008	0.068	dna	21.1	0	4.17	0.004	0.003
	0.249	***0.08***	***0.100***	***0.214***	***dna***	***68.5***	***0.114***	***57.1***	***0.040***	***0.081***
DOC	dna	dna	6.84	dna	dna	162	5.6	20	0.5	2.6
	dna	***30***	***8.07***	***dna***	***dna***	***382***	***23.1***	***60***	***77.6***	***1.5***
Type	Endogenic IV	Endogenic IV	Ectogenic II	Endogenic IV	Endogenic V	Ectogenic I	Ectogenic I	Ectogenic I	Ectogenic III	Ectogenic III
A	3.1	26.8	75	25.8	464	1.27	dna	150	16	120
V	0.29	2.58	9.2	7.2	37.7	0.01	dna	42,1	1.2	2.7
z_{max}	16.2	22.8	21.4	52.5	14.3	2.5	358	64	17.5	10.5

Concentrations in mg L^{-1}; data for monimolimnia in *bold italics*; *dna* data not available; *Type* type of meromixis according to Walker and Likens (1975); *A* surface area in ha; *V* volume in 10^6 m^3; z_{max} maximal depth in m; used references are given in Table 3.2

Table 3.2 Reference sources for the data presented in Table 3.1

Lake	References
Hall Lake	Culver (1977)
Nordbytjernet	Hongve (1974, 1980, 2002)
Längsee	Monitoring data (September 16, 2013) provided by the Kärntner Institut für Seenforschung, Klagenfurt, Austria
Fayetteville Green Lake	Brunskill and Ludlam (1969) and Brunskill (1969)
Waldsea Lake	Hammer (1986, 1994)
Hot Lake	Zachara et al. (2016) (detailed chemical data provided by Zachara et al. for October 23, 2012)
Powell Lake	Perry (1990)
Big Soda Lake	Kharaka et al. (1984)
Moritzteich	Stellmacher (2004) and von Rohden et al. (2010)
Camp Lake	Moncur et al. (2006)

also rather high concentrations of methane (up to 25.2 mg L^{-1}) in its monimolimnion (see Chap. 10).

Lake Nordbytjernet (Norway; Table 3.1) also has relatively low concentrations of dissolved substances. Its most interesting features are the concentrations of dissolved manganese and iron, their biogeochemical cycling and their contributions to the stability of meromixis (see Sect. 3.3.1 and Chap. 2).

Whereas the lakes mentioned above could be called "soft water" lakes, the third and fourth lakes in Table 3.1 are fresh but "hard water" lakes. Lake Längsee (Austria) was selected based on its TDS levels and for its historical significance, as it was one of the Carinthian lakes that were used by Findenegg (1935) to define the term "Meromixis". Later, Längsee became well known for several palaeolimnological investigations (e.g. Frey 1955; Schmidt et al. 2002; Huber et al. 2010).

Fayetteville Green Lake (NY, USA; Table 3.1) is characterised by a high level of sulphate, supported by the local catchment geology and the correspondingly high concentrations of hydrogen sulphide in the monimolimnion. The monimolimnion waters are close to saturation with respect to gypsum ($CaSO_4{\cdot}2H_2O$), and pyrite precipitates because of iron and sulphate reduction in the monimolimnion and the resulting formation of ferrous iron and of high concentration of hydrogen sulphide (Suits and Wilkin 1998; see also Sect. 3.3.2).

Of the naturally meromictic lakes described in this book, lakes La Cruz (Spain), Cadagno (Switzerland) and Kivu (Democratic Republic of the Congo, Rwanda; Chaps. 7, 8 and 10, respectively) are fresh hard water lakes. Lake La Cruz's chemistry is dominated by the presence of limestone in its catchment and by the biogenic calcite precipitation in its mixolimnion (Chap. 8). In Lake Cadagno, inputs of dissolved gypsum from a small part of its catchment assist the lake to sustain meromixis (Chap. 7). Lake Kivu (Chap. 10; TDS ca. 1100 mg L^{-1} in the mixolimnion) receives carbon dioxide from volcanic sources beneath the lake floor (Descy et al. 2012).

Concentrations of biogenic methane (up to 270.4 mg L^{-1}) are high enough for commercial exploitation (Tietze et al. 1980; Wüest et al. 2012).

Characterised by high concentrations of TDS, Waldsea Lake (SK, Canada; Table 3.1) is considered a saline lake (Hammer 1978, 1994; Last and Schweyen 1985; Last et al. 2002; Das 2007). Aragonite ($CaCO_3$) is permanently supersaturated throughout the water column, whereas gypsum is supersaturated in the monimolimnion. Other minerals may become temporarily supersaturated due to freeze-out in winter (Last et al. 2002). High concentrations of sulphate and sodium sustain meromixis (see Sect. 3.3.1).

The two Siberian lakes Shira and Shunet (Russia; Chap. 5) are similar to Waldsea Lake in their salinity and the dominance of sulphate in their mixolimnia. In contrast, Lake Isabel (Mexico; Chap. 12) has higher salinity dominated by sodium chloride due to its location on an island and the influence of guano from colonies of breeding sea birds in the lake catchment, which causes very high phosphorus concentrations in the mixolimnion (ca. 1 mg L^{-1}) and monimolimnion (up to 6 mg L^{-1}; Kienel et al. 2013).

Hot Lake (WA, USA; Table 3.1) is a shallow, highly saline lake whose waters are close to 100 % saturation for many soluble salts. Water chemistry is dominated by magnesium and sulphate. It is one of the most saline, meromictic lakes on Earth (Hammer 1986). Crusts of precipitated salts cover the banks of the lake when the water level is low in autumn. For this lake, Anderson (1958) reported 3.5 m maximal depth, whereas Lindemann et al. (2013) observed a much lower maximum depth range, 1.5–2.5 m, depending on season.

The mixolimnion of Powell Lake (BC, Canada; Table 3.1) has very low concentrations of solutes in the water, even lower than in the mixolimnetic water of Hall Lake, because of the geological and hydrological conditions in the lake's catchment. The monimolimnetic water is saline, and dominance of sodium and chloride indicates it is trapped sea water. Sulphate in the lake plays a role in the microbial decay of organic matter and biogeochemical cycles of heavy metals, particularly iron (see Sect. 3.3.1).

Seawater also has an important influence on the biogeochemistry of Rogoznica Lake (Croatia; Chap. 6), Nohoch Hol (Mexico; Chap. 12) and the Island Copper Mine pit lake (Canada; Chap. 9). Island Copper Mine pit lake is a former open cast mine, which was intentionally filled with seawater and then capped by a fresh to brackish top layer (see Chap. 9 for details), while the other lakes received sea water through groundwater connections.

Big Soda Lake (CO, USA; Table 3.1) is alkaline due to high concentration of sodium carbonate (soda). Mono Lake (CA, USA; Chap. 11) and Lake Rincon de Parangueo (Mexico; Chap. 12) are also alkaline soda lakes. Soda lakes are generally both alkaline and saline.

Lake Moritzteich (Germany; Table 3.1) and Cueva de la Mora (Spain; Chap. 9) are examples of acidic meromictic lakes. Such lakes originate from surface mining, and acidification is caused by pyrite oxidation. The acidification results in partial dissolution of minerals along the flow paths from the catchment and high concentrations of TDS. Camp Lake (MB, Canada; Table 3.1) is a natural lake receiving mining-impacted groundwater (Moncur et al. 2006). This mining-influenced groundwater is the reason for both acidification and meromixis in this lake. Concentrations of metals in lakes that receive waters from metal mining areas may reach grams per litre of iron and several hundred milligrams per litre of other (heavy) metals. We are unaware of examples of meromictic lakes acidified by acid rain, volcanic activities or because of sea spray, which are the other causes for strong acidification of lakes (Geller and Schultze 2010). Mildly acidic conditions in meromictic lakes have been reported for lakes with very low TDS and, accordingly low buffering capacities, with pH between 5.1 and 7 (Parkin et al. 1980; Weimer and Lee 1973). The low pH is due to the carbon dioxide that is produced during the decay of organic matter (see Sect. 3.3.1). Mildly acidic conditions (pH 4.6–6.2) may also result from high concentrations of humic substances, as in some Finnish meromictic lakes (Arvola 1983; Salonen et al. 1984; Kortelainen 1999).

A plot of all TDS concentrations and pH values of the lakes listed in Table 3.1 indicates that the lake water pH is not correlated to TDS values (Fig. 3.1).

3.3 Biogeochemical Processes in Meromictic Lakes

3.3.1 Biogeochemical Processes that May Influence the Stability of Stratification

Most biological processes that contribute to meromixis involve microbial decomposition of settling particulate organic matter and the release of soluble compounds, such as bicarbonate, that significantly contribute to water density (e.g. Boehrer and Schultze 2008). Other substances released into the water, such as ammonia and phosphate, affect the water density to a lesser degree, due to their much lower concentrations. Equation 3.1 shows the organic matter oxidation reaction that uses

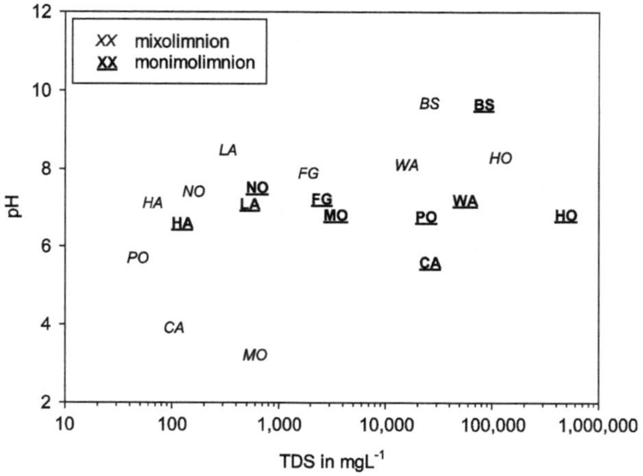

Fig. 3.1 Total dissolved solids and pH of mixolimnia and monimolimnia of the lakes presented in Table 3.1. *BS* Big Soda Lake, *CA* Camp Lake, *FG* Fayetteville Green Lake, *HA* Hall Lake, *HO* Hot Lake, *LA* Längsee, *MO* Moritzteich, *PO* Powell Lake

oxygen. This reaction applies only to oxygenated layers, i.e. the mixolimnion and the upper part of the chemocline.

$$\{CH_2O\} + O_2 \rightarrow HCO_3^- + H^+ \tag{3.1}$$

$\{CH_2O\}$ represents organic matter (as one carbohydrate unit) in Eq. 3.1 and below. Bicarbonate is used in most equations because it is the most abundant species of dissolved inorganic carbon (DIC) in circum-neutral lake water. At lower pH values (<6.5), this changes gradually to dissolved carbon dioxide and under alkaline conditions (pH>8.5) to carbonate. However, using carbon dioxide in equations is not uncommon in literature. All dissolved species of DIC basically increase the density of water and thus can lead to or stabilise meromixis. In rare cases where the partial pressure of carbon dioxide exceeds the absolute pressure (atmospheric pressure plus hydrostatic pressure at a given depth), formation of bubbles may result in destabilisation of meromixis due to advective transport and mixing by the rising bubbles (see Chap. 2).

In deeper, anoxic layers, other oxidising agents may replace oxygen. For example, nitrate, supplied from the mixolimnion, is usually only available in the chemocline. Import of considerable amounts of nitrate into the monimolimnia of meromictic lakes by groundwater has not yet been reported. Denitrification therefore is of lesser importance to the production of bicarbonate in monimolimnia. The reductions of manganese, iron and sulphate (Eqs. 3.2, 3.3 and 3.4, respectively) are more important:

$$\{CH_2O\} + 2MnO_2 + 3H^+ \rightarrow HCO_3^- + 2Mn^{2+} + 2H_2O \tag{3.2}$$

$$\{CH_2O\} + 4Fe(OH)_3 + 7H^+ \rightarrow HCO_3^- + 4Fe^{2+} + 10H_2O \tag{3.3}$$

$$2\{CH_2O\} + SO_4^{2-} \rightarrow 2HCO_3^- + H_2S \tag{3.4}$$

These three reactions are well known to occur in the chemoclines as well as in the monimolimnia and sediments of meromictic lakes (e.g. see Chaps. 5–10). Particulate oxidized manganese and iron enter deeper waters predominantly through sedimentation. Sulphate mainly originates from groundwater inflows but may also enter anoxic layers via diffusion from the mixolimnion or sedimentation of sulphate containing minerals that were formed in the mixolimnion or imported from allochthonous sources. Such sulphate-bearing minerals may include gypsum (occurring in many formations of sedimentary rock and thus also in the catchment of meromictic lakes like Lake Cadagno; see Chap. 8), mirabilite ($Na_2SO_4 \cdot 10\,H_2O$; occurring in highly saline, sulphate-rich lakes, e.g. Waldsea Lake; Hammer 1994) or hydroxysulphates of iron and aluminium in acidic pit lakes (e.g. see Sect. 9.2; Madison et al. 2003; Peiffer et al. 2013).

Other important pathways of bicarbonate production are fermentation (e.g. Eq. 3.5; Conrad 1999) and acetotrophic methanogenesis (Eq. 3.6; Conrad 1999).

$$C_6H_{12}O_6 \rightarrow 2CH_3CH_2OH + 2CO_2 \tag{3.5}$$

$$CH_3COOH \rightarrow CO_2 + CH_4 \tag{3.6}$$

All reaction products from reactions in Eqs. 3.1–3.5 increase the density of water. Conversely, methane as one of the reaction products in Eq. 3.6 lowers the density of water (Dietz et al. 2012). This is a specific property of methane and, to a lesser extent, also of dissolved nitrogen gas (Dietz et al. 2012). Another pathway of methane production is the reduction of carbon dioxide (Eq. 3.7).

$$CO_2 + 4H_2 \rightarrow CH_4 + 2H_2O \tag{3.7}$$

According to Conrad (1999), based on thermodynamic considerations, this pathway can only contribute one-third of the methane production except where there are external sources of hydrogen such as volcanic activity in the subsurface. The isotopic signature of methane indicated the contribution of such sources to the methane production in Lake Kivu (Pasche et al. 2011). Methane accumulation has been reported from many other meromictic lakes, e.g. Lake La Cruz (Chap. 8), Lake Matano (Chap. 10) and Mono Lake (Chap. 11).

The above-described processes are crucial for creating and sustaining meromictic conditions. Formation of bicarbonate in the chemocline, monimolimnion and sediment of meromictic lakes depends on settling organic matter, both allochthonous and autochthonous. Eutrophication, therefore, can cause meromixis through increased primary production in the mixolimnion of lakes and the subsequent greater decay of the produced biomass in the deeper parts of the lakes. All meromictic lakes in Carinthia, for example, are reported as biogenically meromictic (Frey 1955). For Lake Weinfelder Maar (Germany), biogenic meromixis was even found to be reversible by a reduction of nutrient inputs into the lake (Scharf and Oehms 1992).

If high primary production in the mixolimnion and the correspondingly high uptake of carbon dioxide cause calcite precipitation (Eqs. 3.8–3.10), the settling calcite may be dissolved again in the monimolimnion, where the concentrations of dissolved carbon dioxide are high and the pH is low due to decay of organic matter (Eq. 3.11; e.g. Lake La Cruz in Spain, Chap. 8). Relatively high mixolimnetic concentrations of calcium are a prerequisite for this combination of processes.

$$HCO_3^- \leftrightarrow CO_2 + OH^- \tag{3.8}$$

$$HCO_3^- + OH^- \leftrightarrow CO_3^{2-} + H_2O \tag{3.9}$$

$$Ca^{2+} + CO_3^{2-} \leftrightarrow CaCO_3 \tag{3.10}$$

$$CaCO_3 + CO_2 + H_2O \rightarrow Ca^{2+} + 2HCO_3^- \tag{3.11}$$

Precipitation of calcite is not the only process related to the decay of organic matter that may increase the water density. Ferrous iron resulting from reduction of ferric iron (see Eq. 3.3), after upward diffusion into the oxygenated mixolimnion, may be recycled via re-oxidation (Eq. 3.12), precipitation (Eq. 3.13) and sedimentation.

$$4Fe^{2+} + O_2 + 4H^+ \rightarrow 4Fe^{3+} + 2H_2O \tag{3.12}$$

$$Fe^{3+} + 3H_2O \leftrightarrow Fe(OH)_3 + 3H^+ \tag{3.13}$$

In some lakes, the iron cycle consisting of the reactions described in Eqs. 3.3, 3.12 and 3.13 is essential for the formation and stability of meromixis. Such lakes were frequently called iron-meromictic (e.g. Kjensmo 1967; Campbell and Torgersen 1980; Nixdorf and Boehrer 2015). Lake Nordbytjernet (Table 3.1) is also a lake where the iron cycle plays an essential role for the stability of meromixis (Hongve 1974, 1980, 2002). For the pit lake Waldsee (Germany), Dietz et al. (2012) found that solutes contributed 0.51 g L^{-1} to the density difference between the mixolimnion and monimolimnion; dissolved iron contributed 44 % of that difference. The overall density of the mixolimnion varied between 997.8 g L^{-1} in summer and 1000.2 g L^{-1} in winter, while the density of the monimolimnion was about 1000.7 g L^{-1} in all seasons (Dietz et al. 2012). This indicates the importance of the iron cycle in the lake. The reactivity of iron (precipitation and redissolution) was responsible for sustaining meromixis, as shown in the numerical model by Moreira et al. (2011) and as concluded from the conductivity budget (Nixdorf and Boehrer 2015), as well as the double diffusive circulation pattern (Boehrer et al. 2009). Due to the often high concentrations of iron in pit lakes, the iron cycle often contributes considerably to the stability of meromixis in such lakes (see e.g. Cueva de la Mora in Chap. 9).

Another option for iron recycling is oxidation by photo-ferrotrophic microorganisms in the anoxic part of the water body. This pathway for primary production utilises light energy like other photosynthetic processes, as has been observed in Lake Matano (Chap. 10). Equation 3.14 (from Crowe et al. 2008a) shows the combined oxidation of ferrous iron and formation of organic matter.

$$HCO_3^- + 4Fe^{2+} + 10H_2O \rightarrow \{CH_2O\} + 4Fe(OH)_3 + 7H^+ \tag{3.14}$$

This pathway is rare, as it requires the redoxcline to be shallow enough in the water column to be illuminated, and its contribution to sustaining the meromixis is unclear.

If oxidation and precipitation of iron take place in the mixolimnion, the settling iron hydroxides may also transport co-precipitated and adsorbed substances from the mixolimnion into the chemocline and the monimolimnion. Adsorbed substances are released into the monimolimnetic water when ferric hydroxides are dissolved via microbial iron reduction (Eq. 3.3). Even if iron is already oxidised and precipitated in the chemocline or the uppermost monimolimnion, adsorption and co-precipitation of other substances are still important for the cycling of chemicals in lakes. These processes interrupt upward diffusion of solutes enriched in the monimolimnion (e.g. phosphate, organic matter, arsenic, trace metals) and cause their transport back into deeper layers, possibly even their burial in the sediment. For simplification, only $Fe(OH)_3$ was used in the above Eqs. 3.3, 3.13 and 3.14. This does not necessarily mean that only ferrihydrite may precipitate or be the substrate for iron reduction. Goethite and lepidocrocite (FeOOH; Kappler and Newman 2004) or hematite (Fe_2O_3; from allochthonous sources) may also occur. Also the formation of green rust (a ferric-ferrous hydroxide containing an additional anion as carbonate or sulphate) has been observed in the chemocline and found to be important for the cycling of trace elements in meromictic lakes (Zegeye et al. 2012; Usman et al. 2012). However, the transport of co-precipitates like phosphate, organic matter, arsenic or trace metals usually does not substantially influence the water density in the monimolimnion and thus contributes little to the stability of meromixis.

Manganese can also be recycled via oxidation (Eq. 3.15) and precipitation (Eq. 3.16). In Lake Nordbytjernet, the manganese cycle obviously plays a role in the stability of meromixis comparable with that of iron cycling (see Table 3.1; Hongve 1974, 1980, 2002). However, Lake Nordbytjernet is the only lake known to us where manganese is essential for the stability of meromixis.

$$2Mn^{2+} + O_2 + 4H^+ \rightarrow 2Mn^{4+} + 2H_2O \tag{3.15}$$

$$Mn^{4+} + 2H_2O \leftrightarrow MnO_2 + 4H^+ \tag{3.16}$$

Equations 3.2, 3.15 and 3.16 are written with Mn(IV) as the oxidised state of manganese. Depending on conditions, Mn(III) may also occur (Davison 1993; Trouwborst et al. 2006; Pakhomova et al. 2009; Dellwig et al. 2012). The redox cycling of manganese has important implications for the cycling of phosphate and trace metals via co-precipitation (including adsorption) and redissolution of precipitates of manganese (Dellwig et al. 2010; Jones et al. 2011). This is basically comparable with the above-discussed processes related to iron. The transport of phosphate and trace metals usually does not have a substantial influence on the water density in the monimolimnion.

Another way of how dissolved manganese can be recycled from the chemocline to lower layers in the monimolimnion was observed by Cristiani et al. (2014) in Italy. They found that Mn(II) precipitated as hydroxide due to increasing pH in the lower part of the chemocline, forming a zone of white turbidity (Eq. 3.17).

$$Mn^{2+} + 2OH^- \leftrightarrow Mn(OH)_2 \tag{3.17}$$

The data presented in Table 3.1 suggest that high concentrations of iron on one hand and of sulphide on the other hand do not occur simultaneously. The reason is the low solubility and, thus, the precipitation of metal sulphides. Eq. 3.18 shows the respective reaction for iron which usually is more important for the monimolimnetic water density than other heavy metals because iron is typically much more abundant in aquatic systems. The sulphide ions result from the dissociation of hydrogen sulphide (Eqs. 3.19 and 3.20).

$$Fe^{2+} + S^{2-} \rightarrow FeS \tag{3.18}$$

$$H_2S \leftrightarrow HS^- + H^+ \tag{3.19}$$

$$HS^- \leftrightarrow S^{2-} + H^+ \tag{3.20}$$

Over a period of time, and with a sufficient supply of organic matter, precipitation of iron sulphides may substantially change the chemical composition of the monimolimnion. In Lake Rørholtfjorden (Norway), the first lake where trapped old seawater was identified to form the monimolimnion, all initially present sulphate was reduced to sulphide and precipitated as iron sulphide since the lake was estranged from the sea (Strøm 1957). Dissolved iron (390 mg L^{-1}; Strøm 1957) accumulated in the monimolimnion, partly substituting the initially present sulphate. However, this substitution must have taken thousands of years.

Because the precipitation of sulphides transfers two dissolved ions to particulate matter, which settles to the sediment, formation of sulphides and their removal from dissolved phase decreases the water density. Precipitation of iron sulphides has been observed in many meromictic lakes (e.g. Powell Lake, Perry and Pedersen 1993; Hall Lake, Balistrieri et al. 1994). However, the balance over the entire process from oxidation of particulate organic matter by reduction of particulate iron hydroxide and (usually) dissolved sulphate to the precipitation of iron sulphide results in a surplus of dissolved bicarbonate in the water. It implies that only part of dissolved matter is precipitated as iron sulphide, and there is a net increase in the concentration of dissolved matter and, thus, in water density in the monimolimnion, imparting extra stability to meromixis.

The formation of mirabilite ($Na_2SO_4 \cdot 10\ H_2O$) may occur in saline lakes with high concentrations of sulphate and sodium. Hammer (1994) described the precipitation of mirabilite from ice exclusion and its dissolution in the warmer monimolimnion as the process maintaining meromixis in Lake Waldsea (see also Table 3.1). Precipitation of mirabilite caused by formation of ice was observed in several saline lakes in southern Saskatchewan, Canada, in winter (Last and Schweyen 1983). The process described by Hammer (1994) consists of three steps: (1) Salts accumulate immediately below the ice during its formation; (2) mirabilite precipitates due to the increased concentrations and its decreased solubility at temperature close to 0 °C (Marion and Farren 1999); and (3) the mirabilite crystals sink into the monimolimnion where they dissolve due to higher temperatures, 5–7 °C year-round (Hammer et al. 1978). Whereas the solubility of mirabilite is about 0.35 mol $Na_2SO_4 \cdot 10\ H_2O$ per kg water at 0 °C, it increases to about 0.48 mol $Na_2SO_4 \cdot 10\ H_2O$ per kg water at 6 °C (Marion and Farren 1999).

3.3.2 *Characteristic Biogeochemical Processes in Water Columns of Meromictic Lakes*

In this section, we discuss processes observed almost exclusively in the water column of meromictic lakes. Such processes are only rarely reported in water columns of holomictic lakes. They are known to also occur in lake sediments where redox conditions are comparable with those in the chemocline and in the monimolimnion of meromictic lakes. Therefore, Perry and Pedersen (1993) called lakes Powell and Sakinaw (BC, Canada) which are both meromictic "convenient natural laboratories for studies of diagenesis", in their case, for the formation of pyrite.

So-called plates of anoxygenic, phototrophic purple and green sulphur bacteria located in or immediately below the chemocline are often seen as typical features of meromictic lakes. However, such plates do not occur in all meromictic lakes and have been also reported for some holomictic lakes (e.g. Plußsee, Germany, Oberhäuser-Nehls et al. 1994; Lake Dudinghausen, Germany, Selig et al. 2004; Lake Verevi, Estonia, Noges and Solovjova 2005) where these plates were found to occur in or below the thermocline. These layers are sites of anoxygenic photosynthesis (e.g. Overmann et al. 1996; Imhoff 2006; Overmann 2006; Musat et al. 2008). For more examples of meromictic lakes having plates of anoxygenic, phototrophic purple and green sulphur bacteria, see Chaps. 5–8.

Phototrophic sulphur bacteria use sulphur in hydrogen sulphide instead of oxygen in water molecules as electron donor for the reduction of carbon dioxide. The sulphur is oxidised first to its elemental form, which bacteria can store in their cells. When hydrogen sulphide gets depleted, the bacteria can also use the stored sulphur for its oxidation to sulphate (Overmann 2006; Imhoff 2006). Equations 3.21 and 3.22 summarise the described photosynthetic processes. Not all steps of the complex biochemical processes are described by Eqs. 3.21 and 3.22 but only the net reaction from the initial precursors to the final products.

$$2H_2S + CO_2 \rightarrow 2S + \{CH_2O\} + H_2O \tag{3.21}$$

$$2S + 3CO_2 + 5H_2O \rightarrow 2\,SO_4^{2-} + 3\{CH_2O\} + 4H^+ \tag{3.22}$$

Instead of sulphide and elemental sulphur, some sulphur bacteria can also use other reduced sulphur compounds like thiosulphate or sulphite. In addition to elemental sulphur and sulphate, also polysulphides and polythionates may be formed as a result of sulphur oxidation (Overmann 2006; Imhoff 2006). A further anoxygenic photosynthetic process, which is facilitated by particular green sulphur bacteria, is the above-mentioned phototrophic oxidation of ferrous iron (Overmann 2006; see Sect. 3.3.1 and Eq. 3.14). The organisms facilitating phototrophic oxidation of ferrous iron can cope with low light (Hegler et al. 2008). They were detected, e.g. in a meromictic Arctic lake (Comeau et al. 2012) and in tropical Lake Matano (Crowe et al. 2008a; see also Chap. 10).

Pyrite (FeS_2) is often formed in lake sediment (Manning et al. 1999; Holmer and Storkholm 2001; Thomsen et al. 2004; Yin et al. 2008). In the water column, its formation seems to be restricted to meromictic lakes (e.g. Perry and Pedersen 1993; Suits and Wilkin 1998). Rickard and Luther (2007) described two pathways of pyrite formation: (1) reaction of dissolved ferrous iron with dissolved polysulphides (Eqs. 3.23a, 3.23b) and (2) reaction of iron monosulphide (FeS) with hydrogen sulphide (Eq. 3.24):

$$Fe^{2+} + S(-II) \rightarrow [FeS] \tag{3.23a}$$

$$[FeS] + S_5^{2-} \rightarrow FeS_2 + S_4^{2-} \tag{3.23b}$$

[FeS] represents an aqueous ferrous iron sulphide reaction intermediate. Whereas reaction 3.23a is rapid, reaction 3.23b is the rate determining step. $S_5{}^{2-}$ may be replaced also by other polysulphide ions with a number of sulphur atoms other than five. S(−II) stands for chemical species containing sulphur in the oxidation state −2 like H_2S, HS^- or S^{2-} (Rickard and Luther 2007).

$$FeS + H_2S \rightarrow FeS_2 + H_2 \tag{3.24}$$

The hydrogen sulphide acts as an oxidant in reaction described by Eq. 3.24. If hydrogen sulphide represents S(−II) in the reaction shown in Eq. 3.23a, it does not act as oxidising agent. The second important difference between both pathways is that polysulphides are not needed in the reaction shown in Eq. 3.24. For details concerning the mechanisms of pyrite formation, see also Rickard and Luther (2007).

Perry and Pedersen (1993) found pyrite in monimolimnetic water samples from lakes Powell and Sakinaw. Pyrite occurred already where oxygen was just depleted, while iron monosulphide was found only in water samples from deeper waters. This correlated well with calculated saturation level for the two mineral phases. Suits and Wilkin (1998) deduced the pelagic formation of pyrite from the morphology of pyrite particles in the sediment of Fayetteville Green Lake.

Under the special conditions of the meromictic acidic pit lake in the former Berkeley open cast mine, pyrite oxidation in the anaerobic monimolimnion was observed (Madison et al. 2003). Schwertmannite ($Fe_8O_8(OH)_6SO_4$) settling from the mixolimnion into the monimolimnion was dissolved there due to different chemical, but still acidic, conditions while sinking jarosite ($KFe_3[(OH)_6(SO4)_2]$; potassium (K) may be replaced by other cations) was oversaturated also in the monimolimnion and, thus, settled to the sediment. The seepage of groundwater caused these differences. Dissolved ferric iron was not rapidly consumed by iron-reducing bacteria for the decay of organic matter. Therefore, it could oxidise the sulphur of pyrite which was present in the rock that forms the lake basin (Madison et al. 2003). Under neutral or alkaline conditions, this would not be possible since ferric iron would precipitate immediately as hydroxide if liberated by dissolution of other ferric iron-bearing minerals, which settle from the mixolimnion (e.g. iron hydroxo-sulphates).

The permanent redoxcline provides conditions for the anaerobic ammonia oxidation (usually called "anammox"; Mulder et al. 1995) according to Eq. 3.25: the simultaneous availability of ammonia and nitrite and relatively stable conditions for longer periods.

$$NH_4^+ + NO_2^- \rightarrow N_2 + 2H_2O \quad (3.25)$$

In lacustrine systems, this process has been suggested for sediments (Hu et al. 2011) or for the chemocline of a few meromictic lakes: Lake Tanganyika (Schubert et al. 2006), Lake Rassnitz (Germany; Hamersley et al. 2009) and Lake Lugano (Switzerland, Italy; Wenk et al. 2014). The fact that both ammonia and nitrite have to be present and that the involved microorganisms grow slowly (Ding et al. 2013) seems to prevent the occurrence of anammox at seasonal redoxclines in holomictic (eutrophic) lakes.

3.4 Palaeolimnological Aspects of Sediment Chemistry in Meromictic Lakes

Lake sediments preserve records of environmental change that enable palaeolimnological reconstructions. These reconstructions of past conditions are carried out using multiple proxies. For example, biomarkers are well-preserved chemical compounds that carry signatures of the water column or sediment processes at the time of deposition (e.g. Castañeda et al. 2007; Powers et al. 2004). The provenance and breakdown dynamics of organic matter can be reflected by the records of ^{13}C and ^{15}N stable isotopes. Other proxies include algal assemblages (Smol and Cumming 2000), bulk sediment composition, grain size, microfossils, pollen, fish bones, and others (Meyers 2003; Schmidt et al. 2002; Birks 1998). For many of these proxies, interpretations do not differ between meromictic and non-meromictic lakes. Climate reconstructions that rely on biomarker proxies, which are recalcitrant compounds, for example, do not significantly differ in meromictic lakes from those in lakes that experience regular full overturn. Below we discuss several aspects of the sediment chemistry that can affect the palaeolimnological record interpretations that are specific to meromictic lakes.

The composition and vertical distributions of substances within aquatic sediments are modified after their deposition by a suite of chemical, physical and biological processes collectively known as early diagenesis (Berner 1980). These processes recycle substances such as essential nutrients, i.e. carbon, phosphorus and nitrogen, back to the water column and determine the mineral phases that become buried into the deep sediments. Understanding such processes thus plays an important role in the interpretations of palaeolimnological sediment records. In most sedimentary environments, diagenetic transformations are driven by a sequence of redox (reduction-oxidation) reactions fuelled by oxidation of the deposited organic matter. These microbially catalysed oxidation reactions proceed by a cascade of electron acceptors that are typically used sequentially from the sediment surface downward in approximately the following sequence: oxygen, nitrate (NO_3^-), oxides of Mn and Fe, sulphate (SO_4^{2-}) and CO_2 (fermentation).

In non-meromictic lakes, where oxygen is regularly transported into the deeper water layers by mixing, the most intense redox cycling takes place within the sediment column, near the sediment oxic-anoxic boundary (Katsev and Dittrich 2013; Li and Katsev 2014). In meromictic lakes, permanent anoxia effectively shifts these reactions into the water column, where oxic-anoxic boundary lies typically at or below the depth of the thermocline or the most pronounced pycnocline (Crowe et al. 2008b; Katsev et al. 2010). The biogeochemical cycling in sediment in such environments, as a result, becomes simpler, with no recycling of redox-sensitive substances. A weak redox gradient provides less chemical energy as well as supports less biological activity. Decomposition of organic matter is dominated by methanogenesis (fermentation), with minor contributions from sulphate reduction and metal reductions. Permanently anoxic conditions also lead to marked differences in the rates of sediment nutrient recycling compared with those in mixed lakes. Sediment-binding capacity for phosphorus, in particular, is significantly diminished, whereas the preservation of organic carbon is enhanced (Li 2014).

As anoxic conditions favour preservation of organic substances, sediments in meromictic lakes are particularly important as repositories of palaeoenvironmental information (D'Andrea et al. 2012; Cohen et al. 2006; Schmidt et al. 2002; Laird et al. 1996; Sect. 5.4). The absence of bioturbation (mixing by benthic macrofauna) caused by the lack of oxygen further minimises post-depositional disturbance. Varved or annually layered sediments (Zolitschka et al. 2015; Brauer 2004) provide particularly high-resolution records. Seasonal varves can be preserved, for example, when particle settling occurs fast enough to transmit the seasonal signal to the lake floor (e.g. facilitated by formation of zooplankton faecal pellets), bottom currents are minimal to preclude resuspension and gas formation in the sediments occurs slowly enough as to not destroy the laminations (Kelts and Hsü 1978; Cohen 2003). Ancient lakes, such as those of the East African Rift Valley, accumulate kilometres of sediment that contain valuable records of past climatic conditions (Owen et al. 1990; Finney and Johnson 1991; Johnson et al. 2002; Scholz et al. 2007; Johnson et al. 2011). The absence of bioturbation during the meromictic (anoxic) phases of lake histories facilitates preservation of evidence for high-resolution episodic events such as volcanic eruptions (Lane et al. 2013). Well-preserved ash layers can be found as discrete layers identifiable by magnetic susceptibility or X-ray fluorescence (XRF) profiles. Detection of cryptotephra requires more complex methods (Davies et al. 2007; de Fontaine et al. 2007). The onset or disappearance of laminations in sediments can be viewed as evidence for shifts in the lake's hydrodynamic regime, through changes in deepwater oxygen concentrations. Preservation of high-resolution signals in the anoxic sediments of ancient lakes also provides opportunities for evaluating the historical behaviours of high frequency climate signatures, such as El Niño/Southern Oscillation (ENSO; Boës and Fagel 2008; Cohen et al. 2006).

The pH of the bottom waters and of sediment porewaters in meromictic lakes is generally lower than in mixed lakes, as CO_2 produced during organic matter mineralisation becomes trapped in the monimolimnion. This affects the precipitation and dissolution of calcium carbonate (Brunskill 1969). Carbonate-rich layers and shifts in carbonate precipitation/deposition are often used as indicators of environmental changes. For example, detrital carbonates can point out to sources of clastic (weathered rocks) inputs and resedimentation (Kelts and Talbot 1990). Precipitates from surface waters may vary seasonally and with temperature changes. In anoxic sediments, diagenetic carbonates record the carbon isotopic signatures of organic matter mineralisation processes and the oxygen isotopic signatures of the bottom waters at the time of sedimentation (Kelts and Talbot 1990). Precipitation of calcium carbonates (primarily as calcite or aragonite) is regulated by factors such as the dissolved CO_2 (DIC) levels, Ca^{2+} concentrations and pH (Hammes and Verstraete 2002; Pasche et al. 2010). Their enrichment in sediments is frequently regarded as an indicator for higher biological activity, as photosynthesis by primary producers raises the pH. It may also serve as indicator of salt inputs (Votava 2014) and shifting water balance. Calcium carbonates in meromictic lakes may precipitate in saline monimolimnia where concentrations of major cations (including Ca) and DIC may be significantly elevated. In lakes where changes in lake level may cause the lake to alternate between open- and close-basin regimes, calcite enrichment in sediments is typically thought to reflect a closed-basin condition. In iron-rich sediments where reduction of iron oxides releases sufficient quantities of Fe^{2+} and CO_2 (lowering the pH), calcite can be replaced in sediments by siderite ($FeCO_3$) (Scholz et al. 2011; Wittkop 2014). The accumulation of reduced products of sediment redox reactions in monimolimnia may also alter the traditional sequence of diagenetic reactions. Accumulation of reduced iron (Fe^{2+}), for example, may allow methanogens to outcompete the normally more adapted microbes that facilitate the reduction of iron oxides (Crowe et al. 2011). Similarly, differences in the availability of organic or inorganic substrates can determine which of the several possible reaction pathways dominate, e.g. acetoclastic vs. hydrogenotrophic methanogenesis (Pasche et al. 2011; reactions 3.6 and 3.7). These effects of water and porewater chemistry on the kinetics of diagenetic reactions, however, have received relatively little attention.

Evidence of past changes in lake mixing regime, including episodes of deepwater circulation and lake overturns, may be preserved in sediment records. Markers of such dramatic changes may include, for example, changes in microbial and algal assemblages or abrupt precipitation of oxidised phases of redox-sensitive metals. For example, oxygenation of an anoxic monimolimnion containing dissolved Fe^{2+} or Mn^{2+} would be expected to trigger a massive sedimentation of solid oxide particles. Such layers have been suggested as evidence of shifting in the depth of mixing in Lake Malawi (Brown et al. 2000; Finney and Johnson 1991).

Variations in sediment concentrations of organic carbon have been often viewed as proxies for primary productivity, or at least as records of organic sedimentation. Separating the depositional signal from post-depositional alteration, however, is not straightforward. Anoxic conditions foster preservation of organic material (Burdige 2007; Katsev and Crowe 2015), as anaerobic microbes lack the enzymes to break down some of the more complex organic molecules (Jørgensen 2000). Thus, switches between oxic and anoxic conditions in bottom waters may affect organic carbon preservation rates. For example, the "burn down" of organic-rich sediment layers upon reoxygenation is well documented in marine environments (Jung et al. 1997). Numerous processes determine the amount of organic carbon that becomes buried into the deep sediment; hence, the use of organic carbon content as a palaeoproductivity indicator requires careful consideration. Organic matter mineralisation decreases the organic content of the sediment, so observations of higher organic carbon content near the sediment surface cannot be directly interpreted as evidence of increased organic sedimentation. Additional methods, such as ^{13}C and ^{15}N isotopic characterisations (O'Beirne et al. 2015; Castañeda et al. 2009; Schelske and Hodell 1995), or biomarker evidence need to be used. The carbon mineralisation rate decreases roughly as a power function of carbon age (Middelburg 1989; Katsev and Crowe 2015), but mineralisation never stops, so technically no depth within the sediment can be designated as a "no diagenesis" zone. Variations in concentrations of organic material below the upper few decimetres of sediment have been generally interpreted as records of productivity (Castañeda et al. 2011). Though corrections related to the history of the mineralisation conditions (e.g. exposure to oxygen) may be needed in some cases, no easy quantitative methods have been suggested (Zimmerman and Canuel 2000; Katsev and Crowe 2015).

Over geologic time scales, preservation and diagenetic alteration of organic material can lead to formation of petroleum products (Cohen 1989; Talbot 1988). Traditionally, sediments in extant lakes have not been considered as sources of economically exploitable hydrocarbons. In recent years, however, sediments of several ancient East African lakes, including several meromictic lakes, were shown to contain hydrocarbons, sparking a wave of oil and gas explorations across the region. Hydrocarbon deposits in Lake Albert (Democratic Republic of the Congo, Uganda) are already being commercially exploited for oil. Being only 3 million years old, the sediments of Lake Albert are geologically young for oil-bearing rocks. The hydrothermal heat associated with the rift system, however, accelerates maturation reactions, allowing them to occur at shallower depths.

Acknowledgements We thank Michael Moncur (Calgary, Canada) for providing detailed data from figures published in Moncur et al. (2006), Pierangela Cristiani (Milano, Italy) for unpublished data completing the dataset for Lake Idro in Table 3.1, John Zachara and his colleagues (Richland, WA, USA) for providing detailed data for Hot Lake (USA) in addition to those available in Zachara et al. (2016) and the Kärntner Institut für Seenforschung, in particular Roswitha Fresner (Klagenfurt, Austria), for providing unpublished data for Längsee from the state monitoring programme of Austria.

References

Anderson GC (1958) Some limnological features of a shallow saline meromictic lake. Limnol Oceanogr 3:259–270

Arvola L (1983) Primary production and phytoplankton in two small, polyhumic forest lakes in southern Finland. Hydrobiologia 101:105–110

Balistrieri LS, Murray JW, Paul B (1994) The geochemical cycling of trace elements in a biogenic meromictic lake. Geochim Cosmochim Acta 58:3993–4008

Berner RA (1980) Early diagenesis: a theoretical approach. Princeton University Press, Princeton, NJ

Birks HJB (1998) DG Frey and ES Deevey review 1: numerical tools in palaeolimnology–progress, potentialities, and problems. J Paleolimnol 20:307–332

Boehrer B, Schultze M (2008) Stratification of lakes. Rev Geophys 46:RG2005. doi:10.1029/2006RG000210

Boehrer B, Dietz S, von Rohden C, Kiwel U, Jöhnk KD, Naujoks S, Ilmberger J, Lessmann D (2009) Double-diffusive deep water circulation in an iron-meromictic lake. Geochem Geophys Geosyst 10:Q06006. doi:10.1029/2009GC002389

Boës X, Fagel N (2008) Relationships between Southern Chilean varved lake sediments, precipitation and ENSO for the last 600 years. J Paleolimnol 39:237–252

Brauer A (2004) Annually laminated lake sediments and their palaeoclimatic relevance. In: Fischer H, Kumke T, Lohmann G, Flöser G, Miller H, von Storch H, Negendank JFW (eds) The climate in historical times. Springer, Berlin, pp 109–127

Brown ET, Le Callonnec L, German CR (2000) Geochemical cycling of redox-sensitive metals in sediments from Lake Malawi: a diagnostic paleotracer for episodic changes in mixing depth. Geochim Cosmochim Acta 64:3515–3523

Brunskill GJ (1969) Fayetteville Green Lake, New York. II. Precipitation and sedimentation of calcite in a meromictic lake with laminated sediments. Limnol Oceanogr 14:830–847

Brunskill GJ, Ludlam SD (1969) Fayetteville Green Lake, New York. I. Physical and chemical limnology. Limnol Oceanogr 14:817–829

Burdige DJ (2007) Preservation of organic matter in marine sediments: controls, mechanisms, and an imbalance in sediment organic carbon budgets? Chem Rev 107:467–485

Campbell P, Torgersen T (1980) Maintenance of iron meromixis by iron redeposition in a rapidly flushed monimolimnion. Can J Fish Aquat Sci 37:1303–1313

Castañeda IS, Werne JP, Johnson TC (2007) Wet and arid phases in the Southeast African tropics since the last glacial maximum. Geology 35:823–826

Castañeda IS, Werne JP, Johnson TC, Filley TR (2009) Late quaternary vegetation history of Southeast Africa: the molecular isotopic record from Lake Malawi. Palaeogeogr Palaeoclimatol Palaeoecol 275:100–112

Castañeda IS, Werne JP, Johnson TC, Powers LA (2011) Organic geochemical records from Lake Malawi (East Africa) of the last 700 years, part II: biomarker evidence for recent changes in primary productivity. Palaeogeogr Palaeoclimatol Palaeoecol 303:140–154

Cohen AS (1989) Facies relationships and sedimentation in large rift lakes and implications for hydrocarbon exploration: examples from lakes Turkana and Tanganyika. Palaeogeogr Palaeoclimatol Palaeoecol 70:65–80

Cohen AS (2003) Paleolimnology: the history and evolution of lake systems. Oxford University Press, New York

Cohen AS, Lezzar KE, Cole J, Dettman D, Ellis GS, Gonneea ME, Plisnier P-D, Langenberg V, Blaauw M, Zilifi D (2006) Late Holocene linkages between decade–century scale climate variability and productivity at Lake Tanganyika, Africa. J Paleolimnol 36:189–209

Comeau AM, Harding T, Galand PE, Vincent WF, Lovejoy C (2012) Vertical distribution of microbial communities in a perennially stratified Arctic lake with saline, anoxic bottom waters. Sci Rep 2:Article 604. doi:10.1038/srep00604

Conrad R (1999) Contribution of hydrogen to methane production and control of hydrogen concentration in methanogenic soils and sediments. FEMS Microbiol Ecol 28:193–202

Corella JP, Moreno A, Morellón M, Rull V, Giralt S, Rico MT, Pérez-Sanz A, Valero-Garcés BL (2011) Climate and human impact on a meromictic lake during the last 6,000 years (Montcortès Lake, Central Pyrenees, Spain). J Paleolimnol 46:351–367

Cristiani P, Balordi E, Storni G, Guerrini E, Trasatti SP, Bertini A (2014) Manganese cycle in the meromictic lake Idro. In: Biscarini C, Pierleoni A, Naselli-Flores L (eds) Lakes: the mirrors of the earth—balancing ecosystem integrity and human wellbeing. Proceedings of the 15th world lake conference, vol 2. Science4Press, Perugia, pp 289–292

Crowe SA, Jones CA, Katsev S, Magen C, O'Neill AH, Sturm A, Canfield DE, Haffner GD, Mucci A, Sundby B, Fowle DA (2008a) Photoferrotrophs thrive in an Archean Ocean analogue. PNAS 105:15938–15943

Crowe SA, O'Neill AH, Katsev S, Hehanussa P, Haffner GD, Sundby B, Mucci A, Fowle DA (2008b) The biogeochemistry of tropical lakes: a case study from Lake Matano, Indonesia. Limnol Oceanogr 53:319–331

Crowe SA, Katsev S, Leslie K, Sturm A, Magen C, Nomosatryo S, Pack MA, Kessler JD, Reeburgh WS, Roberts JA, González L, Haffner GD, Mucci A, Sundby B, Fowle DA (2011) The methane cycle in ferruginous Lake Matano. Geobiology 9:61–78

Culver AC (1977) Biogenic meromixis and stability in a soft-water lake. Limnol Oceanogr 22:667–686

D'Andrea WJ, Vaillencourt DA, Balascio NL, Werner A, Roof SR, Retelle M, Bradley RS (2012) Mild little ice age and unprecedented recent warmth in an 1800 year lake sediment record from Svalbard. Geology 40:1007–1010

Das B (2007) Reconstruction of historical productivity using visible-near-infrared (VNIR) reflectance properties from boreal and saline lake sediments. Aquat Ecol 41:209–220

Davies SM, Elmquist M, Bergman J, Wohlfarth B, Hammarlund D (2007) Cryptotephra sedimentation processes within two lacustrine sequences from west central Sweden. Holocene 17:319–330

Davison W (1993) Iron and manganese in lakes. Earth Sci Rev 34:119–163

de Fontaine CS, Kaufman DS, Anderson RS, Werner A, Waythomas CF, Brown TA (2007) Late Quaternary distal tephra-fall deposits in lacustrine sediments, Kenai Peninsula, Alaska. Quatern Res 68:64–78

Dellwig O, Leipe T, März C, Glockzin M, Pollehne F, Schmetger B, Yakushev EV, Böttcher ME, Brumsack H-J (2010) A new particulate Mn-Fe-P-shuttle at the redoxcline of anoxic basins. Geochim Cosmochim Acta 74:7100–7115

Dellwig O, Schnetger B, Brumsack H-J, Grossart H-P, Umlauf L (2012) Dissolved reactive manganese at pelagic redoxclines (part II): hydrodynamic conditions for accumulation. J Mar Syst 90:31–41

Descy J-P, Darchambeau F, Schmid M (eds) (2012) Lake Kivu—limnology and biogeochemistry of a tropical great lake. Springer, Dordrecht

Dietz S, Lessmann D, Boehrer B (2012) Contribution of solutes to density stratification in a meromictic lake (Waldsee/Germany). Mine Water Environ 31:129–137

Ding S, Zheng P, Lu H, Chen J, Mahmood Q, Abbas G (2013) Ecological characteristics of anaerobic ammonia oxidizing bacteria. Appl Microbiol Biotechnol 97:1841–1849

Findenegg I (1935) Limnologische Untersuchungen im Kärntener Seengebiete. Int Rev Gesamten Hydrobiol Hydrogr 32:369–423

Finney BP, Johnson TC (1991) Sedimentation in Lake Malawi (East Africa) during the past 10,000 years: a continuous paleoclimatic record from the southern tropics. Palaeogeogr Palaeoclimatol Palaeoecol 85:351–366

Frey DG (1955) Längsee: a history of meromixis. Mem Ist Ital Idrobiol Suppl 8:125–135

Geller W, Schultze M (2010) Acidification. In: Likens GE (ed) Biogeochemistry of inland waters (a derivative of Encyclopedia of inland waters). Elsevier, Amsterdam, pp 557–568

Hamersley MR, Woebken D, Boehrer B, Schultze M, Lavik G, Kuypers MMM (2009) Water column anammox and denitrification in a temperate permanently stratified lake (Lake Rassnitzer, Germany). Syst Appl Microbiol 32:571–582

Hammer UT (1978) The saline lakes of Saskatchewan III. Chemical characterization. Int Rev Gesamten Hydrobiol 63:311–335

Hammer UT (1986) Saline lake ecosystems of the world. Dr W. Junk Publishers, Dordrecht

Hammer UT (1994) Life and times of five Saskatshewan saline meromictic lakes. Int Rev Gesamten Hydrobiol 79:235–248

Hammer UT, Haynes RC, Lawrence JR, Swift MC (1978) Meromixis in Waldsea Lake, Saskatchewan. Verh Int Ver Limnol 20:192–200

Hammes F, Verstraete W (2002) Key roles of pH and calcium metabolism in microbial carbonate precipitation. Rev Environ Sci Biotechnol 1:3–7

Hegler F, Posth NR, Jiang J, Kappler A (2008) Physiology of phototrophic iron(II)-oxidizing bacteria: implications for modern and ancient environments. FEMS Microbiol Ecol 66:250–260

Holmer M, Storkholm P (2001) Sulphate reduction and sulphur cycling in lake sediments: a review. Freshw Biol 46:431–451

Hongve D (1974) Hydrographical features of Nordbytjernet, a manganese-rich meromictic lake in SE Norway. Arch Hydrobiol 74:227–246

Hongve D (1980) Chemical stratification and stability of meromictic lakes in the upper Romerike district. Schweiz Z Hydrol 42:171–195

Hongve D (2002) Seasonal mixing and genesis of endogenic meromixis in small lakes in Southeast Norway. Nord Hydrol 33:189–206

Hu B-I, Shen L-D, Xu X-Y, Zheng P (2011) Anaerobic ammonium oxidation (anammox) in different natural ecosystems. Biochem Soc Trans 39:1811–1816

Huber K, Weckström K, Drescher-Schneider R, Knoll J, Schmidt J, Schmidt R (2010) Climate changes during the last glacial termination inferred from diatom-based temperatures and pollen in a sediment core from Längsee (Austria). J Paleolimnol 43:131–147

Hutchinson GE (1957) A treatise on limnology, Geography, physics and chemistry, vol 1. John Wiley, New York

Imhoff JF (2006) The Chromatiaceae. In: Dworkin MM, Falkow S, Rosenberg E, Schleifer K-H, Stackebrandt E (eds) The Prokaryotes. Proteobacteria: gamma subclass. vol 6, 3rd edn. Springer, Heidelberg, pp 846–873

Johnson TC, Brown ET, McManus J, Barry S, Barker P, Gasse F (2002) A high-resolution paleoclimate record spanning the past 25,000 years in southern East Africa. Science 296:113–132

Johnson TC, Brown ET, Shi J (2011) Biogenic silica deposition in Lake Malawi, East Africa over the past 150,000 years. Palaeogeogr Palaeoclimatol Palaeoecol 303:103–109

Jones C, Crowe SA, Sturm A, Leslie KL, MacLean LSW, Katsev S, Henny C, Fowle DA, Canfield DE (2011) Biogeochemistry of manganese in ferruginous Lake Matano, Indonesia. Biogeosciences 8:2977–2991

Jørgensen BB (2000) Bacteria and marine biogeochemistry. In: Schulz HD, Zabel M (eds) Marine geochemistry. Springer, Berlin, pp 173–207

Jung M, Ilmberger J, Mangini A, Emeis KC (1997) Why some Mediterranean sapropels survived burn-down (and others did not). Mar Geol 141:51–60

Kappler A, Newman DK (2004) Formation of Fe(III)-minerals by Fe(II)-oxidizing phototrophic bacteria. Geochim Cosmochim Acta 68:1217–1226

Katsev S, Crowe SA (2015) Organic carbon burial efficiencies in sediments: the power law of mineralization revisited. Geology 43:607–610

Katsev S, Dittrich M (2013) Modeling of decadal scale phosphorus retention in lake sediment under varying redox conditions. Ecol Model 251:246–259

Katsev S, Crowe SA, Mucci A, Sundby B, Nomosatryo S, Haffner GD, Fowle DA (2010) Mixing and its effects on biogeochemistry in the persistently stratified, deep, tropical Lake Matano, Indonesia. Limnol Oceanogr 55:763–776

Kelts K, Hsü KJ (1978) Freshwater carbonate sedimentation. In: Lerman A (ed) Lakes. Springer, New York, pp 295–323

Kelts K, Talbot M (1990) Lacustrine carbonates as geochemical archives of environmental change and biotic/abiotic interactions. In: Tilzer MM, Serruya C (eds) Large lakes. Springer, Berlin, p 288–315

Kharaka YK, Robinson SW, Law LM, Carothers WW (1984) Hydrochemistry of Big Soda Lake, Nevada: an alkaline meromictic desert lake. Geochim Cosmochim Acta 48:823–835

Kienel U, Plessen B, Schettler G, Weise S, Pinkerneil S, Böhnel H, Engelbrecht AC, Haug GH (2013) Sensitivity of a hypersaline crater lake to the seasonality of rainfall, evaporation, and guano supply. Fundam Appl Limnol 183:135–152

Kjensmo J (1967) The development and some main features of "Iron-meromictic" soft water lakes. Arch Hydrobiol Suppl 32:137–312

Kortelainen P (1999) Acidity and buffer capacity. In: Keskitalo J, Eloranta P (eds) Limnology of humic waters. Backhuys Publishers, Leiden, pp 95–115

Laird KR, Fritz SC, Grimm EC, Mueller PG (1996) Century scale paleoclimatic reconstruction from Moon Lake, a closed basin lake in the northern Great Plains. Limnol Oceanogr 41:890–902

Lane CS, Chorn BT, Johnson TC (2013) Ash from the Toba supereruption in Lake Malawi shows no volcanic winter in East Africa at 75 ka. PNAS 110:8025–8029

Last WM, Schweyen TH (1983) Sedimentology and geochemistry of saline lakes of the Great Plains. Hydrobiologia 105:245–263

Last WM, Schweyen TH (1985) Late Holocene history of Waldsea Lake, Saskatchewan, Canada. Quatern Res 24:219–234

Last WM, Delequiat J, Greengras K, Sukhan S (2002) Re-examination of the recent history of meromictic Waldsea Lake, Saskatchewan, Canada. Sediment Geol 148:147–160

Li J (2014) Sediment diagenesis in large Lakes Superior and Malawi: geochemical cycles and budgets and comparisons to marine sediments. PhD Thesis, University of Minnesota, Duluth, USA

Li J, Katsev S (2014) Nitrogen cycling in deeply oxygenated sediments: results in Lake Superior and implications for marine sediments. Limnol Oceanogr 59:465–481

Lindemann SR, Moran JJ, Stegen JC, Renslow RS, Hutchinson JR, Cole JK, Dohnalkova AC, Tremblay J, Singh K, Malfatti SA, Chen F, Tringe SG, Beyenal H, Fredrickson JK (2013) The epsomitic phototrophic microbial mat of Hot Lake, Washington: community structural responses to seasonal cycling. Front Microbiol 4:Article 323. doi:10.3389/fmicb.2013.00323

Madison JP, Gammons CH, Poulson SR, Jonas JP (2003) Oxidation of pyrite by ferric iron in the Berkeley pit lake, Montana, USA. In: Farell T, Taylor G (eds) Proceedings of the sixth international conference on acid rock drainage, 14–17 Jul 2003, Cairs, Queensland. The Australian institute of mining and metallurgy publication series No 3/2003. The Australian institute of mining and metallurgy, Carlton Victoria, pp 1073–1078

Manning PG, Prepas EE, Serediak MS (1999) Pyrite and vivianite intervals in the bottom sediments of eutrophic Baptiste Lake, Alberta, Canada. Can Mineral 37:593–601

Marion GM, Farren RE (1999) Mineral solubilities in the Na-K-Mg-Ca-Cl-SO_4-H_2O system: a re-evaluation of the sulfate chemistry in the Spencer-Møller-Weare model. Geochim Cosmochim Acta 63:1305–1318

Meyer KM, Macalady JL, Fulton JM, Kump LR, Schaperdoth I, Freeman KH (2011) Carotenoid biomarkers as an imperfect reflection of the anoxygenic phototrophic community in meromictic Fayetteville Green Lake. Geobiology 9:321–329

Meyers PA (2003) Applications of organic geochemistry to paleolimnological reconstructions: a summary of examples from the Laurentian Great Lakes. Org Geochem 34:261–289

Meyers PA, Ishiwatari R (1995) Organic matter accumulation records in lake sediments. In: Lerman A, Imboden DM, Gat JR (eds) Physics and chemistry of lakes. Springer, Berlin, pp 279–328

Middelburg JJ (1989) A simple rate model for organic matter decomposition in marine sediments. Geochim Cosmochim Acta 53:1577–1581

Moncur MC, Ptacek CJ, Blowes DW, Jambor JL (2006) Spatial variations in water composition at a northern Canadian lake impacted by mine drainage. Appl Geochem 21:1799–1817

Moreira S, Boehrer B, Schultze M, Dietz S, Samper J (2011) Modeling geochemically caused permanent stratification in Lake Waldsee (Germany). Aquat Geochem 17:265–280

Mulder A, van de Graaf AA, Robertson LA, Kuenen JG (1995) Anaerobic ammonium oxidation discovered in a denitrifying fluidized bed reactor. FEMS Microbiol Ecol 16:177–184

Musat N, Halm H, Winterholle B, Hoppe P, Peduzzi S, Hillion F, Horreard F, Amann R, Jørgensen BB, Kuypers MMM (2008) A single-cell view on the ecophysiology of anaerobic phototrophic bacteria. PNAS 105:17861–17866

Nixdorf E, Boehrer B (2015) Quantitative analysis of biogeochemically controlled density stratification in an iron-meromictic lake. Hydrol Earth Syst Sci 19:4505–4515

Noges T, Solovjova I (2005) The formation and dynamics of deep bacteriochlorophyll maximum in the temperate and partly meromictic Lake Verevi. Hydrobiologia 547:73–81

O'Beirne MD, Strzok LJ, Werne JP, Johnson TC, Hecky RE (2015) Anthropogenic influences on the sedimentary geochemical record in western Lake Superior (1800–present). J Great Lakes Res 41:20–29

Oberhäuser-Nehls R, Anagnostidis K, Overbeck J (1994) Phototrophic bacteria in the Plußsee: ecology of the sulfuretum. In: Overbeck J, Chrost RJ (eds) Microbial ecology of Lake Plußsee. Ecological studies, vol. 105. Springer, New York, pp 287–325

Overmann J (2006) The family Chlorobiaceae. In: Dworkin MM, Falkow S, Rosenberg E, Schleifer K-H, Stackebrandt E (eds) The prokaryotes. Proteobacteria: delta and epsilon subclasses. vol 7, 3rd edn. Springer, Heidelberg, pp 359–378

Overmann J, Beatty JT, Krouse HR, Hall KJ (1996) The sulfur cycle in the chemocline of a meromictic salt lake. Limnol Oceanogr 41:147–156

Owen RB, Crossley R, Johnson TC, Tweddle D, Kornfield I, Davison S, Eccles DH, Engstrom DE (1990) Major low levels of Lake Malawi and their implications for speciation rates in cichlid fishes. Proc R Soc Lond Biol Sci 240:519–553

Pakhomova SV, Rozanov AG, Yakushev EV (2009) Dissolved and particulate forms of iron and manganese in the redox zone of the Black Sea. Oceanology 6:773–787

Parkin TB, Winfrey MR, Brock TD (1980) The physical and chemical limnology of a Wisconsin meromictic lake. Trans Wis Acad Sci Arts Lett 68:111–125

Pasche N, Alunga G, Mills K, Muvundja F, Ryves DB, Schurter M, Wehrli B, Schmid M (2010) Abrupt onset of carbonate deposition in Lake Kivu during the 1960s: response to recent environmental changes. J Paleolimnol 44:931–946

Pasche N, Schmid M, Vazquez F, Schubert CJ, Wüest A, Kessler JD, Pack MA, Reegurgh WS, Bürgmann H (2011) Methane sources and sinks in Lake Kivu. J Geophys Res 116:G03006. doi:10.1029/2011JG001690

Peiffer S, Knorr K-H, Blodau C (2013) The role of iron minerals in the biogeochemistry of acidic pit lakes. In: Geller W, Schultze M, Kleinmann R, Wolkersdorfer C (eds) Acidic pit lakes. Springer, Heidelberg, pp 57–62

Perry KA (1990) The chemical limnology of two meromictic lakes with emphasis on pyrite formation. PhD Thesis, University of British Columbia, Vancouver, Canada

Perry KA, Pedersen TF (1993) Sulphur speciation and pyrite formation in meromictic ex-fjords. Geochim Cosmochim Acta 57:4405–4418

Powers LA, Werne JP, Johnson TC, Hopmans EC, Damsté JSS, Schouten S (2004) Crenarchaeotal membrane lipids in lake sediments: a new paleotemperature proxy for continental paleoclimate reconstruction? Geology 32:613–616

Rickard D, Luther GW III (2007) Chemistry of iron sulfides. Chem Rev 107:514–562

Salonen K, Arvola L, Rask M (1984) Autumnal and vernal circulation of small forest lakes in Southern Finland. Verh Int Ver Limnol 22:103–107

Sanger JE (1988) Fossile pigments in paleoecology and paleolimnology. Palaeogeogr Palaeoclimatol Palaeoecol 62:343–359

Scharf BW, Oehms M (1992) Physical and chemical characteristics. Arch Hydrobiol Beih 38:63–83

Schelske CL, Hodell DA (1995) Using carbon isotopes of bulk sedimentary organic matter to reconstruct the history of nutrient loading and eutrophication in Lake Erie. Limnol Oceanogr 40:918–929

Schmidt R, Psenner R, Müller J, Indinger P, Kamenik C (2002) Impact of late glacial climate variations on stratification and trophic state of the meromictic lake Längsee (Austria): validation of a conceptual model by multi proxy studies. J Limnol 61:49–60

Scholz CA, Johnson TC, Cohen AS, King JW, Peck JA, Overpeck JT, Talbot MR, Brown ET, Kalindekafe L, Amoako PYO, Lyons RP, Shanahan TM, Castañeda IS, Heil CW, Forman SL, McHargue LR, Beuning KR, Gomez J, Pierson J (2007) East African megadroughts between 135 and 75 thousand years ago and bearing on early-modern human origins. PNAS 104:16416–16421

Scholz CA, Talbot MR, Brown ET, Lyons RP (2011) Lithostratigraphy, physical properties and organic matter variability in Lake Malawi Drillcore sediments over the past 145,000 years. Palaeogeogr Palaeoclimatol Palaeoecol 303:38–50

Schubert CJ, Durisch-Kaiser E, Wehrli B, Thamdrup B, Lam P, Kuypers MMM (2006) Anaerobic ammonium oxidation in a tropical freshwater system (Lake Tangayika). Environ Microbiol 8:1857–1863

Selig U, Hübner T, Heerkloss R, Schubert H (2004) Vertical gradient of nutrients in two dimictic lakes—influence of phototrophic sulfur bacteria on nutrient balance. Aquat Sci 66:247–256

Smol JP, Cumming BF (2000) Tracking long-term changes in climate using algal indicators in lake sediments. J Phycol 36:986–1011

Stellmacher R (2004) Der Moritzteich bei Döbern—Analogon für die Sanierung saurer Bergbauseen?—Hydro- und Sedimentgeochemie eines meromiktischen Restsees. Diploma Thesis, TU Bergakademie Freiberg, Freiberg, Germany

Strøm K (1957) A lake with trapped sea-water? Nature 180:982–983

Suits NS, Wilkin RT (1998) Pyrite formation in the water column and sediments of a meromictic lake. Geology 26:1099–1102

Talbot MR (1988) The origins of lacustrine oil source rocks: evidence from the lakes of tropical Africa. Geol Soc Lond Spec Publ 40:29–43

Thomsen U, Thamdrup B, Stahl DA, Canfield DE (2004) Pathways of organic carbon oxidation in a deep lacustrine sediment, Lake Michigan. Limnol Oceanogr 49:2046–2057

Tietze K, Geth M, Müller H, Schröder L, Stahl W, Werner H (1980) The genesis of the methane in Lake Kivu (Central Africa). Geol Rundsch 69:452–472

Trouwborst RE, Clement BG, Tebo BM, Glazer BT, Luther GW III (2006) Soluble Mn(III) in suboxic zones. Science 313:1955–1957

Usman M, Hanna K, Abdelmoula M, Zegeye A, Faure P, Ruby C (2012) Formation of green rust via mineralogical transformation of ferric oxides (ferrihydrite, goethite and hematite). Appl Clay Sci 64:38–43

von Rohden C, Boehrer B, Ilmberger J (2010) Evidence for double diffusion in temperate meromictic lakes. Hydrol Earth Syst Sci 14:667–674

Votava JE (2014) The holocene history of Lake Kivu (East Africa): new perspectives from new cores. MSc Thesis, University of Minnesota

Walker KF, Likens GE (1975) Meromixis and a reconsidered typology of lake circulation patterns. Verh Int Ver Limnol 19:442–458

Wanchun L, Shijie L, Yu Y, Jiang J, Peimin P (2001) Meromixis in Zige Tangco, central Tibetan Plateau—discovery and significance. Sci China Ser D Earth Sci 44:338–342

Weimer WC, Lee FF (1973) Some considerations of the chemical limnology of meromictic Lake Mary. Limnol Oceanogr 18:414–425

Wenk CB, Zopfi J, Blees J, Veronesi M, Niemann H, Lehmann MF (2014) Community N and O isotope fractionation by sulphide-dependent denitrification and anammox in a stratified lacustrine water column. Geochim Cosmochim Acta 125:551–563

Wittkop C (2014) Are siderite varves deepwater microbialites? In: The Geological Society of America (GSA) (ed) 2014 GSA annual meeting in Vancouver, British Columbia, 19–22 Oct 2014. https://gsa.confex.com/gsa/2014AM/finalprogram/abstract_248376.htm. Accessed 29 Jan 2016

Wüest A, Jarc L, Brüggemann H, Pasche N, Schmid M (2012) Methane formation and future extraction in Lake Kivu. In: Descy J-P, Darchambeau F, Schmid M (eds) Lake Kivu—Limnology and biogeochemistry of a tropical great lake. Springer, Dordrecht, pp 165–180

Yin HB, Fan CX, Ding SM, Zhang L, Zhong JC (2008) Geochemistry of iron, sulphur and related heavy metals in metal-polluted Thaihu Lake sediments. Pedosphere 18:564–573

Zachara JM, Moran JJ, Resch CT, Lindemann SR, Felmy AR, Bowden ME, Cory AB, Fredrickson JK (2016) Geo- and biogeochemical processes in a heliothermal hypersaline lake. Geochim Cosmochim Acta 181:144–163

Zegeye A, Bonneville S, Benning LG, Sturm A, Fowle DA, Jones C, Canfield DE, Ruby C, MacLean LC, Nomosatoryo S, Crowe SA, Poulton SW (2012) Green rust formation controls nutrient availability in a ferruginous water column. Geology 40:599–602

Zimmerman AR, Canuel EA (2000) A geochemical record of eutrophication and anoxia in Chesapeake Bay sediments: anthropogenic influence on organic matter composition. Mar Chem 69:117–137

Zolitschka B, Francus P, Ojala AE, Schimmelmann A (2015) Varves in lake sediments—a review. Quat Sci Rev 117:1–41

Chapter 4
Biological and Ecological Features, Trophic Structure and Energy Flow in Meromictic Lakes

Egor S. Zadereev, Ramesh D. Gulati, and Antonio Camacho

4.1 Introduction

A lake is meromictic when vertical mixing is insufficient to homogenise the waterbody and remove chemical gradients over the course of years (see Chap. 2). As the vertical water exchange between mixolimnion and monimolimnion is relatively weak, strong chemical gradients of salinity, dissolved gases (oxygen, hydrogen sulphide, methane and others) and nutrients are conserved at the depth of the chemocline, that is, the transition zone between the mixolimnion and the monimolimnion. Physical and chemical properties of meromictic lakes affect their biology and food web structure. Some biological processes affect the stability of meromixis. For example, microbiologically controlled chemical reactions, e.g. cycling of iron by the bacterial consortia, including iron oxidisers and iron reducers, contribute to the higher density of monimolimnetic waters due to dissolved iron (Dietz et al. 2012). For these and similar biogeochemical processes, see Chap. 3 under "Chemical Setting and Biogeochemical Reactions in Meromictic Lakes". In this chapter, we will focus mainly on the biological aspects and ecological responses of organisms and food webs to meromixis.

Studies on the biology and ecology of meromictic lakes are often fragmentary, though there are several investigations describing the microbial ecology of the chemocline of meromictic lakes (e.g. Overmann et al. 1991, 1999a; Ovreas et al.

E.S. Zadereev (✉)
Institute of Biophysics SB RAS, Akademgorodok, Krasnoyarsk, Russia
e-mail: egor@ibp.ru

R.D. Gulati
Netherlands Institute of Ecology (NIOO), 6708 PB Wageningen, The Netherlands

A. Camacho
Cavanilles Institute for Biodiversity and Evolutionary Biology & Department of Microbiology and Ecology, Universitat de Valencia, E-46100 Burjassot, Valencia, Spain

© Springer International Publishing AG 2017

R.D. Gulati et al. (eds.), *Ecology of Meromictic Lakes*, Ecological Studies 228,
DOI 10.1007/978-3-319-49143-1_4

1997; Bosshard et al. 2000a, b; Humayoun et al. 2003; Lehours et al. 2007). The gradients of water density, salinity and dissolved gases in the chemocline offer habitats for certain microbial communities. The cycles of nutrients (carbon, sulphur and nitrogen) in meromictic lakes are very often coupled in the vicinity of chemocline. This coupling includes various processes, mostly microbially driven, e.g. the co-occurring denitrification and nitrogen fixation (Halm et al. 2009), photoassimilation of inorganic carbon and dark carbon fixation by the phototrophic and chemolithotrophic sulphur organisms (Camacho et al. 2001a). The chemocline depth determines both the size of the oxic and the anoxic zone and consequently influences the net primary production of autotrophic and chemoautotrophic organisms. Thus, the existence of a chemocline and the ecological conditions that prevail in this zone are, indeed, among the specific features of meromictic lakes.

Another aspect of the biological studies in meromictic lakes is the mechanism that drives the spatial heterogeneity in the distribution of different organisms. The biological research in meromictic lakes often relates to the vertical distribution of plankton communities, e.g. that of phytoplankton and zooplankton (Sharples et al. 2001; Degermendzhy et al. 2002, 2010; Zadereev and Tolomeyev 2007). The mixolimnion of the meromictic lakes can also be stratified during a part of the year. In such a case the water column of the lake will have two chemoclines: (1) the annual chemocline that coincides with the thermocline and is located between the epilimnion and the hypolimnion and (2) a permanent chemocline, separating the mixolimnion from the monimolimnion (see Chap. 2). Physical and chemical gradients of water temperature and different solutes affect the vertical distribution of different organisms, but are also a consequence of their biological activities.

The studies focused on the food web structure and dynamics in meromictic lakes are rare. Over the past 30 years (1985–2015), published work focused on meromictic lakes, including their biology comprises almost 40 % of the studies on bacteria and their role in the lake ecosystems. In contrast, the numbers of publications on zooplankton, fish or food webs are scarce (Table 4.1). This information reflects both the relative importance of bacteria in the biogeochemical cycling in meromictic lakes and the bias of researchers towards studies on microbial processes. On the other hand, the food chain in meromictic lakes is often considered to be simple and stunted.

Table 4.1 Web of Science publications in 1985–2015 on different aspects of the biology of meromictic lakes

Search for total number of published papers relating to meromictic lakes	Number of papers
Topic: (merom* and lake*)	ca. 900
Topic: (merom* and lake* and bacteria*)	ca. 350
Topic: (merom* and lake* and phytoplankton)	ca. 150
Topic: (merom* and lake* and zooplankton)	53
Topic: (merom* and lake* and fish)	41
Topic: (merom* and lake* and food web)	24

Many meromictic lakes lack fish as the top organisms in the food chain. In this case, the prevalence of studies focused on bacteria is easy to explain.

4.2 The Ecology of the Chemocline and Monimolimnion

4.2.1 Microbial Communities

The microbial communities of the chemocline are potentially an important food web component of meromictic lakes. Phototrophic sulphur bacteria are common in the chemocline of many meromictic lakes, where they can use the available light as a source of energy and reduced compounds, such as sulphide, as a source of electrons for the photosynthesis (Camacho 2009). Phototrophic sulphur bacteria develop large concentrations below the interface between mixolimnion and monimolimnion. The depth where these phototrophs are present is not uniform as the light intensity at the depth of the chemocline depends on the water transparency. Usually at the depth of the chemocline for lakes with a relatively shallow monimolimnion, different phototrophic sulphur bacteria form narrow bands of bacterial layers, having a visually recognisable pink (purple sulphur bacteria, PSB) or green or brown (green sulphur bacteria, GSB) colour (e.g. Tonolla et al. 2003). Apart from the light and sulphide availability, the differences in salinity with depth can determine the development of multilayers of PSB and GSB that differentially distribute within the vertical profile of lakes according to their requirements and tolerance, as demonstrated for the meromictic Lake El Tobar (Garcia-Gil et al. 1999; Camacho et al. 2002). Even though the ecophysiology of phototrophic sulphur bacteria has been studied in saline and freshwater ecosystems (e.g. Blankenship et al. 1995; Van Gemerden and Mas 1995; Overmann and van Gemerden 2000; Musat et al. 2008; Overmann 2008), the trophic role of these bacterial communities, which often have both high biomass and high production rates, is not yet well documented.

The development of phototrophic sulphur bacteria is often observed in stratified non-meromictic lakes if the hypolimnion becomes anoxic in summer, mainly due to eutrophication (Dressler et al. 2007), or if enough organic matter is present in the deep layers associated with sufficient sulphate availability so that active reduction in the hypolimnion of sulphate to sulphide can occur. However, as these bacteria require reduced sulphur compounds for their development, during seasonal mixing and breakdown of stratification, the whole water column becomes oxygenated, and these organisms fail to develop high biomass. In meromictic lakes due to the regular presence of sulphide in deep waters, phototrophic sulphur bacteria often develop high densities in the chemocline. The concentration of bacteria in coloured (purple or green) layers in the chemocline can be very high. For example, in Lake Shunet (Siberia, Russia), the concentration of PSB in the chemocline is up to 10^8 cells ml^{-1} (Rogozin et al. 2012). The maximum PSB concentration so far recorded in the chemocline is 4×10^8 cells ml^{-1} in Lake Mahoney (Overmann et al. 1991). These dense bacterial layers can be observed year-round including the winter and ice periods (Overmann 1997; Rogozin et al. 2009).

Despite the high biomass accumulation potentially reached by these phototrophic bacterial blooms, the trophic importance as well as the fate of this abundant source of organic carbon is not clear. Laboratory experiments demonstrate that different zooplankton species can feed on these bacteria (Gophen 1977; Temerova et al. 2002). However, in the field, the toxicity of sulphide should prevent zooplankton from active consumption of this food, as the zooplankton would need to penetrate the anoxic layers to obtain this food. Since the metazooplankton is sensitive to the effect of hydrogen sulphide, the phototrophic sulphur bacteria in the anaerobic zone are not a preferred food. For example, in Lake Ciso (Spain), a well-structured metalimnetic community composed mainly of phototrophic purple bacteria, *Cryptomonas* and a few species of ciliates and rotifers were regularly found for several years at the oxic-sulphide interface (Massana et al. 1994). The presence of *Daphnia pulex* induced a clear-water phase in the epilimnion and reduced the densities of metalimnetic bacteria, cryptomonads and ciliates. The impact of *Daphnia* was very high on the microaerophilic populations of bacteria and protists, whereas the anaerobic bacterial populations were least affected, since sulphide limited the vertical distribution of *D. pulex* (Massana et al. 1994).

Overmann et al. (1999b) and Camacho et al. (2001a) reported that in situ some zooplankton can migrate to the chemocline and consume phototrophic sulphur bacteria. Sulphur bacteria can reach the oxic zone if there is turbulent mixing, especially during autumnal, or even daily caused by the diel changes in the oxygen/sulphide boundaries in lakes with high biogeochemical activity (Camacho and Vicente 1998; Camacho et al. 2000). In many meromictic lakes, during mixing in the mixolimnion, the PSB enter the aerobic zone and become available to the zooplankton (Overmann et al. 1999b). Because of this mixing, substantial bacterial biomass enters the carbon cycle of the conventional food web and links the sulphur and carbon cycles in upper layers (Overmann 1997). The role of PSB in the secondary production remains unclear. Short-term experiments with *Diaptomus connexus* in Lake Mahoney revealed that the carbon from sulphur bacteria contributed only a negligible 0.6 % to the production of copepods in the water column. However, estimates also show that this percentage may reach as high as 85 % (Overmann et al. 1994). Also, based on stable carbon isotope analysis, Camacho et al. (2001a) suggested that some zooplankters may obtain as much as half of their carbon at the chemocline, though the proportion of bacteria and algae in this carbon is not clear. This question will be discussed further in Sect. 4.3.

Many studies consider the bacterial diversity at the chemocline and the monimolimnion of meromictic lakes (e.g. Bosshard et al. 2000a; Overmann et al. 1999a; Lehours et al. 2007). Case-study chapters presented in this book (Part II) cover in detail the wide range of meromictic lake types, each with its specific microbial communities. Lake Rogoznica (typical example of a euxinic marine lake in Croatia) is a system with both meromictic and holomictic conditions alternating between years (see Chap. 6). During the meromictic periods, anoxygenic phototrophic green sulphur bacteria dominate in both the chemocline and the hypolimnion (up to 90 % of total 10^7 cells ml^{-1}), while during and after the holomictic anoxic events, the homogeneous, anoxic water column is characterised by a bloom of *Gammaproteobacteria* sulphur oxidisers (GSO)/SUP05 clade-related *Gammaproteobacteria* (Pjevac et al. 2015). Such a shift in the microbial community after holomixis has been also observed in other stratified lakes (Cohen et al. 1977; Casamayor et al. 2000). However, many other lakes during and after holomixis are usually oxygenated. In such cases, the bacterial community in the mixed water column will be similar to the epilimnetic community (Camacho et al. 2000; Hollibaugh et al. 2001). The anoxic water column of Lake Rogoznica also harbours sulphate-reducing bacteria, which are closely related to *Desulfonema/Desulfococcus/Desulfosarcina* group and *Desulfovibrio/Desulfomicrobium* group, the main sulphate reducers and producers of the hydrogen sulphide in the lake water column (see Chap. 6).

In the iron-rich karstic meromictic lake Laguna de La Cruz (Chap. 8), oxygenic unicellular cyanobacteria and anoxygenic sulphur phototrophic bacteria develop and form relatively dense populations around the oxic-anoxic interface located at the bottom of the metalimnion during thermal stratification (see Chap. 8). A dense plate of *Synechococcus*-like picocyanobacteria develops during the thermal stratification period in the upper part of the oxic-anoxic interface (Camacho et al. 2003a, b), whereas more sparse (up to 3×10^6 cells ml^{-1}) populations of photosynthetic sulphur bacteria, both purple (PSB) and green (GSB), appear below the depth where oxygen is zero (Rodrigo 1997; Rodrigo et al. 2000), both during the thermal stratification (in the hypolimnion as well as the monimolimnion) and only in the monimolimnion during the mixing period. Sulphur bacteria generally link the carbon and sulphur cycles in such lakes. Two main species form these populations in Lake La Cruz (Rodrigo et al. 2000; Casamayor et al. 2012), the PSB *Lamprocystis purpurea* (formerly called *Amoebobacter purpureus*) and the GSB *Chlorobium clathratiforme* (formerly called *Pelodyction clathratiforme*). The former species has a selective advantage, having the ability to accumulate elemental sulphur, which can serve as alternative electron donor in the low-sulphide environment of the Lake La Cruz chemoclines. In addition to the phototrophic bacteria, photo- and chemolithotrophic primary producers, and their activity, have been recently described (Walter 2011; Walter et al. 2014; Camacho et al. 2016) with regard to the carbon, iron and nitrogen cycles.

New methods of genetic analysis and availability of mass sequencing techniques have stimulated progress in studies on bacterial diversity. Meromictic lakes, even if they are geographically nearby, may have a distinct bacterial species composition. For example, Baatar et al. (2016) who studied Lakes Shira and Shunet in Siberia and Lake Oigon in Mongolia, located in landlocked steppes of Central Asia (~2500 km inland), used genetic analyses to demonstrate that bacterial communities not only differ and are dissimilar among these lakes, but they also differ between oxic and anoxic layers within the same lake.

Also studies in several other meromictic lakes show that bacterial communities differ between oxic to anoxic layers and that both the bacterial diversity and abundance are consistently greater in the anoxic than in the oxic waters. For example, microbial diversity in hypersaline alkaline Mono Lake (Chap. 11), both in the chemocline and monimolimnion, was higher than in epilimnetic communities and with a broader phylogenetic representation, including a novel clade of *Bacteriodetes*. Similar results were reported for prokaryotic assemblages from hypersaline meromictic lakes Ursu and Fara Fund (Transylvanian Basin, Romania) (Andrei et al. 2015) and for holomictic and meromictic karstic lakes from Central Spain (Casamayor et al. 2012). The lakes harboured diverse prokaryotic communities. Genetic sequencing of prokaryotic communities revealed a large proportion of operational taxonomic units, OTU, an operational definition of a species or group of species often used when only DNA sequence data are available, with no close identified microbial species, showing that these environments are potential sources of novel microorganisms. It is interesting that microbial populations indigenous to each lake pointed to similarity in carbon degradation and sulphate reduction. Thus, more or less close cycles of nutrients in different meromictic lakes can be driven by different microorganisms but playing similar biogeochemical roles. High bacterial diversity has been reported also for the meromictic Ace Lake (Lauro et al. 2010) and Lake Mahoney (Klepac-Ceraj et al. 2012). The causes for such a diversity are the multiple biogeochemical microbial niches, abundant nutrients, lack of mixing in the anoxic layers, downward metabolic fluxes, sulphur and nitrogen cycles that occur in the anoxic layer and multistep mineralisation processes that need a diverse and complex bacterial community (Yurkova et al. 2002; Biderre-Petit et al. 2011; Tiodjio et al. 2014; Andrei et al. 2015).

With the development of genomics, it appears that even characteristic microorganisms such as phototrophic sulphur bacteria appear to be taxonomically variable among meromictic lakes. Both PSB and GSB typically represent phototrophic sulphur bacteria in the anoxic layers. The aforementioned study of Central Asian meromictic lakes (Baatar et al. 2016) identified three discrete dominant PSB in the chemocline of the three Asian meromictic lakes. Species most closely related to *Halochromatium roseum* dominated in Lake Oigon. Whereas in the Siberian Lake Shunet, the highly abundant species is most closely related to *Thiocapsa rosea*; and, in Lake Shira, PSB species was identified to comprise novel species (Yarza et al. 2014). Even though the two Siberian lakes just mentioned lie within a proximity of 8 km, their dominant PSB species clearly differ. Similarly, dominant species of phototrophic sulphur bacteria differ between Lake Ciso and Lake Vilar in Spain, while these lakes lie only 1 km apart (Casamayor et al. 2000). Such findings imply that even geographically close meromictic lakes may have particular physical, chemical and ecological conditions that promote development of quite diverse bacterial communities.

Phototrophic sulphur bacteria are not only variable among lakes; their diversity within the same lake can be high. For example, in the sulphur-rich Lake Cadagno in Switzerland, a diverse community of microorganisms develops in the chemocline (Chap. 7). This community is dominated by several anoxygenic phototrophic sulphur bacteria, including seven species of PSB and two species of the GSB (Bosshard et al. 2000a, b; Tonolla et al. 1999, 2003, 2005; Peduzzi et al. 2011,

2012). An important question is about the coexistence in the presumably same ecological niche of diverse, but closely related, populations of PSB (Chap. 7).

Comprehensive genetic sequencing studies mentioned below demonstrate that many meromictic lakes contain many unclassified prokaryotes (bacteria and archaea) in their monimolimnia. For example, tag pyrosequencing studies showed almost one-fourth of unclassified bacterial groups and candidate divisions (UBG/CD) in the anoxic layer of Lake Shunet (Baatar et al. 2016). In the saline meromictic Ocnei Lake (Central Romania), archaeal diversity in the water column was studied using microbiological methods and molecular techniques based on the sequence analysis of the 16S rRNA gene (Baricz et al. 2014). Archaeal diversity in the water column increased with increasing depth and salinity. This Romanian lake seems to represent a relatively stable extreme habitat and contains a diverse and novel archaeal community—about 30 % of the OTUs could not be classified at the genus level. For the meromictic Sakinaw Lake (British Columbia, Canada), diversity, abundance and co-occurrence patterns of uncultivated microbial communities were studied using sequencing of the small-subunit rRNA gene (Gies et al. 2014). The monimolimnion was characterised by a rapid increase in methanogen-affiliated and unassigned archaeal sequences as well as bacterial OTUs affiliated with *Chloroflexi* and lineages of prokaryotic organisms for which no cultured representatives have been found. Methanogens are also very abundant in the methane-rich monimolimnia of other meromictic lakes, such as Lake Matano, in Indonesia (Crowe et al. 2011); Lake Kivu, located in the border between the Democratic Republic of Congo and Rwanda (Llirós et al. 2015); and the French Lake Pavin (Lehours et al. 2005, 2007; Biderre-Petit et al. 2011). Also, high proportions of bacteria recorded for Lake Mahoney (Klepac-Ceraj et al. 2012), Lake Alinen Mustajarvi (Peura et al. 2012) and Lake A (Comeau et al. 2012) belonged to UBG/CD.

Based on numerous demonstrations of high numbers of unclassified bacteria, the term *microbial dark matter* was introduced by Gies et al. (2014) to characterise the hidden diversity of microbial communities in meromictic lakes, specifically under anoxic conditions. The challenge in microbial ecology is to both investigate the uncultivated microbial diversity and the metabolic potential of these complex communities. NGS and metagenomic analyses are promising tools to unmask both the taxonomic and genetic diversity in these environments.

The hidden bacterial diversity and metabolic potential can be discovered studying total rather than one nucleic acid pool. Inceoglu et al. (2015) studied the microbial community composition in meromictic Lake Kivu using 16S rDNA and ribosomal RNA (rRNA) pyrosequencing. High proportion of rare species was detected only in either an active (RNA-based) or bulk (DNA-based) community, indicating the possible underestimation of microbial diversity when using only one nucleic acid pool. Research also reveals the co-occurrence of the potentially active sulphur-oxidising and sulphate-reducing bacteria in the anoxic zone that may suggest the presence of an active yet cryptic sulphur cycle (Inceoglu et al. 2015).

The ecological role of microbial guilds in meromictic lakes can be traced by their corresponding activities (Fig. 4.1). Biogeochemical processes are the key for

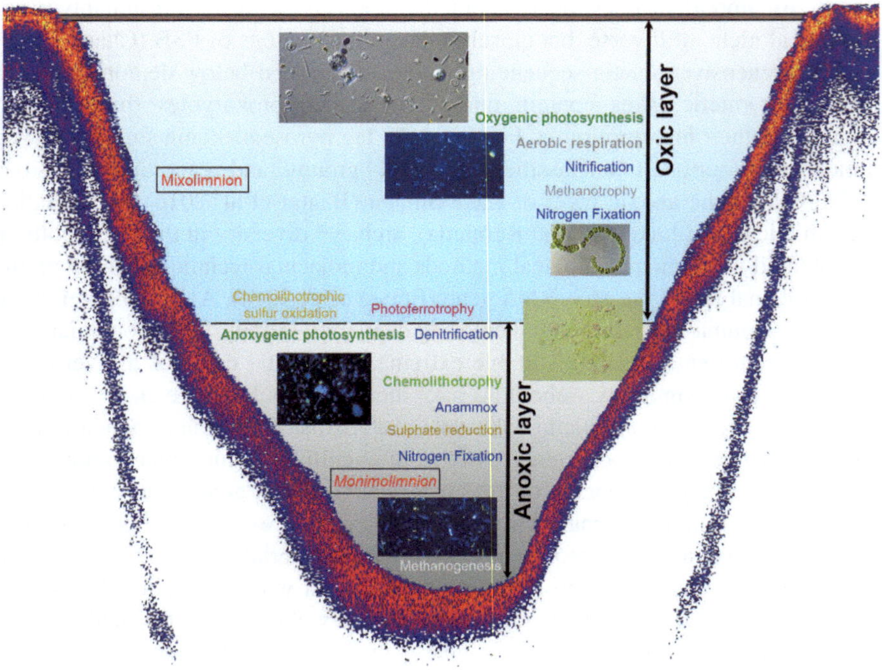

Fig. 4.1 Some of the main biogeochemical processes occurring in the different layers of a meromictic lake. Main processes mainly linked to the uptake of inorganic carbon are in *green*, respiratory processes producing (aerobic respiration) or consuming CO_2 (methanogenesis) are in grey. Main activities linked to the sulphur and nitrogen cycles are in *yellow* and *blue*, respectively. Finally photoferrotrophy is also an autotrophic activity that uses reduced iron as electron donor. Other respiratory processes (e.g. with Fe^{3+}) are not represented

explaining microbial diversity and the interplay between microbial activities and the abiotic characteristics of meromictic lakes. Even if many of these processes are shared also by non-meromictic lakes, some of the activities, especially those enhanced by the stability of steep physical and chemical gradients, feature meromictic lakes, as the existence of these permanent gradients allow these biogeochemical processes to continue through the year and make significant contributions to the elemental cycles.

As in other types of lakes, oxygenic photosynthesis in the oxic waters is the main inorganic carbon incorporation process, although meromictic lakes are featured by the increased relative importance of both anoxygenic photosynthesis, performed by phototrophic sulphur bacteria in anoxic layers, and chemolithoautotrophy, both as inorganic carbon fixing processes. Chemolithoautotrophy can occur in aerobic (e.g. nitrification), microaerobic and anoxic layers (Fig. 4.1). In the steep gradients of the chemocline, where reduced compounds (e.g. sulphide, Fe^{2+}) can coexist with oxygen, sulphur-oxidising bacteria, like the *Epsilonproteobacteria* of the meromictic basin of Lake Banyoles (Noguerola et al. 2015), can achieve high carbon fixation rates in the narrow depth range where sulphide and oxygen coexist.

The same is true for the coexistence of Fe^{2+} and oxygen with enough light availability in very narrow layers of the chemocline of iron meromictic lakes allowing both photoferrotrophy (Walter et al. 2014; Llirós et al. 2015; Camacho et al. 2016) and iron-based chemolithotrophy. Also, in the anoxic waters, the anaerobic oxidation of ammonia (anammox) by obligately anaerobic chemolithoautotrophs contributes to inorganic carbon fixation. Overall, all these carbon fixation activities link the carbon cycle with the sulphur (chemolithotrophic sulphur oxidisers and anoxygenic photosynthetic bacteria), nitrogen (nitrifiers and anammox bacteria) and iron (photoferrotrophs and chemolithoautotrophs) cycles.

Respiratory activities relate to the biogeochemical cycles of several elements, as different electron acceptors can be used for the respiration of organic matter. Aerobic respiration (also including methanotrophy) is an energetically more profitable respiratory process, and consequently it occurs if oxygen is available. However, when oxygen is exhausted, other oxidised compounds can be used as electron acceptors in anaerobic respiratory processes and link carbon cycle with other elemental cycles. The main anaerobic respiratory processes are (1) denitrification, for which nitrate is used as electron acceptor, and (2) sulphate reduction and release of sulphide and (3) methanogenesis, for which CO_2 acts as electron acceptor and methane is released. Since the occurrence of these respiratory processes depends on the relative availability of the potential electron acceptors and the energetic yields of each respiratory type, the relative importance of each of these processes also varies with depth (Fig. 4.1). Accordingly, the most oxidised compounds progressively disappear from the well-oxygenated upper layers, where only aerobic respiration occurs, to the bottom. Chemocline denitrification and even iron (III) respiration can be the main respiratory processes, but in deeper layers, sulphate reduction and methanogenesis become the dominant respiratory processes.

Overall, all the carbon fixation and respiratory processes (Fig. 4.1) occurring with different intensity in different lake layers of meromictic (and also other stratified) lakes make the link between the carbon cycle and nitrogen, sulphur and iron cycles.

4.2.2 *Eukaryotic Primary Producers and Microzooplankton Communities*

The transition redox gradient zone in the deeper layers of meromictic lakes supports eukaryotic primary producers and microzooplankton communities that participate in the deepwater microbial loop. Certain primary producers that produce oxygen, including autotrophic picocyanobacteria and also eukaryotic algae, can accumulate in deep layers, often around the chemoclines of stratified lakes, forming deep chlorophyll maxima (DCM). Although some of the mechanisms leading to the formation of DCM near the chemoclines of stratified lakes are passive (due to sedimentation: see Chap. 5), others involve active in situ growth (Camacho 2006).

Among the most frequently occurring populations of eukaryotic primary producers, thriving in the upper part of the chemoclines, are those of cyanobacteria and cryptomonads (Camacho et al. 1996; Gervais 1998; Camacho et al. 2001b), which grow there and use their mobility or buoyancy changes to move through the physical-chemical gradients of the lake's chemocline to meet their ecological requirements.

Although some eukaryotic algae are mixotrophic (Gervais 1997), microzooplankton are commonly composed of heterotrophic protists, comprising flagellated and ciliated protozoa which are often symbiotic algae or/and bacteria (e.g. Finlay et al. 1991; Miracle et al. 1992). These microzooplankton communities may feed on dense anaerobic bacterial communities in the chemocline and couple deepwater anaerobic microbial production with the grazer food web in the upper aerobic waters.

The typical taxonomic composition of such microzooplankton communities comprises species of heterotrophic protists, mainly ciliates and choanoflagellates. Several meromictic lakes described in details in this book harbour such diverse communities. In the meromictic Lake Cadagno (Switzerland), ciliates and choanoflagellates feed on anaerobic phototrophic sulphur bacteria (see Chap. 7). Metazooplankton (rotifers) and microaerophilic and anaerobic ciliates also form deepwater populations in Lake La Cruz (see Chap. 8). Complex food web interactions between the heterotrophic bacterioplankton and microaerophilic and anaerobic ciliated protozoa that develop in the anoxic hypolimnion and rotifers thriving around the seasonal chemocline during stratification are typical for this lake. Also, such ciliates as *Coleps hirtus* endosymbiotic algae are abundant close to the seasonal chemocline of Lake La Cruz. The coupling of the deepwater microbial loop with the classical grazer food web in the upper waters in Lake La Cruz is, however, quite weak. Reasons for this are the strong changes in density in the water column, the physiological restrictions (e.g. microaerobic or anoxic conditions) for most consumers to graze in these parts of the water column and the lack of biological transport of nutrients to upper layers by the upward migrating zooplankton.

In contrast, in Lake Rogoznica (see Chap. 6), ciliates are important in the carbon transfer throughout the lake's food web, and they also control the microbial loop in the lake. *Strombidium purpureum*, a ciliate, is abundant in the bottom waters of the lake in years when the oxygen saturation % is low and concentrations of sulphur compounds are high. These ciliates are purple in colour, due to the presence of numerous endosymbiotic purple non-sulphur bacteria important for the survival of the ciliates under unfavourable oxygen conditions (Fenchel and Bernard 1993). The ciliates, by grazing on the dense populations of phototrophic sulphur bacteria (Pjevac et al. 2015), transfer large quantities of anoxically produced organic carbon to higher trophic levels.

Several ciliate genera adapted to anoxia and hydrogen sulphide coexist with a population of *Cryptomonas* sp., in the chemocline zone of Lake Shunet (see Chap. 5). The protozooplankton maximum is observed in the 10–35 cm layer of the chemocline region, above the layer of PSB or sometimes in it, where hydrogen sulphide concentrations vary from 0 to 5 mg l^{-1} with little or no oxygen. Here, the ciliates feed on the bacterial community and on *Cryptomonas* spp.

The species diversity of protistan communities in the chemoclines is not well studied. High-throughput sequencing of the 18S rRNA gene provided an insight into the spatio-temporal protistan diversity along the oxygen gradient of the freshwater meromictic Lake Alatsee, Germany (Oikonomou et al. 2015), with the highest diversity mainly owing to mainly to the presence of Euglenozoa. The dominant taxa in the eukaryotic communities in the anoxic monimolimnion belonged Chrysophyta and Bicosoecida. However, it is not clear whether these species can grow under anoxic conditions or they accidently enter the anoxic waters but are not active there. Molecular analytical analyses of the genetic diversity suggest a high degree of undescribed OTUs.

Summarising, highly diverse and specific microbial communities develop at the depth of the oxic-anoxic interface in meromictic lakes, at the top layer of the chemocline. In some cases, such as the moderately productive meromictic Lake La Cruz and the holomictic lakes Arcas and Lagunillo del Tejo (Casamayor et al. 2012), the oxic-anoxic interface may be located above the chemocline, where it separates the metalimnion from the hypolimnion. This transition redox gradient zone can range in thickness from few metres to a few centimetres. Anoxygenic phototrophic bacteria, both PSB and GSB, are characteristic microbial groups for the transition redox gradient zone between the mixolimnion and the monimolimnion. These groups that vary in their relative abundance and diversity have been described from many lakes. The reduction of nitrate, manganese, iron and sulphate takes place at the depth of oxic-anoxic interface or below. Microbial communities take advantage of chemical gradients and share specific metabolic capacities. Biogeochemical cycles of C, N and S in this zone may be coupled due to the activities of various microbial taxa, many of which are still uncultured/unidentified. Transition redox gradient zone also supports specific eukaryotic primary producers and microzooplankton communities. These communities form microbial loop. Diverse eukaryotic species that are potentially mixotrophic, such as *Cryptomonas* sp., may feed on dense anaerobic bacterial communities in the chemocline, as a complementary diet source, but in turn can be a food source for micro- and macrozooplankton. This link can couple deepwater anaerobic microbial production with the grazer food web in the upper waters.

Finally, it may be mentioned that even though we considered the anoxic part of lakes, in many meromictic lakes, most of the water volume does not correspond to the anoxic waters but to the oxic waters. In the oxic waters, there is no role of phototrophic sulphur bacteria in the food web, and for most cases, phototrophic sulphur bacteria are not consumed by metazooplankton. Often the relative contribution of phototrophic bacteria to carbon fixation is relatively low, although it is true that in meromictic lakes in general carbon fixation is greater than that in non-meromictic lakes due to prolonged stratification and relatively stable conditions at the depth of oxic-anoxic interface.

4.3 Food Web Studies

While the structure and ecology of microbial communities of the chemocline in meromictic lakes has received considerable attention, the studies dealing with the higher food web components in the meromictic lakes, i.e. with complex food web, are scarce. Below we will review mostly research papers focused on food web studies in meromictic lakes, which are not included in the case-study chapters. We also support these studies with some examples presented in case-study chapters (Part II of this book). We propose a schematic simplified diagram of the pelagic food web in meromictic lakes and describe different food web components and interactions between them (Fig. 4.2).

4.3.1 Food Web Structure: From Bacteria to Fish

Perhaps the most comprehensively studied meromictic lake in the world is Lake Mahoney. Canada, (Overmann et al. 1991, 1999c). Lake Mahoney has several

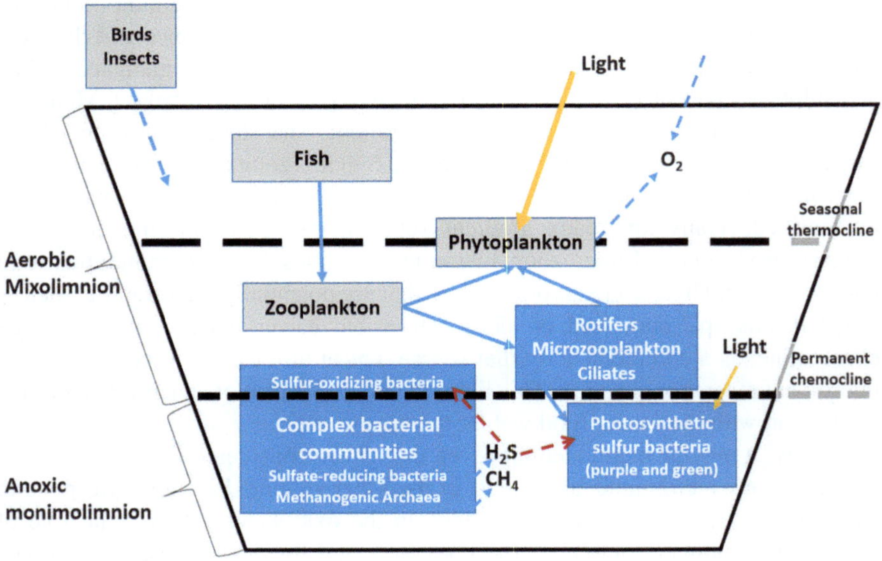

Fig. 4.2 A schematic and simplified diagram of the pelagic food web in a meromictic lake. The more specific features for meromictic lake food web components are coloured *blue*. This picture can be different in the meromictic lakes that develop an anoxic hypolimnion below the seasonal thermocline during thermal stratification. In this case, processes and microorganisms usually occurring in the monimolimnion are also extending to the anoxic hypolimnion. The structure of the food web (components, abundance of species, species diversity and composition) may vary between meromictic lakes depending on various factors (size, salinity, chemical composition, climate and geography, etc.)

typical features of meromictic lakes. It is a small and relatively shallow saline lake. A dense population of purple sulphur bacteria inhabits the chemocline of the lake. Overman et al. (1999b) investigated the biomass and seasonal variability of the planktonic organisms. They observed pronounced vertical differences in the water column for heterotrophic flagellates, ciliates, rotifers, calanoid copepods and growth rates of chemotrophic bacterioplankton. Calanoid copepods represent the highest trophic level in the planktonic food web of this fishless Mahoney Lake. Interestingly, the carbon flux to higher trophic levels through the microbial loop was comparable with the input from the conventional grazing food chain. The carbon supply and demand data revealed that for zooplankton food sources other than phytoplankton must exist. Concluding, it can be surmised that the autumnal massive upwelling of the purple sulphur bacterium *Amoebobacter purpureus* from the chemocline into the mixolimnion is a direct, significant food source for higher trophic levels in the oxic part of Lake Mahoney.

Northcote and Hall (2010) also studied zooplankton in Mahoney Lake and in the nearby meromictic Blue Lake (Canada). They presented the results of an 8-year study, when lake level was decreasing during a period of dry climatic conditions. The research indicates how the zooplankton in these two meromictic Canadian lakes differing in their salinities responded to the decrease in water level and increase in salinity. Moreover, different zooplankton species preferred to inhabit different depths: the cladocerans were found in moderate numbers in upper part of the water column in waters with salinity up to 10 g l^{-1}. The decrease in lake levels and increases in the salinity above this threshold salinity concentration seem to have purged the cladocerans from Mahoney Lake. On the other hand, the abundance of the rotifer *Brachionus plicatilis* greatly increased at a depth close to the chemocline and above the PSB layer. A similar increase was obtained for the calanoid copepod *Diaptomus connexus*. Moderate to high numbers of the late copepodite stages and adults of *D. connexus* inhabited water layers adjacent to the PSB plate, but occasionally their habitats overlapped (Northcote and Hall 2010). Northcote and Hall (2010) proposed that vertical distribution maximum of calanoids within a few cm or less from the PSB plate suggests that these abundant PSB serve as food for the calanoid *D. connexus*. Overmann et al. (1999b) also demonstrated using stable carbon isotope analyses of such copepods and the pigments from *A. purpureus* that these PSB formed between 75 % and 85 % of the copepod diet. On the other hand, the significantly higher numbers of adult *D. connexus* in Blue Lake than that in Mahoney Lake could be related to the importance of the quality of mixolimnetic phytoplankton and its mass as food in the diet of the calanoid copepods. Overmann et al. (1999c) reported that the efficiency of carbon transfer from phytoplankton to zooplankton (*D. connexus*) was much higher than that from chemoheterotrophic bacteria to copepods. Thus, lake Mahoney as a typical meromictic lake shows that the food web in meromictic lake can be driven by the salinity of the lake water (e.g. salinity controls the abundance of cladocerans), by the specific features of the meromictic lake (e.g. phototrophic sulphur bacteria production and microbial loop) and by the overall productivity that is important for the conventional grazing food chain.

The trophic importance of bacterioplankton as a food source for the top consumers in the food web of meromictic lakes is controversial. Studies of Fry (1986) using stable carbon and sulphur isotopic measurements in three lakes near Syracuse, New York, show that the higher level of the food web fish derives their C from a mixture of terrestrial detritus, phytoplankton and littoral vegetation, rather than from bacterioplankton and detritus produced in the lakes. Whereas carbon isotopic values for zooplankton matched with those of the bacterioplankton, those of the sulphur isotopic values did not. Carbon and sulphur isotopic values observed in zooplankton samples can be explained by the mixture of isotopic values of phytoplankton and bacterioplankton. Fry (1986) concluded that the large standing stocks of bacterioplankton and zooplankton present at the top of the monimolimnion and epilimnetic communities in meromictic lakes they studied may exist as relatively isolated communities. Low oxygen level and the presence of sulphide at the edge of monimolimnion may deter fish to predate on zooplankton (Fry 1986).

Murtaugh (1985) also observed a low or little effect of the deepwater production of organic matter on the zooplankton development for small meromictic Lake Roi in central Alberta (Canada). In Lake Roi, a huge peak of chlorophyll was detected in the chemocline, but this seems to be a poor predictor of the zooplankton densities. The zooplankton genera differed in their depth preference. Cladocerans (mostly *Daphnia* sp.) were most abundant in the top several metres of the water column in early summer, but later their numbers declined but increased in deeper-water layers. The copepods showed a similar descent pattern during the summer, when their population size increased with time. *Chaoborus* larvae showed a similar trend, but were mostly found close to chemocline by the end of summer. At least for *Daphnia* sp. fecundity and per capita birth rates were usually the highest in the surface waters of Lake Roi. The decline of *Daphnia* by late summer when ambient concentrations of chlorophyll were high suggests that factors other than temperature and phytoplankton availability were important in influencing the *Daphnia* distribution and dynamics in Lake Roi (Murtaugh 1985). Probably the predation reduced the zooplankton abundances and shaped the vertical distribution. It should be noted that the increase of the zooplankton (biomass) with depth observed in the course of the summer is most probably typical for the stratified waterbodies. Similar trends have been observed for zooplankton populations in oceans, although at different scales (Rudyakov 1986). Such trends have also been observed for saline meromictic Lake Shira in Siberia (Zadereev and Tolomeev 2007).

Many meromictic lakes are fishless. However, the absence of predators and truncated food web is not necessarily a characteristic of the meromictic lake only. Many meromictic lakes do have fish as a part of the food web structure, although the fish are limited to the oxic environments. For example, meromictic Lake Cadagno (Switzerland) supports high fish productivity (see Chap. 7), which was estimated at 25–30 kg ha^{-1} $year^{-1}$ (Tonolla et al. 1988). These values are relatively high for an alpine oligomesotrophic lake. The presence of fish in the lake is due to man-made introductions for exploitation fishery. The species present are rainbow trout (*Oncorhynchus mykiss*), brown trout (*Salmo trutta*), lake charr (*Salvelinus umbla*), brook trout (*Salvelinus fontinalis*), lake trout (*Salvelinus namaycush*), bullhead (*Cottus gobio*), *Phoxinus lumaireul* and Italian chub (*Squalius squalus*). Brook trout and lake trout were introduced to the lake in the 1950s and fry of the commercial fish species are introduced annually (see Chap. 7).

Lake La Cruz (Spain) was naturally fishless; however, the fish are now relatively abundant. Fish were introduced mainly by man to the lake from the nearby mountain streams (see Chap. 8). The main species, *Achondrostoma arcasii* and *Luciobarbus guiraonis*, are benthivorous in their older stages: they probably rely on the abundant chironomids in littoral sediments, on other littoral insect larvae and on detritus from macrophytes mainly in the littoral area. However, in this small lake, some fish are also pelagic and may feed to some extent on *Chaoborus* larvae. Mosquito fish (*Gambusia* sp.) are restricted to littoral areas, though they are mostly planktivorous. In contrast to Lake Cadagno, the coupling of deepwater microbial loop with the classical grazer food web in the upper waters in Lake La Cruz is very weak. The food web in the oxic waters is based mainly on primary production of phytoplankton and, to a much lesser extent, on macrophytes. However, the food web is also linked to the terrestrial system since it incorporates both terrestrial organic matter (pollen, detritus) and flying insects (see Chap. 8).

When fish are present in a meromictic lake, the food web and predator-prey interactions in the aerobic layers of the lake can be quite typical and similar to other inland freshwater lakes. For example, in the meromictic Lake El Tobar (Spain), the size structure of population of species complex formerly classified as *Daphnia longispina*, as well as the vertical distribution and migration of different size classes of this species, was basically controlled by the effect of size-selective predators (Boronat and Miracle 1997). When *Daphnia* mortality by visually hunting predators dominated, a typical nocturnal (upward) migration was observed. If mortality was mainly attributed to the nocturnally migrating *Chaoborus*, *Daphnia* switched to a downward migration at night. When the population of *Daphnia* responded to both visual (fish) and non-visual predators (*Chaoborus*), Daphnia shifted to a pattern of twilight migration. By the end of the season, the vertical distribution of *Daphnia* size classes was markedly segregated between juveniles and adults. While the juveniles were found closer to the surface, the adults were concentrated predominantly in the deeper-water layers just above the oxycline. Thus, the depth with the low concentration of oxygen can be used by zooplankton as a refuge from the predators like fish that are quite sensitive to low oxygen.

Some meromictic lakes (e.g. large tropical meromictic lakes; see Chap. 10) demonstrate even high fish abundance. For example, there are over 250 species of cichlids and 50 species of non-cichlid fish in Lake Tanganyika. Most of these cichlid fish species are endemic (Verburg and Bills 2007). Lake Malawi has been known as a hotspot of biodiversity. While the lake has only four established zooplankton species, there is a great diversity of fish species: there are about 1000 fish species, most of which are endemic (Snoeks 2000). Both these tropical lakes are large deep old lakes with a priori diverse habitats and a long history of fish colonisation, adaptation, local evolution and endemism. Thus, the absence of fish is not a characteristic feature of the meromictic lake.

The causes for the absence of fish in small meromictic lakes are most probably specific. The chemical composition of water; the size of the lake and, specifically, the volume of habitable oxic water; and the absence of appropriate habitats, as well as the landlocked location of many lakes, can be attributed as the most obvious cause that does not allow the fish population from establishing in meromictic lakes. A typical example of the effect of established meromixis on the fish population was presented by McNaughton and Lee (2010). One hundred eighty-metre-deep Caland Pit Lake (Canada) was used for the culture of rainbow trout (*Oncorhynchus mykiss*). After 10 years of operation, because of increased trophic level due to aquaculture, meromictic conditions have occurred in the pit lake with a deterioration of water quality with depth. This regime shift reduced the volume of usable water for aquaculture purposes. The production in the fish farm decreased and was later closed (McNaughton and Lee 2010).

4.3.2 The Effect of Anoxic Conditions on the Food Web in Meromictic Lakes

The idea that meromictic conditions can impact the food web of the lake can be tested on a lake that switches from meromictic to holomictic conditions and back. Lake Sonachi (Kenya) is a small alkaline-saline water crater lake that over the past 200 years has experienced alternation of meromictic and holomictic episodes lasting from a few years to several decades (Verschuren et al. 1999). Reconstruction of the long-term dynamics of algal and invertebrate communities in Lake Sonachi revealed a weak correlation between the distributions of fossil algal pigments, diatoms and chironomid larvae in sediment cores and the documented or reconstructed variation in lake depth, mixing regime and surface-water conductivity (Verschuren et al. 1999). Studies demonstrated the complexity of long-term community dynamics in lakes and revealed how phytoplankton community structure can exert direct control on both benthic and planktonic invertebrate communities. For example, the reduction of algal biomass will favour benthic and planktonic invertebrates by reducing the prevalence of complete water column anoxia associated with intense night-time respiration of cyanobacterial blooms. In Lake Sonachi, Verschuren et al. (1999), through a reconstruction based on pigments, observed planktonic cladocerans in high densities only during two short meromictic phases in the 1970s when conductivity was modest and algal abundance was low.

Another example of the food web that was affected by the transition from dimictic to meromictic conditions is the Third Sister Lake (USA), an urban glacial lake. In this lake, Judd et al. (2005) studied the effect of mixing regime on nutrient dynamics and community structure. Chloride concentrations have dramatically increased in the bottom waters of this lake during the last three decades, preventing complete turnover in the spring. Meromixis affected the composition of both phytoplankton and zooplankton in the lake. The diatom *Asterionella* was the most common genus of phytoplankton in surface sediments and in water column samples. Its abundance, which generally occurs in mesotrophic freshwater to moderately brackish environments, increased during the last 30 years. A second notable shift in phytoplankton community composition was that of the increase in the relative abundance of the filamentous cyanobacterium *Oscillatoria*. The major change in the zooplankton community was a decrease in numbers of large-bodied *Daphnia* sp. The authors attributed such changes to an increase in the anoxic zone. This thus may have restricted the vertical (downward) migration of the daphnids, preventing their escape from predation by fish in the upper layers. Changes in mixing regimes will also alter benthic community composition. Bridgeman et al. (2000), who also studied the Third Sister Lake, noted declines in the abundance of benthic invertebrates (Insecta, Oligochaeta, Sphaeriidae and Gastropoda) and of diversity in recent years. Concluding, the lake has, due to the salt inputs, undergone a transition from dimictic to meromictic conditions, altering the structure and functioning of this lake's ecosystem.

The hypersaline Mono Lake (USA) (see Chap. 11) also alternates between meromictic and monomictic states. Three episodes of meromixis occurred in this lake between 1979 and 2013. The major effect of mixing on the food web in this lake is the interannual variation in standing algal biomass. Low values of chlorophyll are typical for meromictic episodes that contrast with periods of monomixis in which winter periods of complete mixing occur and spring chlorophyll levels are high. At the same time, seasonally averaged adult zooplankton abundances have less interannual variation than phytoplankton abundances.

The effect of anoxic conditions in meromictic lakes on phytoplankton and zooplankton is not well studied. There is general understanding that hydrogen sulphide is toxic for higher animals and eukaryotic algae, as it readily reacts with the iron-containing cytochromes. However, cyanobacteria and some other organisms can cope with high sulphide levels (Camacho et al. 1996). Also, there are reports that many species can be found in waters with low-oxygen or anoxic conditions. Studying the phytoplankton and zooplankton and certain chemical parameters in the meromictic Piaseczno opencast sulphur mine (Poland), Wilk-Wozniak and Zurek (2006) concluded that the meromictic environment is the main cause of low biodiversity of the phytoplankton and zooplankton communities. However, they observed that both phytoplankton and zooplankton were found alive in anoxic layers despite the presence in low concentrations of hydrogen sulphide. Quantitative zooplankton sampling in meromictic Lake Nyahirya (Western Udanda) by Kizito & Nauwerck (1995) revealed that all the 24 species of rotifers observed could withstand the prevailing hypoxic conditions in the lake.

A more detailed study of zooplankton and the food web in the Piaseczno sulphur mine (Zurek 2006) demonstrated a complex vertical structure of food web components. The epilimnion, inhabited by small-size fish, was dominated by rotifers, but there were no *Chaoborus flavicans* and other invertebrate predators (copepods and rotifer *Asplanchna)* inhabiting the cold hypolimnion. Zurek (2006) stated that a few individuals descend into the monimolimnion, near the bottom where diatoms, rotifers and copepods were present. However, the graphical data clearly demonstrated that the maximal densities of zooplankters were located above the monimolimnion, near the thermocline.

The effect of anoxia on the non-adapted populations of zooplankton can be dramatic. Gasol et al. (1995) studied the death and decay of a population of *Daphnia pulex* that had entered the cladoceran-free epilimnion of Lake Ciso (Spain). The cladocerans were found in high concentrations (up to 1900 ind l^{-1}) in the oxygenated epilimnion of the lake but were absent in the anoxic hypolimnion with sulphide concentrations of up to 1.2 mM. On the other hand, most of the available food was concentrated in metalimnion. Daily migrations of *D. pulex* were observed into the anoxic lower part of the metalimnion. The animals which spent most of the daytime at the surface descended at dawn to water layers where sulphide was present. Because the fish predation in the lake was absent, the authors attributed the *D. pulex* migration to the food availability and oxygen conditions (Gasol et al. 1995).

Thus, there are several effects of meromixis on the food web. First, the lack of mixing favours the deposition of nutrients in the monimolimnion. It controls the development of the phytoplankton. Specifically, spring algal blooms can be less pronounced in meromictic lakes because of long-lasting nutrient sequestration in deep layers. Second, the anoxic zone can prevent zooplankton from vertical migrations that change the nature of food web interactions. Zooplankters in this case are more vulnerable to the effect of predators. Third, both the size of the monimolimnion and anoxic conditions affect the biota. The size of the photic zone decreases, benthic community is altered, and zooplankton and fish have less available habitats.

4.3.3 Meromictic Lakes with Extreme Locations and Conditions

Meromixis is very often considered as a stress factor for the functioning of food web in lakes. One can assume that a meromictic lake located in harsh environmental conditions is an example of even more stressful environment. Ace Lake in Antarctica is a saline meromictic lake that has been investigated since the 1970s (Laybourn-Parry and Bell 2014). The lake has a truncated food web, and it supports four main biological communities: (1) a planktonic community in the oxygenated euphotic zone of the mixolimnion dominated by some prokaryotic and eukaryotic microorganisms, (2) a microbial community in the anoxic monimolimnion, where methanogenic archaea and sulphate-reducing and sulphur-oxidising bacteria occur, (3) a benthic algal mat community supported within the littoral zone above the depth of the chemocline and (4) a benthic community both on and in the sediment below the chemocline. In Ace Lake, the only planktonic crustacean observed is the marine copepod, *Paralabidocera antarctica*. It is interesting that copepods are present in the water column in winter and that they do not appear to enter a resting stage or use its eggs to overwinter. In general, communities of the food webs in permanently ice-covered Antarctic lakes are functional in winter at subzero temperatures, and mixotrophy plays an important role in survival of dominant photosynthetic eukaryotic microorganisms in the mixolimnion. Thus, similar typical features for food webs in meromictic lakes also appear in a meromictic lake located in an extreme environment: the food web is truncated; the photosynthetic sulphur bacteria in the chemocline play important role in the overall production budget; the diverse chemoautotrophic bacterial community prevails in the anoxic monimolimnion; and mixotrophic protistan species are an important part of the food web.

Additional stress to the food web in meromictic lakes can be due to the acidity of water over the entire water column. The highly acidic meromictic Hromnice Lake (Western Bohemia) was formed due to the mining of pyritic shales. Two chemoclines separate the extremely acidic (pH about 2.6) mixolimnion from a metal-rich anoxic monimolimnion. Extreme mixolimnetic oxygen maxima in this phosphorus-rich lake are well correlated with upsurges of phytoplankton, which consisted of several acid-tolerant species of the genera *Coccomyxa*, *Lepocinclis*, *Chlamydomonas* and *Chromulina*. The only animals inhabiting the open water were first-instar larvae of the midge *Chironomus* gr. *plumosus* and small numbers of rotifer *Cephalodella acidophila* (Hrdinka et al. 2013). Meromictic lakes that usually have highly acidic water are generally the artificial pit lakes. These lakes are formed in voids of former surface mines deep enough to intersect the local water table. For example, the pit lake Cueva de la Mora (Spain) has pH values of 2.6–3.1. Due to the acidic nature of the lake water, higher organisms are absent in this lake, and, therefore, the food web is restricted to microorganisms (see Chap. 9). It is clear that the extremely truncated food web in such lakes is mainly due to chemical properties of the water rather than to other variables of the meromictic conditions.

Many meromictic lakes are saline lakes. Salinity of water can be another factor that shapes the food web in meromictic lakes. The hypersaline meromictic Mono Lake (USA) has the grazer food web consisting of phytoplankton and the brine shrimp (*Artemia monica*) as the dominant and often sole species of zooplankton (see Chap. 11). No fish occur in the lake. The food web in this and many other saline lakes is completed with the abundant seasonally in near-shore habitats alkali fly (*Ephydra hians*). The brine shrimp and alkali fly larvae provide food for large numbers of birds including California gull (*Larus californicus*) and migratory eared grebes (*Podiceps nigricollis*), Wilson's phalaropes (*Phalaropus tricolor)* and red-necked phalaropes (*Phalaropus lobatus*).

Another type of meromictic lakes with specific conditions are lakes located close to the sea shore where meromixis is supported by the deep subterranean infiltration of the sea water (anchialine lakes, e.g. Lake Rogoznica, see Chap. 6). These marine meromictic lakes are often a refuge for landlocked populations of marine organisms. Thus, the food web and biota in such lakes are a mixture of marine and inland species. For example, the food web in marine meromictic Lake Mogilnoe (the Barents Sea, Russia) consists of both marine and inland species including Atlantic cod, large flocks of sea anemones, scyphomedusae and suberitid sponges, which are quite typical for many anchialine lakes (Strelkov et al. 2014). The isolation of marine species in this meromictic lake affected them in terms of their ecophysiology. Normally Atlantic cod is a saltwater, bottom-dwelling species. In Mogilnoe Lake, however, the cod fish inhabits the saline and freshwater layers. Genetic data show that landlocked population of the Atlantic cod greatly differs from its marine counterpart; in terms of its reproduction, it got isolated from its parental marine counterpart ca. 1800 years ago (Zhivotovsky et al. 2016). Thus, species composition and probably the food web structure in anchialine meromictic lakes will be determined not only by the meromictic nature but also by the interactions with the nearby sea.

4.4 Conclusive Summary

An assessment of research focused on biology and food webs in meromictic lakes demonstrates that the ecology of chemocline is the most studied aspect. At the same time, the link between the conventional grazing food web and the microbial loop in the deep waters is not well defined. The biogeochemistry of the monimolimnion and diversity of deepwater microbial communities are also not fully understood.

Data on the vertical distributions of phytoplankton and zooplankton in meromictic lakes are neither detailed nor available universally. The chlorophyll maxima in meromictic lakes often coincide with the depth where bacterial communities in the oxic-anoxic interface develop or it is located at the depth of the thermocline. The zooplankton community varies in species composition and abundance. Depending on salinity and chemical composition of the mixolimnion, the zooplankton may include certain typical cladocerans and copepods. The general tendency is that with the increase of salinity the cladocerans-dominated communities shift to those dominated by copepods. When the salinity increases, the zooplankton can shift to an *Artemia*-dominated community (e.g. Mono Lake; see Chap. 11) that is typical for hypersaline lakes. Also, rotifers often play an important role in the zooplankton of meromictic lakes. Many meromictic lakes are fishless, with mostly the vertical distribution of zooplankton being stable, and diel vertical migrations are absent (e.g. Lake Shira, see Chap. 5). However, when the food web of the lake has fish or other predators (e.g. Boronat and Miracle 1997; Judd et al. 2005), the typical escape reactions of zooplankton are discernible. The trade-off in this case is on one hand to decrease the predation risk but on the other to avoid deep anoxic waters. There is a limited number of papers that reveal the ability of zooplankters to spend relatively long time in anoxic waters (e.g. DeMeester and Vyverman 1997). However, the question arises whether this phenomenon is universally widespread in lakes. Usually in meromictic lakes, the vertical distribution and migration of zooplankton are restricted by the oxic-anoxic interface.

The vertical distribution of zooplankton for many meromictic lakes is similar. Very often adult animals form the deepwater maximum of biomass, while juveniles and younger animals prefer the warmer surface waters. This size-based distribution pattern has been reported for both cladoceran and copepod species (Boronat and Miracle 1997; Zadereev and Tolomeev 2007). The accepted reason for this type of distribution is the high mass of organic matter near the chemocline; both sinking and in situ production of organic matter play a role for this increase in mass. This is especially important during the end of the vegetative period when the primary production of phytoplankton in the surface waters is nutrient limited. However, direct estimates demonstrate that the efficiency of assimilation of organic matter produced in the chemocline region by the conventional zooplankton grazers is not very high (e.g. Lake La Cruz, see Chap. 8). Both hydrogen sulphide and low oxygen levels are lethal for zooplankton and thus greatly reduce the efficiency of energy transfer. The intermediate link between the chemocline and food web in the mixolimnion is the assemblages of ciliates and other protists. That these communities can reach high densities at the edge of the chemocline and actively consume bacterial biomass there has been reported for several lakes. Ciliates and other protists, in contrast to many bacteria, are quite motile, enabling them to transfer the organic matter from the anoxic to the oxic zone and thus making it available to the metazooplankton. Considering that at the chemocline the gradients of both salinity and oxygen concentration are often quite sharp, even a small spatial transfer can be sufficient to reallocate biomass from unfavourable depth to favourable ones for zooplankton. Finally, the efficiency and intensity of the coupling of microbial loop in chemocline and conventional grazing food web are not well studied yet.

Other not well-studied aspects for meromictic lakes are (1) the trophic links and energy transfer between the littoral and pelagic parts of the lakes and (2) the role of benthic communities on the ecology of meromictic lakes. It is clear that the benthic community in meromictic lakes can thrive only at the depth where the lake is mixed to the bottom. Depending on the depth and morphometry of the lake and the depth of monimolimnion, this bottom habitat still rich with oxygen can be found either only in the littoral or in a relatively large area of the lake. Also, it has been shown for some lakes and species (e.g. *Gammarus lacustris* in meromictic Lake Shira) that some littoral and benthic species can switch to the bentho-planktonic habitat and thus considerably affect the food web structure and energy transfer. This question needs to be studied more intensively.

Finally, we can specify what is typical for the biology and ecology of the meromictic lakes. Features such as the complex bacterial communities and partially coupled sulphur, carbon and nitrogen cycles in anoxic conditions in monimolimnion; the high rate of anoxygenic primary production at the chemocline; the complex communities of the chemocline (bacteria, protists, ciliates, rotifers) that constitute a localised in space microbial loop; and the probable link of anoxic and oxic food webs via the chemocline communities are often considered as specific for meromictic lakes. However, we should remember that most of these features are equally typical for holomictic lakes if they are thermally stratified and the oxygen is depleted in the hypolimnion. It is true that all these activities and communities exist through the year in meromictic lakes, whereas they mostly disappear during mixis in holomictic lakes. Thus, rates of processes and abundances of certain organisms can be very high under some circumstances due to the meromictic conditions.

There are several other features that might be attributed to meromictic conditions but indeed are typical for any stratified lake or are determined by the physical or chemical factors. For example, the complex vertical distributions of zooplankton and phytoplankton are typical for deep stratified lakes with the thermocline. The absence of fish and truncated food web is most probably related to the chemical composition of the lake. There are also some controversial questions, such as the ability of zooplankton to cope with anoxic conditions. Also, some other questions are typical for lake ecology in general. For example, the link between littoral and pelagial of the lake has been in general weakly investigated. Nevertheless, meromictic lakes provide an opportunity to study spatially structured habitats with contrasting physical and chemical conditions. Complex trophic links, coupling of nutrient cycles and anoxic and oxic food web components are peculiar features that make meromictic lakes natural laboratories to study the complexity of the food webs and biological interactions. Understanding of these regularities finally should add to our knowledge on the aquatic ecosystems.

Acknowledgement EZ was partially supported by the Council on grants from the President of the Russian Federation for support of leading scientific schools (grant NSh-9249.2016.5).

References

Andrei A-S, Robeson MS, Baricz A et al (2015) Contrasting taxonomic stratification of microbial communities in two hypersaline meromictic lakes. ISME J 9:2642–2656. doi:10.1038/ismej.2015.60

Baatar B, Chiang P-W, Rogozin DY et al (2016) Bacterial communities of three saline meromictic lakes in Central Asia. PLoS ONE 11:e0150847

Baricz A, Coman C, Andrei AS et al (2014) Spatial and temporal distribution of archaeal diversity in meromictic, hypersaline Ocnei Lake (Transylvanian Basin, Romania). Extremophiles 18:399–413

Biderre-Petit C, Jézéquel D, Dugat-Bony E et al (2011) Identification of microbial communities involved in the methane cycle of a freshwater meromictic lake. FEMS Microbiol Ecol 77:533–545

Blankenship RE, Madigan MT, Bauer CE (eds) (1995) Anoxygenic photosynthetic bacteria. Kluwer Academic Publishers, Dordrecht

Boronat MD, Miracle MR (1997) Size distribution of *Daphnia longispina* in the vertical profile. Hydrobiologia 360:187–196

Bosshard PP, Santini Y, Gruter D et al (2000a) Bacterial diversity and community composition in the chemocline of the meromictic alpine Lake Cadagno as revealed by 16S rDNA analysis. FEMS Microbiol Ecol 31:173–182

Bosshard PP, Stettler R, Bachofen R (2000b) Seasonal and spatial community dynamics in the meromictic Lake Cadagno. Arch Microbiol 174:168–174

Bridgeman TB, Wallace CD, Carter GS et al (2000) A limnological survey of Third Sister Lake, MI with historical comparisons. Lake Reserv Manag 16:253–267

Camacho A (2006) On the occurrence and ecological features of deep chlorophyll maxima (DCM) in Spanish stratified lakes. Limnetica 25:453–478

Camacho A (2009) Sulfur bacteria. In: Likens GE (ed) Encyclopedia of Inland waters, Elsevier—Academic Press, Oxford, NY, pp 261–278

Camacho A, Vicente E (1998) Carbon photoassimilation by sharply stratified phototrophic communities at the chemocline of Lake Arcas (Spain). FEMS Microbiol Ecol 25:11–22

Camacho A, Garcia-Pichel F, Vicente E et al (1996) Adaptation to sulfide and to the underwater light field in three cyanobacterial isolates from Lake Arcas. FEMS Microbiol Ecol 21:293–301

Camacho A, Vicente E, Miracle MR (2000) Spatio-temporal distribution and growth dynamics of phototrophic sulphur bacteria populations in the sulphide-rich Lake. Arcas Aquat Sci 62:334–349

Camacho A, Erez J, Chicote A et al (2001a) Microbial microstratification, inorganic carbon photoassimilation and dark carbon fixation at the chemocline of the meromictic Lake Cadagno (Switzerland) and its relevance to the food web. Aquat Sci 63:91–106

Camacho A, Vicente E, Miracle MR (2001b) Ecology of *Cryptomonas* at the chemocline of a karstic sulphate-rich lake. Mar Freshw Res 52:805–815

Camacho A, Vicente E, García-Gil LJ et al (2002) Factors determining changes in the abundance and distribution of nano- and picoplanktonic phototrophs in Lake El Tobar (Central Spain). Verh Int Ver Limnol 28:613–619

Camacho A, Picazo A, Miracle MR et al (2003a) Spatial distribution and temporal dynamics of picocyanobacteria in a meromictic karstic lake. Algol Stud 109:171–184

Camacho A, Miracle MR, Vicente E (2003b) Which factors determine the abundance and distribution of picocyanobacteria in inland waters? A comparison among different types of lakes and ponds. Arch Hydrobiol 157:321–338

Camacho A, Walter XA, Picazo A et al (2016) Photoferrotrophy: modern remains of an ancient photosynthesis. Front Microbiol (in press)

Casamayor EO, Schafer H, Baneras L et al (2000) Identification of and spatio-temporal differences between microbial assemblages from two neighboring sulphurous lakes: comparison by microscopy and denaturing gradient gel electrophoresis. Appl Environ Microbiol 66:499–508

Casamayor EO, Llirós M, Picazo A et al (2012) Contribution of deep dark fixation processes to overall CO_2 incorporation and large vertical changes of microbial populations in stratified karstic lakes. Aquat Sci 74:61–75

Cohen Y, Krumbein WE, Shilo M (1977) Solar Lake (Sinai). Distribution of photosynthetic microorganisms and primary production. Limnol Oceanogr 22:609–620

Comeau AM, Harding T, Galand PE et al (2012) Vertical distribution of microbial communities in a perennially stratified Arctic lake with saline, anoxic bottom waters. Sci Rep 2:604. doi:10.1038/srep00604

Crowe SA, Katsev S, Leslie K, Sturm A, Magen C, Nomosatryo S et al (2011) The methane cycle in ferruginous Lake Matano. Geobiology 9:61–78. doi:10.1111/j.1472-4669.2010.00257.x

Degermendzhy AG, Belolipetsky VM, Zotina TA, Gulati RD (2002) Formation of the vertical heterogeneity in the Lake Shira ecosystem: the biological mechanisms and mathematical model. Aquat Ecol 36:271–297

Degermendzhy AG, Zadereev YS, Rogozin DY et al (2010) Vertical stratification of physical, chemical and biological components in two saline lakes Shira and Shunet (South Siberia, Russia). Aquat Ecol 44:619–632

DeMeester L, Vyverman W (1997) Diurnal residence of the larger stages of the calanoid copepod *Acartia tonsa* in the anoxic monimolimnion of a tropical meromictic lake in New Guinea. J Plankton Res 19:425–434

Dietz S, Lessmann D, Boehrer B (2012) Contribution of solutes to density stratification in a meromictic lake (Waldsee/Germany). Mine Water Environ 31:129–137

Dressler M, Huebener T, Goers S et al (2007) Multi-proxy reconstruction of trophic state, hypolimnetic anoxia and phototrophic sulphur bacteria abundance in a dimictic lake in northern Germany over the past 80 years. J Paleolimnol 37:205–219

Fenchel T, Bernard C (1993) Endosymbiotic purple non-sulphur bacteria in an anaerobic ciliated protozoon. Microbiol Lett 110:21–25

Finlay BJ, Clarke KJ, Vicente E et al (1991) Anaerobic ciliates from a sulphide rich solution lake in Spain. Eur J Protistol 27:148–159

Fry B (1986) Sources of carbon and sulphur nutrition for consumers in three meromictic lakes of New York State. Limnol Oceanogr 31:79–88

García-Gil LJ, Vicente E, Camacho A et al (1999) Vertical distribution of photosynthetic sulphur bacteria linked to saline gradients in Lake El Tobar (Cuenca, Spain). Aquat Microb Ecol 20:299–303

Gasol JM, Jurgens K, Massana R et al (1995) Mass development of *Daphnia pulex* in a sulphide-rich pond (Lake Ciso). Arch Hydrobiol 132:279–296

Gervais F (1997) Light-dependent growth, dark survival, and glucose uptake by cryptophytes isolated from a freshwater chemocline. J Phycol 33:18–25

Gervais F (1998) Ecology of cryptophytes coexisting near a freshwater chemocline. Freshw Biol 39:61–78

Gies EA, Konwar KM, Beatty JT, Hallam SJ (2014) Illuminating microbial dark matter in meromictic Sakinaw Lake. Appl Environ Microbiol 80:6807–6818

Gophen M (1977) Feeding of *Daphnia* on *Chlamydomonas* and *Chlorobium*. Nature 5:271–273

Halm H, Musat N, Lam P et al (2009) Co-occurrence of denitrification and nitrogen fixation in a meromictic lake, Lake Cadagno (Switzerland). Environ Microbiol 11:1945–1958

Hollibaugh JT, Wong PS, Bano N et al (2001) Stratification of microbial assemblages in Mono Lake, California, and response to a mixing event. Hydrobiologia 1:45–60

Hrdinka T, Sobr M, Fott J, Nedbalova L (2013) The unique environment of the most acidified permanently meromictic lake in the Czech Republic. Limnologica 43:417–426

Humayoun SB, Bano N, Hollibaugh JT (2003) Depth distribution of microbial diversity in Mono Lake, a meromictic soda lake in California. Appl Environ Microbiol 69:1030–1042

Inceoglu O, Lliros M, Crowe Sean A et al (2015) Vertical distribution of functional potential and active microbial communities in meromictic Lake Kivu. Microb Ecol 70:596–611

Judd KE, Adams HE, Bosch NS et al (2005) A case history: Effects of mixing regime on nutrient dynamics and community structure in Third Sister Lake, Michigan during late winter and early spring 2003. Lake Reserv Manag 21:316–329

Kizito YS, Nauwerck A (1995) Temporal and vertical distribution of planktonic rotifers in a meromictic crater lake, Lake Nyahirya (Western Uganda). Hydrobiologia 313:303–312

Klepac-Ceraj V, Hayes CA, Gilhooly WP et al (2012) Microbial diversity under extreme euxinia: Mahoney Lake, Canada. Geobiology 10:223–235

Lauro FM, DeMaere MZ, Yau S et al (2010) An integrative study of a meromictic lake ecosystem in Antarctica. ISME J 5:879–895

Laybourn-Parry J, Bell E (2014) Ace Lake: three decades of research on a meromictic, Antarctic lake. Polar Biol 37:1685–1699

Lehours AC, Bardot C, Thenot A, Debroas D, Fonty G (2005) Anaerobic microbial communities in Lake Pavin, a unique meromictic lake in France. Appl Environ Microbiol 71:7389–7400. doi:10.1125/AEM.71.11.7389-7400.2005

Lehours A-C, Evans P, Bardot C et al (2007) Phylogenetic diversity of archaea and bacteria in the anoxic zone of a meromictic lake (Lake Pavin, France). Appl Environ Microbiol 73:2016–2019

Llirós M, García–Armisen T, Darchambeau F et al (2015) Pelagic photoferrotrophy and iron cycling in a modern ferruginous basin. Sci Rep. doi:10.1038/srep13803

Massana R, Gasol JM, Jürgens K, Pedrós-Alió C (1994) Impact of *Daphnia pulex* on a metalimnetic microbial community. J Plankton Res 16:1379–1399

McNaughton KA, Lee PF (2010) Water quality effects from an aquaculture operation in a meromictic Iron Pit Lake in Northwestern Ontario, Canada. Water Qual Res J Can 45:13–24

Miracle MR, Vicente E, Pedrós-Alió C (1992) Biological studies of Spanish meromictic and stratified karstic lakes. Limnetica 8:59–77

Murtaugh PA (1985) Vertical distributions of zooplankton and population dynamics of Daphnia in a meromictic lake. Hydrobiologia 123:47–57

Musat N, Halm H, Winterholler B et al (2008) A single-cell view on the ecophysiology of anaerobic phototrophic bacteria. PNAS 105:17861–17866

Noguerola I, Picazo A, Llirós M et al (2015) Diversity of freshwater Epsilonproteobacteria and dark inorganic carbon fixation in the sulphidic redoxcline of a meromictic karstic lake. FEMS Microbiol Ecol. doi:10.1093/femsec/fiv086

Northcote TG, Hall KJ (2010) Salinity regulation of zooplanktonic abundance and vertical distribution in two saline meromictic lakes in South Central British Columbia. Hydrobiologia 638:121–136

Oikonomou A, Filker S, Breiner H-W, Stoeck T (2015) Protistan diversity in a permanently stratified meromictic lake (Lake Alatsee, SW Germany). Environ Microbiol 17:2144–2157

Overmann J (1997) Mahoney Lake: a case study of the ecological significance of phototrophic sulphur bacteria. In: Jones JG (ed) Advances in microbial ecology, vol 15. Springer Science +Business Media, New York, pp 251–288

Overmann J (2008) Ecology of phototrophic sulfur bacteria. In: Hell R, Dahl C, Knaff D et al (eds) Sulfur metabolism in phototrophic organisms. Advances in photosynthesis and respiration, vol 27. Springer, Heidelberg, pp 375–396

Overmann J, van Gemerden H (2000) Microbial interactions involving sulfur bacteria: implications for the ecology and evolution of bacterial communities. FEMS Microbiol Rev 24:591–599

Overmann J, Beatty JT, Hall KJ et al (1991) Characterization of a dense, purple sulphur bacterial layer in a meromictic salt lake. Limnol Oceanogr 36:846–859

Overmann J, Beatty T, Hall K (1994) Photosynthetic activity and population dynamics of *Amoebobacter purpurea* in a meromictic saline lake. FEMS Microbiol Ecol 15:309–320

Overmann J, Coolen MJL, Tuschak C (1999a) Specific detection of different phylogenetic groups of chemocline bacteria based on PCR and denaturing gradient gel electrophoresis of 16S rRNA gene fragments. Arch Microbiol 172:83–94

Overmann J, Hall KJ, Northcote TG, Beatty JT (1999b) Grazing of the copepod *Diaptomus connexus* on purple sulphur bacteria in a meromictic salt lake. Environ Microbiol 1:213–221

Overmann J, Hall KJ, Northcote TG et al (1999c) Structure of the aerobic food chain in a meromictic lake dominated by purple sulphur bacteria. Arch Hydrobiol 144:127–156

Ovreas L, Forney L, Daae FL, Torsvik V (1997) Distribution of bacterioplankton in meromictic Lake Saelenvannet, as determined by denaturing gradient gel electrophoresis of PCR-amplified gene fragments coding for 16S rRNA. Appl Environ Microbiol 63:3367–3373

Peduzzi S, Welsh A, Demarta A et al (2011) *Thiocystis chemoclinalis* sp. nov. and *Thiocystis cadagnonensis* sp. nov., motile purple sulphur bacteria isolated from the chemocline of a meromictic lake. Int J Syst Evol Microbiol 61:1682–1687

Peduzzi S, Storelli N, Welsh A et al (2012) Candidatus "Thiodictyon syntrophicum", sp. nov., a new purple sulphur bacterium isolated from the chemocline of Lake Cadagno forming aggregates and specific associations with *Desulfocapsa* sp. Syst Appl Microbiol 35:139–144

Peura S, Eiler A, Bertilsson S et al (2012) Distinct and diverse anaerobic bacterial communities in boreal lakes dominated by candidate division OD1. ISME J 6:1640–1652

Pjevac P, Korlević M, Berg JS et al (2015) Community shift from phototrophic to chemotrophic sulfide oxidation following anoxic holomixis in a stratified seawater lake. Appl Environ Microbiol 81:298–308

Rodrigo MA (1997) Limnología comparada de las lagunas de dos sistemas cársticos de Cuenca: Bacterias fotosintéticas de la Laguna de La Cruz y la Laguna de Arcas-2. PhD Dissertation, Universitat de València

Rodrigo MA, Vicente E, Miracle MR (2000) The role of light and concentration gradients in the vertical stratification and seasonal development of phototrophic bacteria in a meromictic lake. Arch Hydrobiol 148:533–548

Rogozin DY, Zykov VV, Chernetsky MY et al (2009) Effect of winter conditions on distributions of anoxic phototrophic bacteria in two meromictic lakes in Siberia, Russia. Aquat Ecol 43:661–672

Rogozin DY, Zykov VV, Degermendzhi AG (2012) Ecology of purple sulfur bacteria in the highly stratified meromictic Lake Shunet (Siberia, Khakassia) in 2002–2009. Microbiology 81:727–735

Rudyakov YA (1986) Dynamic of the vertical distribution of pelagic animals. Nauka, Moscow

Sharples J, Moore CM, Rippeth TP et al (2001) Phytoplankton distribution and survival in the thermocline. Limnol Oceanogr 46:486–496

Snoeks J (2000) How well known is the ichthyodiversity of the large East African lakes? Adv Ecol Res 31:17–38

Strelkov P, Shunatova N, Fokin M et al (2014) Marine Lake Mogilnoe (Kildin Island, the Barents Sea): one hundred years of solitude. Polar Biol 37:297–310

Temerova TA, Tolomeyev AP, Degermendzhy AG (2002) Growth of dominant zooplankton species feeding on plankton microflora in Lake Shira. Aquat Ecol 36:235–243

Tiodjio RE, Sakatoku A, Nakamura A et al (2014) Bacterial and archaeal communities in Lake Nyos (Cameroon, Central Africa). Sci Rep 4:6151. doi:10.1038/srep06151. pmid:25141868

Tonolla M, Del Don C, Boscolo P et al (1988) The problem of fish management in an artificially regulated meromictic lake: lake Cadagno (Canton Tessin, Switzerland). Riv Ital Acquacolt 23:57–68

Tonolla M, Demarta A, Peduzzi R et al (1999) In situ analysis of phototrophic sulphur bacteria in the chemocline of meromictic Lake Cadagno (Switzerland). Appl Environ Microbiol 65:1325–1330

Tonolla M, Peduzzi S, Hahn D et al (2003) Spatio-temporal distribution of phototrophic sulphur bacteria in the chemocline of meromictic Lake Cadagno (Switzerland). FEMS Microbiol Ecol 43:89–98

Tonolla M, Peduzzi R, Hahn D (2005) Long-term population dynamics of phototrophic sulphur bacteria in the chemocline of Lake Cadagno, Switzerland. Appl Environ Microbiol 71:3544–3550

Van Gemerden H, Mas J (1995) Ecology of phototrophic sulphur bacteria. In: Blankenship RE, Madigan MT, Bauer CE (eds) Anoxygenic photosynthetic bacteria. Kluwer Academic Publishers, Dordrecht, pp 49–85

Verburg P, Bills IR (2007) Two new sympatric species of *Neolamprologus* (Teleostei: Cichlidae) from Lake Tanganyika, East Africa. Zootaxa 1612:25–44

Verschuren D, Cocquyt C, Tibby J et al (1999) Long-term dynamics of algal and invertebrate communities in a small, fluctuating tropical soda lake. Limnol Oceanogr 44:1216–1231

Walter XA (2011) Anaerobic iron cycling in a neoarchean ocean analogue. Dissertation, Université de Neuchâtel

Walter X, Picazo A, Miracle MR et al (2014) Phototrophic Fe(II)-oxidation in the chemocline of a ferruginous meromictic lake. Front Microbiol 5:713. doi:10.3389/fmicb.2014.00713

Wilk-Wozniak E, Zurek R (2006) Phytoplankton and its relationships with chemical parameters and zooplankton in meromictic Piaseczno reservoir, Southern Poland. Aquat Ecol 40:165–176

Yarza P, Yilmaz P, Pruesse E et al (2014) Uniting the classification of cultured and uncultured bacteria and archaea using 16S rRNA gene sequences. Nat Rev Microbiol 12:635–645

Yurkova N, Rathgeber C, Swiderski J et al (2002) Diversity, distribution and physiology of the aerobic phototrophic bacteria in the mixolimnion of a meromictic lake. FEMS Microbiol Ecol 40:191–204

Zadereev YS, Tolomeyev AP (2007) The vertical distribution of zooplankton in brackish meromictic lake with deep-water chlorophyll maximum. Hydrobiologia 576:69–82

Zhivotovsky LA, Teterina AA, Mukhina NV et al (2016) Effects of genetic drift in a small population of Atlantic cod (*Gadus morhua kildinensis* Derjugin) landlocked in a meromictic lake: genetic variation and conservation measures. Conserv Genet 17:229–238

Zurek R (2006) Zooplankton of a flooded opencast sulphur mine. Aquat Ecol 40:177–202

Part II
Case Studies

Chapter 5
Comparative Study of the Stability of Stratification and the Food Web Structure in the Meromictic Lakes Shira and Shunet (South Siberia, Russia)

Denis Rogozin, Egor Zadereev, Igor Prokopkin, Alexander Tolomeev, Yuri Barkhatov, Elena Khromechek, Nadezhda Degermendzhi, Anton Drobotov, and Andrei Degermendzhi

5.1 General Description of the Lakes and Historical Overview

5.1.1 Lakes Physical Surrounding and Historical Overview of Lakes Level/Salinities

Saline lakes are numerous in Chebakovo-Balakhtinskaya depression of Minusinsk trough in Altai-Sayan Mountain region of South Siberia (Republic of Khakassia, Russian Federation). This is a lowland semiarid steppe region surrounded by mountain ridges (Parnachev and Degermendzhy 2002). In this region, the average annual precipitation (about 300 mm yr^{-1}) is significantly lower than potential annual evaporation (about 600 mm yr^{-1}) (Parnachev and Degermendzhy 2002). Such a marked difference with twice higher evaporation than precipitation is a prerequisite for the formation of saline lakes. Because of the continental climate of this territory, the mean annual temperatures vary greatly, i.e. between +18 °C in July and −19 °C in January.

The saline lakes Shira (90.11 E, 54.30 N, 355 m ASL) and Shunet (90.13 E, 54.25 N, 418 m ASL) are the most studied lakes of this Siberian region (Fig. 5.1).

D. Rogozin (✉) • E. Zadereev
Institute of Biophysics SB RAS, Akademgorodok, Krasnoyarsk, Russia

Siberian Federal University, Svobodny 79, Krasnoyarsk, Russia
e-mail: rogozin@ibp.ru

I. Prokopkin • A. Tolomeev • Y. Barkhatov • E. Khromechek • A. Drobotov • A. Degermendzhi
Institute of Biophysics SB RAS, Akademgorodok, Krasnoyarsk, Russia

N. Degermendzhi
Krasnoyarsk State Medical University, ul. Partizana Zheleznyaka 1, Krasnoyarsk, 660022, Russia

© Springer International Publishing AG 2017

R.D. Gulati et al. (eds.), *Ecology of Meromictic Lakes*, Ecological Studies 228,
DOI 10.1007/978-3-319-49143-1_5

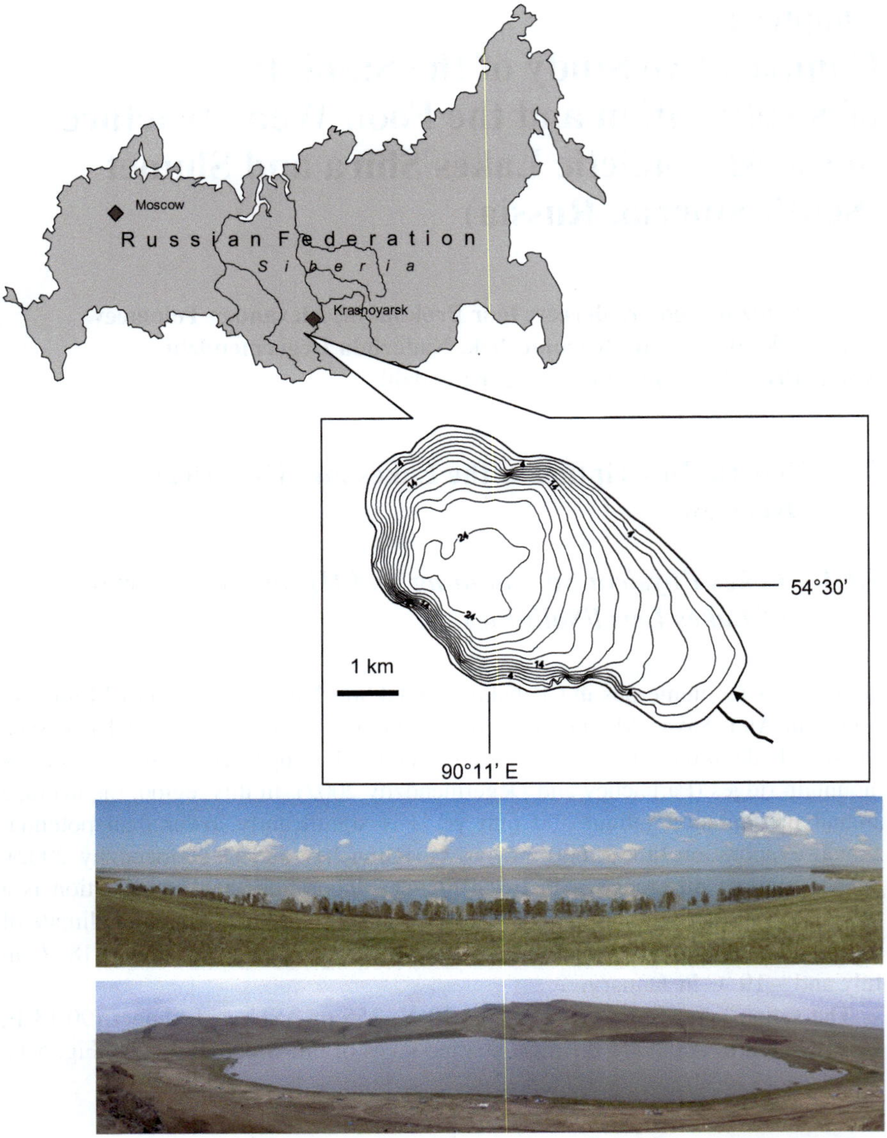

Fig. 5.1 Location of the study area, bathymetry map of Lake Shira and photos of lakes Shira (*upper*) and Shunet (*lower*)

Lake Shira is a closed elliptical (9.3 × 5.3 km) waterbody with an area of 35.9 km^2 and maximum depth of 24 m (2007–2014). The main inflow to the lake is from the River Son, which provides about 42 % of freshwater supply to lake; other sources of water to the Lake Shira are precipitation and bottom seepage. Salinity in summer ranges from 14 g l^{-1} to 15 g l^{-1} in epilimnion and about 18 g l^{-1} in monimolimnion, based on data from 2007–2009 (Rogozin et al. 2010a). Lake Shunet is located about 8 km to the southeast of Lake Shira. This lake is also elliptical (1.2 × 0.4 km) but is both much smaller in area (0.47 km^2) and shallower (6.2 m) than Lake Shira (max. depth 24 m). The lake has no surface outflow, but an inflow from a small stream that enters the lake on its south-west part. The salinity of the Lake Shunet in the mixolimnion (17–20 g l^{-1}) and monimolimnion (up to 66 g l^{-1}) differs markedly (Rogozin et al. 2009). The main anions in Lake Shira are sulphate>chloride>bicarbonate, and main cations are sodium>magnesium>potassium (Kalacheva et al. 2002). In Lake Shunet the chemical composition is similar to that in Lake Shira, but magnesium predominates over sodium. Ice cover persists in both lakes for 5–6 months, from November to early May, and thickness of ice is about 1 m at the middle of March (Rogozin et al. 2009). Both lakes are meromictic: based on studies during 1998–2014 (Zotina et al. 1999; Rogozin et al. 2010a, 2012).

The surface elevation of both lakes has noticeably changed over the observation period. Between the 1920 and 1940, the water level of Lake Shira decreased by about 7 m and salinity increased, reaching a maximum of 27 g l^{-1} in 1926 (Krivosheev and Khasanov 1990). The lake's salinity increase was inversely related to the decrease in water volume (Krivosheev and Khasanov 1990). Also, at the beginning of the twentieth century, Lake Shunet was just 0.5 m deep, and salinity was as high as 380 g l^{-1} (Parnachev and Degermendzhy 2002). Since all the closed lakes in the region were observed to decrease in their water level (Krivosheev and Khasanov 1990; Parnachev and Degermendzhy 2002), it is suggested that fluctuation in balance of regional precipitation-evaporation and predominance of the latter were the cause of water level decreases in the lakes. We have no available information on the lakes' stratification patterns for the period before 1940, and it is also unclear if the lakes were then meromictic or holomictic.

5.1.2 Lakes Stratification

The lakes have been shown to be meromictic with autumnal overturn restricted to mixolimnion (Rogozin et al. 2010a, 2012). The depth of mixolimnion and position of oxic-anoxic interface varied annually in Lake Shira whereas it was stable in Lake Shunet. Spring mixing processes contribute to the autumnal formation and positioning of mixolimnion.

The exceptionally windy spring of 2007 caused the deepening of mixolimnion in Lake Shira in the winter of 2008 (Rogozin et al. 2010a). The winter position of oxic-anoxic interface was affected by the position of lower boundary of mixolimnion. The salinity in the winter mixolimnion increases compared with the autumn because of freezing out of salts from the upper water layers during ice formation and their dissolution in the water below.

During May the salinity in the upper 2-m layer of Lake Shira decreases drastically due to melting of the ice on the lake surface. Therefore, the halocline, i.e. step of salinity represented by the maximum salinity gradient, is formed in the near-surface layers of mixolimnion (Fig. 5.2). The depth of that halocline coincides with the thermocline and pycnocline, indicating the formation of a warmer epilimnion. The oxic-anoxic interface remains near the former mixolimnion depth (Fig. 5.2).

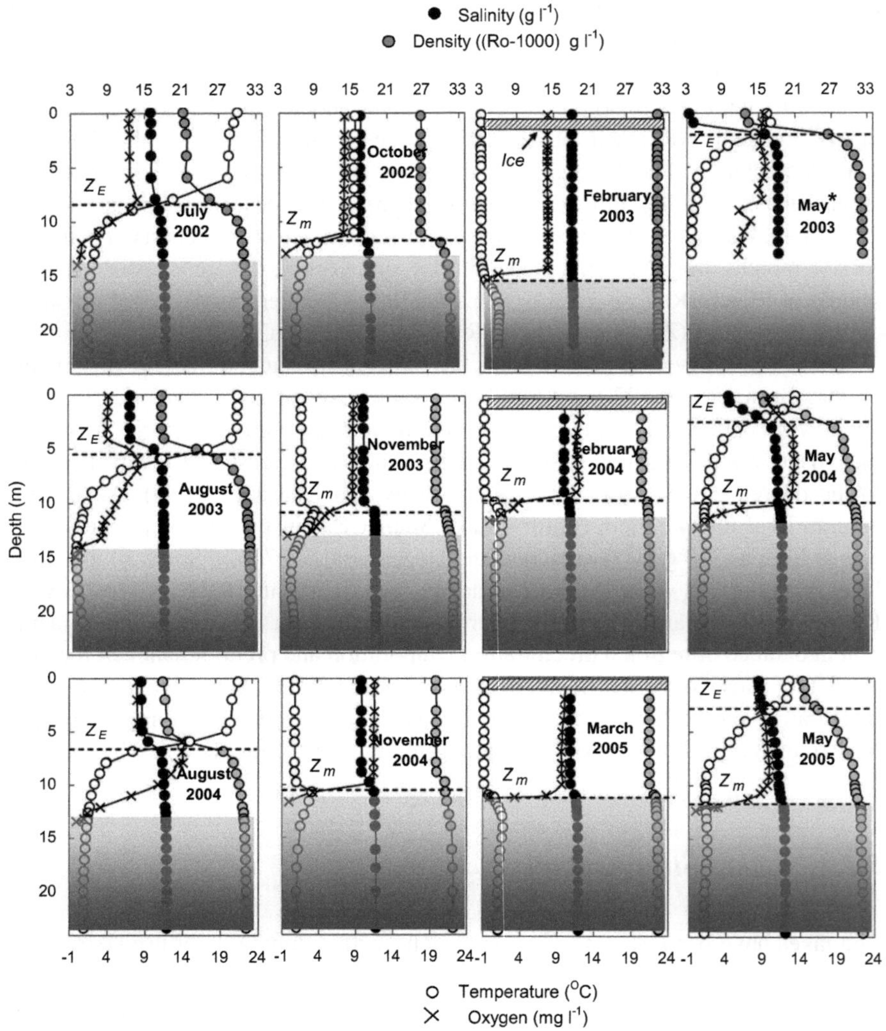

Fig. 5.2 Seasonal dynamics of stratification in Lake Shira. *Grey colour* indicates sulphide-rich water (from Rogozin et al. 2010a)

Even though the thickness of the measured profiles was qualitatively similar, in winter it differed for mixolimnion noticeably from winter to winter. In the years 2003, 2008, 2009, 2013 and 2014, the winter depth of lower boundary of mixolimnion ranged from 15.5 m to 16.7 m. In contrast, in other winters in this period, the mixolimnion depth varied between 9.5 m and 12.5 m (Rogozin et al. 2009; Rogozin et al. 2016).

A similar dynamics of vertical distribution of temperatures and salinity was also observed for Lake Shunet (Rogozin et al. 2012). The vertical distribution of salinity in the depth range from the surface to 4.5 m varied depending on the season; the salinity varied between 10 g l^{-1} and 25 g l^{-1} (Fig. 5.3). The salinity in the lower

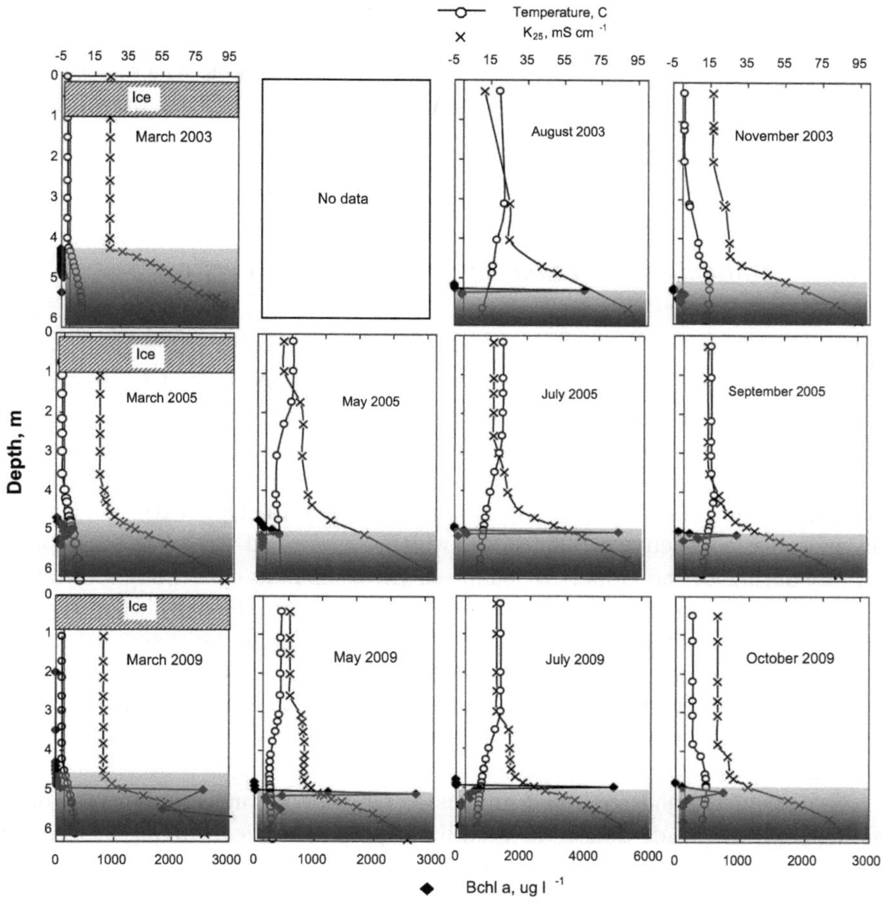

Fig. 5.3 Seasonal dynamics of Lake Shunet stratification. *Grey colour* indicates sulphide-rich water (from Rogozin et al. 2012)

part of the water column hardly changed during the seasons indicating the absence of full circulation of water in Lake Shunet. The sharply heterogeneous distribution of dissolved salts combined with the invariably high sulphide concentration in the bottom layers indicates meromictic character of Lake Shunet. Thus, the water column of the lake is divided into aerobic mixolimnion and anaerobic monimolimnion.

The seasonal dynamics of salinity and temperature in the mixolimnions of both lakes is significantly affected by the formation of the ice cover and its thawing. Thus, the mixolimnion in both lakes was monomictic during the study period, undergoing full circulation in autumn.

Freshwater inflows during the 1930s and 1940s were considerable due to an increase in atmospheric precipitation, and they probably contributed to the formation of the stable salinity gradient observed in Lake Shira nowadays. A similar dramatic increase in water level was observed in 1930–1940s for Lake Shunet (see above). Therefore, lake's meromixis, according to the generally accepted Hutchinson's classification of lake stratification (Hutchinson 1957), has an ectogenic origin (i.e. induced by freshwater covering the saline water) rather than a crenogenic origin (i.e. induced by supply of mineral-rich water from the lake bottom).

5.1.3 Mathematical Modelling of Lake Shira Vertical Structure

The profiles of water temperature, salinity and density in the central deepest part of the lake during an open-water period are based on a 1-D mathematical model (Belolipetskii and Genova 2008). For the subglacial period we used a simplified model of ice formation based on the single-phase Stefan problem (Vuik 1993) with the linear temperature distribution in solid phase (Genova et al. 2010). We isolated an under-ice convective mixing layer, which was formed due to an increase in salinity during ice formation. To obtain analytical solutions concerning the vertical distributions of temperature, we used a vertical structure diagram in the form of several layers (Belolipetskii and Genova 2008). Ice melting processes, from both below and above the ice cover, were modelled for spring period. The calculated profiles of salinity and temperature of Lake Shira were in good agreement with the field data for autumnal and winter periods 2002–2003 and 2003–2004 (Genova et al. 2010).

The calculated and measured profiles of temperature and salinity were quite similar for most study dates. We used the depth of turbulent mixing layer as an indicator of the both calculated and observed mixing patterns (Fig. 5.4). Obviously, the turbulent mixing layer always spread downward to the upper boundary of the pycnocline, i.e. to the depth of maximal salinity gradient and was positioned near thermocline during the warm period and gradually descended to chemocline during autumnal cooling and mixing and winter stratification. Therefore, the sharp maxima indicate the time of ice melting and formation of thermocline in the upper layer (Fig. 5.4). The "shoulders" of solid lines indicate the calculated thermocline position during the stable summer stratification, and lateral straight portions indicate the calculated winter mixolimnion.

Generally, the calculated depth of winter mixolimnion was in good accord with the observed depth for the winters of both 2002–2003 and 2004–2005. In other winters, except 2005–2006, when we did not measure, the agreement between the measures and calculated profiles was not so good. The calculated positions of thermocline in summer 2007 and mixolimnions in 2008 and 2009 were well "shallower" than the observed ones (Fig. 5.4a). The effects of internal currents, wind directions, lake morphology and water level fluctuations cannot be explained by the one-dimensional model. We, therefore, assume that the 2-D and 3-D models provide more adequate prognoses of lake dynamics. In addition, the available data

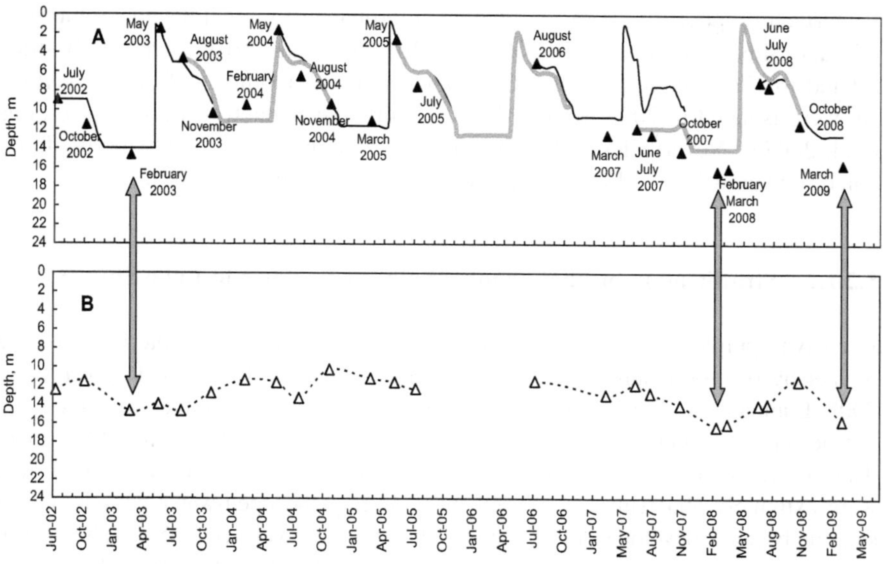

Fig. 5.4 *Upper Panel* (**a**) Depth of maximal gradient of salinity: calculated for odd years (*solid line*), the same for even years (*grey solid line*); measured (*filled triangle*); sharp maxima indicate the moments of ice melting and formation of thermocline in the *upper layer*; (**b**) Depth of observed oxic-anoxic interface (*open triangle*). *Arrows* indicate "deep" mixolimnions (from Rogozin et al. 2010a)

of wind force are based on four preset time measurements per day, but the strong transient storms as observed in spring 2007 will produce noticeable mixing effects between the periods of wind force measurements (Rogozin et al. 2010a).

5.2 Biology and Biogeochemistry

5.2.1 Ecology of Phototrophic Sulphur Bacteria in Lakes Shira and Shunet

5.2.1.1 Species Composition of Phototrophic Sulphur Bacteria

In Lake Shira the purple sulphur bacteria (PSB) dominated in anoxic phototrophic community whereas green sulphur bacteria (GSB) are a minor group. On the contrary, both these bacterial types inhabit the redox zone and monimolimnion of Lake Shunet (Lunina et al. 2007a, b; Rogozin et al. 2009, 2010b). Culture-independent method of denaturing gradient gel electrophoresis of PCR-amplified fragments of 16S rRNA genes with universal eubacterial primers has shown that *Thiocapsa* sp. Shira_1 (*Chromatiaceae*) (AJ633676) isolated from chemocline of Lake Shira was dominant PSB in redox zones of both lakes in 2005 (Rogozin et al. 2010b). According to spectrophotometry analysis, the "green" GSB containing bacteriochlorophyll *c* or *d* and carotenoid chlorobactene inhabit the anaerobic zone of Lake Shunet (Lunina et al. 2007a, b; Rogozin et al. 2010b). One GSB strain related to *Chlorobium limicola* (Chlorobiaceae) and another to *Prosthecochloris* sp. were isolated in 2003 from both lakes Shira and Shunet, respectively (Lunina et al. 2007a). Nevertheless, PCR/DGGE show that another GSB phylotype dominated in chemocline of Lake Shunet in 2005 (Rogozin et al. 2010b).

5.2.1.2 Stratification of Phototrophic Sulphur Bacteria in Lake Shunet

Intensive "purple layer" of PSB was detected at the oxic-anoxic interface in Lake Shunet by precision sampling with multi-syringe sampler (Fig. 5.5) (Rogozin et al. 2005; Lunina et al. 2007a). This layer is observed in 5 cm depth interval at the oxic-anoxic interface, and persisted in all seasons during 2003–2012, though PSB abundance decreased for ice periods (Rogozin et al. 2012; Degermendzhy et al. 2010). PSB were extremely abundant ($>10^8$ cell ml^{-1}) during the open-water periods. Green sulphur bacteria developed mostly under the "purple layer" (Fig. 5.6). The spatio-temporal organization of the bacterial community inhabiting the chemocline of the stratified meromictic Lake Shunet (Khakassia, Russia) was investigated from May to

Fig. 5.5 Multi-syringe thin-layer sampler (Rogozin and Degermendzhy 2008). The photo shows 15 simultaneous, vertical samples taken from Lake Shunet (Siberia, Russia) in August 2004 at 5-cm intervals. The "purple layer" can be easily distinguished (the 9th syringe from the *top*) (From Degermendzhy et al. 2010)

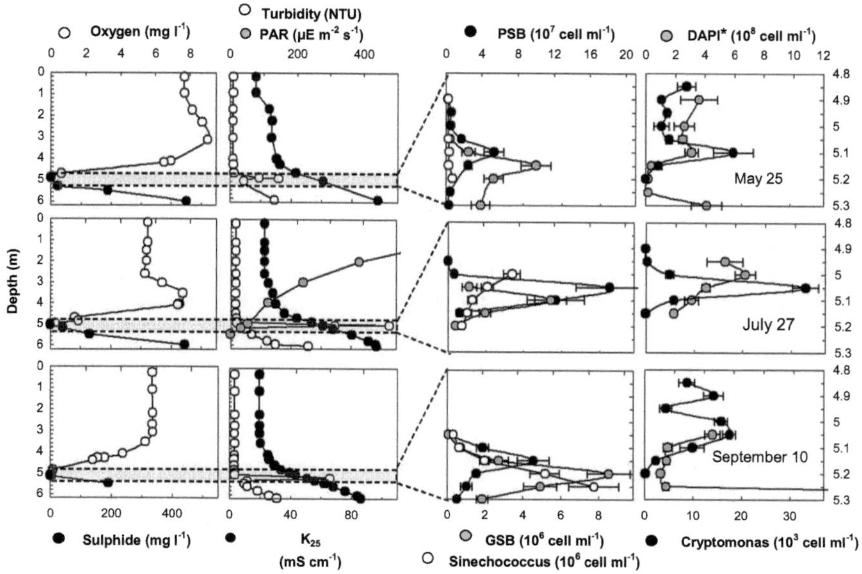

Fig. 5.6 Vertical distribution of the physicochemical characteristics in Lake Shunet and the number of microorganisms in the chemocline in 2005. The total number of bacterial cells (DAPI) does not include the number of PSB cells (from Rogozin et al. 2010b)

September 2005 using microscopical analysis of photosynthetic pigments and PCR-DGGE with subsequent 16S rDNA analysis. The samples were collected with a multi-syringe stratification sampler at every 5 cm (Rogozin and Degermendzhy 2008). There were no large changes observed in the bacterial community of the chemocline, at least among the detected forms (Fig. 5.6). During the entire study, purple sulphur bacteria phylogenetically and morphologically close to *Lamprocystis purpurea* (*Chromatiaceae*) predominated in the chemocline. In the layers below the layer of purple bacteria, green sulphur bacteria were observed. *Cryptomonas* is a phytoflagellate developed in the upper zone of the chemocline (Fig. 5.6). Mahoney Lake (Canada) is the only other lake that is known to have similar abundances of PSB (Overmann 1997).

In chemoclines of both lakes, optimal light conditions and sulphide concentration provide favourable conditions for photosynthetic production, whereas high-density gradients and high hydrophysical stability inhibit the sedimentation and turbulent mixing, respectively. Therefore, the combination of relatively shallow position of chemocline and high-density gradient in both lakes favours dense accumulation of PSB at the chemocline (Rogozin et al. 2012).

5.2.1.3 Effect of Winter Conditions on Phototrophic Sulphur Bacteria in Lakes Shira and Shunet

Anoxic phototrophic bacteria are generally known to persist in chemoclines of temperate meromictic lakes under ice (Lawrence et al. 1978; Overmann et al. 1994; Tonolla et al. 2003). In both lakes we studied vertical distribution and biomass of anoxic phototrophic bacteria and profiles of physical and chemical characteristics during under-ice periods from 2003 to 2008, except in 2006 (Rogozin et al. 2009). The bacterial layers in chemocline of both lakes were sampled with a thin layer using a hydraulic multi-syringe sampler (Rogozin and Degermendzhy 2008). Winter biomass of purple sulphur bacteria in both lakes varied depending on amount of light penetrating into the chemocline through the snow cover and on ice at the lake surface. The quantity of light reaching deeper layers depended on snow cover characteristics. The nature of snow cover could be remotely sensed by reflectance of the lake surface (Rogozin et al. 2009). Therefore, the under-ice light conditions could be roughly assessed by monitoring of snow cover dynamics in studied lakes. In addition, in the relatively less strongly stratified Lake Shira, the vertical position of chemocline in winter can vary resulting in considerable changes in light conditions in chemocline. Hence, the under-ice light conditions of anoxic phototrophic bacteria depend on chemocline position. Interestingly, in Lake Shira due to an increase in transparency of mixolimnion in winter, the combination of shallower chemocline and absence of snow resulted in light intensity exceeding summer values in the chemocline. Therefore, the biomass of purple sulphur bacteria in chemocline of the lake exceeded the summer values under these conditions (Rogozin et al. 2009).

In Lake Shunet, the light intensities in the chemocline and biomasses of PSB were always lower in winter than in summer but the biomass of GSB was similar over the season. In the chemocline of Lake Shira, temperature conditions in winter and summer were similar in most years. The PSB populations developed at temperatures ranging from 0 °C to 1.7 °C for most seasons, except during warm period in June 2007. In contrast, in Lake Shunet, the summer temperatures in chemocline were 5–15 °C higher than in winter.

Previous researchers (Pimenov et al. 2003; Savvichev et al. 2005; Lunina et al. 2007a, b) observed that rates of photosynthesis (^{14}C Method) in these two lakes are lower in winter than in summer. Moreover, contribution of the anoxic photosynthesis to total primary production is rather low in all seasons (Pimenov et al. 2003; Savvichev et al. 2005). In both lakes, the oxic photosynthesis contribute from 93 % to 100 % of total assimilation of photosynthetic inorganic carbon during both summer and winter periods (Pimenov et al. 2003; Savvichev et al. 2005). In each lake the differences in primary production between summer and winter are more due to differences in oxic photosynthesis. However, the winter measurements are based only on data of 2003 (Savvichev et al. 2005) when light conditions in both lakes were very poor. Indeed, in the winters of 2003 and 2008, the light intensity measured in chemocline of Lake Shira was $<0.4\ \mu E\ m^{-2}\ s^{-1}$. This appears to be the lowest light intensity known to support growth of pure PSB cultures (van Gemerden et al. 1989). Thus, anoxic photosynthesis is negligible in the winters in Lake Shira and low in Lake Shunet. Obviously, in winter 2007, the rates of oxic and anoxic photosynthesis in both lakes were higher because of considerably higher under-ice light intensities. We have shown that variation in under-ice light intensities suggest that under-ice photosynthetic production may vary significantly depending on environmental conditions.

5.2.2 Vertical Distribution of Ciliates, Zooplankton and Amphipods

Geographical proximity of the salt lakes Shira and Shunet suggests a similar species composition because they are probably colonized by species from the same regional pool and because of the likely migration of species between the adjacent water bodies. Moreover, both lakes are meromictic having a similar composition of salts and broadly comparable levels of salinities. Both lakes are fishless in pelagic parts. It is not surprising that both lakes are dominated by the same species of pelagic zooplankton. Among the ciliates—representatives of the genera *Cyclidium* sp. and *Strombidium* sp.—and among crustacean zooplankton, both lakes are dominated by calanoid copepod *Arctodiaptomus salinus*. Rotifers are represented by two taxa—*Brachionus plicatilis* and *Hexarthra sp*. The largest invertebrate species in these lakes is the amphipod *Gammarus lacustris*, which inhabits both the littoral and pelagic zones of the lakes. Despite the similarity in species composition of zooplankton, the morphometric differences and the patterns of summer stratification in the lakes affect the vertical structure of zooplankton.

Ciliates The pelagic zone of Lake Shunet comprises several ciliate genera: *Oligotrichida*, *Scuticociliatida*, *Hypotrichida*, *Prostomatida*, *Cyclidium* sp., *Euplotes* sp., *Prorodon* sp., *Balanion* sp., *Strombidium* sp. and *Oxytricha* sp. Most of these genera are adapted to poor light conditions, anoxia and hydrogen sulphide (Khromechek et al. 2010).

The most distinctive feature of ciliate community in the pelagic zone of Lake Shunet is their clear preference for a specific region in the water column: both abundance and biomass were much higher in the chemocline zone than in the mixolimnion (Fig. 5.7a). The protozooplankton maximum was observed in the 10–35 cm layer of the chemocline region, above or sometimes in the layer of PSB, where hydrogen sulphide concentration varied from 0 mg l^{-1} to 5 mg l^{-1} with little or no oxygen. The ciliate abundance in the chemocline (50–400 cells ml^{-1}) was significantly higher than in the mixolimnion (0–36 cells ml^{-1}). Ciliates in this zone do not seem to migrate and their seasonal dynamics in the chemocline is insignificant.

In contrast with Lake Shunet, the species composition of ciliates in the pelagic zone of Lake Shira is extremely poor, being represented by only two species—

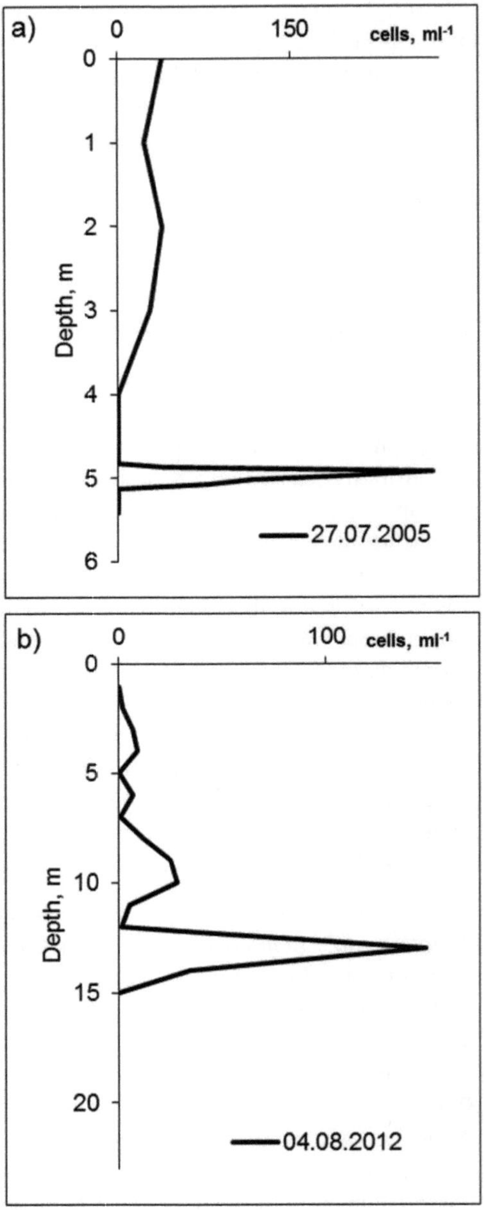

Fig. 5.7 The typical vertical distribution of the total number of ciliates in lakes Shunet (**a**) and Shira (**b**) in summer

Cyclidium sp. and *Strombidium* sp. Unlike Lake Shunet, ciliates in Lake Shira are exposed to significant seasonal fluctuations. In spring and summer, they also form the deep maximum near the chemocline zone. Ciliate maxima were observed at 12–13 m depth, and their concentration was up to 200 ind. ml^{-1} (Fig. 5.7b), with an average of 80 ind. ml^{-1} in the water column. During early fall period, the ciliate maximum drops to about 20 cells ml^{-1}, which is one-tenth the average value in summer when their mean concentration in the water column was only about 10 ind. ml^{-1}; in winter, the ciliates are virtually absent in Lake Shira.

Zooplankton Vertical distribution of different developmental stages and sexes of *Arctodiaptomus salinus*, rotifers and cladocerans, as well as temperature in mixolimnion divided in summer to epi-, meta- and hypolimnion in the two lakes are depicted in Fig. 5.8. The borders of metalimnion are the depths where the water temperature changes by at least 1 °C/m or more (Garneau et al. 2013). The obtained data on the average zooplankton densities (ind. l^{-1}) and temperature distribution are for the four selected periods of the season: ice period (March), period of early stratification of mixolimnion (May–June), the period of stratification of mixolimnion (July–August) and autumn period of extensive mixing of mixolimnion. The data of 32 vertical profiles of zooplankton abundances and water temperature were averaged for the period 1999–2011 (No. of samples annually in parentheses) in Lake Shira (1999 (1), 2001 (6), 2003 (1), 2007 (4), 2008 (6), 2009 (5), 2010 (4), 2011 (5)) and 16 profiles for the period 2003–2011 in Lake Shunet (2003 (1), 2004 (3), 2007 (4), 2008 (4), 2009 (2), 2011 (2)).

March. Under the ice the temperature of both lakes was uniform in the oxygenated water zone. Also copepods *A. salinus* lacked a persistent vertical structure. Copepodites of C1–3 and C4–5 stages dominated. In Lake Shunet the temperature above the chemocline zone was slightly elevated but without an increase of zooplankton. Nauplii were rare in both lakes but the concentration of zooplankton in Lake Shira was higher than in Lake Shunet—average 35 ind. l^{-1} versus 19 ind. l^{-1}, respectively.

May–June. After melting of the ice, there was a mass development of copepodites C4–C5 to adults and increase in reproduction rates, as well as a development of rotifer *Hexarthra* sp. in both lakes. The zooplankton was characterized by a marked predominance of nauplii and rotifers in epi- and metalimnion. In Lake Shira, zooplankton abundance in the hypolimnion declined, unlike in Lake Shunet where all life stages of *A. salinus*, except nauplii, were distributed evenly in the different layers. In this period, concentration of zooplankton in Lake Shunet exceeded that in Lake Shira.

July–August. In mid-summer the epilimnion zone of the lakes expanded. Moreover, metalimnion of Lake Shunet sank to the chemocline zone. The peaks of copepodites C1–C3 and C4–C5 were separated: copepodites C1–C3 were encountered in epilimnion while the C4–C5 mostly occurred in metalimnion. Like in May–June, nauplii preferred to develop in the epilimnion. The high numbers of rotifers in this period were typical for *B. plicatilis* in Lake Shira, but this species was limited in its distribution to the epilimnetic water. In Lake Shunet the *B. plicatilis* massively developed in a thin layer adjacent to the chemocline zone. Other species of zooplankton (cladocera *Moina* sp., rotifers *Hexarthra* sp.) were found in samples occasionally.

In *October*, vertical distribution of temperature dramatically differed between the lakes. In Lake Shira the mixing zone extended almost to the depth where the

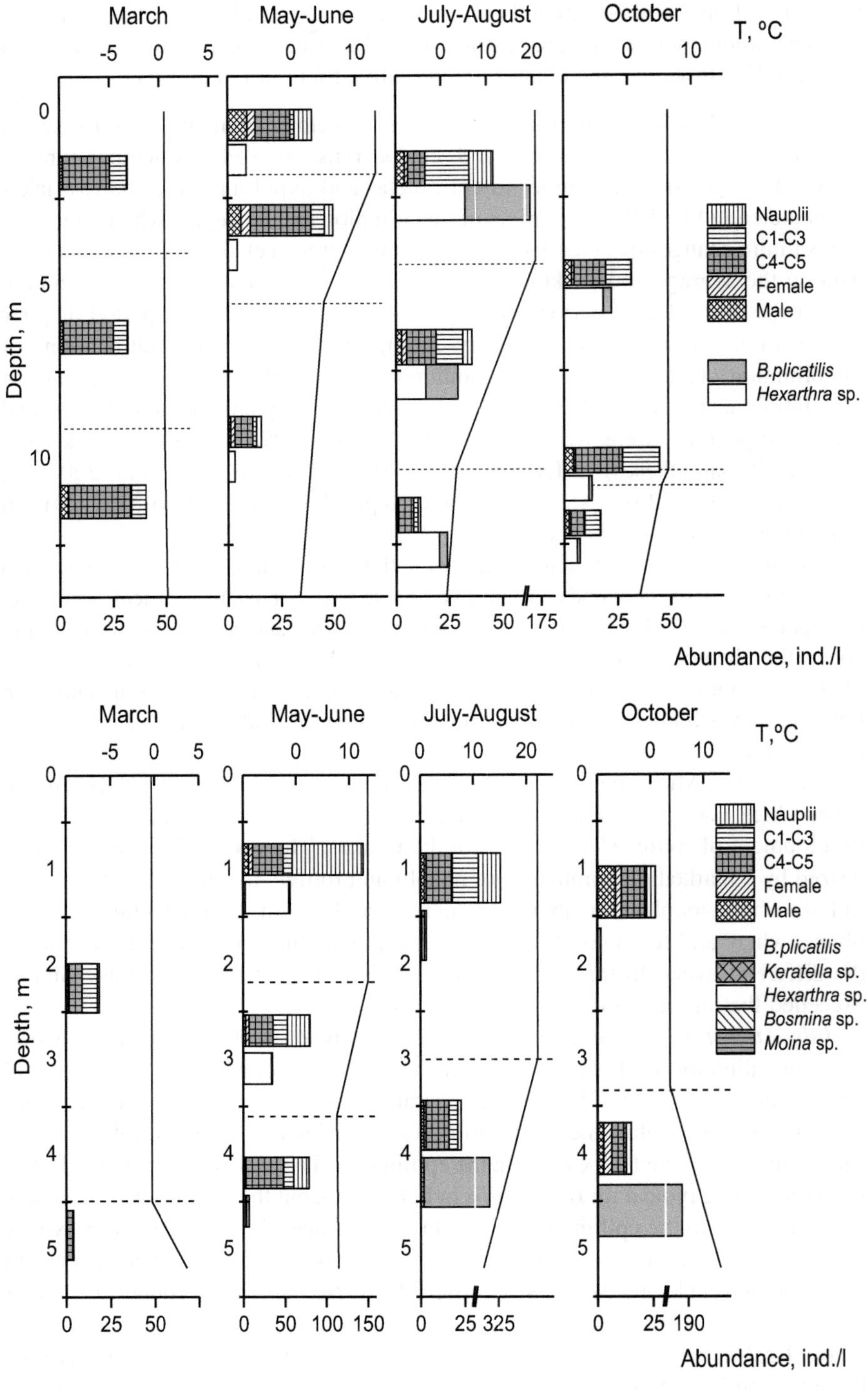

Fig. 5.8 Average concentrations (*bars*) of *A. salinus* and rotifers in different depth strata and temperature profiles (*vertical solid line*) in pelagic part of Lake Shira (1999–2011) and Lake Shunet (2003–2004, 2007–2009 and 2011). Horizontal *dash lines* indicate the borders of epi-,

anoxic zone began. The Lake Shunet had inverse stratification—the surface waters were colder, while the temperature at the depth of chemocline and below was higher (see Fig. 5.3). In Lake Shira concentration of *A. salinus* in a narrow zone of metalimnion was slightly higher than in the epilimnion above. The relative proportions of the adult stages, copepodites C4–5 and C1–3, were similar to that in the period July–August, except for naupliar stages—their contribution to the total number decreased to almost zero. Rotifer *Hexarthra* was still abundant, with a maximum in epilimnion. *B. plicatilis* was also present but in small numbers.

In the Lake Shunet, *A. salinus* was represented mostly by adult stages and copepodites C4–C5. Nauplii and copepodites C1–3 in the population were rare. The concentration of copepods in the metalimnion despite the higher temperature was lower than in the upper mixed layer of water. The numbers of *B. plicatilis* near the chemocline were comparable with summer numbers.

Interestingly, throughout the season the average numbers of *A. salinus*, including nauplii, varied within a rather narrow range, 31–34 ind. l^{-1} from March to October. Throughout the season, the average numbers of copepods *A. salinus* in the Lake Shunet were generally higher than in Lake Shira and were 49 ind. l^{-1} with a maximum in May–June.

The vertical distributions of *A. salinus* in both lakes were similar. Nauplii and copepodites C1–3 preferred to stay in epilimnion. However, copepodites C4–5 in Lake Shira preferred metalimnion, while in Lake Shunet they moved deeper into the hypolimnion. *Hexarthra* sp., a rotifer, in Lake Shira was present in all zones during entire study period. But in Lake Shunet it was limited to the period May–June and to epi- and metalimnion. *B. plicatilis* inhabited the epilimnion in Lake Shira, but in Lake Shunet it developed a rather thin layer only above the chemocline.

Gammarus The amphipod *Gammarus lacustris* is generally considered as a benthic organism in lakes. However, there is growing evidence that the ecological niche of *Gammarus* is not only benthic but benthoplanktonic (e.g. Wilhelm et al. 2000). Lake Shira and Lake Shunet are both meromictic lakes, where *Gammarus* is one of the few dominant species among macro-zooplankton. In meromictic lakes, anoxic deeper layers restrict the vertical distribution of zooplankton (Zadereev and Tolomeyev 2007). It is clear that in the pelagic zone of the meromictic lakes there is no bottom habitat for the benthic animals. Repeated zooplankton sampling in central parts of these both lakes demonstrated that *G. lacustris* is common in samples from the mixolimnion. To understand the ecology of *Gammarus* in meromictic lakes, we studied the vertical distribution and abundance of *Gammarus lacustris* in the pelagic zone of two meromictic lakes during stable thermal stratification (July–August).

Fig. 5.8 (continued) meta- and hypolimnion zones in the thermally and chemically stratified mixolimnion, except March in Lake Shira when the mixolimnion is uniform and lines simply indicate three equal in size zones

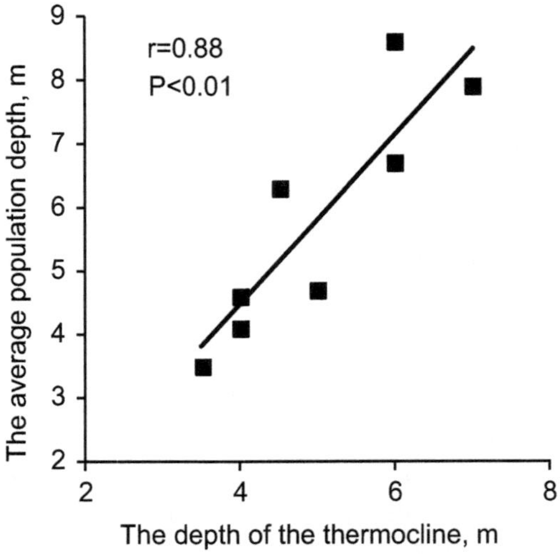

Fig. 5.9 The correlation between the depth of the thermocline and the average depth of *G. lacustris* population in the pelagic zones of Lake Shira and Lake Shunet (Zadereev et al. 2010)

In both lakes during summer stratification, *Gammarus* was distributed non-homogenously, with a stable peak in the metalimnion (Zadereev et al. 2010). The average depth of *Gammarus* population in the pelagic zone was significantly correlated with the depth of the thermocline (Fig. 5.9). Our results shown that *Gammarus* is regularly present in the pelagic zone of two meromictic lakes, Shira and Shunet, during summer stratification. The gammarid exhibited high densities in the thermocline on all sampling dates. Thus, this type of vertical distribution is stable during summer stratification and typical for these lakes. The vertical distribution of *Gammarus* apparently reflects its physiological or ecological preferences, or both. The absence of fish in the pelagic zone, high oxygen concentration, low water temperature, increased seston concentration, elevated water density in the metalimnion and the anoxic hypolimnion can be the most probable combination of factors responsible for the peak abundance of *Gammarus* in the metalimnion of these lakes.

The peak abundance of *Gammarus* in the pelagic zone of both lakes amounted up to 400 ind. m^{-2}, while the peak animal densities in the metalimnion reached 50 ind. m^{-3}. These data are comparable with abundances of the same *Gammarus* species in the littoral of Lake Shira (Yemelyanova et al. 2002). Thus, both littoral and pelagic zones can be equally important habitats for amphipods in meromictic lakes.

It is also important that *Gammarus* summer biomass in the pelagics of both lakes (up to 15 g m^{-2}) was comparable to that of dominant zooplankton grazers (20 g m^{-2} for *Arctodiaptomus salinus and* 15 g m^{-2} for *Brachionus plicatilis*) (Zadereev and Tolomeyev 2007). High pelagic densities of *Gammarus* suggest that this amphipod may have an important effect on the food web in this zone. Gladyshev et al. (2000), who studied the fatty acids in the stomach contents of *Gammarus* from Lake Shira, show that the animals probably feed on freshly sedimenting seston. We determine the role of *Gammarus* in the pelagic woodweb using its diet and growth rates. Based on a survival study of *Gammarus* on the lake's seston in a 20-day mesocosm experiment (Tolomeyev et al. 2006), we can conclude that *Gammarus* can survive for long periods on in situ lake seston particles. We used these assumptions about feeding spectra to simulate with the hydrophysical-ecological model of Lake Shira the effect of *Gammarus* on the pelagic seston in the lake (Degermendzhy et al. 2010). It appears that *Gammarus* does play a role in regulating the seston carbon in lake water. For example, according to model simulations, *Gammarus* present in the water column at the peak densities can control the seston biomass by 30–40 %.

Thus, the presence in abundance of *Gammarus* in pelagic water seems to be an intriguing but an important feature of meromictic lakes, e.g. lakes Shira and Shunet, which lack fish due probably to high salinity. However, a more specific research directed to trophic and population dynamics of *Gammarus*, the role of littoral zone and the coupling of littoral and pelagic trophic budgets in order to examine the role this "predominantly" benthic animal plays in the lake's ecosystem, is needed to quantify its importance in energy flow.

5.2.3 Food Web Structure and Interactions

Food web structure in the lakes Shira and Shunet is quite similar due mainly to similarity of species composition of bacterio-, phyto- and zooplankton (Degermendzhy et al. 2010). This similarity can be explained by the geographical proximity of the two lakes and by their similar salt composition, annual stratification pattern and their simple short food chain structure: mainly because of the absence of fish as a top predator in the pelagial of these both lakes. To understand the food web structure and interactions in the lakes, it is necessary to take into account the vertical stratification of physical, chemical and biological components in these lakes.

5.2.3.1 The Microbial Loop in Mixolimnion and Chemocline

Heterotrophic bacterioplankton is the main constituent of the planktonic microbial community in the mixolimnion of Lake Shira (Kopylov et al. 2002). Both the numbers and biomass of bacteria associated with detrital particles and microcolonies in the water column are considerable (up to 18 % of the total bacterial biomass). They may serve as a source of food for the zooplankton. Heterotrophic flagellates are dominant among the protozoa. Within the microbial loop, a food chain consisting of bacteria and heterotrophic flagellates appears to be a possible pathway transferring bacterial production to higher levels of the planktonic food web, i.e. they facilitate linking the microbial and macrobial food webs in these lakes.

Bacterial community consisting of purple sulphur and heterotrophic bacteria inhabits the chemocline of the Lake Shira. However, we consider the bacterial community in the chemocline as a not essential part of the trophic chain in Lake Shira. Because the depth of the chemocline is variable, the bacterial community does not reach high densities and the densities of ciliates and phytoflagellates are also low (Degermendzhy et al. 2010).

In contrast, the depth of the chemocline in Lake Shunet is stable; the mixolimnion is relatively thin, and the chemocline inhabited by the extremely dense bacterial community, a population of *Cryptomonas* sp. and ciliate community comprising several species (Khromechek et al. 2010). The trophic interactions in Lake Shunet partially differ from those in Lake Shira as both ciliates and zooplankton feed on bacterial community in the chemocline. Ciliates and calanoid copepods (*A. salinus*) also feed on *Cryptomonas* spp. (Tolomeev et al. 2010), which are most probably mixotrophic in the chemocline and consume bacteria.

5.2.3.2 Phytoplankton and Zooplankton

The phytoplankton diversity in both lakes is relatively low. In Lake Shira the dominant species in different seasons and at different depths were cyanobacteria *Lyngbya contorta* Lemm. and *Microcystis pulverea* (H.C. Wood) Forti, diatoms *Cyclotella choctawhatcheeana* (formerly identified as *Cyclotella tuberculate*), green algae *Dictyosphaerium tetrachotomum* Printz and *Oocystis submarina* Lagerheim and cryptophytic algae *Cryptomonas salina* Wislouch. During spring and at the beginning of the summer, the dominant species of phytoplankton in Lake Shira were *L. contorta* and *C. choctawhatcheeana*. They developed in the upper layers of the lake. The peak of *C. salina* near the oxic-anoxic interface was also typical for the spring and early summer periods. From late June to September, the phytoplankton in Lake Shira was stratified, with a maximum in the lower part (8–12 m) of the thermocline. In these strata the chlorophyll concentration reached about 23 $\mu g\ l^{-1}$. This peak value mainly consisted of vertically segregated peaks of *L. contorta*, *M. pulverea*, *D. tetrachotomum* and *O. submarina* var. *schiriensis* (Gaevsky et al. 2002).

In Lake Shunet the composition of phytoplankton slightly differs. The dominant species are cyanobacteria *Synechocystis* sp.; diatoms *Cyclotella* sp. that is observed at all depths of the mixolimnion and *Nitzschia acicularis*, which is typical for surface waters; and green alga *Dictyosphaerium tetrachotomum* Printz which is observed at all depths of the mixolimnion (Degermendzhy et al. 2003). Population density of *Cryptomonas salina* Wislouch in the chemocline is high (Khromechek et al. 2010).

The pelagic zone of these lakes is dominated by identical species of zooplankton. The major constituents of the zooplankton community in both lakes are ciliates represented by the genera *Cyclidium* sp. and *Strombidium* sp., calanoid copepod *Arctodiaptomus salinus*, rotifers *Brachionus plicatilis* and *Hexarthra* sp. and amphipod *Gammarus lacustris*.

The similarity of the vertical structure of the distribution of *A. salinus* in both lakes consists mainly of nauplii and copepodites C1–3 in the epilimnion. Vertical distribution of copepodites C4–5 slightly differs: in Lake Shira copepodites prefer metalimnion zone and in Lake Shunet they move deeper in hypolimnion. Rotifer *Hexarthra* sp. in Lake Shira is present in all lake zones and present in all observation periods. In contrast, in Lake Shunet *Hexarthra* was limited to May–June period alone, only in epi- and metalimnion. Differences in vertical positioning in the lakes showed that whereas *B. plicatilis* inhabited the epilimnion in Lake Shira, in Lake Shunet this rotifer developed as a thin layer above the chemocline.

The feeding spectra of *Arctodiaptomus salinus* (Calanoida, Copepoda) populations inhabiting lakes Shira and Shunet are based on fatty acid (FA) trophic markers measured by Tolomeev et al. (2010). This study revealed the calanoid to feed on *Cryptomonas* and sulphur purple bacteria in Lake Shunet and on ciliates and colonial picoplankton in both lakes. The analysis of markers in storage lipids—triacylglycerols (TAG) of *A. salinus*—reflected the differences in seston composition of the lakes and, consequently, in the feeding preferences. FA composition of seston in the lakes moderately differed: levels of diatom FA markers were higher in Lake Shunet, and those of cyanobacteria and green algae markers were higher in Lake Shira. Moreover, *A. salinus* in Lake Shira had significantly higher concentrations of bacterial FA markers, while in Lake Shunet, contribution of cryptophytes and/or flagellates or both to the diet of *A. salinus* was higher. The analysis of markers in storage lipids also revealed high food selectivity by *A. salinus*. This conclusion is based on the detection in the storage lipids of the trophic markers of the minor components of seston, which must be selectively ingested by the animals in order that it can be detected in the storage lipids.

The presence of *Gammarus lacustris* in the pelagic part of both lakes is a specific feature of their food webs. The study of gut contents of *G. lacustris* and fatty acid trophical markers revealed that gammarids ingest primarily fresh seston (Gladyshev et al. 2000). The 20-day mesocosm experiments demonstrated that *G. lacustris* could grow on in situ concentrations of seston in the lakes. Moreover, the mesocosm experiments revealed that gammarids and copepods are most probably the other competitors for seston in the pelagic part of Lake Shira (Tolomeyev et al. 2006).

Thus, the difference in food web structure and trophic interactions (Fig. 5.10) is mainly due to differences in mean and maximum depths of the two lakes and in their salinity levels that control the stability of permanent stratification. Lake Shira, which is deeper and less saline, has a deeper redox zone with variable depth. Depending on the weather conditions, the oxic-anoxic interface can sink to about16 m or ascend to 11 m depth. The depth of the chemocline is variable and the bacterial densities in the chemocline are not high. The mixolimnion in summer also creates different habitats for various species. The distribution of phytoplankton is non-uniform with peak of biomass in the metalimnion. The distribution of zooplankton is also patchy with rotifers and juvenile copepods in the warm epilimnion and older copepods in the cold oxic hypolimnion (Zadereev

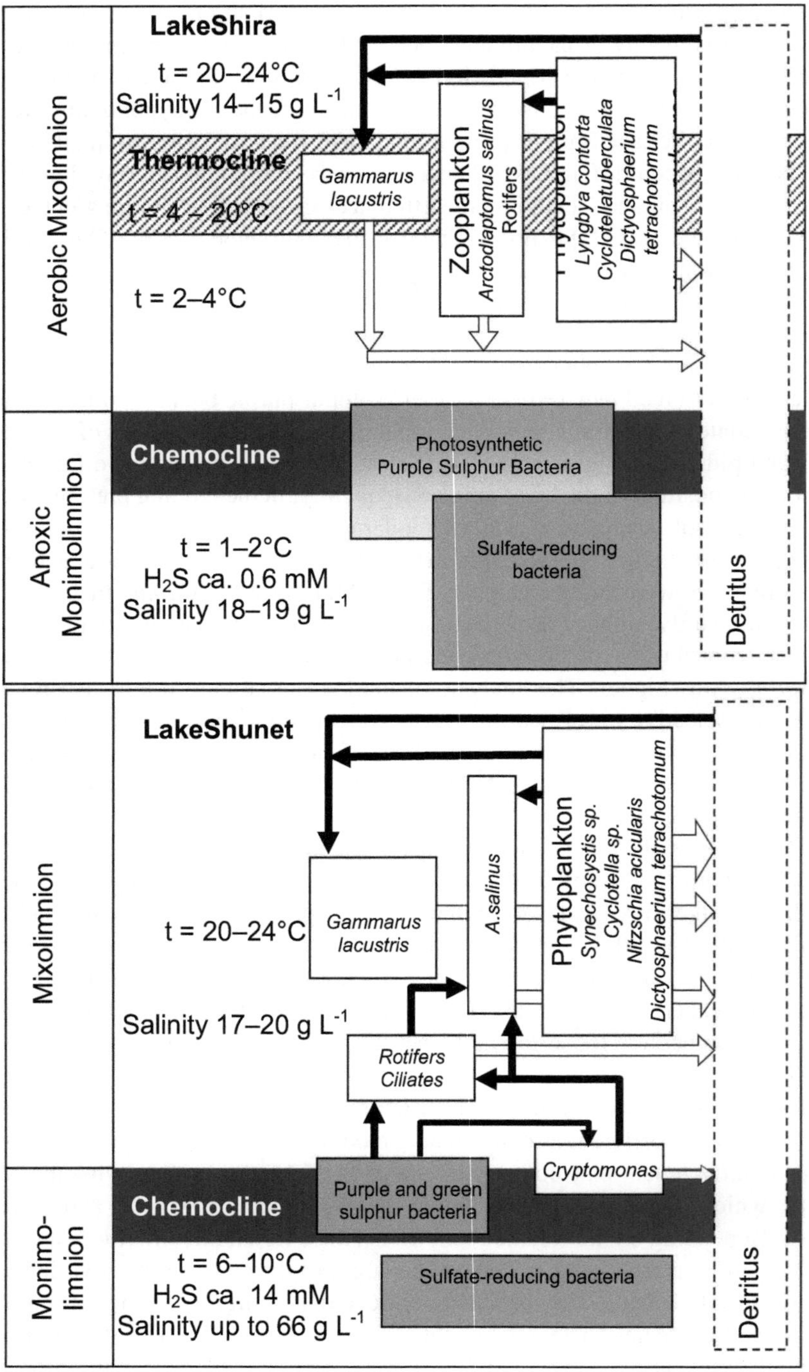

Fig. 5.10 The main food web components (*boxes*) and trophic interactions (*solid arrows*) in two saline lakes Shira and Shunet, (South Siberia, Russia) in summer. The position of boxes demonstrates the location of the given species in the water column. Exact depths, physical and chemical

and Tolomeyev 2007). The zooplankton that is comprised of calanoids *A. salinus* and rotifers *B. plicatilis* and *H. oxiuris* feed mostly on phytoplankton, while the bacterial community in the chemocline does not significantly contribute to the food web.

The chemocline in Lake Shunet is located at 5 m depth, which unlike in Lake Shira is very stable, with the annual variations of <20 cm. This stability of the chemocline depth is based on the sharp salinity gradient between the mixolimnion (salinity ranges from 17 g l^{-1} to 20 g l^{-1}) and monimolimnion (salinity up to 66 g l^{-1}). Because the depth of the chemocline in Lake Shunet is stable and the mixolimnion is relatively shallow, the chemocline is inhabited by an extremely dense bacterial community, and by a population of *Cryptomonas* sp. and several ciliate species. As the mixolimnion of Lake Shunet is thermally not stratified for long periods, the vertical distributions of phytoplankton and zooplankton are also not vertically well defined. The trophic interactions in Lake Shunet partially differ from those in Lake Shira; ciliates, rotifers and zooplankton in the lake feed on bacterial community in the chemocline. Both ciliates and *A. salinus* also feed on *Cryptomonas* sp., which is most probably mixotrophic in the chemocline and consumes bacteria.

We conclude that the food web in the pelagic part of the Lake Shira is mainly based on the transfer of organic matter from the phytoplankton to zooplankton, which can be considered as a classical food web based on the autotrophic production of organic matter by the phytoplankton. In contrast, in Lake Shunet the food web is a combination of classical phytoplankton-zooplankton trophic link and trophic interactions based on the production of organic matter in the chemocline of the lake.

5.3 Mathematical Modelling to Understand the Ecosystem Functioning

5.3.1 General Information on the Lake Shira Model

Lake Shira was selected to develop coupled hydrophysical-biochemical model as the lake has a long history of observations and is of considerable economic value for the region. We developed model of Lake Shira to simulate and explain the thermal structure (the position of thermocline and halocline, vertical mixing rates and the vertical distributions of temperature, salinity and density) and the vertical spatial distribution of the main components of ecosystem (Fig. 5.11). The model was developed to simulate the processes in the pelagic part of the lake during the

Fig. 5.10 (continued) values and concentrations of biological components are described in the other parts of this chapter (Degermendzhy et al. 2010)

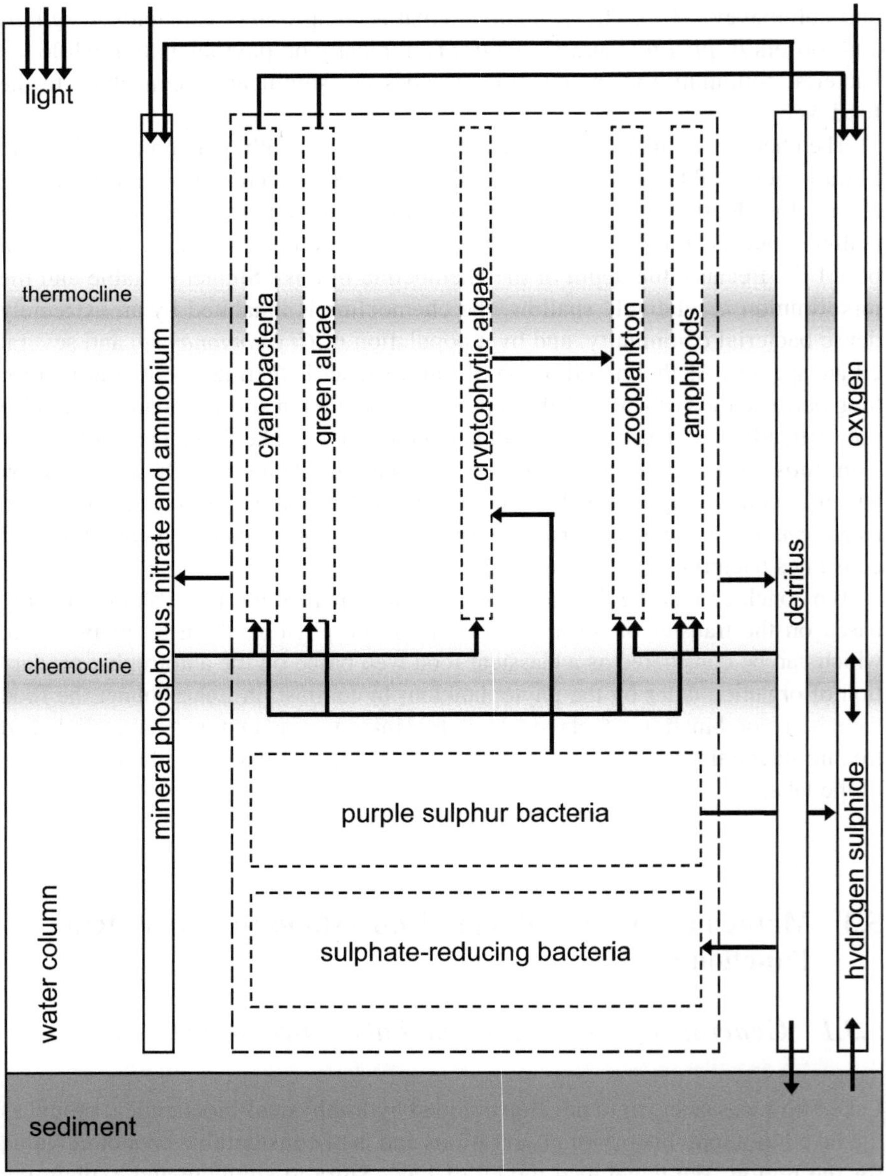

Fig. 5.11 Schematic overview of the structure of Lake Shira one-dimensional imitation model (Prokopkin et al. 2014). The *arrows* show the main processes and interactions in ecosystem

summer season. The thermal structures of the lake were calculated using the meteorological data, % cloud cover, air temperatures, wind speed and direction and vapor pressure. The model was developed taking into account general principles and approaches which were used in developing similar imitation instruments as well as on the basis of the extensive lake model sexperience (Mooij et al. 2010).

Phytoplankton in the model is represented by the dominating species of green alga, *Dictyosphaerium tetrachotomum*, and cyanobacterium, *Lyngbya contorta* (Zotina et al. 1999). Also, the supplementary module describing the cryptophytic alga population was added recently (Prokopkin et al. 2014). Zooplankton is represented by a population of calanoid copepod *Arctodiaptomus salinus*, a species that dominates the biomass of pelagic zooplankton (Zadereev and Tolomeyev 2007), which is divided into two age classes: young nauplii and juvenile copepodite stages and old copepodite stages and adult copepods. The amphipod *G. lacustris* that occupies benthic-planktonic niche in the lake was also introduced in the model. The transformation of sulphur in the model is carried out by two groups of microorganisms: light-dependent purple sulphur bacteria oxidizing hydrogen sulphide (H_2S) and sulphate-reducing bacteria decomposing detritus in the water column. The processes calculated for all biological components of the model are growth and death rates, sedimentation, consumption by other components (if any) and optionally respiration (excretion rates).

The model simulates organic matter cycling in the water column in units of dry weight, phosphorus and nitrogen and processes of diffusion and transformation (e.g. oxidation of hydrogen sulphide and nitrification). The model was calibrated and validated with the field data for the time period 2000–2010. Hydrophysical and biochemical modules of the model were described in detail by Belolipetsky et al. (2010) and Prokopkin et al. (2010, 2014) and compared with other lake models, e.g. the one of Mooij et al. (2010).

The model has wide range of applications—from general simulation of ecosystem development to the analysis of specific processes as well as scenarios of lake behaviour under the effect of different driving forces. Below we provide several examples of how the model was used to understand the stratified meromictic lake ecosystem.

5.3.2 Analysis of the Mechanisms of Formation of Deepwater Peak of Phytoplankton

In summer, the deepwater peak of phytoplankton biomass is typical for Lake Shira. The model simulates the general trends of the development of the phytoplankton (Fig. 5.12a) in agreement with the field observations: the dominating species of green alga *Dictyosphaerium* peaked in July at 8 m depth and in August at 10–12 m depth (Gaevsky et al. 2002); during the second part of summer, cyanobacterium *Lyngbya contorta* begins to dominate in phytoplankton community with the biomass peak above the peak of green alga (Zotina et al. 1999).

To understand the driving forces for the changes in phytoplankton community, we selected several factors and processes that control phytoplankton dynamics:

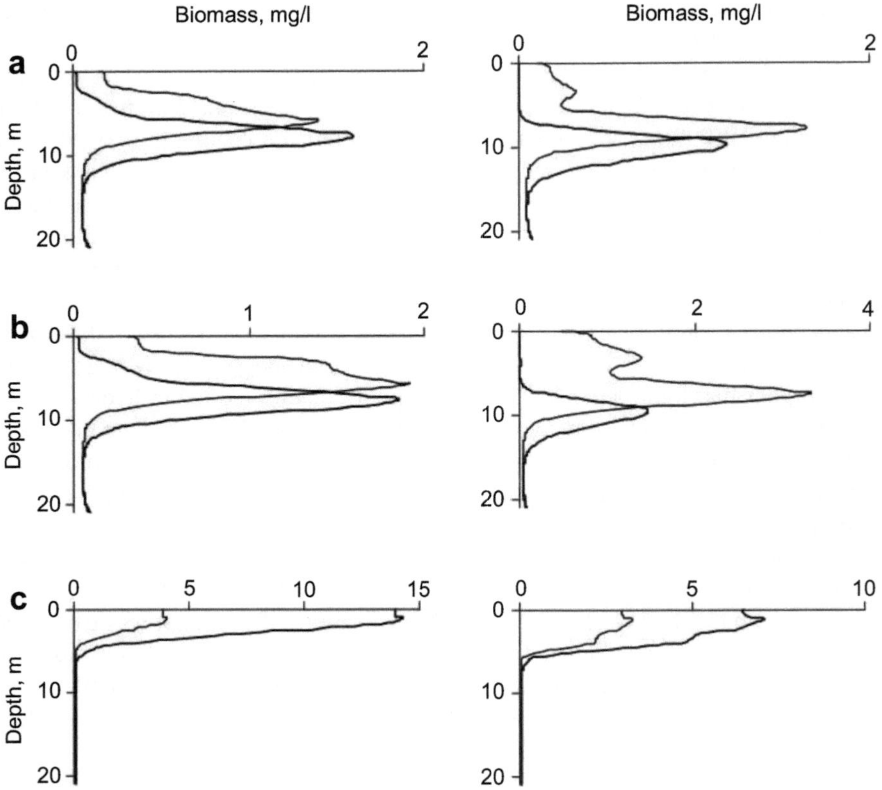

Fig. 5.12 Model calculations (Prokopkin et al. 2010) of the vertical distribution of green alga (*thick line*) and cyanobacteria (*thin line*) biomasses. *Left panels* show the results for July 15; the *right panels* show the results for August 06. (**a**) Calculations under standard conditions; (**b**) results obtained in the absence of trophic pressure exerted by zooplankton; (**c**) calculations in the absence of phytoplankton sedimentation

irradiance; water temperature; nutrient concentration; growth, mortality and respiration; consumption of zooplankton and amphipods; and sedimentation rates of organic matter. Sensitivity analysis of the model based on the different calculations, where these factors and processes were switched on/off or their effects were increased/decreased, was performed. This analysis allows ranking the model parameters, factors and processes according to degree of their influence on the phytoplankton community.

The model calculations show that filter-feeding calanoid *A. salinus* considerably affects the vertical distribution of phytoplankton and change of the dominant species (Fig. 5.12b). The analysis revealed that the sedimentation rate of phytoplankton seems to be the most important factor that determines the deep peak of phytoplankton biomass in the water column, i.e. the biomass would be concentrated in the subsurface water layers if sedimentation is absent (Fig. 5.12c) (Prokopkin et al. 2010).

We conclude that the changes in dominant species in phytoplankton community during summer occur in response to temperature and nutrient factors and grazing activity of zooplankton. The model demonstrates that the change of the vertical positions of the biomass of different phytoplankton species first of all is the result of sedimentation of phytoplankton that is either sinking of live phytoplankton based on gravity or senescence connected sedimentation (Prokopkin et al. 2010).

5.3.3 The Effect of Weather Parameters on the Thermal Stratification of the Lake

To understand the relative importance of the effect of weather parameters on the thermal stratification of Lake Shira, we obtained the temperature profiles of the water column using different weather scenarios. The reference scenario was calculated using the meteorological data set for the year 2002. Then, we compared the output of the reference scenario with the outputs of the calculations performed with altered meteorological data sets. To obtain the altered data sets, the values of wind speed or air temperature or % cloud cover in the reference year weather data set were increased or decreased by several values equal to the unit of measurement of corresponding weather parameter (by 1 m s^{-1} for wind speed, 1 °C for air temperature and 1 relative unit based on % for cloud cover).

The characteristic parameters of the temperature profile of the water column were (a) the depth of the thermocline, (b) the average temperature of water in the epilimnion and (c) the temperature of water at 10 m depth (hypolimnion) (Table 5.1). The values of these parameters were calculated as of June 1 and 15, July 1 and 15 and August 1, 15 and 31. The average of these values was used for analysis.

Changes in the values of weather parameters had strong effect on the temperature of the epilimnion, while for the hypolimnion this effect was less pronounced. We observed similar to Arvola et al. (2010) that in thermally stratified lakes, the increase in summer air temperatures will only result in an increase in hypolimnetic temperatures if winds are strong enough to trigger mixing. Also, Liu et al. (2014) demonstrated that increasing the air temperature warms up the epilimnion but does not effect the hypolimnion temperature. Our calculations support these observations.

The greatest variations in the depth of the thermocline in our calculations were caused by variations in the wind speed and air temperature. Opposite variations in the values of these parameters (increase in air temperature and decrease in wind intensity or vice versa) initiated two scenarios in the water column: (1) the increase in the temperature in the epilimnion together with an increase in the depth of thermocline or (2) a decrease in the temperature of the epilimnion with a decrease

Table 5.1 The effect of weather parameters on the thermal stratification of Lake Shira

		The variation of weather parameters		
		The change of the output with an increase in the wind speed by 1 m s^{-1}	The change of the output with an increase in the cloud cover by one relative unit	The change of the output with an 1 °C increase in the air temperature
Outputs of the thermal stratification	Water temperature at the epilimnion, °C (± S.D.)	−1.77 ± 0.45	−0.73 ± 0.22	0.73 ± 0.01
	Water temperature at the hypolimnion, °C (± S.D.)	0.51 ± 0.40	−0.20 ± 0.07	0.010 ± 0.004
	The depth of the thermocline, m (± S.D.)	0.29 ± 0.13	−0.08 ± 0.04	−0.18 ± 0.03

of the thermocline depth. The second scenario predicts less stable thermal stratification.

De Stasio et al. (1996) and some others state that in response to the ongoing global climate change, the lakes will become warmer and stratification will occur earlier and will be more pronounced. These hypotheses support our calculations if we associate the climate change with an increase in air temperature. The model study by Robertson and Ragotzkie (1990) showed that when the air temperature is increased by 1 °C, the temperature of the epilimnion rises by 0.4–0.85 °C, but the temperature of the hypolimnion in the middle of the summer season would not change significantly. Hondzo and Stefan (1993) showed that because of climate change (scenario of a 100 % increase of CO_2 content in the atmosphere), the temperature of epilimnion will also increase, but to a lesser extent than the air temperature, and hypolimnetic temperature would be determined by the lake morphometry and spring mixing rather than by weather in summer alone. Based on our calculations, we estimated the increase in the water temperature in the epilimnion by 0.73 °C and in the hypolimnion by 0.01 °C, when the air temperature increases by 1 °C, which is in very good agreement with above-mentioned published results.

5.3.4 The Effect of Weather Parameters on the Lake Shira Ecosystem

The model analysis revealed the link between the variations of weather parameters and related changes in temperature profiles in the water column and response of ecosystem components both in the mixolimnion and the monimolimnion (Fig. 5.13).

Response of mixolimnion and monimolimnion components to the weather changes can be depicted by two sets of correlations that influence the content of hydrogen sulphide in the monimolimnion and the depth of the chemocline. What is essential for the ecosystem behaviour is the response of two light-dependent and separated in space primary producers (phytoplankton and purple sulphur bacteria) to the effect of weather parameters and interrelation between them.

The first interrelated set of links is between phytoplankton biomass and detrital mass in the water column and hydrogen sulphide content in the monimolimnion. Phytoplankton is the main source of dead organic matter (detritus) sinking to the monimolimnion, where it participates in the sulphur cycle and production of hydrogen sulphide.

The second interrelated set of links is between the biomass of phytoplankton in the water column, and of purple sulphur bacteria in the chemocline, and the content of hydrogen sulphide in the monimolimnion. Phytoplankton and purple sulphur bacteria are spatially separated light-dependent components; there is a negative correlation between their biomasses. A decrease in phytoplankton biomass

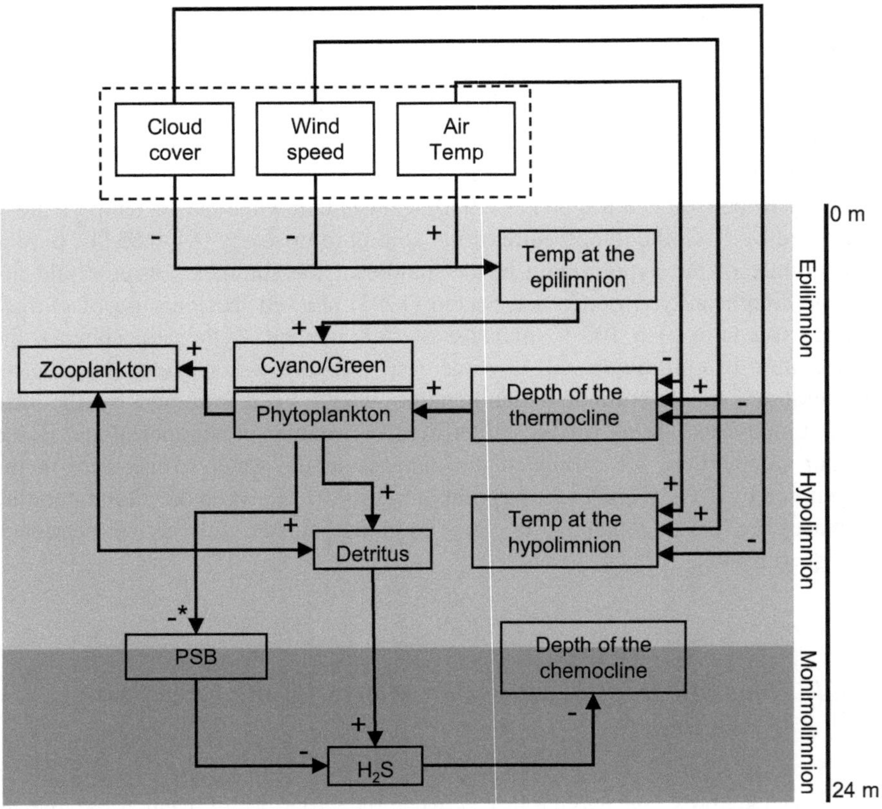

Fig. 5.13 A diagram of weather induced changes in the stratification of the water column and related effects on the food web of Lake Shira. Sign "+" denotes the significant positive correlations between parameters, sign "–"—significant negative correlations. *—the correlation is significant only for the values calculated with the variations in air temperature and wind speed

probably leads to an increase in water transparency, more favourable underwater light conditions, and an increase in the biomass of purple sulphur bacteria. To support this finding, we note that the maximal biomasses of bacteria in winter were observed in winters of 2006 and 2007 when the light intensity in chemocline was highest because of snow-free weather conditions (Rogozin et al. 2009). As hydrogen sulphide is the main substrate for the growth of purple sulphur bacteria (Rogozin et al. 2012), an increase in their biomass leads to a decrease in the content of hydrogen sulphide in the monimolimnion. The content of hydrogen sulphide in the monimolimnion determines the depth of the chemocline.

Thus, we demonstrated how biological processes control the depth of the chemocline during the growing season. In our calculations, we did not use the depth of the chemocline at the beginning of the season as a factor influencing the ecosystem properties. Our studies on the lake during 2002–2009 showed (Rogozin et al. 2010a) that the depth of the chemocline in spring differs from year to year.

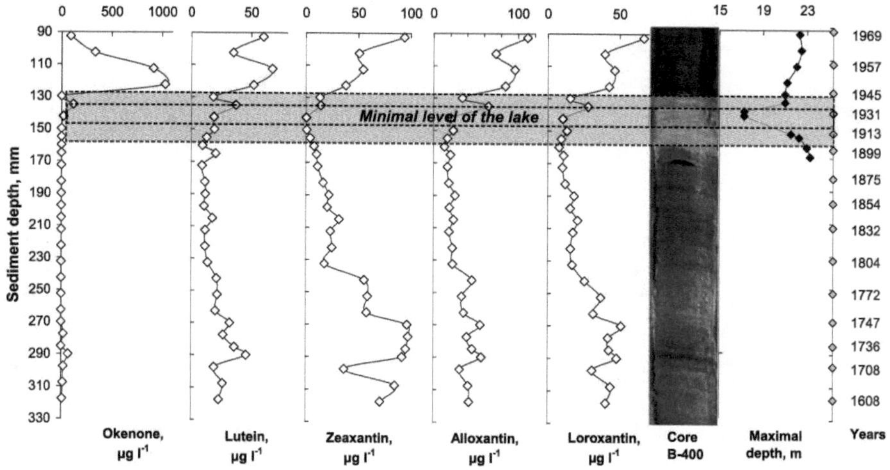

Fig 5.14 Carotenoids in upper sediments of Lake Shira. Right diagram shows the dynamics of the water level (maximum depth) in the lake. *Shaded area* shows carbonate (*white*) layer from Zykov et al. (2012)

This is because this depth in the years to come is likely to be affected by the weather conditions in the preceding year. Moreover, this depth in spring largely depends on the mixing depth of the mixolimnion in the preceding autumn period (Zadereev et al. 2014).

5.4 Paleolimnology. Reconstruction of Local Climate Lake Properties Based on Bacterial Pigments and Other Markers

5.4.1 Geochemical Characteristics of Lake Shira Sediments

We did not study the composition of Lake Shunet sediment because of its non-laminated semi-liquid consistency that prevented detailed analyses and determination of age of the layers. In contrast, Lake Shira sediment is a proper object for paleolimnological studies because of its laminated and consolidated structure (Figs. 5.14 and 5.15).

The water level of a closed saline lake is highly sensitive to climate changes: it depends first of all on humidity. Therefore, reconstruction of the changes in lake's water level may help us understand the climate changes in the past better. The laminated structure of the bottom sediments is best preserved in meromictic lakes, because the stable stratification of the water column prevents sediment stirring and resuspension. In addition, high concentrations of sulphide and anoxia in bottom waters inhibit the activity of benthos, i.e. bioturbation is absent and, hence, the

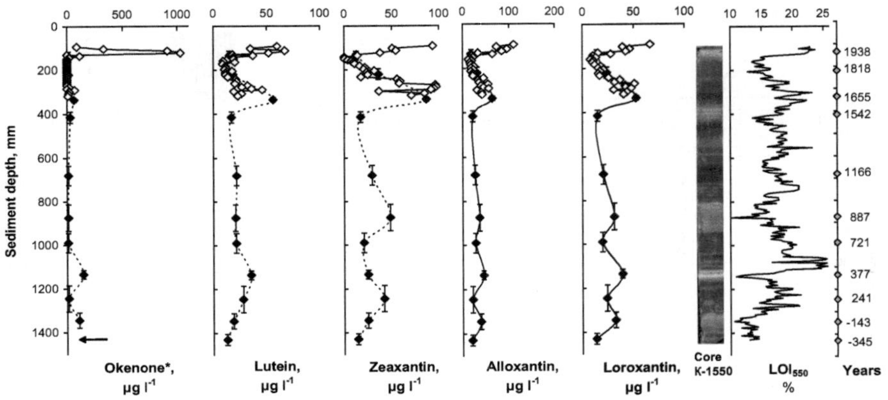

Fig 5.15 Carotenoids and LOI_{550} in bottom sediments of Lake Shira. The absence of okenone in the lower sample is shown by an arrow (from Zykov et al. 2012; see explanation in text)

original progression of the sediment layering remains preserved. Therefore, the bottom sediments in the meromictic lakes are easier to date, and therefore they represent excellent "archives" of the climatic change (Overmann et al. 1993).

In 2009 a paleolimnological study of Lake Shira was first initiated by Rogozin et al. (2011). Isotope dating was carried out using ^{137}Cs, ^{210}Pb and ^{14}C measurements together with direct counting the annual layers (Kalugin et al. 2013). Geochemical analyses show that Lake Shira has two contrasting mixing states: meromictic and holomictic. In upper sediment the period in which the first white layer was formed corresponds to the period of decrease in water level (1900–1930s), indicating that the water level had decreased (1920s) (Fig. 5.14). The content of organic matter is lower in the white sediment layers than in the dark sediment layers (Fig. 5.15), perhaps because of accelerated decomposition rate in the white layers because white layers indicate the oxic conditions in near-bottom waters (Kalugin et al. 2013)

This accelerated decomposition rate of organic matter was probably caused by oxic conditions prevailing in near-bottom waters during the low water-level period of the lake (1900–1930s). Therefore, Lake Shira was presumably holomictic during this period.

Meromictic periods, as in more recent time, are comprised of black thin laminated mud coloured by organics and hydrotroilite (FeS_nH_20), which are predominant in the core. For holomictic conditions, white organic matter-free mud appeared in the periods BC 220–75, AD 480–565, AD 900–1045, AD 1437–1603 and AD 1853–1947 (Fig. 5.15) (Kalugin et al. 2013). Presumably, most organic matter produced in the waterbody and sediment was decomposed by dissolved atmospheric oxygen due to mixing at that time. Simultaneously, the biomass of phototrophic sulphur bacteria seems decreased or disappeared due to lack of hydrogen sulphide.

5.4.2 Photosynthetic Pigments in Bottom Sediments

The fossil molecular remains of APB (such as carotenoids, bacteriochlorophylls and their derivatives (i.e. phaeopigments and DNA) serve as indicators of anaerobic conditions in the photic zone of the lake at some stage of its existence (Overmann et al. 1993). For the first time, in 2011, we analyzed the composition of carotenoids in the bottom sediments of Lake Shira. Okenone, the main carotenoid of purple sulphur bacteria inhabiting Lake Shira, was common in all sediment layers except from deepest layers located at a depth of 1430 m (about 2350 years old) (Fig. 5.14). Other carotenoids were found in all layers without exception. Clear maximum of okenone and local maxima of other carotenoids were detected near the upper limit of the carbonate layer at 13 cm in the sediment core. The concentration maximum of okenone was much higher than the maxima of all other carotenoids.

The presence of okenone allows one to construe that the anaerobic zone existed in the lake from fourth century BC to the present. Nevertheless, the absence of okenone in the earlier sample indicated that during the earlier period, there was no anaerobic activity in the near-bottom water layers. As known from studies on other lakes, okenone is found in bottom sediments only during meromixis/holomixis only if anoxia develops in summer hypolimnion (Schmidt et al. 2002; Dressler et al. 2007). As a rule, meromixis favours the high production of PSB and the better preservation of carotenoids (Overmann et al. 1993; Schmidt et al. 2002; Leavitt 1993). Therefore, a peak in the concentration of okenone in the layers forming 110–130 mm stratum (1945–1970) indicates pronounced meromictic properties of the lake in this period. The most probable (Rogozin et al. 2010a) cause for the current meromixis of Lake Shira is the influence of an ectogenic factor (freshwater layer lying above the saline water), which is in agreement with Hutchinson's lake typology based on type of meromixis in lakes (Hutchinson 1957). Due to the inflow of large amounts of freshwater with the surface run-off into the saltier lake, which was the case during the 1920s and 1930s, in this period the water level was minimum, but a difference in salinity probably favoured a stable density gradient and permanent meromixis (Rogozin et al. 2010a). The low content of okenone in the sediment of Lake Shira until the 1920–1930s indicates that meromixis at a high water level either was quite weak or absent.

In the 1950s when the concentration of okenone was higher, the water level of the lake was about 2 m lower than at present (Fig. 5.14). Thus, it is quite possible that the depth of the mixolimnion (i.e. the depth position of the chemocline) was lower than it is now. Consequently, light conditions in chemocline were better than now. At present, very low light intensity (about 2 $\mu E\ m^{-2}\ s^{-1}$) in the chemocline of the lake limits light conditions for PSB (Rogozin et al. 2009).

5.5 Conclusions

Both lakes Shira and Shunet are examples of ectogenic meromictic lakes (meromixis is induced by freshwater lying on top of the saline water). This condition is sustained annually by ice formation in winter. In spring due to melting of ice, the second halocline is formed at the near-surface layers of mixolimnion that prevents the lakes from full circulation in spring. The mixolimnion in the both lakes is monomictic (based on the regular observations from year 2000 to year 2014) and undergoes full circulation in autumn.

Both these lakes because of their geographical proximity, similar chemical composition, stable stratification and the absence of fish and other plankton predators have quite similar species compositions of bacterio-, phyto- and zooplankton. However, there are also notable differences in the food web structure and interactions because of the differences in morphometry (mean and maximum depths) of these lakes and in the salinities that control the stability of permanent stratification. In Lake Shira, the zooplankton feed mostly on phytoplankton, while the bacterial community in the chemocline does not significantly contribute to the food web. In the shallower Lake Shunet, the extremely dense bacterial community, a population of *Cryptomonas* sp. and several species of ciliates inhabit the chemocline. The food spectra of zooplankton in Lake Shunet and Lake Shira differ. Ciliates, rotifers and zooplankton in Lake Shunet feed on bacterial community in the chemocline. Both ciliates and the calanoid *A. salinus* also feed on *Cryptomonas* sp. which is most probably mixotrophic in the chemocline and consumes bacteria.

We used the coupled biochemical-hydrophysical model of Lake Shira to study the effect of weather parameters on the ecosystem of this meromictic lake. We show that although different weather parameters have opposite effects, i.e. quantitatively their effects on the different ecosystem components are similar but opposite. Thus, even for estimating the consequences of global climate change on a specific ecosystem, local characteristics of weather variations should be taken into account, because they may considerably alter the ecosystem structure and functioning.

The major factors influencing the stability of stratification of the saline meromictic lakes studied by us are the salinity and water level. The maximal depth of Lake Shira in the last 120 years has increased from 16 m to 24 m presumably due to an increase in freshwater inflow into the lake due to a negative annual water budget based on precipitation-evaporation balance. With the increase in the depth of the lake, water salinity decreased. The differences between salinity of upper and bottom waters (i.e. salinity gradient), maximal depth and lake area influence the stability of permanent stratification. According to sediment composition, Lake Shira was holomictic about 90 years ago when the water level was low, but it became strongly meromictic just after the water level increase in the1940s. Therefore, the transition between holomictic and meromictic regimes can be caused by changes in the lake's water level.

Acknowledgement The research on lakes Shira and Shunet was partially supported by the Council on grants from the President of the Russian Federation for support of leading scientific schools (grant NSh-9249.2016.5).

References

Arvola L, Glen G, Livingstone DM, Järvinen M, Blenckner T, Dokulil MT, Jennings E, Aonghusa CN, Nõges P, Nõges T, Weyhenmeyer GA (2010) The impact of the changing climate on the thermal characteristics of lakes. In: Glen G (ed) The impact of climate change on European lakes, Aquatic ecology series, vol 4. Springer, Dordrecht, New York, pp 85–102

Belolipetskii VM, Genova SN (2008) Calculation of vertical profiles of temperature and salinity in Shira Lake. Computational Technologies. Bull KazNU (mathematics, mechanics and informatics issue) 3(58):261–266 (in Russian)

Belolipetsky PV, Belolipetskii VM, Genova SN, Mooij WM (2010) Numerical modelling of vertical stratification of Shira Lake in summer. Aquat Ecol 44:561–570

De Stasio B, Hill D, Kleinhans J, Nibbelink N, Magnuson J (1996) Potential effects of global climate change on small north-temperate lakes: physics, fish and plankton. Limnol Oceanogr 41:1136–1149

Degermendzhy AG, Gaevsky NA, Belonog NP, Ivanova EA, Rogozin DYu, Koltahsev AA, Gribalev ES (2003) Izychenie physico-chimicheskix I biologicheskix charakteristik dvux balneologicheskix ozer (Matarak, Shunet, Khakassia). Vestnik Krasnoyarskogo gosudarstvennogo universiteta 5:107–115 (In Russian)

Degermendzhy AG, Zadereev YS, Rogozin DY, Prokopkin IG, Barkhatov YV, Tolomeev AP, Khromechek EB, Janse JP, Mooij WM, Gulati RD (2010) Vertical stratification of physical, chemical and biological components in two saline lakes Shira and Shunet (South Siberia, Russia). Aquat Ecol 44:619–632

Dressler M, Hubener T, Gors S, Werner P, Selig U (2007) Multi-proxy reconstruction of trophic state, hypolimnetic anoxia and phototrophic sulphur bacteria abundance in a dimictic lake in Nothern Germany over past 80 years. J Paleolimnol 37:205–219

Gaevsky NA, Zotina TA, Gorbaneva TB (2002) Vertical structure and photosynthetic activity of Lake Shira phytoplankton. Aquat Ecol 36:165–178

Garneau M-E, Posch T, Hitz G, Pomerleau F, Pradalier C, Siegwart R, Pernthaler J (2013) Short-term displacement of *Planktothrix rubescens* (cyanobacteria) in a pre-alpine lake observed using an autonomous sampling platform. Limnol Oceanogr 58:1892–1906

Genova SN, Belolipetskii VM, Rogozin DY, Degermendzhy AG, Mooij WM (2010) A one-dimensional model of vertical stratification of Lake Shira focused on winter conditions and ice cover. Aquat Ecol 44:571–584

Gladyshev MI, Emelianova AY, Kalachova GS, Zotina TA, Gaevsky NA, Zhilenkov MD (2000) Gut content analysis of *Gammarus lacustris* from a Siberian lake using biochemical and biophysical methods. Hydrobiologia 431:155–163

Hondzo M, Stefan HG (1993) Regional water temperature characteristics of lakes subjected to climate change. Clim Chang 24:187–211

Hutchinson GE (1957) A Treatise on Limnology, Geography, Physics and Chemistry, vol 1. Wiley, New York

Kalacheva GS, Gubanov VG, Gribovskaya IV, Gladchenko IA, Zinenko GK, Savitsky SV (2002) Chemical analysis of Lake Shira water (1997–2000). Aquat Ecol 36:23–141

Kalugin I, Darin A, Rogozin D, Tretyakov G (2013) Seasonal and centennial cyclesof carbonate mineralization during the past 2500 years from varved sediment in Lake Shira, South Siberia. Quat Int 290–291:245–252

Khromechek EB, Barkhatov YV, Rogozin DY (2010) Densities and distribution of flagellates and ciliates in the chemocline of saline, meromictic Lake Shunet (Siberia, Russia). Aquat Ecol 44:497–511

Krivosheev AS, Khasanov AP (1990) Therapeutic lakes of Krasnoyarsk Region. Krasnoyarsk Publishing House, Krasnoyarsk (In Russian)

Kopylov AI, Kosolapov DB, Romanenko AV, Degermendzhy AG (2002) Structure of planktonic microbial food web in a brackish stratified Siberian lake. Aquat Ecol 36:179–204

Lawrence JR, Haynes RC, Hammer UT (1978) Contribution of photosynthetic green sulphur bacteria to total primary production in a meromictic saline lake. Verh Int Verein Limnol 20:201–207

Leavitt PR (1993) A review of factors that regulate carotenoids and chlorophyll deposition and fossil pigment abundance. J paleolimnol 9:109–127

Liu W, Bocaniov SA, Lamb KG, Smith REH (2014) Three dimensional modeling of the effects of changes in meteorological forcing on the thermal structure of Lake Erie. J Great Lakes Res 40:827–840

Lunina ON, Bryantseva IA, Akimov VN, Rusanov II, Rogozin DY, Barinova ES, Lysenko AM, Pimenov NV (2007a) Seasonal changes in the structure of the anoxygenic photosynthetic bacterial community in Lake Shunet, Khakassia. Microbiology (Translated from Mikrobiologiya) 76:368–379

Lunina ON, Bryantseva IA, Akimov VN, Rusanov II, Barinova ES, Lysenko AM, Rogozin DY, Pimenov NV (2007b) Anoxigenic phototrophic bacteria community of Lake Shira (Khakassia). Microbiology (Translated from Mikrobiologiya) 76:469–479

Mooij WM, Trolle D, Jeppesen E, Arhonditsis G, Belolipetsky PV, Chitamwebwa DBR, Degermendzhy AG, DeAngelis DL, De Senerpont Domis LN, Downing AS, Elliott JA, Fragoso CR, Gaedke U, Genova SN, Gulati RD, Håkanson L, Hamilton DP, Hipsey MR, 't Hoen J, Hülsmann S, FH L, Makler-Pick V, Petzoldt T, Prokopkin IG, Rinke K, Schep SA, Tominaga K, Van Dam AA, Van Nes EH, Wells SA, Janse JH (2010) Challenges and opportunities for integrating lake ecosystem modelling approaches. Aquat Ecol 44:633–667

Overmann J, Beatty JT, Hall KJ (1994) Photosynthetic activity and population dynamics of Amoebobacter purpureus in a meromictic saline lake. FEMS Microbiol Ecol 15:309–320

Overmann J (1997) Mahoney Lake: a case study of the ecological significance of phototrophic sulfur bacteria. Adv Microb Ecol 15:251–288

Overmann J, Sandmann G, Hall KG, Northcote T (1993) Fossil carotenoids and paleolimnology of meromictic Mahoney Lake, British Columbia, Canada. Aquat Sci 55:1015–1621

Parnachev VP, Degermendzhy AG (2002) Geographical, geological and hydrochemical distribution of saline lakes in Khakasia, Southern Syberia. Aquat Ecol 36:107–122

Pimenov NV, Rusanov II, Karnachuk OV, Rogozin DY, Bryantseva IA, Lunina ON, Yusupov SK, Parnachev VP, Ivanov MV (2003) Microbial processes of the carbon and sulphur cycles in Lake Shira (Khakasia). Microbiology (Translated from Mikrobiologiya) 72:221–229

Prokopkin IG, Barkhatov YV, Khromechek EB (2014) A one-dimensional model for phytoflagellate distribution in the meromictic lake. Ecol Model 288:1–8

Prokopkin IG, Mooij WM, Janse JH, Degermendzhy AG (2010) A general one-dimensional vertical ecosystem model of Lake Shira (Russia, Khakasia): description, parametrization and analysis. Aquat Ecol 44:585–618

Robertson DM, Ragotzkie R (1990) Changes in the thermal structure of moderate to large sized lakes in response to changes in air temperature. Aquat Sci 52:360–380

Rogozin DY, Pimenov NV, Kosolapov DB, Chan'kovskaya YV, Degermendzhy AG (2005) Thin layer vertical distributions of purple sulfur bacteria in chemocline zones of meromictic Lakes Shira and Shunet (Khakassia). In: Doklady Biological Sciences. Proceedings of the Russian Academy of Sciences, vol 400, pp 54–56 (Translated from Doklady Akademii Nauk (2005) 400: 426–429)

Rogozin DY, Degermendzhy AG (2008) Hydraulically-operated thin-layer sampler for sampling heterogeneous water columns. J Sib Fed Univ 1:111–117

Rogozin DY, Genova SV, Gulati RD, Degermendzhy AG (2010a) Some generalizations on stratification and vertical mixing in meromictic Lake Shira, Russia, in the period 2002–2009. Aquat Ecol 44:485–496

Rogozin DY, Trusova MY, Khromechek EB, Degermendzhy AG (2010b) Microbial community of the chemocline of meromictic Lake Shunet during summer stratification. Microbiology (Translated from Mikrobiologiya) 79:253–261

Rogozin DY, Zykov VV, Chernetsky MY, Degermendzhy AG, Gulati RD (2009) Effect of winter conditions on distributions of anoxic phototrophic bacteria in two meromictic lakes in Siberia, Russia. Aquat Ecol 43:661–672

Rogozin DY, Zykov VV, Degermendzhi AG (2012) Ecology of purple sulfur bacteria in the highly stratified meromictic Lake Shunet (Siberia, Khakassia) in 2002–2009. Microbiology 81:727–735

Rogozin DY, Zykov VV, Kalugin IA, Daryin AV, Degermendzhy AG (2011) Carotenoids of phototrophic organisms in bottom sediments of meromictic Lake Shira (Siberia, Russia) as an indicator of past stratification. Dokl Biol Sci (Proc Russ Acad Sci) 439:228 (Translated from Doklady Akademii Nauk 439:282–285)

Rogozin DY, Zykov VV, Tarnovsky MO (2016) Dynamics of purple sulfur bacteria in saline meromictic Lake Shira (Khakasia, Siberia) for the period 2007–2013. Microbiology (Translated from Mikrobiologiya) 81:93–101

Savvichev AS, Rusanov II, Rogozin DY, Zakharova EE, Lunina ON, Yusupov SK, Pimenov NV, Degermendzhy AG, Ivanov MV (2005) Microbiological and isotopic–geochemical investigations of meromictic lakes in Khakasia in winter. Microbiology (Translated from Mikrobiologiya) 74:477–485

Schmidt R, Psenner R, Muller J, Indinger P, Kamenik C (2002) Impact of late glacial variations on stratification and trophic state of the meromictic lake Landsee (Austria): validation of a conceptual model by multi proxy studies. J Limnol 61:49–60

Tolomeev AP, Sushchik NN, Gulati RD, Makhutova ON, Kalacheva GS, Zotina TA (2010) Feeding spectra of *Arctodiaptomus salinus* (Calanoida, Copepoda) using fatty acid trophic markers in seston food in two salt lakes in South Siberia (Khakasia, Russia). Aquat Ecol 44:513–530

Tolomeyev AP, Zadereev ES, Degermendzhy AG (2006) Fine stratified distribution of *Gammarus lacustris* Sars (Crustacea: Amphipoda) in the pelagic zone of the meromictic Lake Shira (Khakassia, Russia). Dokl Biochem Biophys 411:346–348

Tonolla M, Peduzzi S, Hahn D, Peduzzi R (2003) Spatio-temporal distribution of phototrophic sulphur bacteria in the chemocline of meromictic Lake Cadagno (Switzerland). FEMS Microbiol Ecol 43:89–98

Van Gemerden H, Tughan CS, de Wit R, Gerbert RA (1989) Laminated microbial ecosystems on sheltered beaches in Scapa Flow, Orkney Islands. FEMS Microbiol Ecol 62:87–102

Vuik C (1993) Some historical notes about the Stefan problem. Nieuw Archief voor Wiskunde 11 (2):157–167

Wilhelm FM, Schindler DW, McNaught AS (2000) The influence of experimental scale on estimating the predation rate of Gammarus lacustris (Crustacea: Amphipoda) on Daphnia in an alpine lake. J Plankton Res 22:1719–1734

Yemelyanova AY, Temerova TA, Degermendzhy AG (2002) Distribution of *Gammarus lacustris Sars* (Amphipoda, Gammaridae) in Lake Shira (Khakasia, Siberia) and laboratory study of its growth characteristics. Aquat Ecol 36:245–256

Zadereev ES, Tolomeev AP, Drobotov AV, Kolmakova AA (2014) The effect of weather variability on the spatial and seasonal dynamics of dissolved and suspended nutrients in the water column of meromictic Lake Shira. Contem Probl Ecol 7:384–396

Zadereev ES, Tolomeyev AP (2007) The vertical distribution of zooplankton in brackish meromictic lake with deep-water chlorophyll maximum. Hydrobiologia 576:69–82

Zadereev ES, Tolomeyev AP, Drobotov AV, Emeliyanova AY, Gubanov MV (2010) The vertical distribution and abundance of *Gammarus lacustris* in the pelagic zone of the meromictic lakes Shira and Shunet (Khakassia, Russia). Aquat Ecol 44:531–539

Zotina TA, Tolomeyev AP, Degermendzhy NN (1999) Lake Shira, a Siberian salt lake: ecosystem structure and function. 1. Major physico-chemical and biological features. Int J Salt Lake Res 8:211–232

Zykov VV, Rogozin DY, Kalugin IA, Dar'in AV, Degermendzhi AG (2012) Carotenoids in bottom sediments of Lake Shira as a paleoindicator for reconstruction of lake states in Khakssiya, Russia. Contemp Probl Ecol 5:434–442

Chapter 6
Rogoznica Lake, a Euxinic Marine Lake on the Adriatic Coast (Croatia) that Fluctuates Between Anoxic Holomictic and Meromictic Conditions

Irena Ciglenečki, Zrinka Ljubešić, Ivica Janeković, and Mirna Batistić

6.1 Introduction

Meromixis is a condition of persistent chemical stratification of lakes with incomplete mixing of the water column over the course of a year. This reduced vertical mixing usually results in anoxia in deeper water layers and accumulation of nutrients in the monimolimnion (Romero and Melack 1996). Several marine basins exhibit a permanent anoxic layer. The most studied such water bodies are the Black Sea (Izdar and Murray 1991), Baltic Sea (Prange and Kremling 1985), Norwegian Fjords (Millero 1991), anoxic brines of Eastern Mediterranean (van der Weijden et al. 1990), Orca Basin in Gulf of Mexico (van Cappellen et al. 1998) and Cariaco Basin in Venezuela (Scranton et al. 2006). On the Mediterranean coast, there are several anoxic marine lake systems with meromictic characteristics: Laguna de la

I. Ciglenečki (✉)
Division for Marine and Environmental Research, Rudjer Bošković Institute, Bijenička 54, 10001 Zagreb, Croatia
e-mail: irena@irb.hr

Z. Ljubešić
Faculty of Science, Division of Biology, University of Zagreb, Rooseveltov trg 6, 10001 Zagreb, Croatia
e-mail: zrinka.ljubesic@biol.pmf.hr

I. Janeković
Division for Marine and Environmental Research, Rudjer Bošković Institute, Bijenička 54, 10001 Zagreb, Croatia

School of Civil, Environmental, and Mining Engineering & UWA Oceans Institute, The University of Western Australia, M470, 35 Stirling Highway, Crowley, WA 6009, Australia
e-mail: Ivica.Janekovic@irb.hr

M. Batistić
Institute for Marine and Coastal Research, University of Dubrovnik, Kneza D. Jude 12, 20000 Dubrovnik, Croatia

© Springer International Publishing AG 2017

R.D. Gulati et al. (eds.), *Ecology of Meromictic Lakes*, Ecological Studies 228,
DOI 10.1007/978-3-319-49143-1_6

Cruz, Lake Ciso and Lake El Tobar are the karstic environments in Spain (Miracle et al. 1992); Aitoliko coastal lake in western Greece (Dassenakis et al. 1994; Kehayias et al. 2013) and Salty lake in Crete, Greece (Mihopoulos et al. 2000); Lake Faro in Italy (Zagami and Brugnano 2013); deep-sea hypersaline anoxic lakes such as Bannock, Discovery, L'Atalante and Urania in the Mediterranean Ridge in the Eastern Mediterranean Sea (Wielen et al. 2005; Daffonchio et al. 2006; Yakimov et al. 2007; Henneke and De Lange 1990). Rogoznica Lake is one of such unique example on the Adriatic Coast, Croatia.

Karst features are widespread in Croatia: karst cover about 50 % of the country's total area. The most important part of the Croatian karstic area lies in the coastal Adriatic region. Croatia's coast consists of Cretaceous limestone in which Pleistocene/Holocene cave systems have developed. Croatia has more than 6000 caves; rich in different formations (stalactites, stalagmites and other speleothems). Collapsing and flooding of these caves have created several seasonally stratified, saline water systems such as marine lakes and anchialine ponds (from Greek *anchialos*, land-locked body of water with a subterranean connection to the ocean, Stock et al. 1986), which are main peculiarity of the karst environment. Rogoznica Lake (local name "Dragon Eye") is one of the most extensively studied of these water bodies on the Adriatic and the Mediterranean coast (Mihelčić et al. 1996; Ciglenečki et al. 2005, 2015; Bura-Nakić et al. 2009; Helz et al. 2011; Žic et al. 2013). The lake is situated in the middle of the eastern Adriatic Coast, close to the village of Rogoznica, on a transect between the cities of Šibenik and Split (43°32′ N 15°58′ E; see map in Fig. 6.1a). The lake is a karstic depression having a circular shape. It is filled with the seawater and covers an area of 10.276 m^2. The lake has maximum length of 143 m, with maximum depth of 15 m and volume derived from the bathymetric data of ca. 85,700 m^3 (Mihelčić et al. 1996; Stipaničev and Branica 2006). The lake is sheltered from the wind by 4–23 m high cliffs that prevent wind shear mixing. Since the lake has no obvious surface connection to the sea, or any known extensive subsurface connection, except visible tides on the rocks, the water exchange between the lake and the open sea is delayed through the karstified carbonate rock. There is a phase shift between the tides in the lake and tides outside the bay for about 2 h.

Rogoznica Lake is thermally and chemically stratified into oxic and anoxic water layers throughout the year. Due to high phytoplankton activity, the upper part of the water column is well oxygenated, while decomposition processes cause enhanced deposition of organic matter and nutrients to deeper waters, leading to hypoxia/anoxia and microbial production of H_2S (Kamyshny et al. 2011; Pjevac et al. 2015; Čanković et al. 2015). The depth of the mixolimnion varies seasonally, influenced greatly by meteorological conditions (temperature, wind, rainfall), which also influence exchange between the water layers.

In its physico-chemical characteristic, the Rogoznica Lake can be considered as a unique system with both meromictic and mixed water conditions which alternate seasonally. Depending on the meteorological conditions, the lake fluctuates between meromictic and holomictic conditions with mainly anoxic water.

A

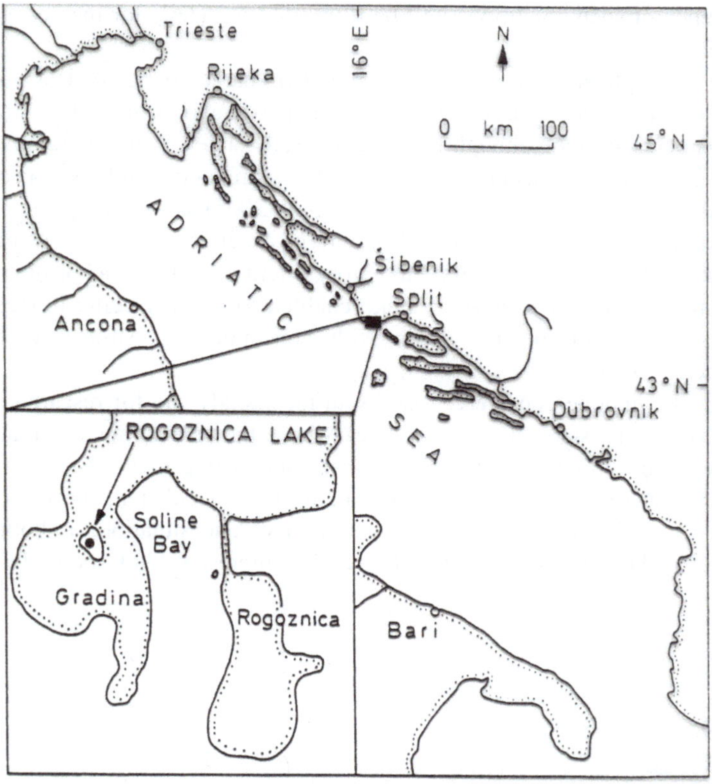

B

Fig. 6.1 (**a**) Map and (**b**) photo of Rogoznica Lake, Croatia (north-west side of the lake)

Prevalence of such conditions will be illustrated in details for the period between 1994 and 2012.

The main physico-chemical and ecological characteristics of Rogoznica Lake are relatively stable with seasonally constant oxic/anoxic interface, stratification that has persisted long enough to produce steady-state conditions relative to alkalinity and sulphide concentrations, reduced phytoplankton and zooplankton diversity (Kršinić et al. 2000, 2013; Ciglenečki et al. 2005) as well as specific microbial community (Pjevac et al. 2015; Čanković et al. 2015). All these characteristics make this lake as unique natural laboratory. Field measurements of biological and physical attributes in Rogoznica Lake enabled to study specific biogeochemical processes that control meromixis with appearance of strong sulphidic and holomictic conditions.

The lake may be considered also as an ideal study site for research on marine redox chemistry since it provides sites where the redoxcline is situated in the euxinic water column (anoxic waters that contain free hydrogen sulphide in the water column) and in the sediment (Fig. 6.2b) (Ciglenečki et al. 2006). Comparison between those two different redox sites is a unique situation, as both sites are situated in the same water body. These sites represent "hot spots" for studying

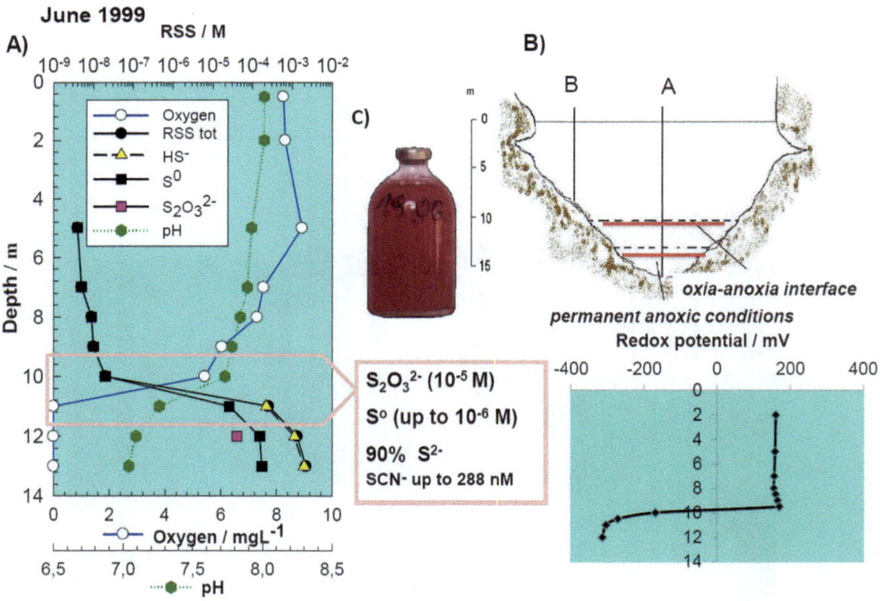

Fig. 6.2 Rogoznica Lake. (**a**) Vertical distribution of main physico-chemical parameters: water temperature, salinity, dissolved oxygen and reduced sulphur species (RSS), including sulphide, elemental sulphur and thiosulphate in the stratified water column of Rogoznica Lake. (**b**) Vertical profile of the water column in Rogoznica Lake with marked sampling locations at permanent anoxic (site A, 14 m depth) and oxic (site B, 8 m depth) environmental conditions, with indicated distribution of the water column redox potential at site A. (**c**) Photo of *pink*-coloured RL sample taken at the chemocline layer

permanently oxic and anoxic conditions in the lake that exist during the periods of stratification and meromixis in the water column or sediment, or both.

In general, Rogoznica Lake is a typical example of anoxic and euxinic water system (Helz et al. 2011). It differs from other well-known anoxic basins such as the Black Sea (Izdar and Murray 1991), Framvaren Fjord (Norway) (Lending and Westerlund 1988) and Lac Pavin (France) (Michard et al. 1994) in its biological, geochemical, geomorphological and hydrological characteristics, including small size and very easy access to anoxic conditions, which usually are almost permanent below 10 m depth (Fig. 6.2).

6.2 Basic Geochemical and Morphological Characteristics of Rogoznica Lake

Rogoznica Lake is located on a narrow limestone peninsula, surrounded on three sides by the Adriatic Sea a few hundred meters away (Fig. 6.1a). The peninsula is tree covered and sparsely populated for the most part, although a hotel and marina were built near the lake in mid-1990. A village of about 1000 permanent residents lies on an adjacent peninsula about 0.5 km from the lake. The local economy is tourism based, but significant industrial activity can be found at about 40–50 km away, around the cities of Šibenik and Split.

The sampling site A in the lake has a depth of 14 m (Fig. 6.2b). It is characterized by permanent anoxic conditions that exist in the overlying water column enriched with relatively high concentrations of reduced sulphur species (compounds) (RSS, range between 8 and 10 mM, mainly in the form of sulphide), dissolved organic matter (DOC, range between 1 and 6 mg L^{-1}) and nutrients (range between 0.23 and 315 μM for NH_4^+, 0.11 and 53 μM for PO_4^{3-} and 1.13 and 680 μM for SiO_4^{4-}) (Ciglenečki et al. 2005, 2015) including iodine species (Stipaničev and Branica 1996, 2006). The sampling site B always lies within the oxic water column, and the redox transition zone lies below the sediment–water interface (Ciglenečki et al. 2006). Changes in redox conditions of the water column and sediments with depth affect the distribution and speciation of major redox-sensitive elements (Fe, Mn, Mo, V) and reduced sulphur species. In the anoxic sediment conditions, enrichment of trace metals occurs, especially those that accumulate sulphides (Fe, Mo) (Ciglenečki et al. 2006; Helz et al. 2011).

Redox potential in the water column (Fig. 6.2b) and surface sediment at site A reaches negative values, up to −420 mV, while in surface sediments at site B, it is around 100 mV; pH at both sites is around 7.00. Sediment at both sites consists of silt and clayey silt. The latter prevails at site A, where the specific surface area with finer particles is much higher (A = 7.97–12.50 $m^2\ g^{-1}$) than at B (=2.57–3.85 $m^2\ g^{-1}$) (Mihelčić et al. 1996; Ciglenečki et al. 2006). Deposition of fine-grained sediments with higher specific surface areas caused the accumulation of both organic matter and trace metals, of which the sedimentary accumulation is about 50 % higher at site

Table 6.1 Concentrations of Fe, Mn and Mo in porewater and sediments of Rogoznica Lake at sites A and B

	A						B					
	Porewater (μg L^{-1})			Sediment (mg kg^{-1})			Porewater (mg kg^{-1})			Sediment (mg kg^{-1})		
Depth cm^{-1}	Fe	Mn	Mo	Fe	Mn	Mo	Fe	Mn	Mo	Fe	Mn	Mo
Bottom Water/sediment	16.8	89	2.4				1740	9	5.8			
0–5	204	278	2	5700	97	38.4	1890	597	2015	3900	59	5.71
5–10	5	142	3	5400	103	36.4	2280	162	3950	2800	26	8.05
10–15	5	149	5	4800	81	34	2295	30	5635	3100	23	17.2
15–20	5	153	5	4800	92	41.3	–	–	–	4000	121	15.9
20–25	5	147	47	4100	91	81.4	–	–	–	–	–	–

Adopted from Ciglenečki et al. (2006)

A than at site B (Table 6.1). Carbonates that are biologically generated in the lake are the most important component of the lake's sediment. Major sediment mineral is calcite (65–92 %), which is the main regulator of water composition and behaviour in karstic environment, followed by aragonite, quartz, dolomite and pyrite (Ciglenečki et al. 2006). The sediment accumulation rate, calculated by using the 137 Cs method, is relatively high (0.45 cm year^{-1}) at the sampling site A, the deepest and permanently anoxic point (Fig. 6.2b). This implies that the sediment core of 25 cm length, taken at site A (Table 6.1), must have required 55 years to deposit. The sediment is laminated, indicating no bioturbation or any mixing caused by physical forces.

Despite its relative isolation from the surroundings, Rogoznica Lake has experienced small but detectable anthropogenic influences. Increase of Cu, Zn and Pb fluxes to the lake's sediments began in the 1950s (Mihelčić et al. 1996), possibly due to increase in the deposition of airborne contaminants, particularly of Pb. The enrichment in the sediment of Mo (up to 81 mg kg^{-1}) can be used as an indicator of paleo-anoxic and sulphidic conditions (Helz et al. 1996; Erickson and Helz 2000; Zheng et al. 2000). Thus, the lake can be characterized as a typical anoxic environment which is geochemically similar to saline, meromictic and sulphidic systems of the Black Sea and Framvaren Fjord (Pilipchuk and Volkov 1974; Lending and Westerlund 1988; Murray 1991) and freshwater, iron-rich crater lake Pavin in Central Massif in France (Michard et al. 1994; Viollier et al. 1995; Bura-Nakić et al. 2009).

Rogoznica Lake sediments at the sites A and B (Fig. 6.2b) show increased concentration of Mo (up to 81.4 mg kg^{-1} in the layer at 20–25 cm in A and up to 20 mg kg^{-1} in the layer at 15–20 cm in B) (Table 6.1). This possibly indicates (1) that historically the sulphide interface in the lake was higher, and site B was also exposed to the sulphidic, anoxic water in the past and (2) that anoxic surface sediment at site A and overlying bottom waters can stay anoxic for >>50 years, according to the evaluated sediment accumulation rate. This is the main attribute to meromictic character of the lake.

6.3 Physico-Chemical Parameters of the Water Column and Seasonal Variation of Oxic and Anoxic Conditions

6.3.1 Stratified and Meromictic Conditions

Seasonal and vertical variations in salinity (expressed as psu—*practical salinity units*) and temperature indicate the existence of two water layers system: (a) surface layer (0–5 m), characterized by variable and lower salinity (14–30 psu) and temperature (6–33 °C), and (b) bottom layer (7–12 m), characterized with higher salinity (33–38 psu) and temperature (23–31 °C) (Figs. 6.3 and 6.4) (Ciglenečki et al. 2013, 2015). Typical spring–summer water column stratification with clear

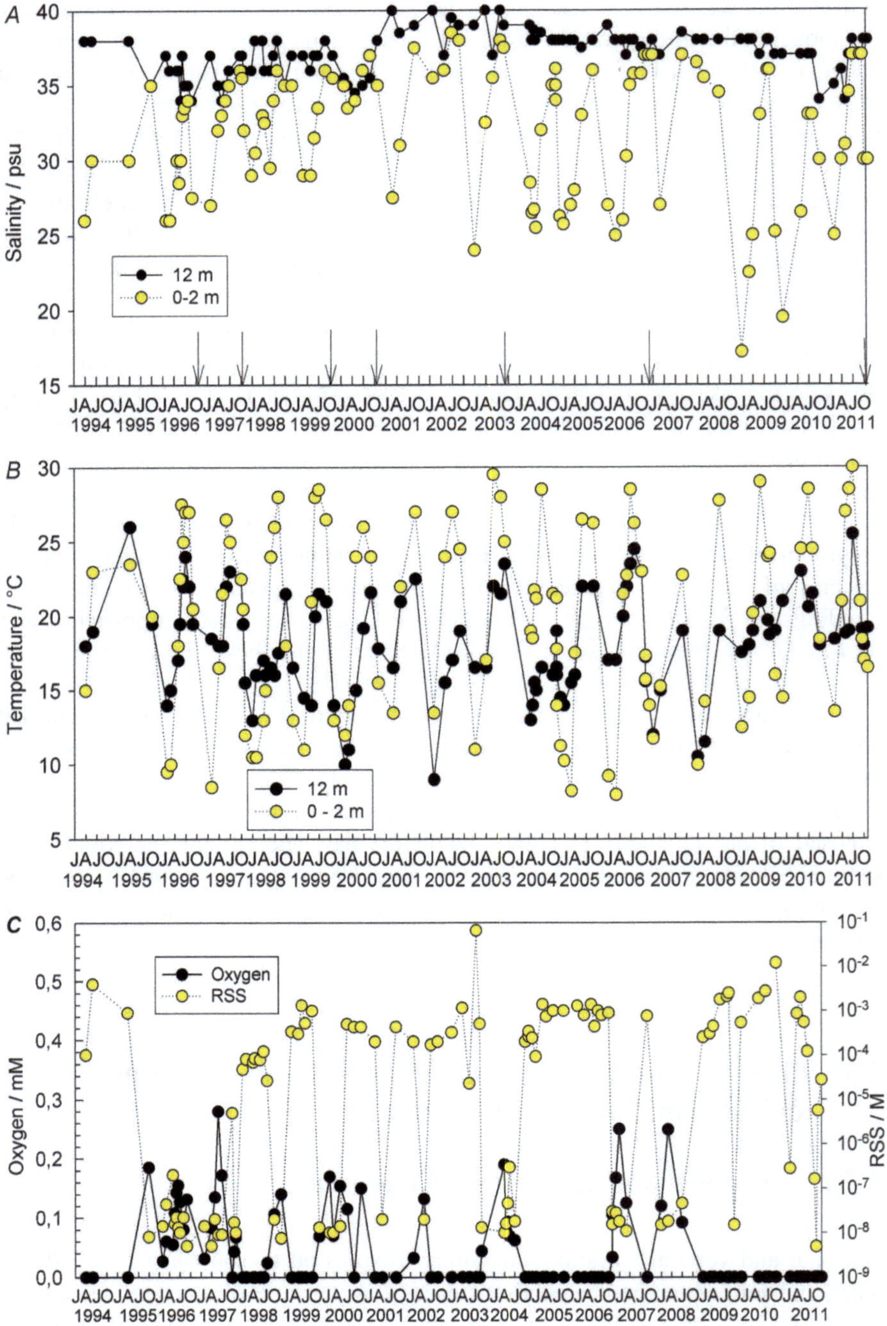

Fig. 6.3 Seasonal variations of (**a**) salinity, (**b**) water temperature and (**c**) dissolved oxygen and total reduced sulphur species (RSS) in the *surface* (0–2 m) and *bottom* (12 m) water column of Rogoznica Lake from 1994 to 2011. For better visibility, RSS and oxygen data are presented only for bottom water at 12 m depth. Recorded mixing between water layers is indicated by *arrows*. On all these mixing occasions, except September 1997 and October 2011, hypoxic conditions

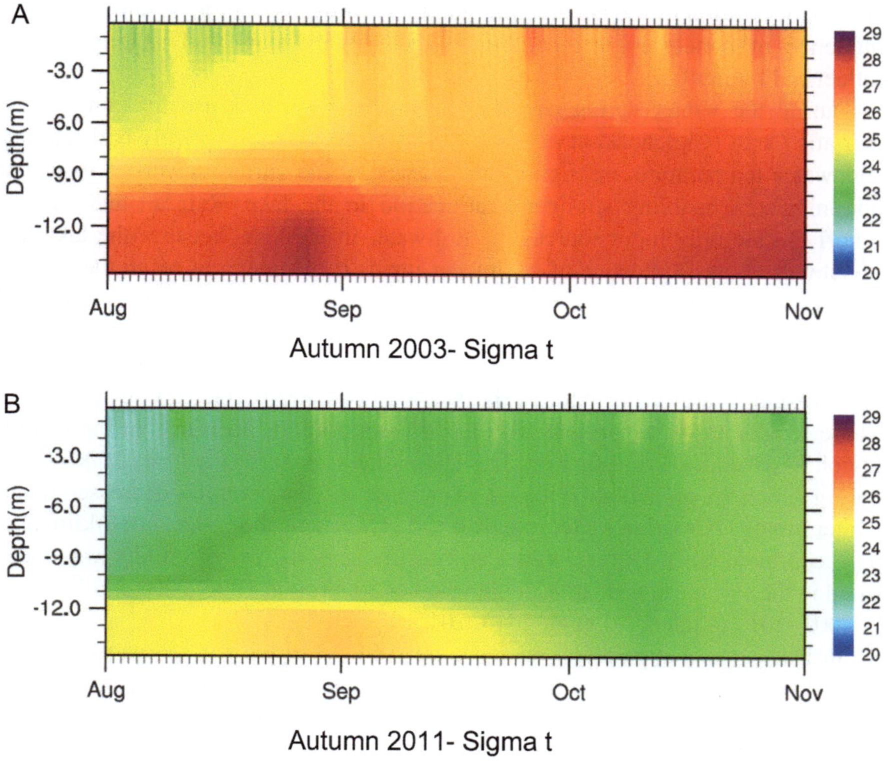

Fig. 6.4 Modelled density (Sigma-t) field in the water column of Rogoznica Lake during the autumn 2003 (**a**) when holomixis with hypoxic conditions occurred, and autumn 2011 (**b**) when holomixis with anoxic conditions were recorded in the water column of Rogoznica Lake (adopted from Ciglenečki et al. 2015)

chemocline situated around 10 m depth is a notable feature of the lake (Fig. 6.2a). At the boundary of oxic–anoxic conditions during spring and summer months, usually a pink-coloured and ca/50 cm thick chemocline layer, enriched with dense population of anoxygenic phototrophs, develops (Pjevac et al. 2015). The chemocline depth (as well as its thickness) usually fluctuates diurnally by about 1 m and thus influences the depth of maximum light intensity and the oxidation and reduction processes that control the concentrations of oxygen and sulphide. Due to intensive biochemical activities in the chemocline, higher concentration of thiosulphate (up to 80 uM) and elemental sulphur (up to 150 uM) are found (Ciglenečki et al. 1996, 1998; Bura-Nakić et al. 2009; Kamyshny et al. 2011). Stratification of the Rogoznica Lake water column supports meromictic conditions

Fig. 6.3 (continued) (1–4 mg L^{-1} of oxygen) were recorded. In 1997 and 2011, anoxia persisted for more than 1 week in the whole water column of the lake and depth position of the chemocline. In other words, redoxcline is under a strong influence of meteorological conditions (both temperature and salinity greatly influenced by rainfall).

During the winter–spring period (December–May), it is usually rainy at the Adriatic Coast (Ciglenečki et al. 2005; Žic et al. 2013); consequently both salinity and water temperature are significantly lower in the surface layers than in the bottom water, enabling stronger stratification in the lake. Rainfall and surface runoff are the only input sources of freshwater in the lake, because the lake has no lateral inflows. Influence of freshwater inputs can sometimes reach the bottom layers, as in the autumns of 1996, 2009, 2010 and in the spring–summer period of 2011 (Fig. 6.3a). Generally, salinity decreases in the upper 0–2 m layer. The influence of fresh water input after 2000 was reduced in the bottom layers, indicating stronger stratification than in the years 1994–2000 (Ciglenečki et al. 2015). These results are in agreement with the average rainfall measured in the nearby region of Šibenik—Primošten (middle Dalmatian Coast) for the years 2000–2006, during which the precipitation tended to increase and temperatures decreased in the winter months (Croatian Meteorological and Hydrological Service, http://klima.hr/ocjene_arhiva.html). During winter and spring seasons, lower salinity was measured in the surface layers, an average value of 30.5 psu in 1996–2000 and 28.4 psu in 2001–2012 (Ciglenečki et al. 2013, 2015).

Surface water temperatures increase by 0.64 °C in the surface waters (depth 0–5 m) and 1.1 °C in the bottom water (depth 12 m) during summers in 2001–2012, compared with the summer periods of 1996–2000 (Ciglenečki et al. 2015). Linear regression of temperature data measured in years 1996–2012 clearly reveals increasing surface temperatures in the upper 5 m. Positive linear trend of 1.72 °C/10 years was obtained with rather high correlation coefficient ($n = 120$, $R^2 = 0.82$, Fig. 6.5).

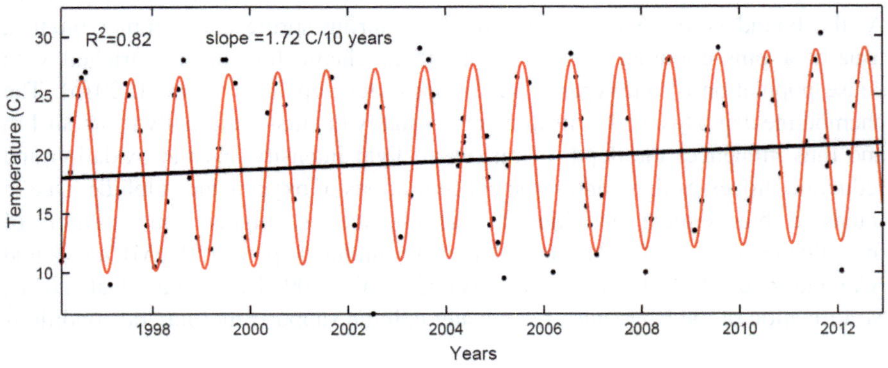

Fig. 6.5 Linear regression and seasonal cycle (evident from the figure) for Rogoznica Lake surface temperature (0–5 m), measured during 1996–2012 period. There is a linear trend of increasing surface T (slope of the regression line) of 1.72 °C/10 years (adopted from Ciglenečki et al. 2015)

This increase of the average summer temperature after 2001 follows the measured higher air temperatures during the dry summer months, especially in the 2003 in Croatia (Croatian Meteorological and Hydrological Service, http://klima.hr/ocjene_arhiva.html) and in lakes around the world, as well as in both coastal and open sea areas (IPCC 2007; Adrian et al. 2009; Shimoda et al. 2011). Thus, Rogoznica Lake may potentially serve as a valuable sentinel of the climate change in the area.

Thermohaline features of Rogoznica Lake fit quite well with the water column redox and euxinic conditions (Fig. 6.3) (Ciglenečki et al. 2015). Concentrations of dissolved oxygen (O_2), which ranged from 0 to 10 mg L^{-1}, are closely related with that of the total concentration of reduced sulphur species. The highest values of O_2 were detected invariably in the mixolimnion up to 8 m depth, where it was related with higher rates of primary production in this part. Corresponding to O_2 concentrations, concentration of reduced sulphur species ranged between 10^{-9} and 10^{-1} M, with always higher concentration in the bottom. In the years when O_2 concentrations were low in the bottom layers, around 0 (1994, 1998, 1999, 2001, 2003, 2005–2006, 2009–2011), the concentrations of reduced sulphur were extremely high but variable, 10^{-4}–10^{-1} M. The concentration of reduced sulphur in the surface layers during stratification is around 10^{-8} M: it mainly corresponds to non-volatile sulphur species such as elementary sulphur and organic sulphur compounds (Fig. 6.2a) (Bura-Nakić et al. 2009; Ciglenečki et al. 2015). Reduced sulphur species in the bottom water occurs mainly as free sulphide.

In addition to the sulphide, stratified anoxic bottom water is extremely rich in several nutrients (Table 6.2) ($NH4^+$, up to 315 μM; PO_4^{3-}, up to 53 μM; and SiO_4^{4-}, up to 680 μM) (Kršinić et al. 2000; Ciglenečki et al. 2005, 2015; Žic et al. 2010, 2013) and dissolved organic matter or DOC (up to 6 mg L^{-1}) (Ćosović et al. 2000; Ciglenečki et al. 2005). These values indicate strong decomposition of organic matter in the mixolimnion during periods of high primary production, i.e. in spring and early summer.

Table 6.2 Maximum (max), minimum (min) and average (avg) values of phytoplankton (during 1998–2013) and nutrients (during 2000–2008) in Rogoznica Lake

Parameter	Avg	Max	Min	*N*
Total microphytoplankton (cells L^{-1})	461,400	8,138,480	0	382
Diatoms (cells L^{-1})	451,360	7,926,200	0	382
Dinoflagellates (cells L^{-1})	10,040	840,910	0	382
NO_3 (μM)	2.03	44.21	0	476
NO_2 (μM)	0.26	3.7	0	478
NH_4 (μM)	14.84	315	0.23	457
Total inorganic nitrogen (μM)	17.21	315.83	0.31	440
PO_4 (μM)	1.98	33.1	0.01	496
SiO_4 (μM)	5.5	679	0.34	497

N number of samples
Adopted from Malešević et al. (2015)

Depending on the atmospheric conditions (air temperature, wind, heat fluxes, precipitation), the thermocline accompanied by halocline weakens, starting in the summer, and the chemocline descends to the deeper water. Later, usually in autumn, thermohaline stratification together with redoxcline is completely destroyed as indicated by inverse peaks of salinity and temperature, as well as holomictic conditions (Figs. 6.3 and 6.4). Holomixis that enables mixing between water layers in the lake was recorded in the autumn periods of 2003 and 2011 (Žic et al. 2013; Ciglenečki et al. 2015), also denoted by arrows in Fig. 6.3a.

6.3.2 Holomictic Conditions

Mixing between water layers primarily depends on the meteorological conditions: they highly influence vertical dynamics and duration of mixing process within the water column similar to many other saline lakes (Romero and Melack 1996). Due to mixing, relatively high concentrations of sulphide are transported from the bottom to the surface where it is oxidized to sulphide, polysulphide and colloidal sulphur (Bura-Nakić et al. 2009, 2012; Ciglenečki et al. 2015). The oxygen consumption can result in occurrence of anoxic holomictic conditions in the whole water column (Fig. 6.6). Rapid precipitation of colloidal sulphur influences the colour of the surface water. In stratified conditions the water transparency is relatively high, while during holomixis is non-transparent and milky yellowish (Fig. 6.6). If the vertical dynamics is slow, diffusion of well-oxygenated surface water towards the bottom, and of the sulphide-rich bottom water to the surface, can continue for several days, as was recorded in September 2003 and November 2006 (Bura-Nakić et al. 2009; Ciglenečki et al. 2015). While the surface layer remains oxygenated the whole time, bottom layers usually remain hypoxic or anoxic. Under such conditions, upward transport of ammonium facilitates nitrification processes as was evident from higher concentrations of nitrate and nitrite in the surface waters (Žic et al. 2013; Ciglenečki et al. 2015). On the other hand, fast intrusion of the nutrient-rich bottom waters to the surface decreases the N/P ratio because of P release from the bottom waters. In contrast, there is P limitation in the surface water during spring and summer stratification (Ciglenečki et al. 2015). During and after the appearance of anoxic holomixis, the whole water column appears as N limited. This highly influences biology of the lake.

The anoxic and sulphidic holomictic events were studied in September 1997 (Bura-Nakić et al. 2009) and in October 2011 (Ciglenečki et al. 2015). On both the occasions, high sulphide and ammonium concentrations were detected in the entire water column. This led to a mass mortality of all phytoplankton and benthic organisms (Kršinić et al. 2000; Barić et al. 2003; Ciglenečki et al. 2005). The mass mortality contributed to increased nutrient concentrations in the whole lake, including the surface layers (Fig. 6.7). Nutrient concentrations that ranged up to: $NH4^+$, 100 μM; NO_3^-, 25 uM; PO_4^{3-}, 5 μM; and SiO_4^{4-}, 60 μM, remained high for several months after the anoxia conditions due to decomposition of large amount of particulate organic carbon (POC) and low primary production rates.

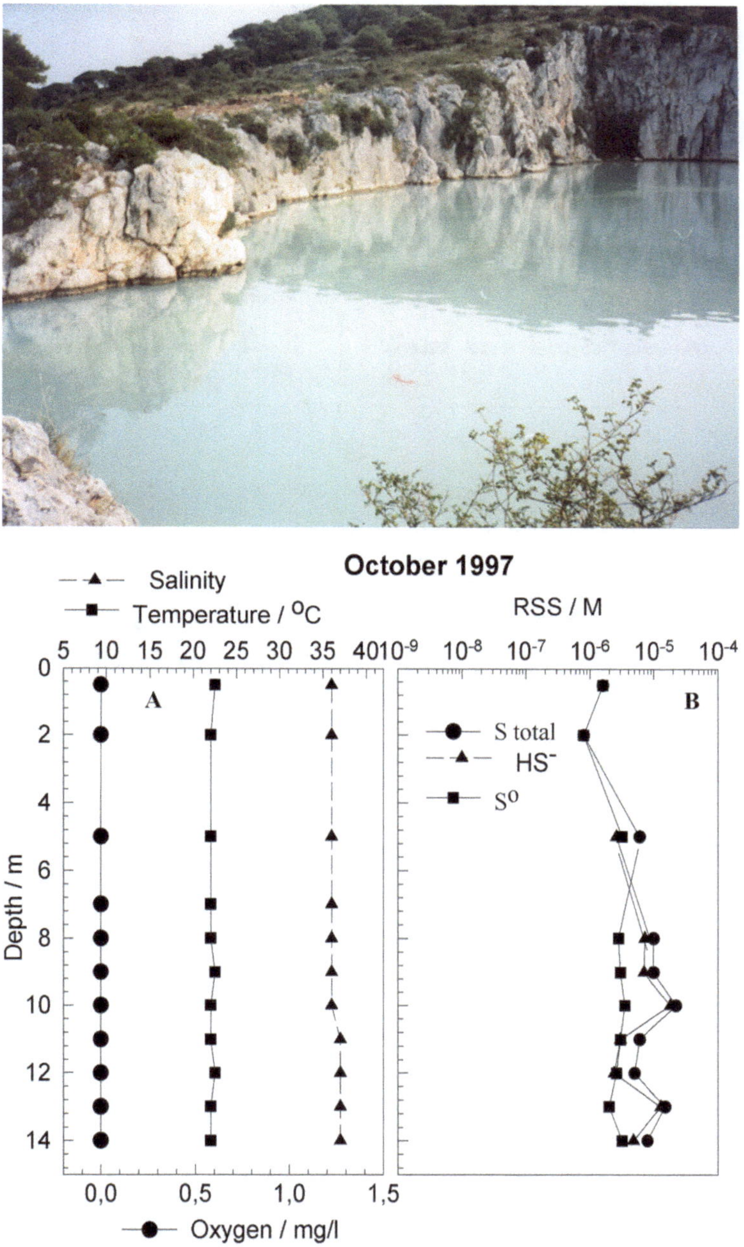

Fig. 6.6 Photo of Rogoznica Lake, south-west side (*upper panel*); panels below show vertical profiles of the oxygen, temperature, salinity and reduced sulphur species in the water column of the lake during the holomictic–anoxic event in September 1997. (*Panel A*) oxygen (note anoxic conditions throughout the depth), temperature and salinity. (*Panel B*) reduced sulphur species, RSS

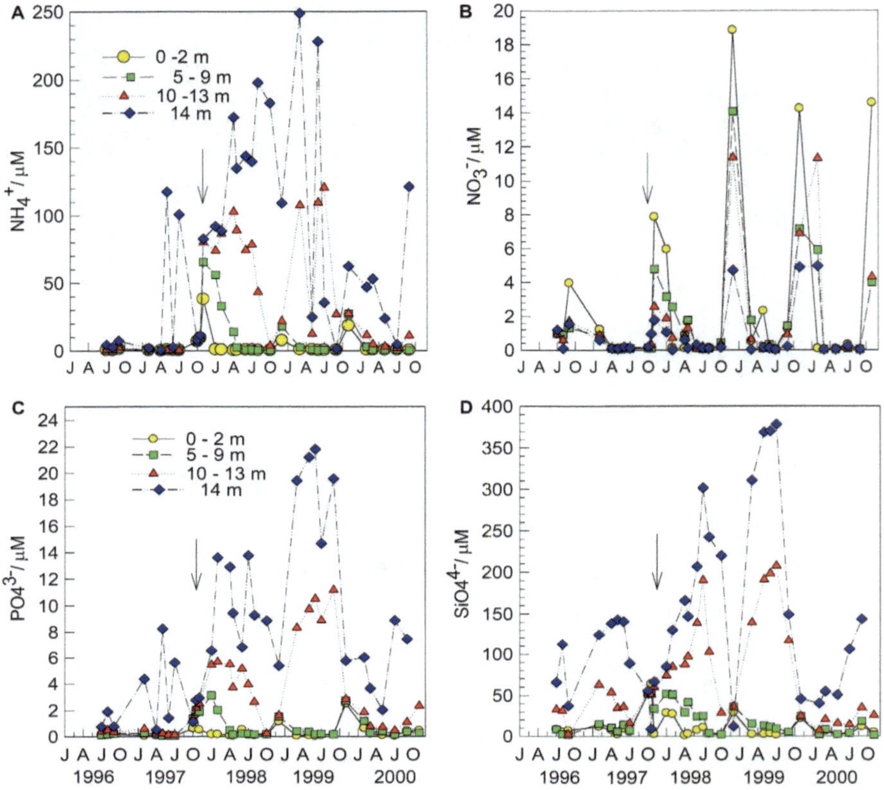

Fig. 6.7 Seasonal variations of (**a**) ammonium, (**b**) nitrate, (**c**) phosphate and (**d**) silicate in the water column of Rogoznica Lake. Turnover of the water layers followed by anoxic holomixis is denoted by *arrows* (adopted from Ciglenečki et al. 2005)

6.4 Mathematical Modelling of Mixing Between Water Layers and Predictions of the Lake's Future

1-D General Ocean Turbulent Model (GOTM; http://www.gotm.net) was used in order to simulate vertical and temporal evolution of physical parameters within the lake (Ciglenečki et al. 2015). Vertical dynamics between water layers and vertical turbulence are modelled using the well-established two-equation k-epsilon turbulence model. In the model, atmospheric forcing with temporal resolution of 6 h, obtained from the closest meteorological station during the study period, was applied. To obtain realistic profiles of temperature and salinity, the model was nudged (with relaxation time of one day) towards the measured temperature and salinity data by multimetric probe. Nudging, or Newtonian relaxation, is a simple form of data assimilation that adjusts dynamical variables of free-running models using measured data to provide a realistic dynamics of the system at a given time. Vertical model grid discretization of 0.5 m (maximum depth of 15 m) insured appropriate representation of the probe measurements, not requiring for observation to model interpolations, as well as smooth vertical mixing and transport dynamics induced with atmospheric forcing.

GOTM model revealed that vertical stability of the water column in Rogoznica Lake differs both annually as well as seasonally. Furthermore, it confirmed that interaction among water layers is driven by the atmosphere conditions and mostly occurs in autumn and early winter due to atmosphere cooling (Figs. 6.3, 6.4 and 6.6). Vertical stability for two different regimes of mixing processes (slow and fast vertical dynamics between water layers) was estimated with squared Brunt–Vaisala frequency (N2). The modelled values of the N2 were larger in the late summer–autumn season (August–September–October) of 2003 when holomictic and hypoxic conditions prevailed, than during the same period of 2011 when anoxic and holomictic conditions were recorded in the entire water column for more than a week (Fig. 6.8a, b). Consequently, in September 2003 slower diffusion and interaction among the water layers lead to development of moderate holomixis with hypoxic conditions, while in the beginning of October 2011 fast vertical dynamics between water layers lead to holomictic, anoxic and sulphidic conditions in the lake. Similar anoxic holomictic events occurred also in the late summer–autumn period of 1997.

The model also showed that if the typical reduction of the water column stability is induced with extreme weather events, such as an abrupt change in the air temperature, with a temperature drop between 10 and 15 °C in <24 h, along with strong wind named Bora wind (strong, dry and cold easterly wind, blowing predominantly along the east coast of the Adriatic Sea) (Fig. 6.8c), then mixing between water layers could be governed by fast turnover. This could result in a rapid oxidation of the sulphide, which depletes the oxygen and results in holomictic–anoxic conditions in the entire water column. Such conditions usually end with mass mortality of all organisms and community shift in the lake. These events nowadays are discussed in relation to climate change impacts (Ciglenečki et al. 2015). It is interesting that the anoxic event with the same consequences as reported for Rogoznica Lake were recorded in the late August in 1997 in Mariager Fjord, Denmark (Fallesen et al. 2000). The long period of calm and unusual warm weather that reduced mixing of the water column in the Mariager Fjord was reported as the main factor triggering anoxic conditions there.

6.5 Food Web Interactions, Phytoplankton and Zooplankton Interactions

6.5.1 Phytoplankton and Zooplankton Populations

Extremely fluctuating environmental conditions, such as high seasonal variations in salinity and temperature, water stratification with strong redox gradient and euxinia accompanied by high concentrations and variability of nutrients on one side and

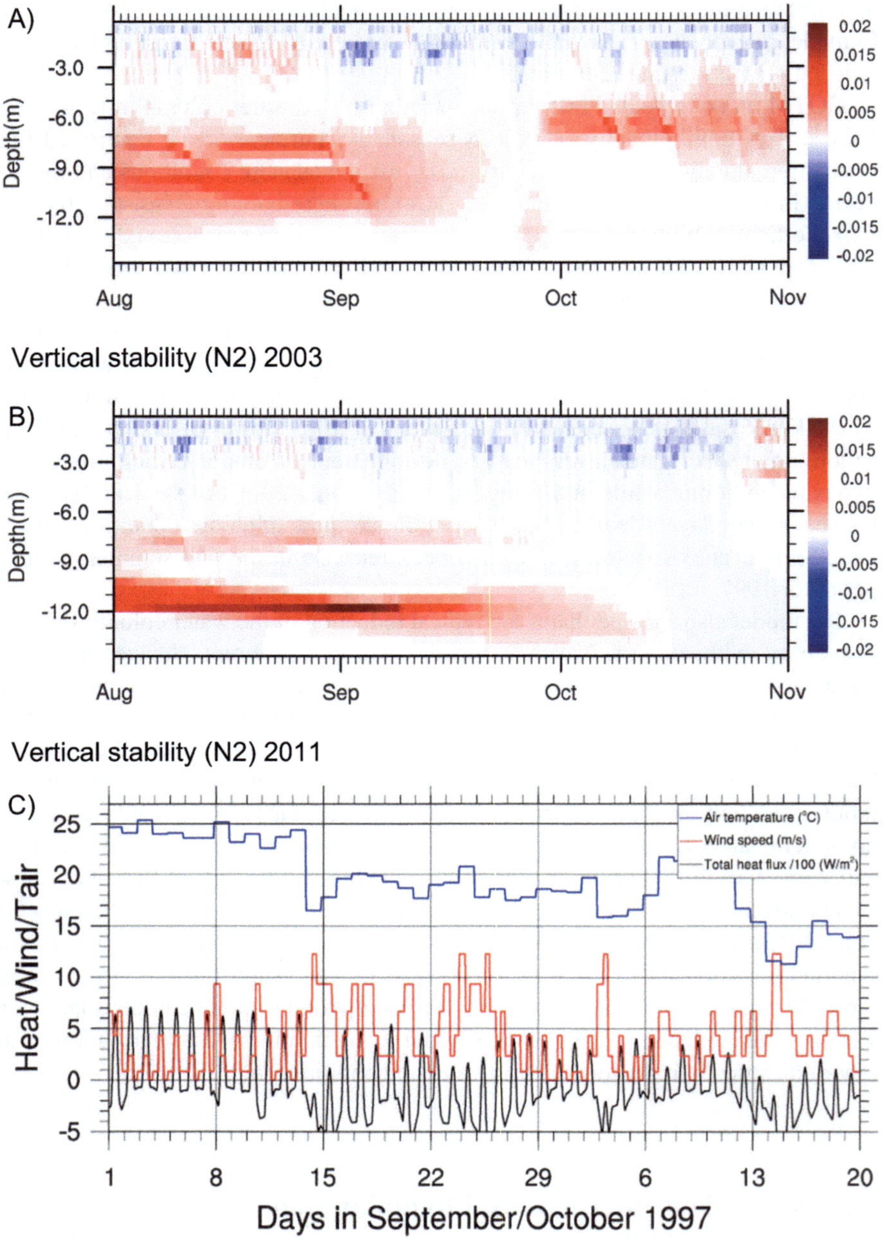

Fig. 6.8 Modelled squared Brunt–Vaisala frequency (N2) of the water column of Rogoznica Lake in the late summer–autumn season of (**a**) 2003 and (**b**) 2011. (**c**) Observed air temperature and wind speed with modelled heat flux in autumn season 1997 (adopted from Ciglenečki et al. 2015)

episodic appearance of holomixis with hypoxia and anoxia conditions on the other side, greatly influence the community structure of phytoplankton and zooplankton in the lake. High concentrations of organic detritus also create a turbid environment below 5 m depth, as indicated by Secchi disc depth (usually 5–6 m). Influenced by these physico-chemical constraints, the few species capable of surviving in Rogoznica Lake must be distributed through a narrow zone of maximum depth of 9–10 m (Svensen et al. 2008). Diatoms (*Chaetoceros curvisetus*, *Thalassionema nitzschioides*, *Dactyliosolen fragilissimus*), which are present in a relatively smaller number of species (0–17 diatom species per sample from a total of 51 phytoplankton taxa recorded during the study) and relative high single species densities (up to 10^8 cells L^{-1}), higher than those in the neighbouring sea (Svensen et al. 2008; Burić et al. 2009; Kršinić et al. 2013; Malešević et al. 2015), are the best adapted to these extreme and highly eutrophic and fluctuating conditions (Tables 6.2 and 6.3). In relation to the physico-chemical parameters within the water column, before and after the year 2000, diatoms became dominant in the phytoplankton assemblage. Before 2001 rare microflagellate *Prorocentrum arcuatum* and the heterotrophic ebridian microflagellate *Hermesinum adriaticum* together with diatoms *Chaetoceros curvisetus* and *Eunotia* sp. dominated the phytoplankton assemblage in the lake (Ciglenečki et al. 2005; Burić et al. 2009); after 2001 diatoms contributed up to 97 % on average (Kršinić et al. 2013; Ciglenečki et al. 2015; Malešević et al. 2015).

Dominant diatoms encountered after 2000 are *Thalassionema nitzschioides* (up to 2.91×10^7 cells L^{-1}, frequency of appearances 29 %), *Cyclotella choctawhatcheeana* (up to 5.61×10^7 cells L^{-1}, frequency of appearances 29 %), *Dactyliosolen fragilissimus* (up to 3.24×10^7 cells L^{-1}, frequency of appearances 30 %) and *Chaetoceros curvisetus* (up to 2.95×10^7 cells L^{-1}, frequency of appearances 51 %). Frequency of appearances is defined as percentage of appearances of some phytoplankton species in analysed samples. Domination of diatoms in the lake varies seasonally (Table 6.3, Fig. 6.9) in view of their adaptation to the environmental conditions.

Main environmental forces recognized to influence the lake's phytoplankton community structure are as follows:

(1) Seasonal variation of thermohaline conditions that strongly control both meromixis and holomixis in the lake
(2) Seasonal fluctuation of nutrients supply and the nutrient ratios within the water column
(3) Seasonal fluctuations of sulphidic conditions
(4) Anoxic stress conditions which develop because of holomixis and oxygen depletion
(5) Grazing pressure of zooplankton (Ciglenečki et al. 2015)

Fast intrusion of nutrient-rich bottom waters to the lake surface, together with decomposition of dead organisms during and after holomictic–anoxic events, drastically change the N/P/Si ratio in the lake water (Ciglenečki et al. 2015). During spring and summer stratification, the surface water layer is potentially P limited

Table 6.3 List of diatoms taxa found in Rogoznica Lake phytoplankton in the 1998–2013 study period

Taxa	Fr (%)	Max
Cerataulina pelagica (Cleve) Hendey+	2	137,960
Chaetoceros affinis Lauder	4	13,300
Chaetoceros compressus Lauder	4	112,2000
Chaetoceros curvisetus Cleve	51	1,079,090
Chaetoceros danicus Cleve	6	112,530
Chaetoceros decipiens Cleve	2	28,120
Chaetoceros diversus Cleve	2	6400
Chaetoceros lauderi Ralfs	2	1600
Chaetoceros perpusillus Cleve	2	560
Chaetoceros rostratus Lauder	2	800
Chaetoceros spp.	11	1,165,510
Cyclotella choctawhatcheeana Prasad	30	7,926,200
Cylindrotheca closterium (Ehrenberg) Reimann & J.C.Lewin	1	760
Dactyliosolen fragilissimus (Bergon) Hasle	30	2,582,210
Eunotia sp.	3	170,808
Guinardia flaccida (Castracane) H.Peragallo	1	272,790
Guinardia striata (Stolter.) Hasle	6	109,730
Hemiaulus hauckii Grunow ex Van Heurck	1	1520
Hemiaulus sinensis Greville	1	380
Leptocylindrus danicus Cleve	6	313,480
Leptocylindrus minimus Gran	2	7090
Licmophora sp.	3	1140
Microtabella interrupta (Ehrenberg) Round	2	760
Nitzschia longissima (Brébisson) Ralfs	10	63,080
Pleurosigma sp.	6	380
Proboscia alata (Brightwell) Sundström	1	181,060
Pseudo-nitzschia delicatissima (Cleve) Heiden	1	2,974,060
Pseudo-nitzschia spp.	6	622,945
Rhizosolenia calcar avis Schultze	1	1890
Rhizosolenia delicatula Cleve	1	6400
Rhizosolenia fragilissima Bergon	1	4000
Rhizosolenia imbricata Brightwell	1	1140
Striatella unipunctata (Lyngbye) C.Agardh	3	22,800
Thalassionema nitzschioides (Grunow) Mereschkowsky	30	1,667,885
Thalassiosira sp.	10	100,475
Unidentified pennate diatoms	25	812,600

Max is for maximum abundance of cells L^{-1}, and Fr is frequency of appearance, i.e. percentage of appearance of some phytoplankton species in analysed samples (382 samples is 100 %)
Adopted from Malešević et al. (2015)

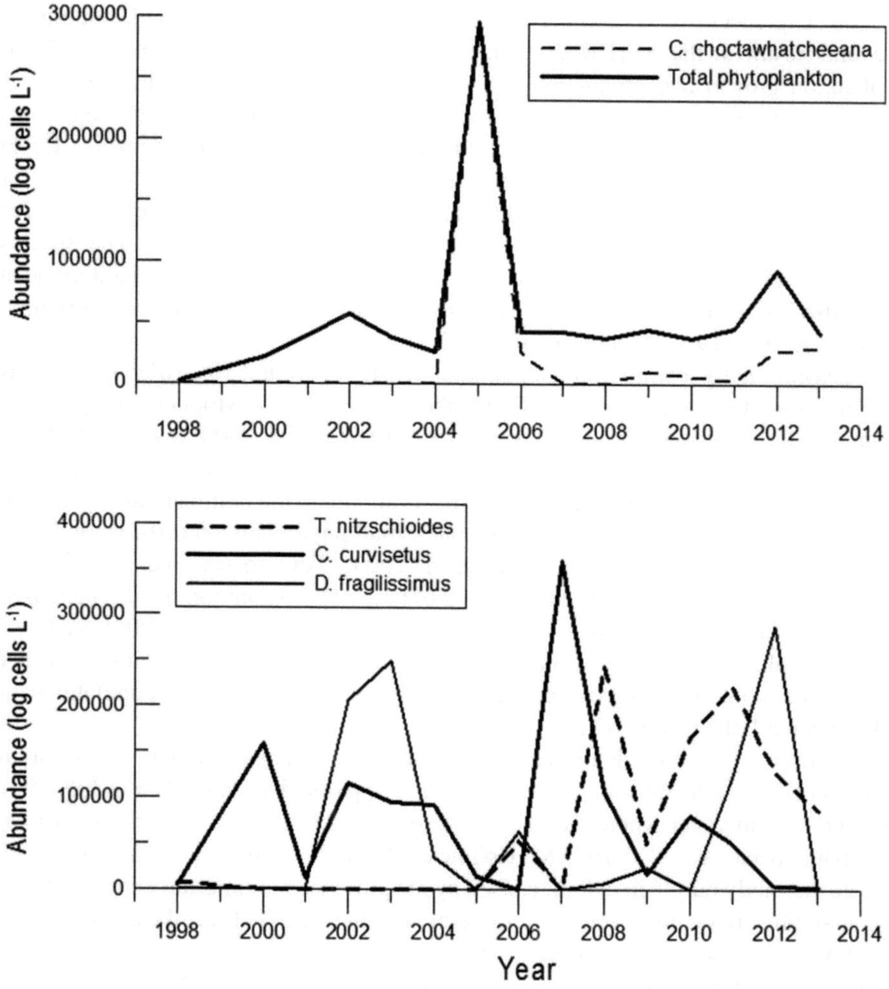

Fig. 6.9 Spatial and temporal distribution (average abundances per year) of microphytoplankton and its dominant diatom species during the period of 1998–2013 in Rogoznica Lake (modified from Malešević et al. 2015)

(Chicharo et al. 2006) with an average N/P molar ratio (N is including ammonium and nitrate) of 60 and average PO_4^{3-} concentration of 0.34 μM, while N/P molar ratio of 26 (with average PO_4^{3-} concentration of 2.42 μM) is calculated as the grand average value for the all seasons (Ciglenečki et al. 2015). During and after anoxic holomixis due to transport of bottom water enriched with phosphate and silicate to the surface, the whole water column becomes potentially N limited (Fig. 6.7c, d). The grand average molar ratios measured for Si/N and Si/ P in the period 2000–2012 was 36 and 81 for the whole water column (taking into account all samples within the water column), respectively.

In most cases when sulphide enters in the upper water layers, both phytoplankton abundance and species diversity decrease due to sulphide toxicity (Kršinić et al.

2000; Ciglenečki et al. 2005; Burić et al. 2009). On the other hand, holomictic events result in the nutrient enrichment of the surface layer, stimulating phytoplankton growth and production, while generally thermohaline stratification prevents nutrients to reach the water layers above the thermocline. Such conditions result in reduced phytoplankton diversity in the lake (51 taxa in Rogoznica Lake) compared with open Adriatic waters, where species' number is more than an order of magnitude higher. Recent insights reveal that the phytoplankton of the eastern Adriatic Sea has 888 phytoplankton species (Viličić et al. 2002). The species composition and diversity in other karstic and saline lakes in the Mediterranean are also limited to the adapted species only, which densities are generally very high (Miracle et al. 1992; Moraies et al. 2003).

Exceptionally low microphytoplankton density of around 10^3–10^4 cells L^{-1} was found after the anoxic event in September 1997, when phytoplankton spp. with small cell dimensions (2–40 μm) were found (Kršinić et al. 2000; Barić et al. 2003). Such a phytoplankton succession was governed by nutrient composition and possible concentration after the anoxic event. Ammonium, the dominant nitrogen form in the post-anoxic period (Fig. 6.7), favours small phytoplankton cells, while larger ones are favoured by nitrate (Barić et al. 2003). Four to six weeks after mass mortality in September 1997, the biomass of larger phytoplankton species increased at the cost of nanoplankton and picoplankton biomass (Barić et al. 2003). Reduced phytoplankton diversity together with extreme environmental conditions further governed both the composition and abundance of zooplankton populations in Rogoznica Lake (Table 6.4).

During 1997, oligotrich ciliates were highly abundant in the lake before total anoxia. The ciliate *Strombidium purpureum* was particularly abundant in July 1997 in the bottom water where oxygen saturation was 55 % and concentration of total sulphur species was 9×10^{-9} M (Kršinić et al. 2000). Individuals of *S. purpureum* were distinctly purple in colour, caused by the presence of numerous endosymbiotic non-sulphur-reducing bacteria important for the survival of these ciliates in unfavourable oxygen conditions (Fenchel and Bernard 1993). We assume that in low oxygen conditions, these ciliates are the important zooplankters in Rogoznica Lake because they allow carbon flow from the anoxic to oxic layers. In general, the ciliates by grazing on bacteria, particularly on phototrophic sulphur bacteria recorded in a high density in the lake (Pjevac et al. 2015), transfer high quantity of organic carbon to higher trophic levels. This is especially true in case of nauplii

Table 6.4 List of calanoid copepods (ind. m^{-3}) and ciliates (ind. L^{-1}) found in Rogoznica Lake

Taxa	Max
Acartia italica	25,677
Tigriopus sp.	26
Unidentified harpacticoid species	3
Ciliates	6000

Data are expressed for the study periods between 1994 and 2004 and between 2011 and 2012 for calanoid copepods and 1994–2004 for ciliates

Max is for maximum abundance of individuals

and copepodites of *Acartia italica* in the lake, for which ciliates serve as an important food source (Kršinić et al. 2013). Therefore, ciliates are very important in the carbon transfer throughout the lake food web, and they also control microbial loop in Rogoznica Lake. On the other hand, calanoid copepods, *Acartia italica*, play an important role and also influence the lake's food web.

Whylie and Currie (1991) have demonstrated that suspended single bacteria cells contribute insignificant amount of carbon to copepod zooplankton. Bacteria bound to detrital and other particles could be ingested and digested by marine copepods (Lawrence et al. 1993), thereby linking the microbial loop directly with the grazing food chain (Overmann et al. 1999). In Mahoney Lake a significant carbon transfer was demonstrated from phototrophic bacteria (*Amoebobacter purpureus*) to planktonic multicellular invertebrates, the calanoid copepod *Diaptomus connexus* (Overmann et al. 1999). *A. purpureus* represents a major bacterial food source of copepods in Mahoney Lake, implying that carbon flow from anoxic to oxic habitats via the biomass of phototrophic sulphur bacteria could be also significant in the stratified and meromictic ecosystems. Similar is reported for Lake Shunet (south Siberia, Russia) where ciliates and calanoid copepods (*Arctodiaptomus salinus*) are shown to feed on *Cryptomonas* spp. and sulphur purple bacteria (Tolomeev et al. 2010; Chap. 5).

The copepod community in Rogoznica Lake is characterized by three species (*Acartia italica* Steuer, *Tigriopus* sp. and one unidentified harpacticoid species). We also found some planktonic stages of benthic species in the plankton (Ciglenečki et al. 2015). However, the most important planktonic copepod species in the lake, found in extreme high abundance (Table 6.4) is *Acartia italica* Steuer, endemic to the Mediterranean (Hure and Kršinić 1998). This species showed great adaptation ability to extreme environmental conditions, especially to higher concentrations of ammonium and sulphide that characterize the bottom waters, but sometimes during the holomixis also the whole water column of the lake (Kršinić et al. 2000, 2013). In autumn 1997 after anoxic holomixis, ammonium maximum occurred in November with average concentration of 66.0 ± 17.37 μM. Only 4 months after anoxic holomixis did the average ammonium concentration decrease to <50.00 μM and by August 1998 to 17.70 μM (Fig. 6.7). The toxic effect of ammonium on many marine organisms is well documented (WHO 1986). Based on laboratory toxicology tests, Buttino (1994) found that *A. clausi* exposed to low ammonium concentrations of 6.6 μM for 9 days showed a considerable increase in egg production, accompanied by a fall in hatching rate to about 50 %. A significant proportion of the copepodites collected in November 1997 exhibited a subcylindrical outgrowth on the anal somite, representing the evaginated hind gut (Kršinić et al. 2000). Crisafi (1974) documented the same phenomenon in specimens of *A. italica* sampled from a harbour and affected by pollution. The adaptation of *A. italica* to chronic toxicity with higher nutrient concentrations (especially of ammonium) in Rogoznica Lake is underscored here. The factors that cause morphological changes on the anal somite require further investigation.

Usually after the anoxia events, the *A. italica* population re-establishes quite quickly in the lake. The first nauplii stages are found in net samples as soon as

2 weeks after anoxic holomixis, and it takes about 40 days after anoxia for the first sexually mature individuals to appear in the plankton again (Kršinić et al. 2000, 2013). Survival strategies for many coastal planktonic copepods rely on the production of resting eggs during unfavourable conditions (Marcus 1996). These eggs sink to the sea bottom and then disappear from the plankton. The *Acartia* population was most likely renewed after anoxic events from subitaneous and resting eggs that had been present on the lake sediment for some time (Kršinić et al. 2000, 2013), but delayed-hatching eggs cannot be excluded (Chen and Marcus 1997).

A maximum for adult copepods recorded almost at the same time as the diatom density maximum as well as the positive significant correlation ($R^2 = 0.242$; $n = 73$; $p < 0.05$) between copepods and *C. curvisetus* distribution suggests that the diatom *C. curvisetus* (length 10–19 μm, cell diameter 5–10 μm) probably forms the diet of adult copepods. At least it seems to be the case during our study from 1997 to 2004 (Kršinić et al. 2013). Congener *Acartia tonsa* preferred feeding on motile prey (flagellates and ciliates), but diatoms comprised a substantial part of *Acartia*'s diet when Si/N ratios are high (Sommer 2009), which is similar to that reported for Rogoznica Lake (Svensen et al. 2008; Kršinić et al. 2000).

It is known that ecosystem functioning can be controlled by bottom-up factors which control species growth (e.g. light intensity, temperature, salinity, availability of nutrients, nutrient ratios, etc.) and top-down factors which control phytoplankton biomass (e.g. predation, viral lyses and parasitism) (Moraies et al. 2003). In Rogoznica Lake, after the anoxic holomictic events, shift from consumption (top-down) to resource limitation (bottom-up) is regularly observed (Kršinić et al. 2000, 2013; Ciglenečki et al. 2015). Before the holomictic anoxia event in 1997, the lake was mainly top-down controlled, while in the post-anoxic period, after September 1997 to the year 2000, bottom-up control became more important. The population structure of *A. italica* totally differed before and after total anoxia in 1997. Basic features of the pre-anoxia population were primarily determined by predator/prey relationship while in the post-anoxia, the nutrient–phytoplankton–copepod relationships were marked. The conditions from before the disastrous anoxia event in September 1997 re-established in about 2 years (Kršinić et al. 2013); however, the system did not recover to attain the same conditions as before.

The copepod *Tigriopus* sp., recorded for the first time in Rogoznica Lake after anoxic event in October 2011, usually inhabits splash pools located in the lake's supralittoral zone. It shows a remarkable adaptation to changes in salinity, temperature and oxygen (McAllen and Block 1997; McAllen 1998, 1999) as the characteristic for Rogoznica Lake. Also, these copepods are omnivorous and feed on bacteria, protozoans, detritus, diatoms and small crustaceans (Lewis et al. 1998; Gallucci and Ólafsson 2007). *Tigriopus* sp. occurred in the lake only in winter, with a maximum in January 2012, i.e. 2 months after anoxic holomictic event; it is very likely that the animals were swept away by wave action from their natural habitat of rock pools. *Tigriopus* sp. could be also an indicator of environmental changes in the case of more extreme conditions potentially favoured by climatic variations already registered for copepod species in the Mediterranean (Molinero et al. 2005).

6.5.2 Microbiological Activity

The purple and green phototrophic sulphur bacteria are one of the best known features of meromictic and stratified lakes. In layers where light intensity and sulphide concentrations are favourable, these bacteria build microstratified plates which are dense enough to colour the water green, brown or purple. Phototrophic sulphur bacteria form dense blooms in numerous lakes, contributing up to 90 % of the total primary productivity (Overmann 1997). Usually in Rogoznica Lake with its pink-coloured chemocline layer, about 50 cm thick (Fig. 6.2b, c), high activity of anoxygenic phototrophs, i.e. green and purple sulphur bacteria (GSB, PSB), was recorded (Pjevac et al. 2015). These organisms are mainly responsible for photo-oxidation of sulphide and other reduced sulphur compounds (e.g. thiosulfate), which were regularly detected in the lake, to elemental sulphur and sulphate. The most dominant oxidized form is elemental sulphur, especially in rapid chemical oxidation, similar to that recorded during holomictic–anoxic events in September 1997 and October 2011. During stratification, anoxygenic phototrophic green sulphur bacteria are dominant at the chemocline and in the hypolimnion (up to 90 % of total 10^7 cells mL^{-1}) (Pjevac et al. 2015). While during and after the anoxic holomictic event in 2011, the anoxygenic phototrophic green sulphur bacteria population entirely disappeared, and homogeneous, anoxic water column was characterized by a bloom of *Gammaproteobacteria* sulphur oxidizers (GSO)/ SUP05 clade-related *Gammaproteobacteria* (Pjevac et al. 2015). Population density of GSB is much higher, but due to the size of individual bacterial PSB cells, chemocline layer in Rogoznica Lake has a purple colour (Ciglenčki et al. 1998; Pjevac et al. 2015).

Microbial community shifts during and after holomixis are known from other stratified lakes (Casamayor et al. 2000; Cohen et al. 1977); however, most other lakes are oxygenated during holomixis and an epilimnion-like community is usually observed (Camacho et al. 2000; Hollibaugh et al. 2001). For example, Lake Ciso (Pedrós-Alió and Guerrero 1993) in Spain is dominated by phototrophic chemotrophic sulphur-oxidizing prokaryotes (SOPs). Anoxic water column in Rogoznica Lake also harbours specific community of sulphate-reducing bacteria closely related to *Desulphonema*/*Desulphococcus*/*Desulphosarcina* group and *Desulphovibrio*/*Desulphomicrobium* group which are main drivers for the sulphate reduction process and production of the toxic hydrogen sulphide in the lake water column (Čanković et al. 2015). Preliminary investigation showed that community structure of these species is very sensitive to physico-chemical conditions in the water column, and accordingly seasonal shifts in structure composition occur.

Although microbiological studies in Rogoznica Lake are not detailed, they do show how well the microbial diversity relates to that recorded for hydrographical and chemical conditions of the stratified and mixed water column. For most of the sampling dates, the microbial communities in the lake can be divided into that from epilimnion and from chemocline/hypolimnion populations (Pjevac et al. 2015). Generally, bacterial community has a high abundance in the chemocline with rather low productivity per cell, while in the surface layer bacteria are low in numbers but have higher rate of productivity (Šestanović et al. 2005; Pjevac et al. 2015). Accordingly, organic matter distribution represented mainly by particulate carbon (POC) and dissolved organic carbon (DOC), and surface-active substances (SAS) as an important fraction of the DOC and POC that possess surface-active properties in natural waters (Ćosović and Vojvodić 1998), has their maxima at the chemocline in the stratified conditions (Fig. 6.10) (Ćosović et al. 2000; Šestanović et al. 2005; Pjevac et al. 2015; Marguš et al. 2015). In the same layer, increased biomass of phytoplankton is evidenced from very high chlorophyll *a* concentrations (up to 8000 ng L^{-1}) compared with that in the upper water layer. The DOC and POC concentrations in the lake range between 1.00–6.0 mg C L^{-1} and 0.50–4.0 mg C L^{-1}, respectively (Ciglenečki et al. 2005; Marguš et al. 2015), reflecting the importance of phytoplankton, zooplankton and their grazing and bacterial productivity. Similar ranges of DOC concentrations have been reported for the anoxic water columns of other meromictic and seasonally anoxic lakes, such as Hall Lake in the USA (Balistrieri et al. 1994), Lake Pavin in France (Alberic et al. 2000) and Esthwaite Water in the UK (Hamilton-Taylor et al. 1996). A major fraction of the DOC in Rogoznica Lake must be a long-lived material that has been produced by the

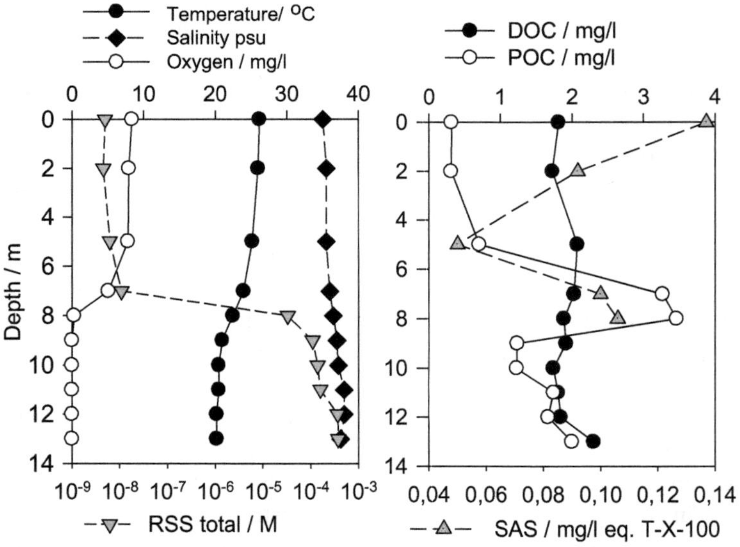

Fig. 6.10 Physico-chemical parameters of the water column in Rogoznica Lake (temperature, salinity, dissolved oxygen, reduced sulphur species (RSS)) (*left panel*) and organic compounds that possess surface-active properties (SAS), DOC and POC (*right panel*) measured during the stratified conditions in May 2012, i.e. 7 months after anoxic holomictic event in October 2011. Please note that recorded situation reflects intensive phytoplankton activity of diatoms *Cyclotella choctawhatcheeana* and *Dactyliosolen fragilissimus*, with maximum cell density, up to 3 × 10^6 cells L^{-1} recorded between 2 and 5 m depth (Malešević et al. 2015; Ciglenečki et al. 2015). Maximum of total bacteria cell counts, up to 6.4 × 10^6 cells mL^{-1} was detected around the chemocline, situated at 8 m depth (Pjevac et al. 2015) (modified from Marguš et al. 2015)

breakdown of organic matter and dispersed through the lake by mixing events (Ćosović et al. 2000; Bura-Nakić et al. 2009). POC and SAS variability are mainly connected with the intensity of primary production (Parszuto and Kaliszewska 2007; Ćosović et al. 2000; Marguš et al. 2015). The contribution of POC to the total organic carbon (TOC) in Rogoznica Lake is very high and typical for highly eutrophic areas (Parszuto and Kaliszewska 2007). In Rogoznica it usually varies between 20 and 60 % of the TOC concentration (Marguš et al. 2015). Seasonal variability in concentration and composition of organic matter in Rogoznica Lake is assumed to highly influence the microbial structure and its activities.

6.6 Conclusions

Rogoznica Lake is a unique karstic marine system in the middle of the eastern Adriatic Coast, lying on a transect between the cities of Šibenik and Split (43°32′ N 15°58′ E) in Croatia. It is a highly eutrophic, anoxic and sulphidic marine lake that depends highly on the meteorological conditions and consequently fluctuates between anoxic holomixis and meromictic conditions. Such fluctuations strongly impact the biogeochemistry and food web interactions in the lake.

Stratification and mixing between the water layers in the lake are directly influenced by the precipitation and evaporation events, heat flux and increased frequency of extreme weather events due to climate change.

Most stratified and sulphidic lakes become oxygenated after annual turnover. However, in contrast, Rogoznica Lake becomes anoxic on establishment of holomixis.

It differs from the other well-known anoxic basins such as the Black Sea, Framvaren Fjord (Norway) and Lac Pavin (France) in its biological, geochemical, geomorphological and hydrological characteristics, including small size and very easy access to anoxic conditions which usually are almost permanent below 10 m. In the Mediterranean area, the ecosystem of Rogoznica Lake is quite similar to the karstic and saline lakes in Spain, Lake Ciso; Italy, Lake Faro; and Greece, Aitoliko coastal lake and Salty Lake in Crete.

High seasonal variations in salinity and temperature, water stratification regarding strong redox gradient, euxinia accompanied by high concentrations and variability of nutrients and episodic holomixis causing hypoxia and anoxia conditions highly affect the community structure of phytoplankton and zooplankton in the lake. Diatoms are the most dominant phytoplankton that is well adapted to these extreme and highly eutrophic fluctuating conditions. They have relatively few species but high single species abundance (up to 10^8 cells L^{-1}). Ciliates and copepods, the heterotrophic zooplankton organisms, play an important role and control the lake's food web. The most important and abundant plankton copepod species in the lake is *Acartia italica* Steuer. This species, which is endemic to the Mediterranean, has a great adaptive ability at higher ammonium and sulphide concentrations as in Rogoznica Lake. Population dynamics of *A. italica* and development of phytoplankton reflect the fluctuation between anoxic holomixis and stratified conditions. Usually these changes are followed by a shift from the top-down to the bottom-up control of the food web in the lake. The lake with its pink-coloured chemocline layer has a very active community of green and purple sulphur bacteria (GSB, PSB).

Rogoznica Lake appears to be an ideal site to examine the processes that control the biogeochemistry of coastal and potentially hypoxic–anoxic marine environment. Organic matter supply, stimulated by recycling of nutrients, leads to eutrophication conditions in the lake. In general, enclosed marine basins are exposed to the increased human activities from the coastal areas. Today eutrophication-related phenomena are closely related to anthropogenic climate changes, as illustrated here for Rogoznica Lake. The consequences could be catastrophic for the benthos because of the development of hydrogen sulphide, which is highly toxic and can lead to extinction of organisms. The integration of the long-term (from 1994 up to the present) data on physico-chemical aspects and biology indicate that the marine lake, Rogoznica Lake, in addition to its meromictic and holomictic characteristics, can potentially serve as a valuable sentinel of the extreme weather events in response to climate change.

Acknowledgements The Croatian Ministry of Science, Education and Sports financially supported this study through the projects: 119-1191189-1228, 098-0982934-2717 and 275-0000000-3186. The financial support of Croatian Academy of Science and Croatian Science Foundation under the projects (1) "Investigation of Rogoznica Lake, a unique anoxic environment on the Adriatic coast related to recovering from catastrophic anoxia occurred in October 2011" and (2) IP-11-2013-5928 (ADAM-ADRA) and (3) IP-11-2013-1205 (SPHERE) is highly acknowledged. The authors thank Z. Zovko, Z. Roman, M. Čanković, E. Bura-Nakić and M. Marguš for the help and assistance during the lake water sampling and analyses. E. B. N. and M. M. are acknowledged for their help in RSS analyses.

References

Adrian R, O'Reilly CM, Zagarese H et al (2009) Lakes as sentinels of climate change. Limnol Oceanogr 54:2283–2297

Alberic P, Viollier E, Jezequel D et al (2000) Interaction between trace elements and dissolved organic matter in the stagnant anoxic deep layer of a meromictic lake. Limnol Oceanogr 45:1088–1096

Balistrieri LS, Murray JW, Paul B (1994) The geochemical cycling in a biogenic meromictic lake. Geochim Cosmochim Acta 58(19):3993–4008

Barić A, Grbec B, Kušpilić G et al (2003) Mass mortality event in a small saline lake (Lake Rogoznica) caused by unusual holomictic conditions. Sci Mar 67:129–141

Bura-Nakić E, Helz GR, Ćosović B et al (2009) Seasonal variations in reduced sulphur species in a stratified lake (Rogoznica Lake, Croatia); evidence for organic carriers of reactive sulphur. Geochim Cosmochim Acta 73:3738–3751

Bura-Nakić E, Marguš M, Orlić S et al (2012) Zmajevo oko—a unique example of anchialine system in the Adriatic coast (Croatia) during spring-summer stratification and autumn mixing period. Nat Croat 21:17–20

Burić Z, Caput Mihalić K, Cetinić I et al (2009) Occurrence of the rare microflagellates Prorocentrum arcuatum Issel and Hermesinum adriaticum Zacharias in the marine Lake Rogoznica (Eastern Adriatic coast). Acta Adriat 50:31–44

Buttino I (1994) The effect of low concentrations of phenol and ammonia on egg production rates, fecal pellet production and egg viability of the calanoid copepod Acartia clausi. Mar Biol 119:629–634

Camacho A, Vicente E, Miracle MR (2000) Spatio-temporal distribution and growth dynamics of phototrophic sulphur bacteria populations in the sulphide-rich Lake. Arcas Aquat Sci 62:334–349

Čanković M, Petrić I Ciglenečki I (2015) Rogoznica Lake—an extreme environment hosting specific sulfate-reducing bacterial community. In: Rene ER, Bhattarai S, Nancharaiah YV, Lens PNL (eds) Proceedings of the 4th international conference on research frontiers in chalcogen cycle science & technology, Delft, The Netherlands, May 2015, p 115

Casamayor EO, Schäfer H, Bañeras L et al (2000) Identification of and spatio-temporal differences between microbial assemblages from two neighboring sulphurous lakes: comparison by microscopy and denaturing gradient gel electrophoresis. Appl Environ Microbiol 66:499–508

Chen F, Marcus NH (1997) Subitaneous, diapause, and delayed-hatching eggs of planktoniccopepods from the northern Gulf of Mexico: morphology and hatching success. Mar Biol 127:587–597

Chicharo L, Chicharo MA, Ben-Hamadou R (2006) Use of a hydrotechnical infrastructure (Alqueva Dam) to regulate planktonic assemblages in the Guadiana estuary: basis for sustainable water and ecosystem services management. Estuar Coast Shelf Sci 70:3–18

Ciglenečki I, Kodba Z, Ćosović B (1996) Sulphur species in Rogoznica Lake. Mar Chem 53:101–110

Ciglenečki I, Kodba Z, Viličić D et al (1998) Seasonal variation of anoxic conditions in the Rogoznica Lake. Croat Chem Acta 71(2):217–232

Ciglenečki I, Carić M, Kršinić F et al (2005) The extinction by sulphide-turnover and recovery of a naturally eutrophic, meromictic seawater lake. J Mar Syst 56:29–44

Ciglenečki I, Pichler S, Prohić E et al (2006) Distribution of redox-sensitive elements in bottom waters, porewaters and sediments of Rogoznica Lake (Croatia) in both oxic and anoxic condition. Water Air Soil Pollut 6:537–545

Ciglenečki I, Bura-Nakić E, Marguš M (2013) (Rogozničko jezero jedinstveni anoksični sustav na Jadranskoj obali) Rogoznica Lake unique anoxic system on the Adriatic coast. Hrvatske Vode 86: 295–302 (on Croatian)

Ciglenečki I, Janeković I, Marguš M et al (2015) Impacts of extreme weather events on eutrophic seawater ecosystem (Rogoznica Lake, Adriatic coast). Cont Shelf Res 108:144–155

Cohen Y, Krumbein WE, Shilo M (1977) Solar Lake (Sinai). Distribution of photosynthetic microorganisms and primary production. Limnol Oceanogr 22:609–620

Ćosović B, Vojvodić V (1998) Voltammetric analysis of surface active substances in natural seawater. Electroanalysis 10:429–434

Ćosović B, Ciglenečki I, Viličić D et al (2000) Distribution and seasonal variability of organic matter in a small eutrophicated salt lake. Estuar Coast Shelf Sci 51:705–715

Crisafi P (1974) Some responses of planktonic organisms to environmental pollution. Rev Int Oceanogr Med 34:145–154

Daffonchio D, Borin S, Brusa T et al (2006) Stratified prokaryote network in the oxic-anoxic transition of the deep-sea halocline. Nature 440:203–207

Dassenakis M, Krasakopoulou E, Matzara B (1994) Chemical characteristics of Aetoliko Lagoon, Greece, after an ecological shock. Mar Pollut Bull 28(7):427–433

der Weijden V, Luther GW III, Bregant D (eds) (1990) Anoxic brines of the Mediterranean Sea. Mar Chem 31(1–3):1–320

Erickson BE, Helz GR (2000) Molybdenum(VI) speciation in sulfidic waters; stability and lability of thiomolybdates. Geochim Cosmochim Acta 64:1149–1158

Fallesen G, Andersen F, Larsen B (2000) Life, death and revival of the hypertrophic Mariager Fjord, Denmark. J Mar Syst 25:313–321

Fenchel T, Bernard C (1993) Endosymbiotic purple non-sulphur bacteria in an anaerobic ciliated protozoon. Microbiol Lett 110:21–25

Gallucci F, Ólafsson E (2007) Cannibalistic behaviour of rock-pool copepods: an experimental approach for space, food and kinship. J Exp Mar Biol Ecol 342:325–331

Hamilton-Taylor J, Davison W, Morfett K (1996) The biogeochemical cycling of Zn, Cu, Fe, Mn, and dissolved organic C in seasonally anoxic lake. Limnol Oceanogr 41:408–418

Helz GR, Miller CV, Charnock JM et al (1996) Mechanism of molybdenum removal from the sea and its concentration in black shales. Geochim Cosmochim Acta 60:3631–3642

Helz GR, Bura-Nakić E, Mikac N et al (2011) New model for molybdenum behavior in euxinic basins. Chem Geol 284:323–333

Henneke E, De Lange GJ (1990) The distribution of DOC and POC in the water column and brines of the Tyro and Bannock basins. Mar Chem 31:113–122

Hollibaugh JT, Wong PS, Bano N et al (2001) Stratification of microbial assemblages in Mono Lake, California, and response to a mixing event. Hydrobiologia 1:45–60

Hure J, Kršinić F (1998) Planktonic copepods of the Adriatic Sea. Spatial and temporal distribution. Natura Croat 7:1–135

IPCC (2007) Contribution of Working Groups I, II and III to the Fourth Assessment Report of the Intergovernmental Panel on Climate Change. In: Pachauri RK, Reisinger A (eds) IPCC. Switzerland, Geneva, p 104

Izdar E, Murray JW (1991) Black Sea oceanography—NATO ASI series C, vol 351. Kluwer, Dordrecht

Kamyshny A, Zerkle ZL, Mansaray ZF et al (2011) Biogeochemical sulfur cycling in the water column of a shallow stratified sea-water lake: speciation and quadruple sulfur isotope composition. Mar Chem 127:144–154

Kehayias G, Ramfos A, Ntzialas P et al (2013) Zooplankton diversity and distribution in a deep and anoxic Mediterranean coastal lake. Mediterr Mar Sci 14(1):179–192

Kršinić F, Carić M, Viličić D et al (2000) The calanoid copepod Acartia italica Steuer, phenomenon in the small saline Lake Rogoznica (Eastern Adriatic coast). J Plankton Res 22:1441–1464

Kršinić F, Ciglenečki I, Ljubešić Z et al (2013) Population dynamics of the calanoid copepod Acartia italica Steurer in a small saline lake. Acta Adriat 54:229–240

Landing WM, Westerlund S (1988) The solution chemistry of iron(II) in Framvaren Fjord. Mar Chem 23:329–343

Lawrence SG, Ahmad A, Azam F (1993) Fate of particle-bound bacteria ingested by *Calanus Pacificus*. Mar Ecol Prog Ser 97:299–307

Lewis AG, Chatters L, Raudsepp M (1998) Feeding structures and their functions in adult and preadult Tigriopus californicus (Copepoda: Harpacticoida). J Mar Biol Assoc U K 78:451–466

Malešević N, Ciglenečki I, Bura-Nakić E et al (2015) Diatoms in the extreme euxinic environment (Rogoznica Lake, Eastern Adriatic coast). Acta Bot Croat 74(2):333–343

Marcus NH (1996) Ecological end evolutionary significance of resting eggs in marine copepods: past, present, and future studies. Hydrobiology 320:141–152

Marguš M, Morales-Reyes I, Bura-Nakić E et al (2015) The anoxic stress conditions explored at the nanoscale by Atomic Force Microscopy in highly eutrophic and sulphidic marine lake. Cont Shelf Res 109:24–34

McAllen R (1998) Life at its limits—the ecophysiology of the high shore rockpool inhabitant Tigriopus brevicornis. Dissertation, University of London

McAllen R (1999) Enteromorpha intestinalis—a refuge for the supralittoral rockpool harpacticoid copepod Tigriopus brevicornis. J Mar Biol Assoc U K 79:1125–1126

McAllen R, Block W (1997) Aspects of the cryobiology of the intertidal harpacticoid copepod Tigriopus brevicornis. Cryobiology 35:309–317
Michard G, Viollier E, Jézéquel D et al (1994) Geochemical study of a crater lake: Pavin Lake, France—identification, location and quantification of the chemical reactions in the lake. Chem Geol 115:103–115
Mihelčić G, Šurija B, Juračić M et al (1996) History of the accumulation of trace metals in sediments of the saline Rogoznica Lake (Croatia). Sci Total Environ 182:105–115
Mihopoulos N, Dassenakis M, Anastasakis G et al (2000) Physical and chemical characteristics of a Mediterranean coastal microenvironment: the "Salty lake" of Aghios Nikolaos (Crete, Greece). Glob NEST Int J 2(2):167–178
Millero FJ (1991) The oxidation of H2S in Framvaren Fjord. Limnol Oceanogr 36(5):1007–1014
Miracle MR, Vicente E, Pedrós-Alió C (1992) Biological studies of spanish meromictic and stratified karstic lakes. Limnetica 8:59–77
Molinero JC, Ibanez F, Nival P et al (2005) North Atlantic climate and northwestern Mediterranean plankton variability. Limnol Oceanogr 50:1213–1220
Moraies P, Chícharo MA, Barbosa A (2003) Phytoplankton dynamics in a coastal saline lake (SE-Portugal). Acta Oecol 24:S87–S96
Murray JWM (ed) (1991) Black Sea oceanography: results from the 1988 Black Sea expedition. Deep Sea Res Part A Oceanogr Res Pap 38(2):S655–S1266
Overmann J (1997) Mahoney Lake: a case study of the ecological significance of phototrophic sulphur bacteria. Adv Microb Ecol 15:251–288
Overmann J, Hall KJ, Northcote TG et al (1999) Grazing of the copepod Diaptomus connexus on purple sulphur bacteria in a meromictic salt lake. Environ Microbiol 1(3):213–221
Parszuto K, Kaliszewska EA (2007) The evaluation of the trophy of selected Olsztyn lakes by means of indices of particulate organic carbon (POC) and dissolved organic carbon (DOC). Limnol Rev 7:29–35
Pedrós-Alió C, Guerrero R (1993) Microbial ecology in Lake Ciso. Adv Microb Ecol 13:155–209
Pilipchuk MF, Volkov II (1974) Behavior of molybdenum in processes of sediment formation and diagenesis in Black Sea. Mem Am Assoc Petrol Geol 20:542–553
Pjevac P, Korlević M, Berg JS et al (2015) Community shift from phototrophic to chemotrophic sulfide oxidation following anoxic holomixis in a stratified seawater lake. Appl Environ Microbiol 81:298–308
Prange A, Kremling K (1985) Distribution of dissolved molybdenum, uranium and vanadium in the Baltic sea waters. Mar Chem 16:259–274
Romero JR, Melack JM (1996) Sensitivity of vertical mixing in a large saline lake to variations in runoff. Limnol Oceanogr 41:955–965
Scranton MI, McIntyre M, Astor Y et al (2006) Temporal variability in the nutrient chemistry of the Cariaco basin. In Neretin LN (ed) Past and present water column anoxia. Nato science series: IV: earth and environmental sciences, vol 64, p 139
Šestanović S, Šolić M, Krstulović N et al (2005) Vertical structure of microbial community in an eutrophic meromictic saline lake. Fresen Environ Bull 14(8):668–675
Shimoda M, Azim E, Perhar G et al (2011) Our current understanding of lake ecosystem response to climate change: what have we really learned from the north temperate deep lakes? J Great Lakes Res 37:173–193
Sommer U (2009) Copepod growth and diatoms: insensitivity of Acartia tonsa to the composition of semi-natural plankton mixtures manipulated by silicon: nitrogen ratios in mesocosms. Oecologia 159(1):207–215
Stipaničev V, Branica M (1996) Iodine speciation in the water column of the Rogoznica Lake (Eastern Adriatic Coast). Sci Total Environ 182:1–9
Stipaničev V, Branica M (2006) The distributions of iodate and iodide in Rogoznica Lake (East Adriatic coast). Estuar Coast Shelf Sci 66:55–66
Stock JH, Iliffe TM, Williams D (1986) The concept 'anchialine' reconsidered. Stygologia 2:90–92

Svensen C, Wexeles Riser C, Cetinić I et al (2008) Vertical flux regulation and plankton composition in a simple ecological system: snapshots from the small marine Lake Rogoznica (Croatia). Acta Adriat 49:37–51
Tolomeev AP, Sushchik NN, Gulati RD et al (2010) Feeding spectra of Arctodiaptomus salinus (Calanoida, Copepoda) using fatty acid trophic markers in seston food in two salt lakes in South Siberia (Khakasia, Russia). Aquat Ecol 44:513–530
Van Cappellen P, Viollier E, Roychoudhury A et al (1998) Biogeochemical cycles of manganese and iron at the Oxic-Anoxic transition of a stratified marine basin (Orca Basin, Gulf of Mexico). Environ Sci Technol 32:2931–2939
Viličić D, Marasović I, Mioković D (2002) Checklist of phytoplankton in the Eastern Adriatic Sea. Acta Bot Croat 61:57–91
Viollier E, Jézéquel D, Michard G et al (1995) Geochemical study of a crater lake (Pavin Lake, France): trace element behavior in the monimolimnion. Chem Geol 125:61–72
WHO (1986) Environmental health criteria for ammonia. World Health Organization, The International Programme on Chemical Safety (IPCS) 54 Vammala, Finland
Whylie JL, Currie DJ (1991) The relative importance of bacteria and algae as food sources for crustacean zooplankton. Limnol Oceanogr 36:708–728
Wielen PW, Bolhuis H, Borin S (2005) Bio Deep Scientific Party: the enigma of prokaryotic life in deep hypersaline anoxic basins. Science 307:121–123
Yakimov MM, La Cono V, Denaro R et al (2007) Primary producing prokaryotic communities of brine, interface and seawater above the halocline of deep anoxic lake L'Atalante, Eastern Mediterranean Sea. ISME J 1:743–755
Zagami G, Brugnano C (2013) Diel, seasonal and man-induced changes in copepod assemblages and diversity, with special emphasis on hyperbenthic calanoid species, in a Mediterranean meromictic system (Lake Faro). Mar Freshw Res 64(10):951–964
Zheng Y, Anderson RF, van Geen A et al (2000) Authigenic molybdenum formation in marine sediments: a link to pore water sulphide in the Santa Barbara Basin. Geochim Cosmochim Acta 64(24):4165–4178
Žic V, Carić M, Viollier E et al (2010) Intensive sampling of iodine and nutrient speciation in naturally eutrophicated anchialine pond (Rogoznica lake) during spring and summer seasons. Estuar Coast Shelf Sci 87:265–274
Žic V, Carić M, Ciglenečki I et al (2013) The impact of natural water column mixing on inorganic iodine and nutrient speciation in a eutrophic anchialine pond (Rogoznica Lake, Croatia). Estuar Coast Shelf Sci 133:260–272

Chapter 7
Lake Cadagno: Microbial Life in Crenogenic Meromixis

Mauro Tonolla, Nicola Storelli, Francesco Danza, Damiana Ravasi, Sandro Peduzzi, Nicole R. Posth, Raymond P. Cox, Mårten F. Jørgensen, Lea H. Gregersen, Niels Daugbjerg, and Niels-Ulrik Frigaard

M. Tonolla (✉)
Laboratory of Applied Microbiology, University of Applied Sciences of Southern Switzerland, Via Mirasole 22a, 6500 Bellinzona, Switzerland

Microbiology Unit, Department of Botany and Plant Biology, University of Geneva, 1211 Geneva, Switzerland

Alpine Biology Center Foundation, Via Mirasole 22a, 6500 Bellinzona, Switzerland
e-mail: mauro.tonolla@supsi.ch

N. Storelli • D. Ravasi
Laboratory of Applied Microbiology, University of Applied Sciences of Southern Switzerland, Via Mirasole 22a, 6500 Bellinzona, Switzerland

F. Danza
Laboratory of Applied Microbiology, University of Applied Sciences of Southern Switzerland, Via Mirasole 22a, 6500 Bellinzona, Switzerland

Microbiology Unit, Department of Botany and Plant Biology, University of Geneva, 1211 Geneva, Switzerland

S. Peduzzi
Alpine Biology Center Foundation, Via Mirasole 22a, 6500 Bellinzona, Switzerland

Section of Earth and Environmental Sciences, Forel Institute, University of Geneva, 1211 Geneva, Switzerland

N.R. Posth • R.P. Cox
Nordic Center for Earth Evolution (NordCEE), Department of Biology, University of Southern Denmark, 5230 Odense M, Denmark

M.F. Jørgensen • L.H. Gregersen • N. Daugbjerg • N.-U. Frigaard
Marine Biological Section, Department of Biology, University of Copenhagen, 1353 Copenhagen and, 3000 Helsingør, Denmark

© Springer International Publishing AG 2017

R.D. Gulati et al. (eds.), *Ecology of Meromictic Lakes*, Ecological Studies 228,
DOI 10.1007/978-3-319-49143-1_7

7.1 Introduction

Due to their permanent stratification, meromictic lakes are interesting model ecosystems. In particular, the chemocline offers many advantages for studying the planktonic life and processes mediated by microorganisms. The physico-chemical gradients across the chemocline support the development of very intense microbial blooms (Overmann et al. 1996; Parkin and Brock 1981; Sorokin 1970; van Gemerden and Mas 1995). Along vertical gradients of light and different electron donors and acceptors, diverse physiological groups of organisms substitute each other (Guerrero et al. 1985; Jørgensen et al. 1979; Overmann et al. 1991) in distinct layers at depth intervals ranging from cm to m. These distinct layers facilitate accurate sampling and measurement of environmental characteristics. Lake Cadagno is a crenogenic meromictic lake located in the southern Swiss Alps (46°33′N, 8°43′E) at 1921 m asl, in the catchment area of a dolomite vein rich in gypsum (Piora-syncline). A permanent chemocline lies between 10 and 14 m depth, stabilized by density differences of salt-rich, high-density water constantly supplied by sub-aquatic springs to the monimolimnion and of electrolyte-poor, low-density surface water that feeds the mixolimnion. Steep sulphide gradients in the chemocline support the growth of large numbers of phototrophic sulphur bacteria (up to 10^7 cells ml^{-1}). This community includes purple sulphur bacteria (PSB, family *Chromatiaceae*) of the genera *Chromatium*, *Lamprocystis*, *Thiocystis* and *Thiodictyon* and green sulphur bacteria (GSB. family *Chlorobiaceae*) of the genus *Chlorobium*. Since the beginning of the last century, Lake Cadagno and the nearby Lake Ritom have been used as models to study water stratification, sulphidogenic bottom waters and blooms of PSB. Since the early 1980s, studies on ecological issues in Lake Cadagno significantly increased, supported by two major developments. The first was the establishment of research facilities at the shore of the lake, the Alpine Biology Center, and the second was the introduction of novel molecular techniques that allowed researchers to analyse microbial populations in Lake Cadagno unaffected by the difficulties in culturing them.

This contribution presents an overview of more than 30 years of studies on Lake Cadagno, including biogeochemical analyses, diversity surveys and ecological studies on the very diverse microbial community developing in it.

The present chapter is structured in the following major parts:

1. Hydrology, and morphology of the lake in brief, including the lake's origin and age
2. Historical perspective and description of early investigations conducted on Lake Cadagno and description of the lake's meromixis and the main physico-chemical features of the lake and its related stratification
3. The organisms inhabiting the lake, with its diversity dominated by microorganisms, particularly in relation to the sulphur cycle

The vast majority of the studies on Lake Cadagno concern the bacterial community developing in the chemocline. The chapter on the biology of the lake reviews comprehensively studies published since 1985 and results of recent studies on prokaryotic diversity (see Sect. 5.1) and on microbial eukaryotic diversity (see Sect. 5.3).

7.2 Origin, Morphology and Hydrology of Lake Cadagno

Lake Cadagno (Fig. 7.1) originated due to glacial erosion during the last glacial period, about 10,000 years ago, and the formation of a moraine acting as a small dam (Boucherle and Züllig 1988; Krige 1918; Stapfer 1991; Wirth et al. 2013). The lake is located in the southern Swiss Alps at 1921 m asl (latitude, 46°33′N; longitude, 08°43′E), in Val Piora, a region characterized by its richness in mountain lakes, waterways, wetlands and ponds (Garwood 1906; Peduzzi 2011). Lake Cadagno is small (maximum length, 830 m; maximum width, 420 m; surface area, 0.26 km^2; maximum depth, 21 m; average depth, 9.27 m) (Del Don et al. 2001), situated among pre-Triassic gneiss and mica-schists of central alpine crystalline rocks just south of the Gotthard Pass in the Canton Ticino (Switzerland). It has a catchment area of 2.71 km^2. The southern lake shore rests on late Triassic and early Jurassic rocks folded into the so-called Piora-syncline, characterized by the presence of tectonized dolomitic limestones (magnesian limestone) and gypsum deposits with a karstic hydrology. The Piora-syncline forms a vein that separates the crystalline rocks of the Gotthard massif to the north and the metamorphic

Fig. 7.1 Lake Cadagno and Piora valley. Photo taken from the west side of the lake

Pennine Lucomagno Nappe, about 10 km to the south, along the whole Piora valley (Otz et al. 2003).

Lake's sediment analysis reveals that permanent stratification, i.e. meromixis, and the presence of phototrophic sulphur bacteria must have been established very early in the history of the lake. This suggestion was first based on the analyses of organic components in the upper sediments (Putschew et al. 1995; Hebting et al. 2006) and of Cladocera (Crustacea) remains (Boucherle and Züllig 1988). Trace pigments, e.g. okenone and isorenieratene, characteristic for purple and green phototrophic bacteria, were detected in the upper 170 cm of the lake's sediment (Züllig 1985). However, the detailed longer-term history of development of meromixis and differential redox conditions in Lake Cadagno, and the timing of the transition from oxic to euxinic (i.e. sulphidic and anoxic) conditions in the water column, remained unknown until 2009, when 10.5-m-long sediment cores covering the complete lacustrine record of Lake Cadagno were investigated using an interdisciplinary approach (Niemann et al. 2012; Ravasi et al. 2012b; Dahl et al. 2013; Wirth et al. 2013). The lake history was characterized by a first oxic-intermediate-sulphidic redox-transition period shortly after glacial retreat, and lake formation was estimated at 12.5 kyr BP (before present). During the 2300-year-long intermediate period from 12.1 to 9.8 kyr BP, strongly fluctuating Mn concentrations in the sediments associated with episodic water column mixing events were recorded (Wirth et al. 2013). At the same time, efficient Mo burial (Dahl et al. 2010) and the presence of remains of phototrophic sulphur bacteria (Ravasi et al. 2012b) provided an evidence for a sulphidic environment. In fact, at the end of this period with intermediate redox conditions, ca. 9800 ± 130 years ago, a pronounced drop in Mn and a simultaneous rise in Mo concentrations in the sediments indicated efficient Mo burial in sulphide-rich water column and thus the onset of euxinic conditions in Lake Cadagno. This redox transition was causally linked to the concurrent climate warming, which led to permafrost melting and the activation of the sub-aquatic springs. The inflow of salt-rich waters to the lake bottom likely initiated the formation of a chemocline and a stable crenogenic meromixis (créno, spring; from the Greek "Krini") in the early Holocene. Recent biogeochemical data provide evidence that Lake Cadagno has remained sulphidic from the start of Holocene to date. Flood and mass movement events, however, generated underflows of turbidity-rich water that sporadically introduced oxygen-rich water to the hypolimnion, thereby only temporarily weakening the sulphidic conditions (Wirth et al. 2013). Such events may have influenced water turbidity and nutrient supply and may thus have caused shifts in the microbial community structure (see Sect. 5.1.1.2.) (Tonolla et al. 2005b; Decristophoris et al. 2009; Gregersen et al. 2009).

Water penetrating and circulating through the karstic dolomite and reappearing as underwater springs in the southern part of the lake has a rather high ionic strength (9–10 mM), and it flows downwards feeding the anoxic lower part of the water column (monimolimnion). Dye tracing experiments (Otz et al. 2003) have revealed that this groundwater originates from a relative small underground catchment area which extends for about 2 km to the southeast up to the so-called Calderoni Sinkhole. The sinkhole causes, uphill, a flow of the underground water towards the Lucomagno, i.e. in the opposite direction to the surface water flow. On the other hand, the upper part of the lake's aerobic water column (mixolimnion) is fed by surface run-off from a small drainage area of about 2 km^2, north of the lake, which encompasses the crystalline rocks of the Gotthard massive. The major source of surface input to Lake Cadagno is a small stream that drains water from Lake Stabbio, a small lake at 2351 m asl. The crystalline rocks of the watershed are rather resistant to chemical weathering, such that the water of the mixolimnion has a relatively low ionic strength and is oligo- to mesotrophic. The density difference between the two water masses in Lake Cadagno causes the development of a very stable stratification with a chemocline at about 10–14 m depth characterized by the rapid change of the concentrations of chemical components (Del Don et al. 2001; Hanselmann and Hutter 1998; Peduzzi et al. 1998).

7.3 Historical Overview: Scientific Studies and Hydroelectric Exploitation

Lake Cadagno has attracted scientists since the end of the nineteenth century. In fact, since the foundation of limnology in 1904, the lakes of the Piora region have created considerable scientific interest (Bourcart 1906; Garwood 1906; Burckhardt 1910); for a review on the historic data sets on lakes Ritom and Cadagno, see Peduzzi and Bianconi (2012). Stratification phenomena, as well as sulphur-rich deep waters and mass developments of phototrophic sulphur bacteria, are among the early observations on the lake. A monograph on the nearby Lake Ritom (Collet et al. 1918) reported results of studies carried out during the construction of the Ritom hydroelectric dam (1915–1920). Already then, the importance of the sulphur cycle and the bacteria involved therein were mentioned (Düggeli 1924). Studies on fish populations (Surbeck 1917) and earlier researches on phytoplankton and sulphur bacteria were carried out on Lake Ritom and Lake Cadagno (Bachmann 1924, 1928; Borner 1928a, b; Burckhardt 1910; Düggeli 1924). These early published works are important documents reporting the loss of meromixis in Lake Ritom following the construction of the Ritom dam. Similar results were reported for a French alpine lake, Lac de la Girotte in Savoie, in 1926 (Delebeque 1898; Barades 1926). Lake Cadagno is probably the most unique among the lakes in the entire alpine range that exhibit crenogenic meromixis. In 1948, a small dam was built raising the lake's surface water level from 1918 to 1921 m asl, and up to the year 2007, the lake was exploited for production of electricity resulting in a 3-m decrease in the lake's water level during the winter. During almost 60 years of exploitation of the lake for hydroelectric generation, it retained its permanent stratification. The lowering of the water level led to a loss of about 720,000 m^3 water, i.e. about 30 % of the lake's total water volume (2.42 million m^3). This comprised 50 % of the water volume with O_2 concentration >4 mg l^{-1} (around the minimum survival limit for Salmonidae). Aerobic organisms in the mixolimnion were subjected to a significant reduction in vital living space. In addition to that, 45,000 m^2 of littoral dried up causing, on the one hand, the death of numerous populations of sediment dwellers (e.g. Ephemeroptera, Plecoptera and Trichoptera), which were a potential source of food for fish species and, on the other hand, the loss of spawning areas for the fish reproduction (Tonolla et al. 1988). At present, the lake is no more actively being exploited for the production of electricity, and therefore its water level remains fairly constant at 1921 m asl.

Studies on Lake Cadagno intensified during the twentieth century, more so since the 1980s. The opening of the Alpine Biology Center (ABC) in Piora in 1994 in cooperation with the Universities of Geneva, Zurich, Lugano and recently the newly opened University of Applied Sciences and Arts of Southern Switzerland (SUPSI), as well as participating cantons, the municipality and the patriciate of Quinto, have led to a substantial increase in scientific studies in the region.

The presence of scientific database is an invaluable source of information for researchers, as it provides an important historical perspective and solid data sets as a basis for future research. For a complete bibliography on scientific studies on the region and on Lake Cadagno before 1998, refer to *Chronological list of publications on Lake Cadagno and Piora Valley* in Peduzzi et al. (1998), and for the period after 1998, check on the website of the Alpine Biology Center, Piora (*http://www.piora.org/*).

7.4 Lake Water Chemistry and Meromixis

The dolomite vein present in the Piora valley is in direct contact with the southern part of the Cadagno basin. The water which percolates through these easily weathered rocks becomes enriched in calcium, magnesium, carbonate and sulphate ions and enters the waterbody through sublacustrine springs. The analyses of seven underwater springs located in the southern part of the lake at depths between 8.0 and 12.5 m provided the following averages for: conductivity, 1359 (±145) μS cm^{-1}; calcium, 171.1 (±22.5) mg l^{-1}; magnesium, 74.7 (±10.1) mg l^{-1}; sulphate, 622.9 (±81.5) mg l^{-1}; and carbonate, 120.7 (±35.6) mg l^{-1} (Del Don et al. 2001). The inflow of this water contributes to the establishment of the crenogenic meromixis in the lake. The chemocline is permanent and stabilized by the difference in density between the two water strata: the upper mixolimnion and the lower monimolimnion (Del Don et al. 2001; Lehmann and Bachofen 1999; Tonolla et al. 1998b).

The concentrations of nutrients in the oxic mixolimnion (between 0 and 10–12 m depth) are low, with phosphate (PO_4_P) close to the detection limit (<0.3 μg l^{-1}), nitrate (NO_3_N) below 11.5 μg l^{-1} and dissolved inorganic carbon (DIC) ca. 10 mg l^{-1} (Tonolla et al. 1998b, 1999). The monimolimnion is anoxic and rich in reduced compounds due to the action of anaerobic sulphur bacteria. Moreover, it is a steady food source indispensable for the development of the dense community of phototrophic sulphur bacteria in the chemocline. High concentrations of sulphur compounds (HS^- up to 30 mg l^{-1}; SO_4^{2-} up to 200 mg l^{-1}) ammonium (NH_4_N up to 3.9 mg l^{-1}), phosphate (PO_4_P up to 0.132 mg l^{-1}), carbonates (HCO_3^- and CO_3^{2-} up to 50 mg l^{-1}) (Del Don 2001; Tonolla et al. 1998b) and dissolved organic carbon (DOC, up to 6 mg C l^{-1}) (Bertoni et al. 1998) are the basis for the massive development of populations of these phototrophic sulphur bacteria, which reach maximum densities of 10^7 cells per ml^{-1} during summers.

With sulphate concentrations ranging between 100 and 200 mg l^{-1} (1–2 mM), Lake Cadagno contains up to ten times more sulphate than most freshwater lakes (Hanselmann and Hutter 1998). In addition, sulphate, bicarbonate, calcium and manganese are the dominant ions in Lake Cadagno (Tonolla et al. 1998a).

For the water stability, the massive development of large-celled, flagellated phototrophic purple sulphur bacteria in the chemocline may also influence the profiles of temperature and salinity by their active movements, contributing to maintain constant values over a depth interval of about 0.5 m. The water mixing caused by the organisms at this level destabilizes the density gradient and the physico-chemical stratification (Wüest 1994).

7.5 Biology of Lake Cadagno

7.5.1 Prokaryotic Diversity

The prokaryotic diversity of Lake Cadagno has been studied by several workers in the last three decades (Züllig 1985; Peduzzi et al. 1998, 2000, 2012; Tonolla et al. 1999, 2000, 2005a). Because of the difficulty in culturing environmental micro-organisms, the study of the bacterial populations was focused, as early as in the 1990s, on direct detection methods such as nucleic acid staining with DAPI or acridine orange (Bensandoun et al. 1998; Tonolla et al. 1998a). Such approaches revealed the presence of several bacterial morphotypes (e.g. *Chromatium okenii*, *Lamprocystis* sp., *Desulfocapsa* and morphotype R) and allowed an evaluation of the bacterial diversity in the lake's water column. The phylogenetic analysis of bacterial communities continued with the amplification and generation of gene clone libraries for 16S ribosomal RNA genes (Bosshard et al. 2000a; Demarta et al. 1998; Schramm et al. 2003). The microbiological analyses did not only focus on the chemocline (Tonolla et al. 1999, 2000, 2005a; Bosshard et al. 2000a, b; Decristophoris et al. 2009; Halm et al. 2009), but also on the monimolimnion, the mixolimnion (Tonolla et al. 2005b; Gregersen et al. 2009; Milucka et al. 2015) and the anoxic sediment (Schubert et al. 2011). They allowed the identification of novel bacterial species, later isolated from the chemocline of the lake (e.g. *Thiocystis chemoclinalis* sp. nov., *Thiocystis cadagnonensis* sp. nov. and *Candidatus "Thiodictyon synthrophicum"* sp. nov.) (Peduzzi et al. 2011; Peduzzi et al. 2012).

Recently a next-generation sequencing technology, 454 pyrosequencing, was also applied to explore the prokaryotic diversity in the layers of the water column (mixolimnion, chemocline, monimolimnion) and the superficial anoxic sediment of Lake Cadagno (unpublished data). This generated 22,700 sequences of the hypervariable V1–V3 regions (average length, 400 bp) of the 16S rRNA gene. Through PICRUSt algorithm (Langille et al. 2013), 3244 sequences (1227 sequences for mixolimnion, 621 for chemocline, 1326 for monimolimnion and 70 for anoxic sediment) could be assigned to 18 known bacterial phyla, 11 of which were dominant (>1 % relative abundance). This number is higher than the 12 phyla identified in the previous studies (Ravasi et al. 2012a). The monimolimnion noticeably differed, with 13 phyla identified by 454 pyrosequencing, compared with eight phyla previously identified. Also, microbial communities differed among the different layers in the water column and the anoxic sediment. The mixolimnion was dominated by the phylum *Verrucomicrobia* (51 %), which was less abundant in the chemocline (8 %) and monimolimnion (4 %) and absent in the anoxic sediment. Other important phyla of mixolimnion were *Actinobacteria* (14%) and *Bacteroidetes* (3 %). *Proteobacteria* were the most represented phylum in the chemocline communities (76 %); they were also well represented in the mixolimnion (31 %), monimolimnion (19 %) and anoxic sediment (18 %). Other important communities in the chemocline were *Chlorobi* (8 %), *Actinobacteria* (5 %) and *Bacteroidetes* (2 %). Other than the presence of *Proteobacteria*, the monimolimnion communities were mainly comprised of *Chlorobi* (66 %), *Actinobacteria* (6 %), *Bacteroidetes* (3 %) and *Chloroflexi* (1 %). The anoxic sediment was the most diverse habitat, with various phyla almost equally represented, in addition to the previously mentioned *Proteobacteria* (*Chloroflexi* 19 %, *Chlorobi* 15 %, *Actinobacteria* 14 %, *Cyanobacteria* 11 %, OP8 8 %, *Bacteroidetes* 7 %, *Firmicutes* 2 % and *Nitrospirae* and *Acidobacteria* 1 %). At a finer taxonomical resolution, *Alphaproteobacteria* were the dominant class of *Proteobacteria* in the mixolimnion, with the family of *Pelagibacteraceae* (aerobic oxygenic bacteria). Alphaproteobacteria also represented a good part of the *Proteobacteria* found in the anoxic sediment (*Hyphomicrobiaceae*, *Pelagibacteraceae* and *Rhodobacteraceae*). The class *Betaproteobacteria* was represented in all samples, mainly by *Comamonadaceae*. *Deltaproteobacteria* were also present in all habitats analysed, except in the mixolimnion; they were best notable in the monimolimnion, with the sulphate reducers *Desulfobulbaceae*, and in the sediment (clade OM27). *Epsilonproteobacteria* were present in chemocline and monimolimnion (*Helicobacteraceae*). *Gammaproteobacteria* were the most dominant *Proteobacteria* class in the chemocline, with the sulphur purple photosynthetic *Chromatiaceae*, and were also prominent in the monimolimnion (*Chromatiaceae*) and the sediment (*Moraxellaceae*, *Pseudoalteromonadaceae*, *Sinobacteraceae*, *Vibrionaceae*).

The diversity of archaea in Lake Cadagno has received less attention than the bacteria. Nonetheless, the few studies that were conducted have highlighted interesting patterns of distribution. Bottinelli (2008) explored the coexistence of sulphate-reducing bacteria and methanogenic archaea in the sediments of the lake.

The analysis of 16S rRNA gene sequences allowed the detection of few phylotypes, with more diversity in the deeper part of the sediments. Schubert et al. (2011) studied the archaeal community involved in methane production or degradation in upper sediments of Lake Cadagno, constructing 16S rRNA gene clone libraries from 2 to 4 cm sediment (42 clones) and 9- to 15-cm (43 clones) sediment depth. These depths represented the peak (2–4-cm sediment) and bottom (9–15-cm sediment) of the AOM (anaerobic oxidation of methane) zone. The archaeal diversity revealed in this study was also low, with only nine different phylotypes detected, based on a 97 % 16S rRNA gene sequence similarity criterion. The majority of 16S rRNA gene sequences belonged to the euryarchaeotal marine benthic group D (MBGD), found in a variety of limnic and marine habitats. In total, six distinct MBGD phylotypes were retrieved, three of which were only found in 2–4-cm sediment depth. The second archaeal lineage detected was the recently described AOM-associated archaea (AAA) clade of Euryarchaeota. As third and fourth archaeal lineages, the crenarchaeotal marine benthic groups B and C (MBGB and MBGC) were identified. Catalysed reporter deposition-fluorescence in situ hybridization (CARD–FISH) analysis revealed a high abundance of *Deltaproteobacteria*, especially of free-living sulphate-reducing bacteria of the *Desulfosarcina/Desulfococcus* branch of *Deltaproteobacteria* in the AOM zone. Here, loose aggregations of AAA cells were found that might be responsible for oxidation of methane in the lake sediments (Schubert et al. 2011). A more recent study by Milucka et al. (2015) on methane oxidation coupled to oxygenic photosynthesis in the chemocline of Lake Cadagno revealed that, despite the geochemical evidence for high methane-oxidizing activity, no anaerobic methane-oxidizing archaea (ANME)-1 or (ANME)-2 were detected in samples from the anoxic zone (i.e. from 10 and 13 m) by CARD–FISH with any of the used probes.

The Anoxygenic Phototrophic Sulphur Bacteria in the Chemocline
One of the main features of sulphur-rich Lake Cadagno is the diverse community of microorganisms which develops in the chemocline especially in the spring and summer months. The community, which is dominated by anoxygenic phototrophic sulphur bacteria, also contains microorganisms that acquire energy via anaerobic respiration, aerobic respiration and fermentation (Bosshard et al. 2000a). Both green sulphur bacteria (GSB) and purple sulphur bacteria (PSB) have been observed in Lake Cadagno, and a number of their strains have been successfully isolated, cultivated in the laboratory and described as new bacterial species (i.e. Eichler and Pfennig 1988; Tonolla et al. 1999; Peduzzi et al. 2011, 2012). *Thiocystis chemoclinalis*, *T. cadagnonensis* and *Candidatus* "*Thiodictyon syntrophicum*" were recently described (Peduzzi et al. 2011, 2012).

Anoxygenic phototrophic sulphur bacteria are phylogenetically diverse, but all harvest light energy using antenna complexes containing bacteriochlorophylls, and accessory pigments, such as carotenoids. The antenna complexes adsorb light radiation and transfer the captured energy to a reaction centre where it is subsequently used to generate ATP and reductant (NAD(P)H). Anoxygenic phototrophic sulphur bacteria utilize either the pheophytin–quinone-type reaction centres (PSB and *Chloroflexus* sp.) or the Fe–S reaction centres (GSB and *Heliobacteria* sp.). In addition to sulphide, some of these bacteria utilize elemental sulphur and thiosulphate as electron donors (Frigaard and Dahl 2009).They can also utilize hydrogen ($H_2(g)$) as a reductant. Many anoxygenic phototrophs can also oxidize ferrous iron to reduce CO_2 (Widdel et al. 1993; Ehrenreich and Widdel 1994; Overmann and Garcia-Pichel 2006; Hegler et al. 2008). The oxidation of reduced inorganic sulphur by sulphide-oxidizing anoxygenic phototrophic bacteria is usually coupled to CO_2 reduction to form cell biomass following a number of possible pathways (Eqs. 7.1–7.5):

$$CO_2 + 2H_2S \rightarrow CH_2O + H_2O + 2S^0 \quad (7.1)$$

$$3CO_2 + 2S^0 + 5H_2O \rightarrow 3CH_2O + 2SO_4{}^{2-} + 4H^+ \quad (7.2)$$

$$2CO_2 + H_2S + 2H_2O \rightarrow 2CH_2O + SO_4{}^{2-} + 2H^+ \quad (7.3)$$

$$CO_2 + 2S_2O_3{}^{2-} + H_2O \rightarrow CH_2O + 2S^0 + 2SO_4{}^{2-} \quad (7.4)$$

$$2CO_2 + S_2O_3{}^{2-} + 3H_2O \rightarrow 2CH_2O + 2SO_4{}^{2-} + 2H^+ \quad (7.5)$$

Taxonomy of Anoxygenic Phototrophs from Lake Cadagno

In Lake Cadagno, the cell concentration maxima of anoxygenic phototrophs (*Chromatiaceae* and *Chlorobiaceae*) in the summer are located between 10.5 and 13.5 m depth in the chemocline (Schanz et al. 1998; Bosshard et al. 2000a). The community of phototrophic bacteria in the chemocline is diverse but dominated by a few species of purple and green sulphur bacteria (Tonolla et al. 1999; Bosshard et al. 2000a). The main anoxygenic phototrophic bacteria identified in Lake Cadagno (Table 7.1) are three flagellated PSB, one large-celled, *Chromatium okenii*, and two small-celled, *Thiocystis chemoclinalis* and *T. cadagnonenensis*; four small-celled PSB *Lamprocystis purpurea*, *Candidatus* "*Thiodictyon syntrophicum*", *Lamprocystis roseopersicina* and *Lamprocystis* sp. strain D; and two GSB *Chlorobium clathratiforme* and *Chlorobium phaeobacteroides* (Bosshard et al. 2000a, b; Tonolla et al. 1999, 2003, 2005b; Peduzzi et al. 2011, 2012). The main sulphide-oxidizing anoxygenic phototrophs are grouped according to their physiology, morphology, pigment composition and membrane structure (Table 7.1).

The PSB belong to the gamma division of the *Proteobacteria*. They have BChl *a* as the main bacteriochlorophyll and contain carotenoids, such as okenone. The reaction centre is of Type II containing pheophytin and quinone. They fix CO_2 using the reductive pentose phosphate cycle. Metabolically, they can be photoautotrophs or photoheterotrophs and in some cases also chemolithotrophs growing in the absence of light (Table 7.1). Two examples of strains that can grow chemolithotrophically are *Lamprocystis purpurea* and *Candidatus* "*Thiodictyon syntrophicum*". PSB are mostly obligate anaerobes, although some may tolerate low-oxygen conditions. The four main small-celled PSB found in Lake Cadagno are between 1.4 and 4 μm in diameter and spherical to oval in shape. The best described of the group, *L. purpurea* and *Candidatus* "T. syntrophicum", utilize

Table 7.1 Characteristics of the main sulphide-oxidizing anoxygenic phototrophs in the chemocline of Lake Cadagno

Organism	Group	Phylogenic position	Electron donor	Cell size [μm]	Shape	Abs max [nm]	Photosynthetic pigments	C-fixation pathway	Other characteristics	References
Chromatium okenii	PSB	γ-*Proteobacteria*	Sulphide, thiosulphate, elemental sulphur	4.5–6 × 8–15	Rod	370, 520, 835	Purple-red colour BChl *a* okenone	Reductive pentose phosphate cycle	Motile, intracellular sulphur globules	Fischer et al. (1996), Tonolla et al. (1999)
Cand. Thiodictyon syntrophicum (*Strain F, Cad16*^{T})				1.4–2.4[a]	Spherical to oval	374, 528, 582, 833	Purple-red colour BChl *a* okenone		Slime capsule, intracellular sulphur globules, gas vacuoles, immotile, aggregation, chemolithotrophic	Peduzzi et al. (2003b, 2012)
Thiocystis chemoclinalis (*Strain CadH11*^{T})				2.3–3.6	Spherical to oval	525, 590, 830	BChl *a*, okenone		Motile by flagella, aggregates, chemolithotrophic growth in the presence of oxygen in the dark	Peduzzi et al. (2011)
Thiocystis cadagnonensis (*Strain Cad448*^{T})				2.3–4.7	Ovoid to rod	525, 590, 830	BChl *a*, okenone		Motile by flagella, chemolithotrophic growth under oxic and micro-oxic conditions in the dark with sulphide and thiosulphate	Peduzzi et al. (2011)
Lamprocystis purpurea				1.9–2.3 × 2.0–3.2	Spherical to oval	369, 523, 580, 770, 805, 831	Purple-red colour BChl *a* okenone		Nonmotile, aggregation, intracellular sulphur globules, gas vacuoles, vesicular-type photosynthetic membranes, chemolithoautotrophic with sulphide and thiosulphate in dark	Eichler and Pfennig (1988), Fischer et al. (1996), Tonolla et al. (1999), Imhoff (2001), Peduzzi et al. (2012)
Lamprocystis roseopersicina				2.0–3.5	Spherical to oval	N/A	Pink-violet colour BChl *a* rhodopinal		Motile by flagella, gas vesicles, irregular aggregates	Tonolla et al. (1999), Bosshard et al. (2000a), Imhoff (2001)

(continued)

Table 7.1 (continued)

Organism	Group	Phylogenic position	Electron donor	Cell size [μm]	Shape	Abs max [nm]	Photosynthetic pigments	C-fixation pathway	Other characteristics	References
Chlorobium clathratiforme	GSB	Chlorobi	Sulphide	~1 × 4–7	Rod	750	Brown colour BChl *e* isorenieratene	Reductive citric acid cycle	Net-forming, immotile, cultivation-dependent gas vacuole formation	Pfennig and Cohen-Bazire (1967), Gregersen et al. (2009), Habicht et al. (2011)
Chlorobium phaeobacteroides			Sulphide, elemental sulphur	0.6–0.8 × 1.3–2.7	Rod	725	BChl *d* isorenieratene		Nonmotile	Pfennig and Cohen-Bazire (1967)

Notes: *Bchl* bacteriochlorophyll, *PSB* purple sulphur bacteria, *GSB* green sulphur bacteria
[a]Cell size depicts diameter as measured for a spherical cell, unless otherwise marked

BChl *a* and the carotenoid okenone to harvest light for photosynthesis. The small-celled PSB in Lake Cadagno are immotile and contain gas vacuoles and intracellular sulphur globules. They also form aggregates, notably together with sulphate-reducing bacteria (SRB) belonging to the *Desulfovibrionaceae* (Tonolla et al. 2003; Peduzzi et al. 2003b; review by Tonolla et al. 2004). This association with the SRB is considered to provide them with a environmental advantage by providing substrate (Descristophoris et al. 2009). The 16S rRNA gene sequence of *Lamprocystis roseopersicina* is very similar to that of *Lamprocystis purpurea*, a finding that supports the reclassification of the latter species from *Amoebobacter* to *Lamprocystis* (Imhoff 2001). The large-celled PSB in the lake, *Chromatium okenii*, also belongs to the gamma division of the *Proteobacteria* (Table 7.1). This bacterium is rod shaped and has large cells (4.5–6 × 8–15 μm). *C. okenii* are motile using flagella and they harbour intracellular sulphur globules (Fig 7.2).

The GSB belong to a separate phylum (*Chlorobi*). They have a Type I reaction centre (Fe–S) and are generally considered strict photoautotrophs fixing CO_2 using the reverse citric acid cycle. The occurrence of the concentrated pigments in chlorosomes allows the GSB to use light at very low intensities (Frigaard and Bryant 2006). The abundance of the brown-coloured GSB, *Chlorobium clathratiforme*, increased in 2000 in Lake Cadagno and thereafter formed a significant population in the chemocline (Tonolla et al. 2005b; Gregersen et al. 2009; Habicht et al. 2011). *Chlorobium clathratiforme* has irregular, rod-shaped cells with typical dimensions of ~1 by 4–7 μm (Fig 7.2). They contain BChl *e* and the carotenoid isorenieratene. The cells are net-forming and nonmotile and display cultivation-dependent gas vacuoles.

Temporal and Spatial Distribution of the Anoxygenic Phototrophic Sulphur Bacteria

The population density of anoxygenic phototrophic sulphur bacteria in the chemocline varies with depth in the anaerobic water column (upper monimolimnion), seasonally and over longer timescales. As expected from the

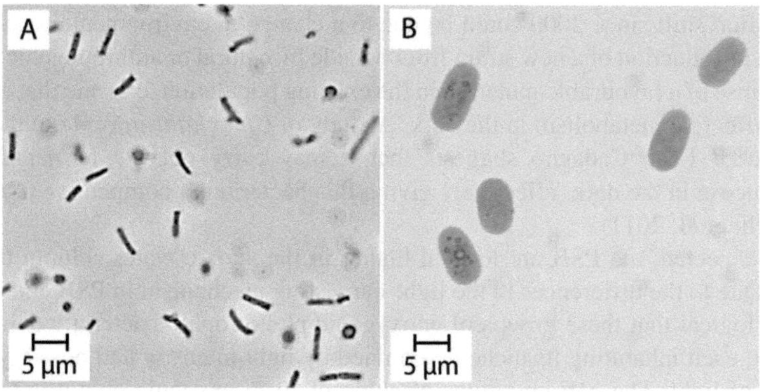

Fig. 7.2 Light microscopy images of (**a**) green sulphur bacterium *Chlorobium clathratiforme* and (**b**) purple sulphur bacterium *Chromatium okenii* showing intercellular sulphur globules

geochemical characteristics of the lake, there is a strong change in community composition between the oxic and anoxic layers (Bosshard et al. 2000b). The seasonal distribution of the anoxygenic phototroph population in the chemocline suggests that the population is adapted to changing geochemical niches (Tonolla et al. 1999). In March and June 1999, bacterial populations were evenly distributed in depth, but a microstratification of the population was observed as the season progressed and the geochemical gradient developed (Tonolla et al. 2003). A shift in dominance among the *Chromatiaceae* from *Chromatium okenii* in the spring and early summer to an increase in *Lamprocystis purpurea* in the late summer and autumn was observed by Bosshard et al. (2000a, b). However, this pattern has not been consistent in recent decades. In another seasonal study, Tonolla et al. (2003) observed high densities of *C. okenii* in the late autumn, but small-celled forms of the PSB were most abundant during spring and summer. Thus, even among the PSB populations, a number of ecophysiological factors contribute to the population composition. These results are further confounded by the motility of *C. okenii*, likely a response to light and nutrient availability.

The changes in the anoxygenic phototroph community structure in Lake Cadagno during the last decades are intriguing. Also, the relative abundance of PSB and GSB has changed dramatically over the past 15 years. A study on in situ hybridization during 1994–2003 shows that up to 2001, PSB were the most prominent phototrophic sulphur bacteria forming 70–95 % of the total bacterial population (review by Tonolla et al. 2004). In 2000, the dominance shifted from PSB to GSB, due to a dramatic increase of *Chlorobium clathratiforme* (Tonolla et al. 2005b). The bacterial biomass increased threefold, due entirely to the increase in abundance of GSB, while PSB densities remained stable possibly because of associations with SRB in the aggregates (Decristophoris et al. 2009). After this change, up to 95 % of the phototrophic sulphur bacteria community was comprised of *Chl. clathratiforme* (Tonolla et al. 2005b). Indeed, subsequent studies revealed that *Chl. clathratiforme* in Lake Cadagno is a clonal population forming up to 70 % of the bacterial 16S rRNA gene sequences obtained in the chemocline at 11–12-m depth, where the cell concentration was the highest (Gregersen et al. 2009). The population shift since 2000 could be due to a change in environmental conditions, i.e. the introduction of a new strain from outside by natural or anthropogenic means, or because of a favourable mutation in the existing population, e.g. one that allows a more effective metabolism in the dark. A study of *Chl. clathratiforme* in the water column of Lake Cadagno suggests that it may carry out the fermentation of polyglucose in the dark, effectively giving this bacterium a competitive advantage (Habicht et al. 2011).

As expected, the PSB are located higher in the anoxic water column than are GSB. Due to the differences in the light-harvesting mechanism in PSB and GSB, it seems logical that these groups of anoxygenic phototrophic bacteria would easily coexist, each inhabiting its niche determined by light intensity and possibly wavelength of light. The concentrations of chlorophyll in oxygenic phototrophs in the oxic upper layers of Lake Cadagno are low. At water depths corresponding to the depth of chemocline, the wavelength distribution of the available light is largely determined by the light absorbance by water itself. The wavelengths reaching a depth of 10 m mainly correspond to that absorbed by carotenoids and the Qx band of BChl *e*. The intensity of near-infrared radiation at wavelengths above 700 nm is much less so the characteristic long-wavelength absorbance of aggregated BChl *e* in the chlorosomes of *Chl. clathratiforme* or protein-associated BChl *a* in the PSB is unlikely to be important in the lake. Nonetheless, long-term studies of the phototrophic population reveal that the changes in light-related limnological properties of the lake, such as turbidity, light availability and light intensity over long timescales (decades), profoundly influence the population composition (Tonolla et al. 2005b).

Role of the Anoxygenic Phototrophic Sulphur Bacteria in the Primary Production of Lake Cadagno

Although the chemocline represents only a small part of the total lake volume, here about half of the total daily photoassimilation of carbon in the lake occurs (Camacho et al. 2001). Photosynthetic ^{14}C-assimilation, measured with radioisotope (^{14}C) technique, generally peaks at 1–3 m (phytoplankton) and at 10–12 m, the location of the chemocline with the anoxygenic phototrophic bacteria community. Very high values of phototrophic and chemotrophic C-assimilation are generally observed below 10 m (Bossard et al. 2001). Indeed, in a study on Lake Cadagno in September 1999, Camacho and coworkers found high rates of dark carbon fixation by chemolithotrophs in the chemocline, suggesting a high contribution of these organisms to the overall carbon cycle of the lake (Camacho et al. 2001). Recently, the rates of CO_2 assimilation of the most abundant phototrophic sulphur bacteria of the chemocline were measured using both nanoscale secondary ion mass spectrometry (nanoSIMS) (Musat et al. 2008; Zimmermann et al. 2015) and $^{14}CO_2$ quantitative assimilation in dialysis bags (Storelli et al. 2013). Thus, the strongest assimilator in the presence of light was the large-celled PSB *C. okenii*, while in the dark the major assimilator was the small-celled PSB *Candidatus* "T. syntrophicum", which was also a strong CO_2 assimilator in the light. While *C. okenii* represents only ca. 0.3 % of the total cell number, it contributes 70 % of the total uptake of carbon in the system (Musat et al. 2008), an assimilation effectiveness theorized by the authors to be based in *C. okenii*'s mobility towards substrate and optimal light conditions. Likewise, although it only contributes just about 2 % of the total bacterial populations in the chemocline, *Candidatus* "T. syntrophicum" appears to also play a greater role in CO_2 fixation in Lake Cadagno, irrespective of if there is light (Storelli et al. 2013). The most abundant population of the chemocline, GSB *Chl. clathratiforme*, showed only low CO_2 fixation rates. These results were supported by a micro-autoradiography analysis which revealed that about 10 % of the cells of *Chl. clathratiforme* from below the photic zone in Lake Cadagno fixed CO_2 in the dark, and 20–55 % of the cells from the photic zone fixed CO_2 in the light (Habicht et al. 2011). Although GSB are considered strictly phototrophic, their low rates of CO_2 photoassimilation are surprising. Another metabolic pathway probably contributes to biomass production. At present, only acetate and pyruvate are known to be used by GSB as organic carbon source in the presence of sulphide and CO_2 (Van Gemerden and Mas 1995).

Non-photosynthetic carbon assimilation is often observed in phototrophic sulphur bacteria but the pathway is not yet fully understood. In order to gain insight into the process of dark CO_2 fixation, two-dimensional differential gel electrophoresis (2D-DIGE) was used to monitor the global changes in the proteome of *Candidatus* "T. syntrophicum" strain Cad16^T in anoxic autotrophic cultures, irrespective of availability of light. Interestingly, three enzymes more abundant in the dark are part of a hypothetical anaerobic dicarboxylate/4-hydroxybutyrate (DC/HB) cycle, an autotrophic CO_2 fixation pathway found essentially in archaea (Berg 2011). The substrates needed for this process [NAD(P)H and acetyl-CoA] should be provided by the degradation of the storage globules of poly (3-hydroxybutyrate) (PHB), whose synthesis was shown to be higher in the presence of light (Storelli et al. 2014).

The Sulphate-Reducing Bacteria (SRB)

Within the chemocline, non-phototrophic bacteria are also present: these include SRB that can account for a significant fraction of the total community (Tonolla et al. 1998a). Their seasonal depth distribution was consistent with the variations of the phototrophic PSB. Over an annual cycle, however, differences in abundance of bacteria detected with probes SRB385 (detecting most *Desulfovibrionaceae*) and SRB385Db (detecting most *Desulfobacteriaceae*) were observed (Peduzzi et al. 2003a). In situ determination of sulphide turnover rates (Luthy et al. 2000) and high-resolution images of sulphide concentrations in the water column (Lehmann and Bachofen 1999) confirmed the presence of active SRB in the bacterial layer dominated by the phototrophic sulphur bacteria and introduced the possibility of a significant sulphur cycling within the chemocline coupled to a rapid turnover of sulphide (Fritz and Bachofen 2000). The vertical distribution profiles of SRB detected with probes SRB385 and SRB385Db were similar to those of the phototrophic sulphur bacteria and amounted on average 24 % of the DAPI-stained bacteria in the chemocline before the rise of the GSB *Chl. clathratiforme* population at the beginning of the twenty-first century. This percentage shows that SRB made up a significant part of the bacterial population in the chemocline of Lake Cadagno (Tonolla et al. 1998a, 1999, 2000). The in situ observations also reveal majority of the SRB consisted of one particular morphotype, which was observed in cell aggregates formed by small-celled phototrophs (Fig 7.3).

Desulfomonile tidjei in the monimolimnion of Lake Cadagno is easily recognizable as it emits a green fluorescence after acridine orange staining (Bensadoun et al. 1998) and is detectable with the two specific probes DsmA445 and DsmB455 (Tonolla et al. 2005a). Its vertical distribution was unusual, compared with other SRB, as it did not produce a maximum in abundance in the chemocline, but its distribution pattern followed that of hydrogen sulphide profiles. It accounted from 17 to 44 % of the total bacterial numbers, especially at 15–17 m depth (Bensadoun et al. 1998; Tonolla et al. 1998a, 2005a). Seasonal maximum was most frequently

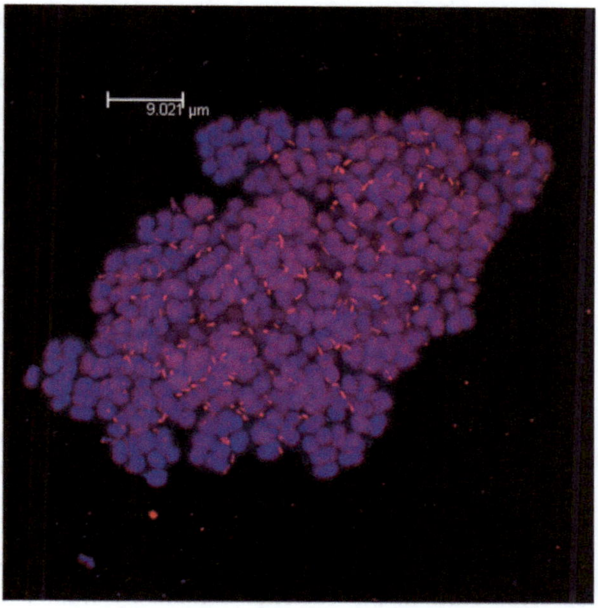

Fig. 7.3 Aggregates of small-celled purple sulphur bacteria (*in purple-violet, coccoid cells*) and their association with sulphate-reducing bacteria of the genus *Desulfocapsa* (*pink-red rod shaped cells*) in samples from the chemocline of Lake Cadagno. Confocal laser scanning microscopy image

around 30 %, the calculated biovolumes reaching values between 10 and 17 % of the total bacterial volume (Tonolla et al. 1998a, 2005a).

7.5.2 Interactions and Functions of the Microbial Populations

Long-term Alternation Between Green and Purple Bacteria

The long-term changes of GSB and PSB in Lake Cadagno were followed by analysing pigments of photosynthetic sulphur bacteria and 16S rRNA gene sequences in long-core sediment samples. The carotenoid pigments okenone and isorenieratene were extracted and identified, and their concentrations measured semi-quantitatively by chromatography (Wirth et al. 2013). Total DNA was extracted, and, for each of the sediment samples, the number of copies of 16S rRNA genes of nine targeted phototrophic sulphur bacterial populations (seven *Chromatiaceae* and two *Chlorobiaceae*) was determined by quantitative PCR (Ravasi et al. 2012b).

Out of the seven *Chromatiaceae* populations analysed using DNA sequences, six contained okenone and only one, *Lamprocystis roseopersicina*, contained rhodopinal (Peduzzi et al. 2011, 2012). During 1990–2012, *L. roseopersicina* had very low numbers in the water column and therefore constituted a minor fraction of the total *Chromatiaceae* population (Tonolla et al. 2005b). Both the investigated *Chlorobiaceae* produce isorenieratene. Therefore, as expected, records from lipid biomarkers (carotenoid pigments) and DNA analysis showed a consistent pattern (Wirth et al. 2013).

The GSB/PSB ratio was found to vary during the evolution of the lake. In the early Holocene (10.5–8 cal kyr BP), GSB weakly dominated over PSB. On the contrary, the middle Holocene (8–5 cal kyr BP) was characterized by a minor prevalence of PSB. In the period <5 cal kyr BP, flood frequencies in the lake area increased. Floods and mass movement events (i.e. mobilization, mixing and redeposition of sediments from shallower parts of the lake to the deepest lake area) generated turbiditic underflows that sporadically introduced O_2-rich water to the hypolimnion, thus briefly weakening the sulphidic conditions. Concomitantly with the rise in flood frequency, the relative abundance of GSB increased again. Here, the two isorenieratene peaks at 3.7 and 3.4 cal kyr BP seemed to correspond to flood-frequency highs with a time lag of roughly 50–100 years (e.g. at 4.3, 3.8, 3.4 and weakly at 2.5 thousand years ago). Enhanced light absorption due to higher suspended particle concentrations and algal/cyanobacterial growth in the surface and subsurface waters during floods likely reduced light availability within the chemocline. Since GSB are known to be more low-light-tolerant than PSB (Biebl and Pfennig 1978; Brocks and Schaeffer 2008), this could have triggered the observed shift in the bacterial community from a PSB- to a GSB-dominated population (Tonolla et al. 2005a; Decristophoris et al. 2009; Gregersen et al. 2009; Wirth et al. 2013). In any case, there appears to be a time lag of 50–100 years between the deposition of the most prominent flood layers and the subsequent restructurings of the phototrophic community. It is thus uncertain what exactly controlled past shifts in the microbial community structure in the water column of Lake Cadagno. Between 1.5 and 0.5 cal kyr BP, flood frequency decreased, while the Mo burial rate again increased and PSB became more abundant. However, with increasing flood activity towards the present time, the bacterial community structure has again shifted to dominance of GSB.

Microbial Cell–Cell Interactions: The Case of Lake Cadagno aggregates
We already showed that in the chemocline, up to 15 % of the cells of the total microbial community were associated in three-dimensional cell–cell aggregates (Peduzzi et al. 2003b; Decristophoris et al. 2009). The isolate Cad16 was recently proposed as *Candidatus "Thiodictyon syntrophicum"* sp. nov. strain Cad16^T, a provisionally novel species within the genus *Thiodictyon* (Peduzzi et al. 2012). This PSB strain Cad16^T together with the SRB *Desulfocapsa thiozymogenes* (isolate Cad626) forms stable aggregates in the lake. We will further refer to these aggregates as *Lake Cadagno aggregates.* Aggregation of cells of *Candidatus* "T. syntrophicum" and *Desulfocapsa thiozymogenes* is not mandatory as it is for other consortia (Overmann and Schubert 2002), since in their natural environment both partners can be encountered as free-living cells although numerically inconspicuous (Peduzzi et al. 2003b; Decristophoris et al. 2009).

Major community shifts and changes have recently been observed not only in the phototrophic sulphur bacteria but also in size and composition of aggregates in Lake Cadagno (see in Sect. 5.1.1.2: Temporal and spatial distribution of the anoxygenic phototrophic sulphur bacteria). Aggregates' particle size decreased, whereas in 2004, compared with 1998, the number of SRB cells in the aggregates increased about three times (Fig 7.3) (Decristophoris et al. 2009). A reduction in size of *Lake Cadagno aggregates*, due to reduced abundance of small-celled PSB, was previously described for winter and spring periods when ice and snow cover had reduced light transmission (Peduzzi et al. 2003b). Although the driving forces that are regulating aggregate formation and composition are still not well understood, overall, these findings suggest that the aggregate formation is a dynamic association responding to shifts in environmental conditions (e.g. changes in sulphide and oxygen concentration as suggested also by Overmann 1997) and to changes in global microbial community structure.

Ecological consequences of aggregate formation may be important for other members of the microbial community not directly involved in aggregation since aggregation might significantly alter competition dynamics among the phototrophic sulphur bacteria community. During periods of intense sulphide photo-oxidation, anoxygenic photosynthesis at the upper boundary of the bacterial layer is limited by the availability of reduced sulphur compounds, a situation encountered in Lake Cadagno (Luthy et al. 2000) and also Lake Mahoney (Overmann 1997) and other lakes (Shira, Shunet, Cisò, etc.) treated in this book (see Chaps. 7 and 9 of part II). Since sulphide and light occur in vertical opposing gradients with depth in the lake, phototrophic sulphur bacteria can find suitable conditions for growth only in a narrow zone of overlap between the two opposing gradients. Under these conditions, *Candidatus* "T. syntrophicum" in aggregates could have a growth advantage over non-associated phototrophic sulphur bacteria since the sulphate-reducing partner could serve as a permanent source of sulphide. Thus, an internal sulphur cycling has the potential to influence competition dynamics among the anaerobic photosynthetic community.

Peduzzi et al. (2003a, b) discussed the potential physiological interactions between the microbial partners in *Lake Cadagno aggregates*. However, from the results presented, it is evident that the role of both sulphur and carbon compounds in the interactions of the bacteria in this association needs to be elucidated in detail. Exploiting nanoSIMS provided more evidence that the metabolic interaction and exchange are not limited to sulphur compounds but most probably also involve organic metabolites (Musat et al. 2008, 2012).

7.5.3 Novel Eukaryotic Microbes in the Anoxic Water Column

Eukaryotic microbial species diversity in freshwater ecosystems is not well understood and especially so in permanently anoxic systems (Saccà 2012). Currently, only a few molecular sequencing studies exist that address this issue in meromictic freshwater lakes (Lefèvre et al. 2007; Triadó-Margarit and Casamayor 2012;

Gies et al. 2014). From a species biodiversity perspective, meromictic lakes are the only freshwater lakes suitable for obligate anoxic organisms and should, at least in theory, hold a higher biodiversity and possibly also higher endemism, the surrounding oxic environment simply acting as a barrier to dispersal.

The anaerobic eukaryotic biodiversity of Lake Cadagno was examined by constructing SSU rRNA clone libraries (Frigaard, unpublished data). Samples for the SSU rRNA clone libraries were collected from three depths: the lower oxic mixolimnion (10 m), the chemocline (11.5 m) and the anoxic monimolimnion (15 m), during 24–28 September 2006. The cell size fraction of the microorganisms analysed was in the range between 0.2 and 149 μm. Between 49 and 98 SSU rRNA sequences per depth were retrieved from the libraries made. Further details about the physico-chemical measurements, sampling and sample preparation are described elsewhere (Gregersen et al. 2009). At the time of sampling, the Chl *a* concentration in the samples from the mixolimnion, chemocline and monimolimnion was approx. 7, 45 and <1 μg L^{-1}, respectively. The ratio of Chl *c* to Chl *a* was about 0.25 in both the mixolimnion and chemocline, indicating that Chl *c*-containing algae were present. Chl *c* was undetectable in the monimolimnion.

The non-parametric richness estimator S_{Chao1} values were about twofold higher in the monimolimnion than in the mixolimnion and about fourfold higher than in the oxycline, which indicates a more complex eukaryotic community at the time of sampling in the monimolimnion than in the oxycline or mixolimnion. Cryptophytes and ciliates dominated the libraries from Lake Cadagno (Fig. 7.4). The dominance of cryptophytes in the chemocline corresponds well to the high concentration of Chl *c* found here (11 μg L^{-1}). The other common groups encountered in the samples included choanoflagellates (related to *Monosiga* and *Desmarella*) and metazoan rotifers (Monogononta), retrieved from all three libraries, and fungi and ichthyosporean sequences retrieved from the mixo- and monimolimnic samples. The absence of Chl *b* in the examined samples corresponds well to the absence of sequences from chlorarachniophytes, green algae and euglenophytes.

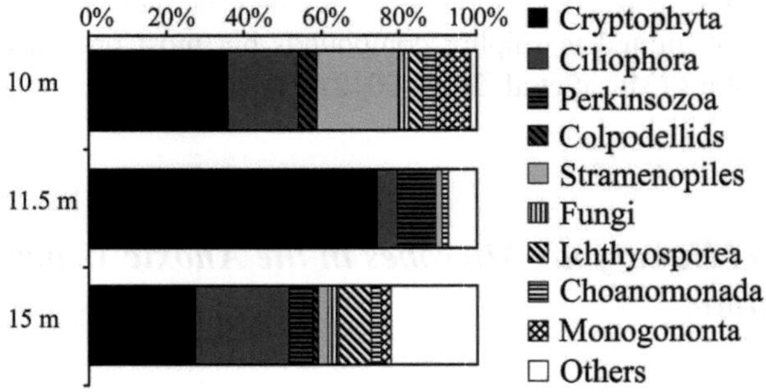

Fig. 7.4 Distribution of the main eukaryotic groups present in Lake Cadagno based on the number of clones as compared with total numbers of clones retrieved (see text for details). "Others" include sequences and OTUs (operational taxonomic units) of uncertain taxonomic origin as well as OTUs of heterolobosean and katablepharidean origin

One cryptophyte operational taxonomic unit (OTU) dominated both the oxycline and monimolimnion clone libraries based on the number of clones retrieved, but no reference sequence in GenBank matched the retrieved SSU rRNA sequences. However, the cloning generated nine nucleomorph SSU rRNA sequences identical to the nucleomorph sequence of *Cryptomonas phaseolus* (Gregersen et al. 2009). This species has been found to be sulphide tolerant and to be the dominant single cryptophyte species in the deep chlorophyll maximum in Lake Cisó, a holomictic lake in Spain that periodically experiences total anoxia with high levels of sulphide in the hypolimnion (Pedrós-Alió et al. 1987). *C. phaseolus* was also dominant in the holomictic Lake Großer Vätersee in Germany (Gervais et al. 2003), which also has sulphide-rich hypolimnic water. It forms stable populations, probably due to the lack of competition from other eukaryotes in the oxycline, one of the normal mechanisms believed to account for succession (Gasol et al. 1992).

Within the alveolates, the dominant group was the Ciliophora (ciliates). As expected, different sequences were retrieved from the oxic and anoxic samples, as ciliates in general are adapted to varying oxygen concentrations. The ciliate sequences found in the monimolimnic sample were most closely related to sequences from species known to be obligate anaerobes: *Caenomorpha uniserialis*, *Cyclidium porcatum*, *Metopus contortus* and *Trimyema compressum*. All four species have been found to harbour endosymbiotic methanogenic bacteria (Wagener et al. 1990) and also other bacteria with unknown function (Clarke et al. 1993; Bruggen et al. 1983).

In the genetic library for the monimolimnion, four sequences retrieved were related to the heterolobosean amoeba *Sawyeria marylandensis* (Fig. 7.5). The latter was isolated from an anoxic salt marsh sediment sample and was found to be microaerophilic (O'Kelly et al. 2003). Together with *Monopylocystis visvesvarai* and *Psalteriomonas lanterna*, it forms a highly supported clade within the heteroloboseans. As the Lake Cadagno OTU was found to be unique (<91 % sequence identity), a water sample from the lake was screened for possible amoeboid cells, but none were encountered. However, in both the oxycline and monimolimnic samples, a round organism was present in large numbers, feeding ferociously on the photosynthetic sulphur bacteria present in large numbers (Fig. 7.6a). Single-cell PCR of this organism revealed that it had an identical SSU rRNA with the heterolobosean OTU. Phase-contrast microscopy revealed two flagella in this organism, one long and thin and the other one short and thick (Fig. 7.6b). The general outline and presence of flagella distinguish this organism from the outline of *Sawyeria marylandensis*, and it is likely that it represents both a new genus and a new species. However, we consider that for a formal description, an ultrastructural examination needs to be conducted.

While the abundance of alveolate ciliates, choanoflagellates and fungi sequences in Lake Cadagno is similar to that found in molecular studies on protists in Lake Pavin in France (Lefèvre et al. 2007), the absence of cercozoan sequences in Lake Cadagno is surprising, unlike in the mixolimnion and in the oxycline of Lake Pavin.

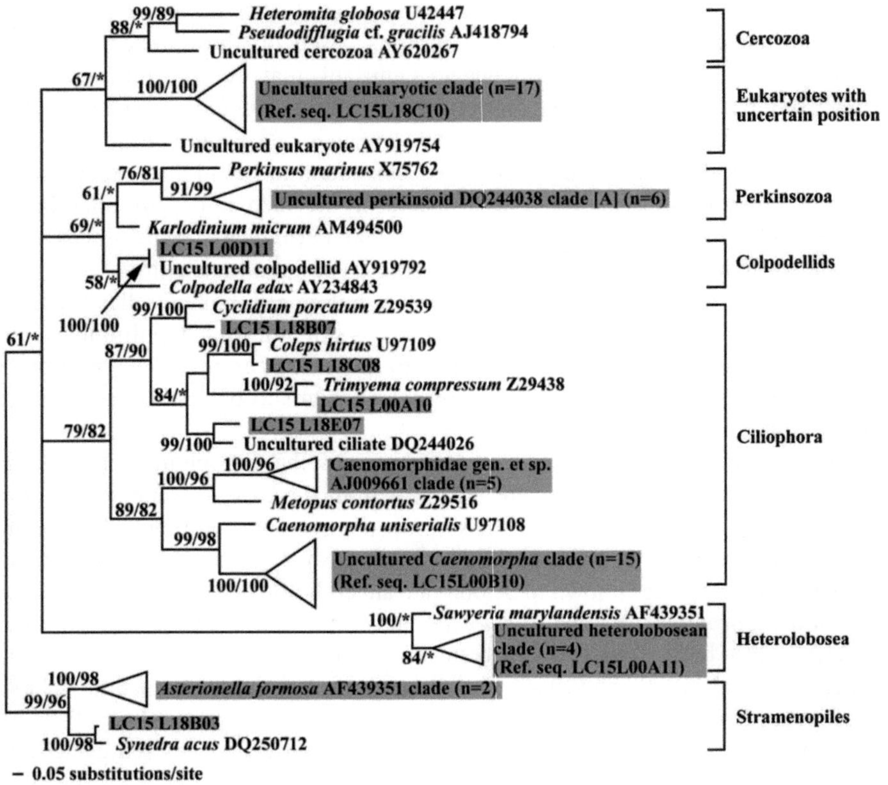

Fig. 7.5 Phylogenetic tree of a subset of monimolimnic SSU rRNA sequences (including the heterolobosean sequences) based on Bayesian analysis. Support values are Bayesian posterior probability values and neighbour-joining bootstrap support values

Lefranc et al. (2005) retrieved cercozoan sequences only from the oligomesotrophic Lake Pavin and the eutrophic Lake Aydat, in France. Thus, the absence of cercozoan organisms might be a consequence of nutrient limitation. Lake Cadagno can be characterized as an oligotrophic to oligomesotrophic lake, and this might partly explain the absence of cercozoans. However, further studies are needed to clarify whether chemical differences or population dynamics could explain the observed difference.

In conclusion, eukaryotic primary producers in the anoxic zones were dominated by *Cryptomonas phaseolus*. Eukaryotic heterotrophs were dominated by ciliates and choanoflagellates, but also included a novel clade of heteroloboseans and two novel clades of uncertain position distantly related to opisthokonts and Cercozoa. Using single-cell PCR, the novel heterolobosean clade was linked to an abundant heterotrophic flagellate in the lake that feeds on anaerobic phototrophic sulphur bacteria.

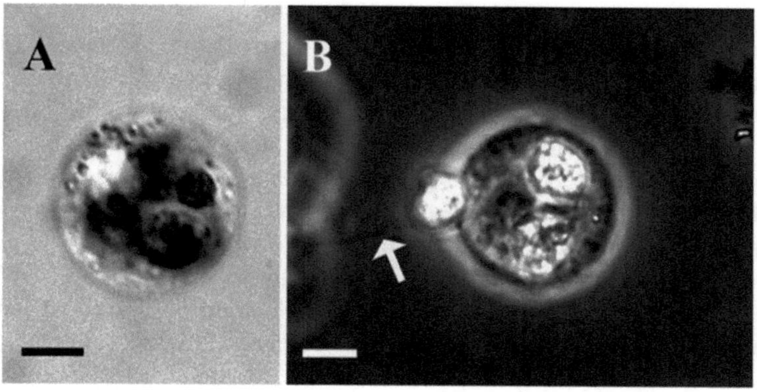

Fig. 7.6 Light micrographs of the novel heterolobosean organism observed in Lake Cadagno. (**a**) Bright-field micrograph showing the general outline of the cell. The cell interior is filled with ingested phototrophic sulphur bacteria. (**b**) Phase-contrast micrograph showing the ingestion of phototrophic sulphur bacteria. *Arrow* points to the longest flagellum; the smaller flagellum is not visible. The bacterium slides along the flagellum and ingestion occurs along close to the flagellar base. Scale bars = 10 μm

7.5.4 Phyto- and Zooplankton

Güttinger and Straub (1998) addressed the question whether the special conditions (meromicticity) of the lake had any influence on the composition of the diatom flora. As expected, live diatoms in Lake Cadagno have been found in the oxic layer only and to a depth of about 10 m. A list of diatom species observed in the water column in earlier (e.g. Schanz et al. 1988) and recent studies is given in Güttinger and Straub (1998). However, in their qualitative study, no species were found, which are not common in other alkaline alpine lakes of non-meromictic character (Güttinger and Straub 1998).

Other authors followed phytoplankton summer dynamics in Lake Cadagno (Schanz and Stalder 1998). At the beginning of summer, phytoplanktonic communities in the upper 10 m are characterized by a uniform vertical distribution of pennate diatoms (above all species of the genus *Fragilaria*, i.e. *F. crotonensis* and *F. capucina*) and centric diatoms (species of the genera *Stephanodiscus*, with *St. hantzschii*, and *Cyclotella*) as well as green algae (mostly *Scenedesmus costato-granulatus* and *Dictyosphaerium subsolitatum*). Pennate and centric diatoms become dominant later in the summer. In autumn, green algae (*Echinocoleum elegans*, *Sphaerocystis schroeterii* and *Oocystis lacustris*) dominate (Camacho et al. 2000; Schanz and Stalder 1998).

Chlorophyll *a* concentrations in the water column vary considerably, e.g. in summer 1988 between 26 mg m^{-2} in late July and 4 mg m^{-2} in early September or in summer 1986 between 18 and 65 mg m^{-2}, thus depending on the period of sampling, while phytoplankton primary production varies between 2 and 17 mg C m^{-3} h^{-1}, from 0 to 9.5 m depth (Schanz and Friedl 1993; Bossard et al.

2001). The comparison of 1987 and 1997 production rates did not show any statistically significant variation (Bossard et al. 2001; Pasini and Schanz 1998).

Dense developments of *Fragilaria capucina* and *Cyclotella comensis* are sometimes noted in the oxic–anoxic transition zone of the upper chemocline. A few centimetres deeper, the massive presence of cryptomonads *Cryptomonas erosa* and *Cryptomonas phaseolus* sometimes gives the water an intense green colour. At the same depth, often by nearly anoxic conditions, an important zooplankton community thrives all year round (Camacho et al. 2001; Tonolla et al. 1988). The dominant species are Cladocera *Daphnia longispina* complex and *Bosmina longirostris*, which often represent up to 40 % of the zooplankton density. Rotifers constitute, however, the largest share of zooplankton diversity (15 species, according to Riccardi et al. 2012), with *Conochilus* sp. (35 % of total zooplankton density) and *Asplanchna priodonta* (10 %) (Camacho et al. 2001). Other zooplankton species encountered in significant quantities are the calanoid copepod *Acanthodiaptomus denticornis* and the cyclopoid *Cyclops abyssorum*. The complete list of zooplankton species of Lake Cadagno is available in Riccardi et al. (2012).

Recently, possible responses of the biological parameters of the Lake to climate changes were investigated through the analysis of the populations of different diatom genera found in the sediments over the last 100 years (Tiffay et al. 2015). A significant shift in relative genus abundance was observed, with a dominance of small-size planktonic diatoms in recent years (1985–2012) compared with periphytic/benthic diatoms dominated in earlier years (1900–1950). Interestingly, taxa previously not observed in the Lake (Güttinger and Straub 1998, Riccardi et al. 2012) were also identified in the sediments: *Aneumastus* sp., *Craticula* sp., *Gyrosigma* sp., *Meridion* sp. and *Reimeria* sp. (Tiffay et al. 2015).

In contrast to Mahoney Lake in Canada (Overmann 1997), Lake Cadagno planktonic community in oxic layer takes advantage of the bacterial development in the chemocline for growth, and zooplankton is a link between bacterial primary production and the higher levels of the food chain (e.g. fish). In situ hybridization has showed the presence of PSB in the intestine of these predators (e.g. *Cyclops abyssorum*). This was further confirmed by the analysis of stable carbon isotopes in bacteria and zooplankton by which the contribution of the bacterial layer in providing carbon to the zooplankton through the latter's predation on bacteria was estimated to be around 50 % (Camacho et al. 2001).

7.5.5 Fish

Meromictic conditions in Lake Cadagno support high fish productivity. The massive development of anaerobic photosynthetic bacteria in the chemocline provides a supplementary form of primary production along with that of the algae. On the whole, bacterial and phytoplankton primary productions are the beginning of a food chain that ensures the development of a fairly large fish population (Tonolla et al.

1988). Fish productivity, estimated at 25–30 kg ha^{-1} (Marrer 1975; Tonolla et al. 1988), represents relatively high values for an alpine oligomesotrophic lake situated at an altitude close to 2000 m asl. In fact, the lakes of the Piora region, Lakes Ritom and Cadagno in particular, have a long history of fish management, documented as far back as the fifteenth century (Surbeck 1917). For the inhabitants of this mountain region, fish represented an important supplementary source of proteins (Peduzzi 2000). Moreover, the fish of both lakes Ritom and Cadagno were renowned for their natural salmon-like colouration supposedly due to the carotenoids of the PSB inhabiting the chemocline. The presence of fish is due to man-made introductions for exploitation of fishery. The species present are rainbow trout (*Oncorhynchus mykiss*), brown trout (*Salmo trutta*), lake charr (*Salvelinus umbla*), brook trout (*Salvelinus fontinalis*), lake trout (*Salvelinus namaycush*), bullhead (*Cottus gobio*), *Phoxinus lumaireul* and Italian chub (*Squalius squalus*), though the last two species are not caught by fishermen. Brook trout and lake trout were introduced to the lake in the 1950s and fry of the commercial fish species are introduced annually (Polli 2012).

7.6 Concluding Remarks

Lake Cadagno represents an ideal model lake for studying the role of microbes in aquatic environments. The scientific projects performed over many decades offer an important source of information supporting ongoing and future scientific projects. Despite its constant meromictic state, Lake Cadagno is unique in that it is responding to the ongoing climatic changes, and man-made changes are influencing the lake itself as well as the catchment area. In addition, continuing studies of the major bacteria involved in the sulphur cycle, as well as studies of microbial interactions and the investigation of the microbes from the mixolimnion and from the sulphidogenic sediments, will contribute to advance our knowledge of this interesting ecosystem.

Meromictic lakes such as Lake Cadagno can be used as a model for studying biogeochemical processes mediated by microorganisms and characterized by permanently anoxic water columns, as earlier reported for Lake Cadagno (Hanselmann and Hutter 1998; Putschew et al. 1995). In this perspective, of great interest are the investigations promoted by the Nordic Center of Earth Evolution focusing on the long history and evolution of life on Earth.

Since anoxic and sulphidic marine conditions may have prevailed during the Precambrian (Canfield et al. 1998; Meyer et al. 2011), Lake Cadagno has the potential to provide insights into the biogeochemical cycling of elements in the Proterozoic oceans and thus can be considered as the present-day analogue to the ancient prokaryote-dominated oceans (Canfield et al. 2010; Honeycutt et al. 2008), considering particularly the bacteria involved in the sulphur cycling (Johnston 2011). Phototrophic sulphur bacteria, especially the GSB *Chlorobium* spp., have been suggested to proliferate in ancient anoxic ocean basins (Crowe et al. 2008;

Summons and Powell 1987; Halm et al. 2009). They are, indeed, believed to have been important for the carbon fixation and chemical development of the anoxic Precambrian ocean, where they may have contributed to the genesis of banded iron formations and the stabilization of the sulphide-rich ocean during the mid-Proterozoic period, Proterozoic aeon 2500–542 million years ago (Brocks et al. 2005; Crowe et al. 2008; Habicht et al. 2011; Johnston et al. 2009).

Further, we applied in Lake Cadagno very innovative investigation techniques, which are among the most promising and powerful in the field of molecular microbial ecology to study of the lake's ecosystem. Thus, Lake Cadagno and its sulphur phototrophic bacteria and associated sulphate-reducing bacteria have proven to represent an excellent model system to test and develop novel investigation techniques in both environmental microbiology and microbial ecology. Particularly interesting are the methods described in Musat et al. (2008), Halm et al. (2009), Habicht et al. (2011) and Storelli et al. (2013) to infer or measure in situ activities of single cells of diverse groups of organisms (e.g. *Chromatium okenii*). These techniques have now the potential to be applied to more difficult environments such as some other aquatic environments, running waters, sediments and microbial mats or soils.

In the studies reviewed here, much of the attention was devoted to population dynamics of phototrophic sulphur bacteria comprising both an apparently monophyletic GSB community and a multiple and diverse PSB community. One question raised but yet unanswered is about the coexistence in the apparently same ecological niche of diverse, but closely related, populations of phototrophic sulphur bacteria. This coexistence and alternation has been present on a long timescale, since the formation of euxinic conditions in the lake, soon after its formation, about 10,000 years BP.

As already pointed out above, other functional groups in the bacterial community such as SRB or, as highlighted in a recent work (Berg et al. 2016), bacteria involved in the iron cycle have a great impact in the functioning of this ecosystem. In particular, sulphate-reducing bacteria in *Lake Cadagno aggregates* may well play a major role in competition dynamics among the phototrophic populations. Thus, the mechanisms underlying PSB–SRB aggregate formation certainly deserve more attention. The study of aggregates, with the tools and techniques described above, potentially integrates all levels of biological analysis from molecular to ecological and holds the promise to give new insights into the metabolic coupling between organisms in aquatic systems.

There are striking parallels and analogies between Lake Cadagno and other stratified meromictic ecosystems worldwide, from microbial mats to ancient oceans. The most interesting analogy is with a low sulphidic and prokaryote-dominated Proterozoic ocean (2500–542 million years ago). Research on this particular ecosystem thus holds the promise to give insights into many fascinating aspects of lake biology and biogeology like life and Earth history and into some of the growing interdisciplinary fields in environmental microbiology: the microbial community dynamics and interactions, as well as the single-cell trait-based biodiversity in microbial communities and its link to ecosystem functioning.

Acknowledgements We are grateful to Prof. R. Peduzzi and the Alpine Biology Foundation, Piora, for the cooperation and support. We also would like to thank K. Habicht, D. Canfield, C. Schubert, X. Perret, J. Zopfi, M. Bueche, J. A. Wuest, A. Bruder, T. Sommer, M. Kuypers and M. Saad for their collaboration and contributions to specific projects.

References

Bachmann H (1924) Der Ritomsee. Schweiz Z Hydrol 2:7–28

Bachmann H (1928) Das Phytoplankton der Pioraseen nebst einigen Beiträgen zur Kenntnis des Phytoplanktons schweizerischer Alpenseen. Schweiz Z Hydrol 5:50–103

Barades J (1926) Le percement du Lac de la Girotte. L'Illus 4338:400–401

Bensadoun J-C, Tonolla M, Demarta A et al (1998) Vertical distribution and microscopic characterization of a non-cultivable microorganism (morphotype R) of Lake Cadagno. Doc Ist Ital Idrobiol 63:45–51

Berg IA (2011) Ecological aspects of the distribution of different autotrophic CO_2 fixation pathways. Appl Environ Microbiol 77:1925–1936

Berg JS, Michellod D, Pjevac P, Martinez-Perez C, Buckner CRT, Hach PF, Schubert CJ, Milucka J, Kuypers MMM (2016) Intensive cryptic microbial iron cycling in the low iron water column of the meromictic Lake Cadagno. Environ Microbiol. doi:10.1111/1462-2920.13587

Bertoni R, Callieri C, Pugnetti A (1998) Dinamica del carbonio organico nel Lago di Cadagno e attività microbiche nel mixolimnio. Doc Ist Ital Idrobiol 63:11–17

Biebl H, Pfennig N (1978) Growth yields of green sulfur bacteria in mixed cultures with sulfur and sulfate reducing bacteria. Arch Microbiol 117:9–16

Borner L (1928a) Die Bodenfauna des Lago Ritom und seines Deltagebiets vor der Absenkung (1916) I. Schweiz Z Hydrol 4:121–162

Borner L (1928b) Die Bodenfauna des Lago Ritom und seines Deltagebiets vor der Absenkung (1916) II. Schweiz Z Hydrol 4:1–30

Bossard P, Gammeter S, Lehman C et al (2001) Limnological description of the lakes Zürich, Lucerne, and Cadagno. Aquat Sci 63:225–249

Bosshard PP, Santini Y, Grüter D et al (2000a) Bacterial diversity and community composition in the chemocline of in the meromictic Lake Cadagno as revealed by 16S rDNA analysis. FEMS Microbiol Ecol 31:173–182

Bosshard PP, Stettler R, Bachofen R (2000b) Seasonal and spatial community dynamics in the meromictic Lake Cadagno. Arch Microbiol 174:168–174

Bottinelli M (2008) Approche moléculaire à l'étude des bactéries sulfato-réductrices et des Archaea méthanogènes dans les sédiments des lacs Cadagno et Rotsee. Dissertation, Université de Genève, Terre et Environnement 71

Boucherle MM, Züllig H (1988) Lago Cadagno: an environmental history. In: Lang G, Schlüchter C (eds) Lake, mire and river environments. Balkema Publishers, Rotterdam, pp 234–295

Bourcart FE (1906) Les lacs alpins suisses, étude chimique et physique. Dissertation, Université de Genève, Georg & Co., Genève

Brocks JJ, Love GD, Summons RE et al (2005) Biomarker evidence for green and purple sulphur bacteria in a stratified Palaeoproterozoic sea. Nature 437:866–870

Brocks JJ, Schaeffer P (2008) Okenane, a biomarker for purple sulfur bacteria (Chromatiaceae), and other new carotenoid derivatives from the 1640 Ma Barney Creek Formation. Geochim Cosmochim Acta 72:1396–1414

Bruggen JJA, Stumm CK, Vogels GD (1983) Symbiosis of methanogenic bacteria and sapropelic protozoa. Arch Microbiol 136:89–95

Burckhardt G (1910) Das Plankton des Lago Ritom unter dem Einfluss der Schwefelquellen. Verh Schweiz Naturforsch Ges 93:302–303

Camacho A, Vicente E, Miracle MM (2000) Spatio-temporal distribution and growth dynamics of phototrophic sulfur bacteria populations in the sulfide-rich lake Arcas. Aquat Sci 62:334–349
Camacho A, Erez J, Chicote A et al (2001) Microbial microstratification inorganic carbon photoassimilation and dark carbon fixation at the chemocline of the meromictic Lake Cadagno (Switzerland) and its relevance to the food web. Aquat Sci 63:91–106
Canfield DE, Thamdrup B, Fleischer S (1998) Isotope fractionation and sulfur metabolism by pure and enrichment cultures of elemental sulfur disproportionating bacteria. Limnol Oceanogr 43:253–264
Canfield DE, Farquhar J, Aubrey L et al (2010) High isotope fractionations during sulfate reduction in a low-sulfate euxinic ocean analog. Geology 38:415–418
Clarke KJ, Finlay BJ, Esteban G et al (1993) *Cyclidium porcatum* n. sp.: a free-living anaerobic scuticociliate containing a stable complex of hydrogenosomes, eubacteria and archaeobacteria. Eur J Protistol 29:262–270
Collet LW, Mellet R, Ghezzi C (1918) Le lac Ritom. Département suisse de l'intérieur. Communication du service des eaux (Dr C Mutzner) Berne Nr. 13
Crowe SA, Jones C, Katsev S et al (2008) Photoferrotrophs thrive in an Archean ocean analogue. Proc Natl Acad Sci USA 105:15938–15943
Dahl TW, Anbar AD, Gordon GW et al (2010) The behavior of molybdenum and its isotopes across the chemocline and in the sediments of sulfidic Lake Cadagno, Switzerland. Geochim Cosmochim Acta 74:144–163
Dahl TW, Chappaz A, Fitts JP et al (2013) Molybdenum reduction in a sulfidic lake: evidence from X-ray absorption fine-structure spectroscopy and implications for the Mo paleoproxy. Geochim Cosmochim Acta 103:213–231
Decristophoris PMA, Peduzzi S, Ruggeri-Bernardi N et al (2009) Fine scale analysis of shifts in bacterial community structure in the chemocline of meromictic Lake Cadagno, Switzerland. J Limnol 68:16–24
Del Don C, Hanselmann KW, Peduzzi R et al (2001) The meromictic alpine Lake Cadagno: orographycal and biogeochemical description. Aquat Sci 63:70–90
Delebeque A (1898) Les lacs français. Chamerot & Renouard, Paris
Demarta A, Tonolla M, Caminada AP et al (1998) Phylogenetic diversity of the bacterial community from the anoxic layer of the mero-mictic Lake Cadagno. Doc Ist Ital Idrobiol 63:19–30
Düggeli M (1924) Hydrobiologische Untersuchungen im Pioragebiet. Bakteriologische Untersuchungen am Ritomsee. Schweiz Z Hydrol 2:62–205
Eichler B, Pfennig N (1988) A new purple sulfur bacterium from stratified freshwater lakes, *Amoebobacter purpureus* sp. nov. Arch Microbiol 149:395–400
Ehrenreich A, Widdel F (1994) Anaerobic oxidation of ferrous iron by purple bacteria, a new tyoe of phototrophic metabolism. Appl Environ Microbiol 60:4517–4526
Fischer C, Wiggli M, Schanz F et al (1996) Light environment and synthesis of bacteriochlorophyll by populations of Chromatium okenii under natural environmental conditions. FEMS Microbiol Ecol 21:1–9
Frigaard NU, Bryant D (2006) Chlorosomes: antenna organelles in photosynthetic green bacteria. Microbiol Monogr 2:79–114
Frigaard NU, Dahl C (2009) Sulfur metabolism in phototrophic sulfur bacteria. Adv Microb Physiol 54:103–200
Fritz M, Bachofen R (2000) Volatile organic sulfur compounds in a meromictic Alpine lake. Acta Hydrochim Hydrob 28:185–192
Garwood EJ (1906) The tarns of Canton Ticino. Q J Geol Soc 62:165–195
Gasol JM, Guerrero R, Pedrós-Alió C (1992) Spatial and temporal dynamics of a metalimnetic Cryptomonas peak. J Plankton Res 14:1565–1579
Gervais F, Siedel U, Heilmann B et al (2003) Small-scale vertical distribution of phytoplankton, nutrients and sulphide below the oxycline of a mesotrophic lake. J Plankton Res 25:273–278

Gies EA, Konwar KM, Beatty JT et al (2014) Illuminating microbial dark matter in meromictic Sakinaw Lake. Appl Environ Microbiol 80:6807–6818

Gregersen LH, Habicht KS, Peduzzi S et al (2009) Dominance of a clonal green sulfur bacterial population in a stratified lake. FEMS Microbiol Ecol 70:30–41

Guerrero R, Montesinos E, Pedrós-Alió C, Esteve I, Mas J, van Gemerden H (1985) Phototrophic sulfur bacteria in two Spanish lakes: vertical distribution and limiting factors. Limnol Oceanogr 30:919–931

Güttinger W, Straub F (1998) Diatoms of Lake Cadagno. Doc Ist Ital Idrobiol 63:57–64

Habicht KS, Miller M, Cox RP et al (2011) Comparative proteomics and activity of a green sulfur bacterium through the water column of Lake Cadagno, Switzerland. Environ Microbiol 13:203–215

Halm H, Musat N, Lam P et al (2009) Co-occurrence of denitrification and nitrogen fixation in a meromictic lake, Lake Cadagno (Switzerland). Environ Microbiol 11:1945–1958

Hanselmann K, Hutter R (1998) Geomicrobiological coupling of sulfur and iron cycling in anoxic sediments of a meromictic lake: sulfate reduction and sulfide sources and sinks in Lake Cadagno. Doc Ist Ital Idrobiol 63:85–98

Hebting Y, Schaeffer P, Behrens A et al (2006) Biomarker evidence for a major preservation pathway of sedimentary organic carbon. Science 312:1627–1631

Hegler F, Posth NR, Jiang J et al (2008) Physiology of phototrophic iron(II)-oxidizing bacteria: implications for modern and ancient environments. FEMS Microbiol Ecol 66:250–260

Honeycutt C, Canfield D, Bjerrum C (2008) One dimensional physico-chemical ecological modelling of sulfidic oceans: applications to proterozoic ocean chemistry and euxinic photoic zones. Geophysical Research Abstracts, vol 10, EGU: 2008-A-08712, 2008 EGU General Assembly 2008

Imhoff JF (2001) Transfer of Pfennigia purpurea Tindall 1999 (Amoebobacter purpureus Eichler and Pfennig 1988) to the genus Lamprocystis as Lamprocystis purpurea comb. nov. Int J Syst Evol Microbiol 51:1699–1701

Johnston DT, Wolfe-Simon F, Pearson A et al (2009) Anoxygenic photosynthesis modulate proterozoic oxygen. Proc Natl Acad Sci USA 106:16925–16929

Johnston DT (2011) Multiple sulfur isotopes and the evolution of earth's surface sulfur cycle. Earth-Sci Rev 106:161–183

Jørgensen BB, Kuenen JG, Cohen Y (1979) Microbial transformations of sulfur compounds in a stratified lake (Solar Lake, Sinai). Limnol Oceanogr 24:799–822

Krige LJ (1918) Petrographische Untersuchungen im Val Piora und Umgebung. Ecol Geol Helv 14:549–654

Langille MGI, Zaneveld J, Caporaso JG et al (2013) Predictive functional profiling of microbial communities using 16S rRNA marker gene sequences. Nat Biotechnol 31:814–823

Lefèvre E, Bardot C, Noël C et al (2007) Unveiling fungal zooflagellates as members of freshwater picoeukaryotes: evidence from a molecular diversity study in a deep meromictic lake. Environ Microbiol 9:61–71

Lefranc M, Thénot A, Lepère C et al (2005) Genetic diversity of small eukaryotes in lakes differing by their trophic status. Appl Environ Microbiol 71:5935–5942

Lehmann C, Bachofen R (1999) Images of concentrations of dissolved sulphide in the sediment of a lake and implications for internal sulphur cycling. Sedimentology 46:537–544

Luthy L, Fritz M, Bachofen R (2000) In situ determination of sulfide turnover rates in a meromictic alpine lake. Appl Environ Microbiol 66:712–717

Marrer H (1975) Zur fischereilichen Bewirtschaftung meromiktischer Alpenseen (Lago di Cadagno, Tessin). Schweiz Z Hydrol 37:213–219

Meyer KM, Macalady JL, Fulton JM et al (2011) Carotenoid biomarkers as an imperfect reflection of the anoxygenic phototrophic community in meromictic Fayetteville Green Lake. Geobiology 9:321–329

Milucka J, Kirf M, Lu L et al (2015) Methane oxidation coupled to oxygenic photosynthesis in anoxic waters. ISME J. doi:10.1038/ismej.2015.12

Musat N, Halm H, Winterholler B et al (2008) A single-cell view on the ecophysiology of anaerobic phototrophic bacteria. Proc Natl Acad Sci 105:17861–17866
Musat N, Foster R, Vagner T et al (2012) Detecting metabolic activities in single cells, with emphasis on nanoSIMS. FEMS Microbiol Rev 36:486–511
Niemann H, Stadnitskaia A, Wirth SB et al (2012) Bacterial GDGTs in holocene sediments and catchment soils of a high Alpine lake: application of the MBT/CBT-paleothermometer. Clim Past 8:889–906
O'Kelly CJ, Silberman JD, Amaral Zettler LA et al (2003) Monopylocystis visvesvarai n. gen., n. sp. and Sawyeria marylandensis n. gen., n. sp.: two new amitochondrial Heterolobosean amoebae from anoxic environments. Protist 154:281–290
Otz MH, Otz HK, Otz I et al (2003) Surface water/groundwater interaction in the Piora aquifer, Switzerland: evidence from dye tracing tests. Hydrgeol J 11:228–239
Overmann J, Beatty T, Hall KJ, Pfennig N, Northcote TG (1991) Characterization of a dense, purple sulfur bacterial layer in a meromictic lake. Limnol Oceanogr 36:846–859
Overmann J, Beatty T, Krouse HR, Hall KJ (1996) The sulfur cycle in the chemocline of a meromictic salt lake. Limnol Oceanogr 41:147–156
Overmann J (1997) Mahoney lake: A case study of the ecological significance of phototrophic sulfur bacteria. In: Jones JG (ed) Advances in microbial ecology, vol 15. Plenum Press, New York, p 251
Overmann J, Schubert K (2002) Phototrophic consortia: model systems for symbiotic interrelations between prokaryotes. Arch Microbiol 177:201–208
Overmann J, Garcia-Pichel F (2006) The phototrophic way of life. In: Dworkin M, Falkow S, Rosenberg E et al (eds) The prokaryotes, Ecophysiology and biochemistry, vol 2. Springer, New York, pp 32–85
Parkin TB, Brock TD (1981) The role of phototrophic bacteria in the sulfur cycle of a meromictic lake. Limnol Oceanogr 26:880–890
Pasini P, Schanz F (1998) Influence of UV-radiation on the primary production of two high mountain lakes in the Piora region. Doc Ist Ital Idrobiol 63:65–70
Pedrós-Alió C, Gasol JM, Guerrero R (1987) On the ecology of a Cryptomonas phaseolus population forming a metalimnetic bloom in Lake Cisó, Spain: annual distribution and loss factors. Limnol Oceanogr 32:285–298
Peduzzi R, Bachofen R, Tonolla M (1998) Lake Cadagno: a meromictic alpine lake. Doc Ist Ital Idrobiol 63
Peduzzi R (2000) Piora e San Gottardo, due secoli di ricerca scientifica. In: Peduzzi R (ed) La ricerca alpina e le trasversali. Atti delle 178a assemblea annuale dell'ASSN 1998, vol 1. Documenta Centro di Biologia Alpina, Piora, pp 17–25
Peduzzi R, Bianconi F (2012) Storia della ricerca biologica e limnologica. In: Rampazzi F, Tonolla M, Peduzzi R (eds) Biodiversità della Val Piora—Risultati e prospettive delle "Giornate della biodiversità". Mem Soc Ticinese Sci Nat 11:13–19
Peduzzi S, Tonolla M, Hahn D (2003a) Vertical distribution of sulfate-reducing bacteria in the chemocline of Lake Cadagno, Switzerland, over an annual cycle. Aquat Microb Ecol 30:295–302
Peduzzi S, Tonolla M, Hahn D (2003b) Isolation and characterization of aggregate-forming sulfate-reducing and purple sulfur bacteria from the chemocline of meromictic Lake Cadagno, Switzerland. FEMS Microbiol Ecol 45:29–37
Peduzzi S (2011) Wasserkraft Val Piora-Piotta. Wege durch die Wasserwelt—5.1. Region Tessin, Hydrologischer Atlas der Schweiz-Geographisches Institut der Universität Bern, Bern: p 42
Peduzzi S, Welsh A, Demarta A et al (2011) Thiocystis chemoclinalis sp. nov. and Thiocystis cadagnonensis sp. nov., motile purple sulfur bacteria isolated from the chemocline of a meromictic lake. Int J Syst Evol Microbiol 61:1682–1687
Peduzzi S, Storelli N, Welsh A et al (2012) Candidatus "Thiodictyon syntrophicum", sp. nov., a new purple sulfur bacterium isolated from the chemocline of Lake Cadagno forming aggregates and specific associations with Desulfocapsa sp. Syst Appl Microbiol 35:139–144

Pfennig N, Cohen-Bazire G (1967) Some properties of the green bacterium Pelodictyon clathratiforme. Arch Mikrobiol 59:226–236

Polli B (2012) Pesci della Val Piora (Cantone Ticino, Svizzera). In: Rampazzi F, Tonolla M, Peduzzi R (eds) Biodiversità della Val Piora—Risultati e prospettive delle "Giornate della biodiversità". Memorie della Società Ticinese di Scienze Naturali 11:237–240

Putschew A, Scholz-Bättcher BM, Rullkätter J (1995) Organic geochemistry of sulfur-rich surface sediments of meromictic lake Cadagno, swiss Alps. In: Vairavamurthy MA, Schoonen MAA (eds) Geochemical transformations of sedimentary sulfur. ACS Symposium Series 612, Oxford Press, pp 59–78

Ravasi DF, Peduzzi R, Tonolla M (2012a). La diversità microbica del Lago Cadagno e degli endoliti della Val Piora (Cantone Ticino, Svizzera). In: Rampazzi F, Tonolla M, Peduzzi R (eds) Biodiversità della Val Piora—Risultati e prospettive delle "Giornate della biodiversità". Mem Soc Ticinese Sci Nat 11:69–78

Ravasi DF, Peduzzi S, Guidi V et al (2012b) Development of a real-time PCR method for the detection of fossil 16S rDNA fragments of phototrophic sulfur bacteria in the sediments of Lake Cadagno. Geobiology 10:196–204

Riccardi N, Austoni M, Kamburska L et al (2012) Phytoplankton and zooplankton species distribution in the high altitude lakes of the Piora valley (Canton Ticino, Switzerland). In: Rampazzi F, Tonolla M, Peduzzi R (eds) Biodiversità della Val Piora—Risultati e prospettive delle "Giornate della biodiversità". Mem Soc Ticinese di Sci Nat 11:77–91

Saccà A (2012) The role of eukaryotes in the anaerobic food web of stratified lakes. In: Altenbach A, Bernhard JM., Seckbach J (eds) anoxia: evidence for eukaryote survival and paleontological strategies. Cell Orig Life Extrem Habitat Astrobiol 21:403–419

Schanz F, Elber F, Huerlimann J (1988) Ricerche algali nella regione di Piora: Confronto dei risultati del periodo 1915–1928 e del 1987. Boll Soc Ticinese Sci Nat LXXVI:35–46

Schanz F, Friedl C (1993) Environmental effects on light utilization by phyto-plankton in a meromictic alpine lake. Verh Int Ver Limnol 25:621–624

Schanz F, Fischer-Romero C, Bachofen R (1998) Photosynthetic production; photoadaptatin of phototrophic sulfur bacteria in Lake Cadagno (Switzerland). Limnol Oceanogr 43:1262–1269

Schanz F, Stalder S (1998) Phytoplankton summer dynamics: sedimentation in the themally stratified Lake Cadagno. Doc Ist Ital Idrobiol 63:71–76

Schramm A, Fuchs BM, Nielsen JL et al (2003) Fluorescence in situ hybridization of 16S rRNA gene clones (Clone-FISH) for probe validation; screening of clone libraries. Environ Microbiol 4:713–720

Schubert CJ, Vazquez F, Lösekann-Behrens T et al (2011) Evidence for anaerobic oxidation of methane in sediments of a freshwater system (Lago di Cadagno). FEMS Microbiol Ecol 76:26–38

Sorokin YI (1970) Interrelations between sulphur and carbon turnover in meromictic lakes. Arch Hydrobiol 66:391–446

Stapfer A (1991) Pollenanalytische Untersuchungen im Val Piora (Tessin). Ein Beitrag zur Klima und Vegetationsgeschichte der Nacheiszeit. Geogr Helv 46:156–164

Storelli N, Peduzzi S, Saad M et al (2013) CO_2 assimilation in the chemocline of Lake Cadagno is dominated by a few types of phototrophic purple sulfur bacteria. FEMS Microbiol Ecol 84:421–432

Storelli N, Saad M, Frigaard N-U et al (2014) Proteomic analysis of the purple sulfur bacterium Candidatus "Thiodictyon syntrophicum" strain Cad16T isolated from Lake Cadagno. EuPA Open Proteom 2:17–30

Summons RE, Powell TG (1987) Identification of aryl isoprenoids in source rocks and crude oils: biological markers for the green sulphur bacteria. Geochim Cosmochim Acta 51:557–566

Surbeck G (1917) Ueber die Fische des Ritom-, Cadagno- und Tomsees in Val Piora. Verh Schweiz Nat Ges 99:264–265

Tiffay M-C, Stoll S, Peduzzi S (2015) Ecosystem responses to climate variations of the high Alpine Lake Cadagno over the last 100 years. RS02. SEFS 9 Symposium for European Freshwater Sciences

Tonolla M, Del Don C, Boscolo P et al (1988) The problem of fish management in an artificially regulated meromictic lake: Lake Cadagno (Canton Tessin, Switzerland). Riv Ital Acquacolt 23:57–68

Tonolla M, Demarta A, Hahn D et al (1998a) Microscopic and molecular in situ characterization of bacterial populations in the meromictic Lake Cadagno. Doc Ist Ital Idrobiol 63:31–44

Tonolla M, Demarta A, Peduzzi R (1998b) The chemistry of Lake Cadagno. Doc Ist Ital Idrobiol 63:11–17

Tonolla M, Demarta A, Peduzzi R et al (1999) In situ analysis of phototrophic sulfur bacteria in the chemocline of meromictic Lake Cadagno (Switzerland). Appl Environ Microbiol 65:1325–1330

Tonolla M, Demarta A, Peduzzi S et al (2000) In situ analysis of sulfate-reducing bacteria related to Desulfocapsa thiozymogenes in the chemocline of meromictic Lake Cadagno (Switzerland). Appl Environ Microbiol 66:820–824

Tonolla M, Peduzzi S, Hahn D et al (2003) Spatio-temporal distribution of phototrophic sulfur bacteria in the chemocline of meromictic Lake Cadagno (Switzerland). FEMS Microbiol Ecol 43:89–98

Tonolla M, Peduzzi S, Demarta A et al (2004) Phototrophic sulfur and sulfate-reducing bacteria in the chemocline of meromictic Lake Cadagno, Switzerland. J Limnol 63:161–170

Tonolla M, Bottinelli M, Demarta A et al (2005a) Molecular identification of an uncultured bacterium ("morphotype R") in meromictic Lake Cadagno, Switzerland. FEMS Microbiol Ecol 53:235–244

Tonolla M, Peduzzi R, Hahn D (2005b) Long-term population dynamics of phototrophic sulfur bacteria in the chemocline of Lake Cadagno, Switzerland. Appl Environ Microbiol 71:3544–3550

Triadó-Margarit X, Casamayor EO (2012) Genetic diversity of planktonic eukaryotes in high mountain lakes (Central Pyrenees, Spain). Environ Microbiol 14:2445–2456

Van Gemerden H, Mas J (1995) Ecology of phototrophic sulfur bacteria. In: Blankenship RE, Madigan MT, Bauer CE (eds) Anoxygenic photosynthetic bacteria. Kluwer Academic, The Netherlands, pp 49–85

Wagener S, Schulz S, Hanselmann K (1990) Abundance and distribution of anaerobic protozoa and their contribution to methane production in Lake Cadagno (Switzerland). FEMS Microbiol Ecol 74:39–48

Wirth SB, Gilli A, Niemann H et al (2013) Combining sedimentological, trace metal (Mn, Mo) and molecular evidence for reconstructing past water-column redox conditions: the example of meromictic Lake Cadagno (Swiss Alps). Geochim Cosmochim Acta 120:220–238

Widdel F, Schnell S, Heising S et al (1993) Ferrous iron oxidation by anoxygenic phototrophic bacteria. Nature 362:834–836

Wüest A (1994) Interaktionen in Seen: Die Biologie als Quelle dominanter physikalischer Kräfte. Limnologica 24:93–104

Zimmermann M, Escrig S, Hübschmann T et al (2015) Phenotypic heterogeneity in metabolic traits among single cells of a rare bacterial species in its natural environment quantified with a combination of flow cell sorting and NanoSIMS. Front Microbiol 6:243

Züllig H (1985) Pigmente phototropher Bakterien in Seesedimenten und ihre Bedeutung für die Seeforschung. Schweiz Z Hydrol 47:87–126

Chapter 8
Lake La Cruz, an Iron-Rich Karstic Meromictic Lake in Central Spain

Antonio Camacho, María R. Miracle, Lidia Romero-Viana, Antonio Picazo, and Eduardo Vicente

8.1 Introduction

Meromixis is a condition of permanent stratification of water masses in lakes undergoing incomplete circulation at the fall overturn. A meromictic lake has some water which partly or wholly is unmixed with the main water mass at circulation periods. Meromixis is favoured in lakes with a very high relative depth, which enhances the physical stability of the water column. Stability further becomes stronger when salt concentrations are higher in deep waters than in the upper water layers.

Karstic lakes appear as the consequence of limestone or gypsum dissolution, which can be followed by structural collapse, having a relatively impermeable basin. Their morphometry is quite characteristic, with one or several circular sinkholes (dolines) showing steep walls and a high relative depth, namely, the ratio of maximum depth to mean diameter. Depending on the water level, the whole basin can be flooded. Then the lake spans through several dolines or a polje (large, flat-floored depression within karst limestone, whose long axis develops in parallel with major structural trends). However, if only part of the basin is flooded, the steep walls help to protect the lake from wind. Such lakes are somewhat associated to the karstic aquifers, which feed the lake, at least partially. This confers on the lake a hydrological inertia that increases water residence time and reduces water movements. These features, together with a higher salt content of the deeper waters due to the dissolution of salts, can favour the stratification of the water layers. Most karstic lakes, however, are not meromictic but monomictic (Hutchinson 1957;

A. Camacho (✉) • M.R. Miracle • L. Romero-Viana • A. Picazo • E. Vicente
Cavanilles Institute for Biodiversity and Evolutionary Biology and Department of Microbiology and Ecology, Universitat de Valencia, Edificio de Investigación "Jeroni Muñoz", Campus de Burjassot, E-46100, Burjassot, Valencia, Spain
e-mail: antonio.camacho@uv.es

© Springer International Publishing AG 2017

R.D. Gulati et al. (eds.), *Ecology of Meromictic Lakes*, Ecological Studies 228,
DOI 10.1007/978-3-319-49143-1_8

Imboden and Wüest 1995). Wind sheltering is not enough by itself to maintain the thermal stratification through the cold periods, as many karstic lakes with similar morphometry and substratum are not meromictic. Instead, the higher salt content in deep waters, of geological or biogenic origin, is actually the major factor in determining the capacity of the lake to maintain stratification through the annual cycle. This primarily determines the trend of a lake to remain meromictic.

Several karstic areas in Spain contain karstic lakes. Among them, the area of "Serrania de Cuenca", located in the Central-Eastern part of the Iberian Peninsula in Castilla-La Mancha region (Fig. 8.1), and that of Banyoles, in the Northeastern part of the Iberian Peninsula, are the most prominent, and both present meromictic lakes. From up to 35 dissolution lakes in the northern area in the province of Cuenca, only two lakes, Laguna de la Cruz (Lake La Cruz) and Laguna del Tobar (Lake El Tobar), are meromictic (Miracle et al. 1992). Lake La Cruz is a biogenic meromictic lake due to its monimolimnetic water being enriched with bicarbonates of calcium, magnesium and iron. Lake El Tobar shows, however, a crenogenic meromixis (due to saline springs delivering dense water into the deep layers), where the monimolimnetic waters contain sodium chloride brine. The karstic lakes of Cañada del Hoyo (Cuenca) were scientifically described already about 140 years ago by Botella y de Hornos (1875). In this area, only Lake La Cruz is permanently meromictic out of the seven dissolution lakes. Near Banyoles (Girona) is the other main Iberian karstic area with meromictic lakes. The bigger lake there, called Estany de Banyoles (Lake Banyoles), lies in a polje. It comprises six basins, but only one of them is permanently meromictic. Moreover, among the 15 small lakes that surround the main lake, only Lake Vilar is meromictic (Miracle et al. 1992). Thus, although most Iberian karstic lakes present morphometric features that favour summer stratification, only a few have a density gradient strong enough to maintain meromixis. Additionally, some Spanish coastal lakes, Estany de Cullera near Valencia and Estany d'Es Cibollar in Majorca, were also meromictic, but alteration of the land–sea interfaces resulted in the disappearance of the meromictic condition.

This chapter presents the physical, chemical and biological features of Lake La Cruz, their main ecological processes and the palaeoecological studies performed in this lake, allowing a reconstruction of historical processes, both on the origin of the meromictic condition and on the climate evolution within the last centuries in the area.

8.2 Physical Features, Climate and Vertical Structure of Lake La Cruz

Lake La Cruz is located in the karst system of Cañada del Hoyo, Guadazaón River (39° 59′ 16.54″ N, 1° 52′ 25.32″ W; UTM 30S X595932 Y4427087), near the city of Cuenca, in Central-Eastern Spain (Fig. 8.1). It is a solution lake on the dolomite

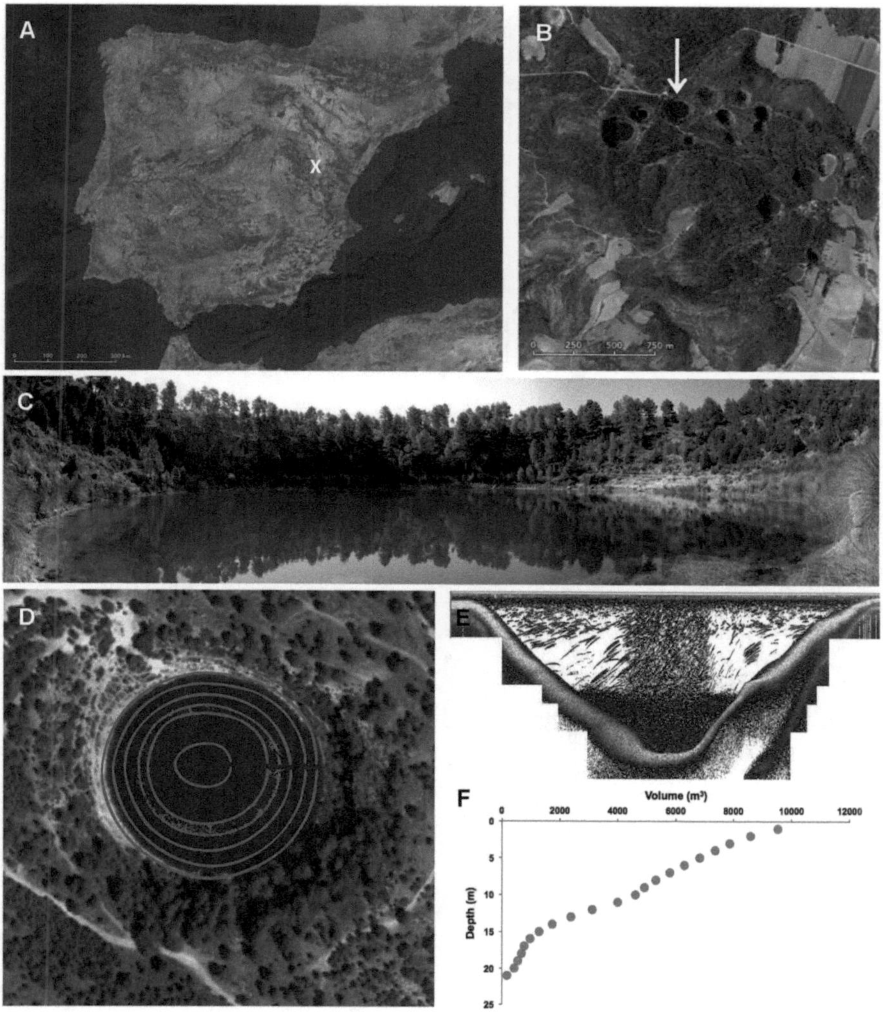

Fig. 8.1 (**a**) Location of Lake La Cruz in the Iberian Peninsula. (**b**). Aerial view of the lacustrine system of Cañada del Hoyo, *arrow* shows Lake La Cruz. (**c**) Lateral view of Lake La Cruz. (**d**) Bathymetric map, (**e**) sonar transect and (**f**) hypsographic curve (depth-area) of Lake La Cruz

limestone of the mountains of the Serrania de Cuenca, at ca. 1000 m a.s.l. The karst lies on Upper Cretaceous (Cenomanian) formations with soluble limestone in the upper series and impermeable green marls in the lower, where the lake bottom lies (Eraso et al. 1979). The lake area has a transitional Mediterranean to Continental climate, with summer as the driest season (Fig. 8.2), cool winters and strong temperature differences between day and night. Regional precipitation is limited to total annual rainfall of around 600 mm, being variable during winter and spring, and mainly related to westerly winds with moist air from the Atlantic Ocean (Fig. 8.2). Eastern maritime winds from the Mediterranean can also occasionally

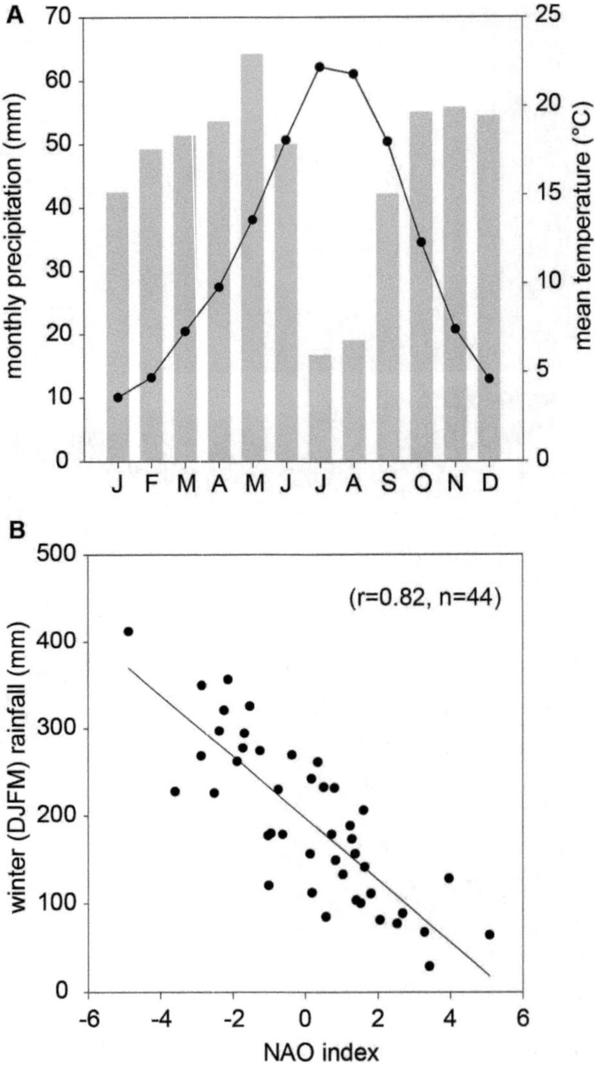

Fig. 8.2 (**a**) Monthly total precipitation (*grey bars*) and mean temperature (*black dots*). Data are averages from the meteorological data set recorded from 1950 to 2003 at Cuenca's meteorological station. (**b**) Relationship between the North Atlantic Oscillation (NAO) winter (DJFM) index and the regional winter (DJFM) rainfall (1950–2003). NAO index is given as normalized sea level pressure differences between Lisbon (Portugal) and Reykjavik (Iceland) based on data of the Climate Analysis section, NCAR, Boulder, USA (http:/www.cgd.ukar.edu/kas/jhurell/indices.html)

reach the lake area promoting episodic events of heavy rainfall (Rodriguez-Puebla et al. 1998). Lake La Cruz has a negligible catchment area with no connection with surface running waters (no inflow, no outflow), but the lake is in hydrostatic equilibrium with the local groundwater table. The lake is mainly fed by lateral groundwater sources located about 4–5 m above the lake bottom, at the level where the dolomite meets the impermeable layer of green marls. It lies in a circular closed doline with steep walls that rise 20–30 m above the lake surface and reduce the wind mixing effect. The lake has a maximum diameter of 136 m and a surface area of 1.4 ha. The bathymetric map (Fig. 8.1) shows a strong slope towards the central flat bottom of the lake. Water level oscillates between 20 and 25 m depending on the multiannual rainfall pattern.

The study of the sediment cores from the central part of the lake shows that the lake became meromictic ca. 300 years ago (Julià et al. 1998). A permanent monimolimnion occurs below 18 m, which is overlaid by the mixolimnetic waters during the mixing period. The dissolved solid gradient, enhanced by the internal loading (biogenic meromixis), is the main cause of the presence of this stable monimolimnion, which has a slightly higher temperature, 6–7 °C, than the overlying water layer, 4–5 °C. Both are separated by a permanent chemocline with strong redox gradient. Thermal stratification additionally occurs in the lake from April to late October, and three water layers are formed from surface to the monimolimnion. Then, the epilimnion spans from surface to 4 m during the early stages of the stratification but reaches up to 10 m later. Below, the metalimnion spans along the thermocline (Fig. 8.3), and then the lower hypolimnion extends from the bottom of the thermal gradient to the permanent chemocline starting at ca. 16 m (Fig. 8.4). During summer, a secondary chemocline (and also with a redox gradient) appears at the bottom of the thermocline, similar to holomictic lakes, but it disappears when the upper water layers are progressively mixed in the fall period.

In general, the vertical structure of the water column of a deep lake reflects differences in water density due to both temperature and chemical gradients. The surface water temperature of the lake is strongly variable over the year because of heat exchange with air and the seasonal variations of incident solar radiation and air temperature (Fig. 8.3). The depth of wind mixing depends upon the surface area and the degree of exposure to wind, which in turn depends on the extent of mountains, trees and on local topography. The situation of Lake La Cruz, in the deepest part of

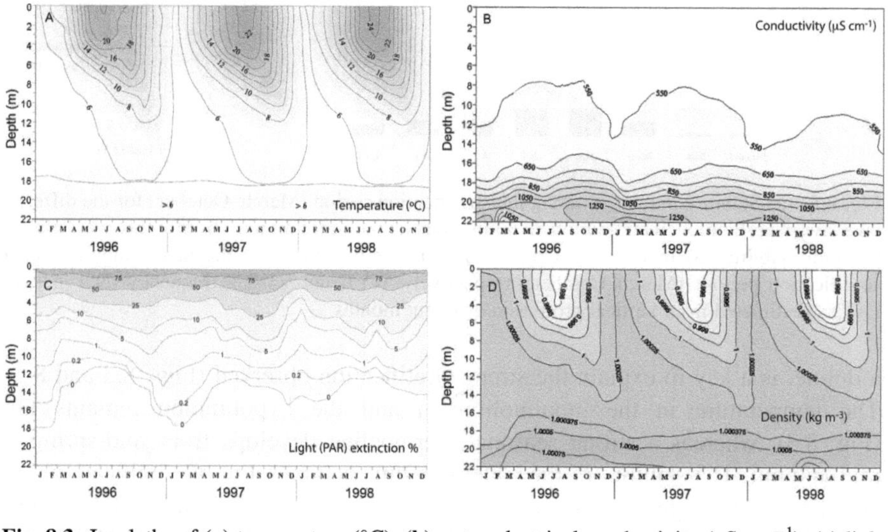

Fig. 8.3 Isopleths of (**a**) temperature (°C), (**b**) water electrical conductivity (μS cm^{-1}), (**c**) light extinction (% of surface irradiance, PAR) and (**d**) water density (Kg m^{-3}) along the years 1996, 1997 and 1998 in Lake La Cruz

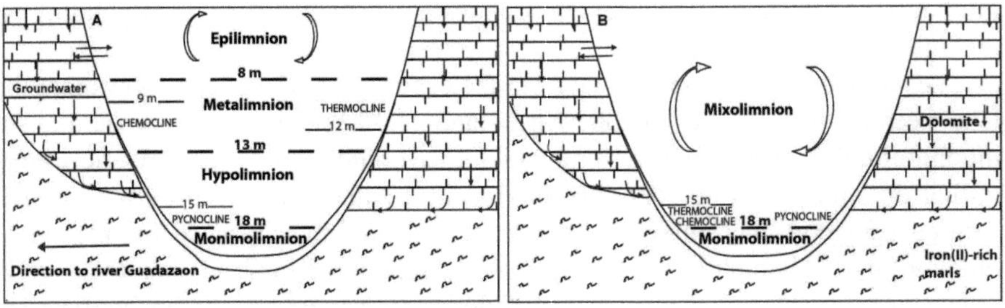

Fig. 8.4 Conceptual model of the vertical structure of Lake La Cruz during (**a**) the summer thermal stratification period and (**b**) the winter period, the latter with an upper mixed layer (mixolimnion) and a permanently anoxic bottom water (monimolimnion). Modified from Walter (2011)

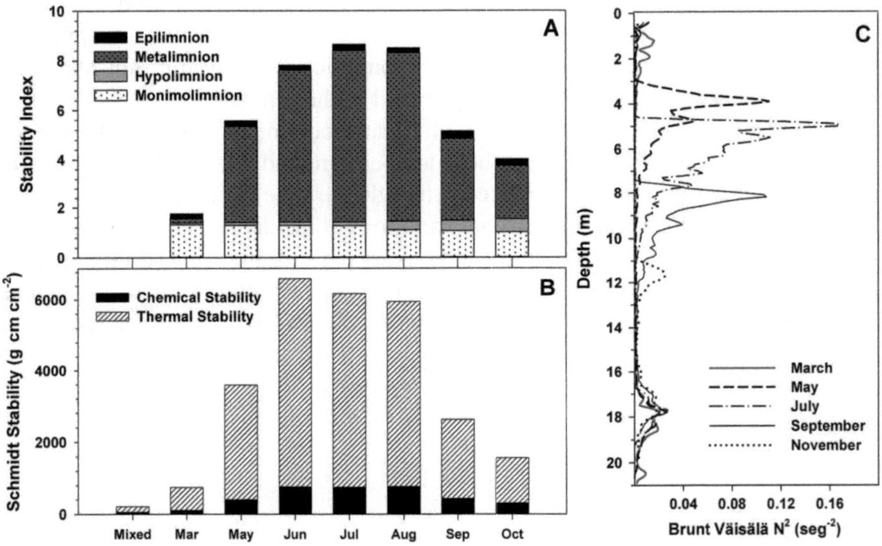

Fig. 8.5 (**a**) Stability index (SI) along the stratification period (March–October) for the different lake layers, epilimnion, metalimnion, hypolimnion and monimolimnion. (**b**) Monthly whole lake's Schmidt stability (S) in Lake La Cruz during the thermal mixis of the mixolimnion and the stratification periods (March–October), respectively. (**c**) Brunt-Väisälä frequency (N^2) in Lake La Cruz estimated from vertical profiles in different months

a doline, is a key to explain the strong stratification observed (Figs. 8.3 and 8.5). The temperature in the monimolimnion and the hypolimnion remains low (Fig. 8.3), whereas a strong seasonal thermocline develops from mid-spring to mid-autumn. The stability of the water column, as a measure of stratification, can be estimated through several parameters (Fig. 8.5), such as Schmidt's stability (S), the stability index (SI) and the Brunt–Väisälä frequency (N^2) (Wetzel 2001). The stability index (SI), calculated to compare different layers in the lake (Fig. 8.5a), shows the stable density gradient in the monimolimnion. With the establishment of the seasonal thermocline, the density gradient in the metalimnion strongly increases with the stratification pattern with maximum values in summer. Maxima of both the stability index (SI) and the Schmidt's stability in Lake La Cruz occur from June to August (Fig. 8.5). In winter, the upper water layers in the lake are mixed, and the lake is in circulation except for the monimolimnion. In this period, both the stability index (SI) and the Schmidt's stability (S) have their minima as the thermal gradient disappears, and the remaining stability is due to the higher salt content in the monimolimnion (Figs. 8.3 and 8.5). During the circulation period, dissolved oxygen and nutrients distribute over the entire mixolimnion. Hence, the circulation pattern is a decisive factor for the changes with time of the water chemistry and the development of planktonic populations in the lake.

The stability of stratification can also be quantified by the stability frequency or Brunt–Väisälä frequency (N^2), which is the maximum frequency for internal waves that can propagate in the stratified water layers. N^2 indicates how much energy is required to exchange water in the vertical profile. In Lake La Cruz, the variation of the Brunt–Väisälä frequency has two main gradients. A primary permanent chemical gradient (permanent chemocline) appears between 16 and 18 m, in the transition to the monimolimnion, determined by both salinity and, to a lesser extent, by a small temperature gradient. A second one is a thermal gradient, which is seasonally determined by the development of a thermocline during the summer thermal stratification period. During the mixing period, the vertical profile of the Brunt–Väisälä frequency shows how the mixolimnion extends from the surface to about 16 m (Fig. 8.5c), the upper border of the monimolimnion. In the early stages of thermal stratification, the vertical profile of the Brunt–Väisälä frequency shows that, at the beginning of the stratification period (May in Fig. 8.5c), the epilimnion extends to about 4 m, the metalimnion is located between 4 and 10 m, and the hypolimnion spans from 10 to 16 m, whereas the chemocline extends down to the upper part of the permanent monimolimnion between 16 and 18 m. In the summer, the depth of the epilimnion increases and the density of stratification (Brunt–Väisälä frequency) keeps on decreasing until the total mixing ensues.

The monimolimnion does not significantly vary in depth and is always located from 18 m to the bottom. This layer is excluded from the gas exchange with the water layers above. The diffusion rate across the chemocline is usually very low. Then, anoxia exists permanently, and reducing conditions regulate the anaerobic microbial oxidation of the settled organic materials. In addition, some chemical compounds that would not be stable in the mixolimnion are produced and accumulate in the monimolimnion. Due to the higher hydrostatic pressure, gases such as CO_2 can accumulate in monimolimnion much more than in the mixolimnion.

Light penetration during the period of normal transparency, with Secchi disc values of 4–6 m, shows that the layer where 10 % of surface light intensity reaches seasonally varies from 4 to 6 m and that of 1 % surface light reaches from 8 to 11 m (Fig. 8.3). Below these depths, the light extinction by the picocyanobacterial and photosynthetic bacterial populations causes complete darkness. Apart from the typical summer whiting event, the minimal transparency of lake waters occurs in spring and early summer, but transparency substantially increases by midsummer after the whiting event caused by precipitation of calcium carbonate.

8.3 Water Chemistry

Dissolved salt content of Lake La Cruz waters is strongly related to the immediate substrate and that of the catchment because the lake is water-fed in equilibrium with the groundwater table. The main rock components in this area are dolomites (calcium-magnesium carbonate), whereas the lake bottom is formed by green marls containing clay, calcium carbonate and reduced iron (Fig. 8.4). Conversely, these substrates are very poor in gypsum and chloride. Therefore, lake waters are rich in bicarbonates of calcium and magnesium, with high pH values of 8.5–8.8, and very poor in sulphate and chloride. The monimolimnion below has similar chemical components, but concentrations are higher, especially that of calcium, and also the concentration of iron here is high (Fig. 8.6).

The mixolimnion contains mainly dissolved bicarbonates, with alkalinity of around 6 meq L^{-1} and much lower concentrations of chloride (0.2–0.3 meq L^{-1}) and sulphate (around 0.1 meq L^{-1}). The ratio among the main anions is approximately Alk/Cl^-/SO_4^{2-} 50/2/1. Cations, calcium (1–2 meq L^{-1}) and, especially, magnesium (4–5 meq L^{-1}) far exceed sodium and potassium concentrations (around 0.2 meq L^{-1}). Mixolimnetic waters of Lake La Cruz are lower in calcium than in magnesium, because of the removal by precipitation of calcium carbonate as calcite. Calcite crystals sink to the monimolimnion, where they partially redissolve

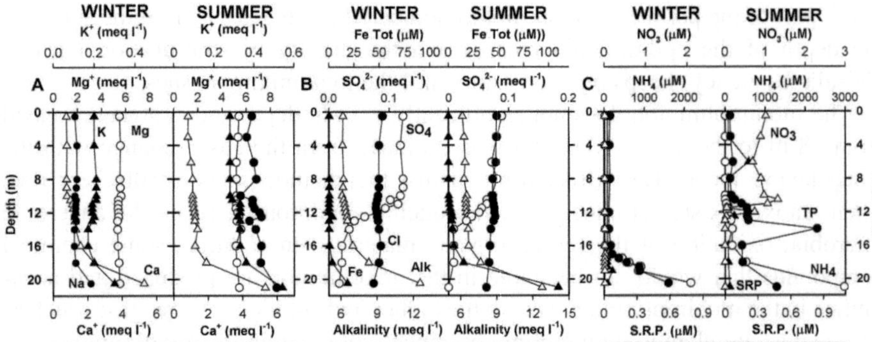

Fig. 8.6 Water concentrations of (**a**) main cations (Ca^{2+}, Mg^{2+}, Na^+, K^+, in meq L^{-1}), (**b**) main anions (alkalinity, SO_4^{2-} and Cl^- in meq L^{-1}) and total iron (in μM) and (C) main inorganic nutrients (NO_3^-, NH_4^+, *SRP* soluble reactive phosphorus, *TP* total phosphorus, in μM) in the vertical profile of Lake La Cruz comparing the winter period with the thermal stratification in summer

because of the lower pH, thus increasing calcium concentration there, whereas the remaining crystals finally integrate in the sediment as a thin white micritic layer.

In the anoxic and highly reduced waters of the permanent chemocline and the monimolimnion, concentrations of most dissolved salts increase (Figs. 8.3 and 8.6). Alkalinity increases to near 15 meq L^{-1}, and among cations, calcium concentrations rise to about 6 meq L^{-1}, whereas magnesium maintains the concentrations of the upper layers of around 5 meq L^{-1}. The relationship Alk/Cl^-/SO_4^{2-} here increases to 200:3:1.

Lake La Cruz is a mesotrophic lake, with low bioavailable dissolved N and P compounds through the photic zone. Ammonia, silicates and, though moderately, orthophosphate accumulate in the monimolimnion (Fig. 8.6). Nitrogen compounds are transformed mainly near the redoxclines. In addition to the low P concentrations, dissolved nitrogen concentrations in upper layers are also low, and even nitrate is depleted through the summer period and almost disappears from the epilimnion. The nitrogen thus becomes a limiting nutrient for phytoplankton growth (Camacho et al. 2003a, b). Studies for 30 years show that Lake La Cruz is a typical meromictic lake with increase of ammonia, phosphate, alkalinity (and Fe^{2+} in this case) in the monimolimnion: their concentrations slightly increase from spring (start of stratification) to autumn (mixing period). These trends suggest that the development of the meromixis is influenced by the supply of organic matter from upper layers, and this is, thus, a case of biogenic meromixis (Walker and Likens 1975).

The lake is also characterized by very low concentration of sulphide in the anoxic, reduced waters (to around 10–15 μM). The sulphate concentrations, which are very low, do not allow a substantial sulphate reduction, which implies that most sulphide comes from the proteolysis. Consequently, the anaerobic consumption of organic carbon is mainly processed through methanogenesis in the deep monimolimnion, the water-sediment interface and the surface sediment. Thus, an important quantity of methane is released from the sediments and rises upwards to the surface forming columns of small bubbles, before they escape to the atmosphere. The released gases analysed by gas chromatography shows about 50 % CH_4 and 30 % CO_2. Not all the produced methane is released to the atmosphere, since in the upper zone of the monimolimnion and along the mixolimnion part of this methane that is dissolved in the water is metabolized by several groups of methane-consuming bacteria that develop sequentially along the vertical profile.

The dynamics of organic matter in Lake La Cruz can also be followed through the chromophoric dissolved organic matter (cDOM), that is, the optically measurable component of the dissolved organic matter in water that is present in all natural aquatic environments. This is also one of the largest biologically active carbon reservoirs made up of a complex mixture of organic molecules with a large variability in the molecular weight. A variety of sources of cDOM allows a high heterogeneity in composition (Bertilsson and Jones 2003). The major sources of cDOM in the lake result from a balance between the extracellular release of organic matter by algae, the release by grazers , the viral lysis of bacteria and algae and, to a lesser degree, the release of the DOM from the sediments to the monimolimnion and its diffusion upwards to the hypolimnion. Due to the high light absorption coefficient, cDOM markedly affects light penetration to the deeper layers, influencing also the biological productivity, the formation of chemical reactants and the chemical speciation, transport and availability of trace elements, thus being an outstanding factor for the understanding of biogeochemical cycles in aquatic ecosystems (Mopper and Schultz 1993). The lake water is, however, relatively clear with respect to the optical effects of cDOM, and light extinction through the water column (Fig. 8.3) is mostly caused by both the absorption by the photosynthetic pigments of phytoplankton and light scattering of suspended particles, especially calcite crystals when precipitation of carbonates is high. Although the cDOM fluorescence properties (Ex/Em: 370/460 nm) show that the concentrations of cDOM in the monimolimnion are an order of magnitude higher than those observed in the rest of the lake, this has no optical effect as light does not reach these depths. On the other hand, a marked temporal seasonally dynamics of cDOM concentrations was observed, with the higher values immediately after mixing process, as a result of the mixing with hypolimnetic water containing higher concentrations of organic matter. From spring to autumn, as the stratification advances, the cDOM concentrations decrease mainly in the epilimnion in a process that occurs simultaneously with the progressive nutrient limitation and the decrease of planktonic primary production in upper layers. In contrast, the metalimnion shows a relative cDOM maximum that matches with the deep chlorophyll maximum (DCM) and with the maximum values of primary production in the stratification period. In the hypolimnion, however, cDOM concentration increases as the stratification proceeds mainly due to the degradation of organic matter that settles down from the water column.

Meromictic lakes, as Lake La Cruz, have a relatively high concentration of dissolved salts in the deepest waters, which allows the permanent stratification by the establishment of a strong density gradient, the so-called pycnocline (Wetzel 2001). In this lake, organic matter, which is mostly autochthonous, allows the reduction of ferric oxides in the upper sediment, spreading iron into the monimolimnion where it also accumulates as ferrous iron and bicarbonate, thus contributing to the stability of the water column. The low sulphate availability prevents an active sulphate reduction, and thus sulphide is relatively scarce in the anoxic waters of the lake. These conditions influence the iron cycle, allowing high soluble concentrations of iron in the anoxic waters, as it occurs in the anoxic layers of other lakes where sulphide concentrations are very low.

Iron-rich meromictic lakes are rare and unusual ecosystems, of which only a few have been discovered worldwide. Lake La Cruz can be considered as one such "iron-rich meromictic lake", though iron does not account for most of the chemical gradients that maintain the meromixis. Instead, the high concentrations of dissolved calcium bicarbonate in the monimolimnion mostly cause the stability of the water column (Rodrigo et al. 2001), although iron concentrations in deep waters also enhance the density gradient. Other meromictic lakes have sulphidic deep waters due to the activity of sulphate-reducing bacteria that use the relative abundant sulphate as electron acceptor for anaerobic respiration and release sulphide (Orem et al. 1991; Wetzel 2001; Canfield et al. 2010). The absence of gypsum in the surrounding rock substrate (Vicente and Miracle 1988; Miracle et al. 1992) determines the low sulphate and sulphur concentrations in Lake La Cruz waters, where iron is much more abundant than sulphur and Fe(II) is concentrated in the monimolimnion (Fig. 8.6) and reaches concentrations as high as 250 μM at the sediment–water interface. The maintenance of these monimolimnetic iron concentrations involves its active recycling at the chemocline described as the iron "wheel" by Campbell and Torgersen (1980).

Fe(II) is soluble at circumneutral pH, whereas Fe(III) only shows high solubility at low pH (Weber et al. 2001). Since Lake La Cruz water column has an alkaline pH in the well-oxygenated waters, oscillating around 8.5–8.8, iron is expected to be found mainly as Fe(III)-oxides particles that likely settle to the lake bottom, thus causing low iron concentrations in upper waters. The particulate Fe(III) fraction constantly has low concentrations along the water column. Soluble Fe(II) is oxidized in the summer redoxcline between the meta- and the hypolimnion, as indicated by two separate Fe(III) peaks. An upper broad peak of about 3 μM Fe (III) can be observed in the lower metalimnion, where picocyanobacteria are the most abundant photosynthetic microorganism and Fe(III) is presumably formed by direct chemical reaction with photosynthetically produced O_2 or by microaerophilic chemotrophs (Lehours et al. 2007; Walter et al. 2014). The concentrations of colloidal Fe(III) and total Fe(III) also vary along the water column, with occasional peaks in the chemocline, being almost absent in the upper hypolimnion, and increasing concentrations towards the bottom of the lake. Most of the total Fe(III) is in colloidal form. During the late summer stratification, an accumulation of iron oxides is found in the water layer just above the sediments, with a higher ferrireduction in winter than in summer. On the contrary, higher Fe(III) concentration in lower layers suggests a higher Fe(II) oxidation in late summer or a lower iron reduction. While iron reduction is a biological process (Walter et al. 2014), Fe (II) oxidation can be either a biotic or an abiotic reaction, or both. Differentiation between both biotic and abiotic Fe(II) oxidations can be inferred on the basis of depths where Fe(III) peaks are detected. Lake La Cruz iron profiles show two different types of situations: on the one hand, Fe(III) peaks are all located within and below the micro-oxic part of the chemocline, while on the other hand, the upper peaks are situated in the more oxygenated layers. The lake exhibited a recurrent peak of Fe(III) in the anoxic euphotic layer of the chemocline associated with an increase of inorganic carbon uptake in presence of Fe(II) (Walter 2011; Walter et al. 2014). The occurrence of light-dependent Fe(II) oxidation in this layer provides a consistent evidence of the photo-ferrotrophic activity in the lake's chemocline. This is the first demonstration of the coexistence of both iron- and sulphur-driven anoxygenic photosynthesis with oxygenic photosynthesis in the same environment (Walter et al. 2014).

Fig. 8.7 Main organisms found in Lake La Cruz. (**a**) Autofluorescence microphotography of unicellular picocyanobacteria *Synechococcus* sp, (**b**) DAPI staining microphotography of bacterioplankton and pico- and nanoflagellates, (**c**) DAPI staining microphotography of a ciliate, (**d**) photosynthetic bacteria and algae forming clumps after thermal mixing, (**e**) microphotography of *Cyclotella* sp., lateral side (**f**) microphotography of *Euglena* sp., (**g**) microphotography of *Cosmarium* sp., (**h**) microphotography of *Tetraedron* sp., (**i**) microphotography of *Monoraphidium* sp., (**j**) microphotography of the rotifer *Anuraeopsis fissa*, (**k**) microphotography of the rotifer *Keratella* sp., (**l**) aquatic helophyte *Scirpus lacustris*, (**m**) aquatic hydrophyte *Chara vulgaris*, (**n**) aquatic hydrophyte *Myriophyllum spicatum*

8.4 Biological Communities

Obviously living beings inhabiting an ecosystem interact within them and with the abiotic environment and cannot be seen as separate entities. However, the differentiation of taxonomic groups or functional guilds helps to describe biological communities. We have used this approach for a better description of lake's biota (Fig. 8.7).

8.4.1 Bacterioplankton

The most striking biological feature of Lake La Cruz is the existence of a multi-layered planktonic community with high abundance of microorganisms in a thin portion of the water column, coincident with the gradient of chemical compounds and redox potential (Fig. 8.8). The depth at which the redoxcline and the associated layers of organisms are found varies seasonally. Both the thermocline, with the associated chemical gradient at its bottom, and the permanent chemocline, which separates the mixolimnion from the monimolimnion, provide a suitable environment for the establishment of stratified populations of planktonic microorganisms (Miracle and Vicente 1983). Total bacterioplankton densities can exceed 5×10^7 cell mL^{-1} (Casamayor et al. 2012), and although autotrophic bacterioplankton accounts for an important part of this community, heterotrophic bacteria are also abundant, especially in the deeper layers of the tropholytic zone where they take advantage of the settled organic matter originating from primary production in upper layers. Casamayor et al. (2012) offer an overview of the main planktonic

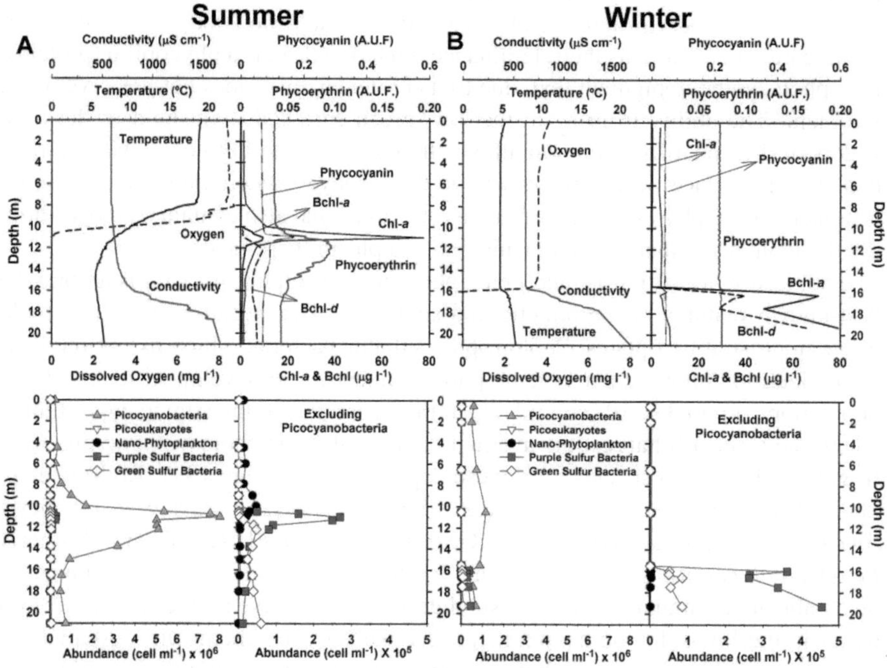

Fig. 8.8 Typical vertical profiles of the physical and chemical variables (temperature, electrical conductivity and dissolved oxygen), photosynthetic pigments (chlorophyll-*a*, phycoerythrin, phycocyanin, bacteriochlorophyll-*a* and bacteriochlorophyll-*d*), and abundance of the main photosynthetic microorganisms (picocyanobacteria, picoeukaryotes, eukaryotic nanoalgae, purple sulphur bacteria and green sulphur bacteria) for the (**a**) summer stratification period and (**b**) winter period, in Lake La Cruz. Scale on *both right–bottom* charts has been expanded by removing the data from picocyanobacteria

bacterial groups dominating Lake La Cruz bacterioplankton, among which the dominant species are of phototrophs.

Oxygenic and anoxygenic phototrophic bacteria develop in the lake and form relatively dense populations that mainly appear at both sides of the oxic–anoxic interface during thermal stratification. A dense plate of *Synechococcus*-like picocyanobacteria develops during the stratification period in the upper part of the oxic–anoxic interface (Camacho et al. 2003a, c), whereas more sparse populations of photosynthetic sulphur bacteria, both purple (PSB) and green (GSB), appear below the depth where oxygen is zero (Rodrigo 1997; Rodrigo et al. 2000), both during the stratification and the mixing periods.

Synechococcus-like picocyanobacteria are spread through the water column, reaching concentrations of around 1×10^6 cell mL^{-1}, but they tend to concentrate at the bottom of the thermocline over the oxic–anoxic interface and can reach very high densities (up to 8×10^6 cell mL^{-1}) in summer (Camacho et al. 2003a). One of the selective advantages of picocyanobacteria over other phototrophs, which allow them to reach these very high numbers, is their high surface-to-volume ratio, which promotes a higher nutrient uptake capacity compared with eukaryotic algae. They can harvest the green light prevailing at the summer chemocline, thanks to the presence of phycoerythrin (Camacho et al. 2003c; Camacho 2006). This is the other main feature that explains their dominance. Phycoerythrin is a phycobilin (protein)-type photosynthetic pigment that, due to its proteic nature, does not preserve in the sediment. The ability to grow in the low, green-enriched, light field in the deep waters of the upper hypolimnion allows these cyanobacteria to cope with nutrient depletion in surface waters by growing in deeper layers where nutrient availability is higher. Cryptophytes, which are eukaryotic algae that also contain phycoerythrin, are the main accompanying oxygenic phototrophs in these layers.

The occurrence of purple and green phototrophic sulphur bacteria is among the best-known features of meromictic and stratified lakes, because there they build dense microstratified plates. Phototrophic sulphur bacteria, which use sulphide as main electron donor for anoxygenic photosynthesis, also develop relatively dense populations (up to 3×10^6 cell mL^{-1}) in the upper part of the anoxic hypolimnion where they can still harvest the available light with their photosynthetic pigments (Fig. 8.8). Two main species form these populations (Rodrigo 1997; Rodrigo et al. 2000; Casamayor et al. 2012), the PSB *Lamprocystis purpurea* (formerly *Amoebobacter purpureus*) and the GSB *Chlorobium clathratiforme* (formerly *Pelodyction clathratiforme*), with the former having a selective advantage as being able to accumulate elemental sulphur, which can serve as alternative electron donor in the low-sulphide environment of Lake La Cruz chemocline. Apart from the phototrophic bacteria, both cyanobacteria and sulphur bacteria, the chemolithotrophic primary producers and their activity have been recently described (Walter 2011; Walter et al. 2014) as related to the iron and nitrogen cycle, and it has been hypothesized that sulphur bacteria could also be involved in these activities.

8.4.2 Nano-phytoplankton and Plants

Phytoplankton studies in Lake La Cruz started in the 1980s (Rojo and Miracle 1987), but detailed systematic studies were undertaken in 1987–1988 (Dasí and Miracle 1991; Miracle et al. 1998) and 1996–1998 (Camacho et al. 2003a, c; Sendra 2009). About 130 nano-phytoplankton species have been identified (e.g. Fig. 8.7). Lake meromixis and the marked physical stability of the water column influence phytoplankton spatio-temporal distributions, depth (with the associated ecological features) being the main structuring factor as indicated by a multivariate statistical analysis performed with phytoplankton species data (Sendra 2009). At the onset of stratification phytoplankton growth consumes nutrients, but most of the generated biomass, including the faecal pellets of zooplankton, sink downwards. Therefore, the well-illuminated surface waters become nutrient depleted (Figs. 8.3 and 8.6), and phytoplankton growth is limited. Recycling to the upper layers is restricted, and phytoplankton production settles to pycnoclines down below and even deeper. Downwelling production is accumulated at the steep density gradient established across the mixo-monimolimnion interface and increases the decay processes, causing a progressive exhaustion of oxygen in the upper layers. In mid-late summer, the anoxic zone is extended up to the lower metalimnion, where the available light is still enough for photosynthesis.

In the mixolimnion, a well-defined phytoplankton seasonal cycle (Fig. 8.9) may be identified in the lake, very similar to that in other karstic lakes in Spain, especially those with high relative depths as for Lake La Cruz (Planas 1973; Ávila et al. 1984; Modamio et al. 1988; Morata et al. 2003). The cycle starts with allogenic changes due to the autumnal overturn causing a turbulent disruption of the stratified system up to the top of the permanent chemocline (ca.16 m). The mixing period lasts through winter, and the mixolimnion depth varies with the meteorological conditions and is responsible for interannual differences in the vernal peak of algae. During mixing, nano-phytoplankton is dominated by *Cyclotella* spp. (diatoms) in the mixolimnion (and by picocyanobacteria). A predominant population of *Cyclotella delicatula* occupied the lake continuously, with a much lower abundance of *Cyclotella distinguenda*, at least for the last 200 years (Kiss et al. 2007). In earlier papers, these species were given inaccurate names. *C. delicatula* responds to mixing, as it develops a peak at the end of winter or early spring depending on the onset and intensity of mixing. It is a small-sized (5–6 μm Ø) centric diatom that can attain a maximum of 4×10^3 cell mL^{-1} at the end of winter-early spring. *C. distinguenda*, a much bigger (8–9 μm Ø) species, grows less and a bit later. When these diatom species peak, the deficiency of silica is very pronounced, and thereafter the diatom population nearly disappears from the epilimnion. They may, however, persist in low numbers in the newly formed thermocline to attain later a higher development at the deeper and well-established thermocline in early summer, until they virtually disappear in midsummer (Assemblage A, Fig. 8.9). *Gymnodinium*, a dinoflagellate, has similar distribution to these diatoms but with a late winter maximum.

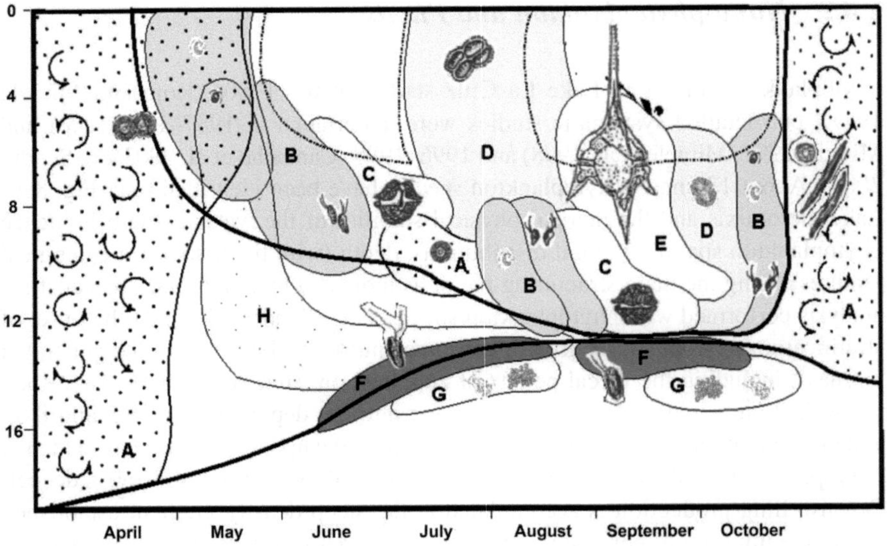

Fig. 8.9 Diagram indicating phytoplankton taxa prevalence in relation to the development of the vertical structure during stratification in Lake La Cruz, based on the annual cycle of 1987, with rains occurring in early July. *Solid* lines represent the bottom of the thermocline (isotherm approx. 12°C) and the situation of the oxycline, respectively. The *dashed line* encloses the nutrient-depleted zone in the epilimnion. (**A**) Mixed water column with prevalence of *Cyclotella* and *Synechococcus*-like picocyanobacteria; (**B**) fast-growing community dominated by *Pedinomonas*, chrysophytes and *Rhodomonas lacustris*; (**C**) small dinoflagellates *Peridinium* sp.; (**D**) community dominated by slow growing colonial and cenobial chlorophytes (*Crucigenia*, *Scenedesmus*, *Oocysts*); (**E**) slow-growing large dinoflagellates, *Ceratium*; (**F**) plate forming algae, *Cryptomonas*; (**G**) photosynthetic sulphur bacteria (*Lamprocystis purpurea*, *Chlorobium clathratiforme*); (**H**) *Synechococcus*-like picocyanobacteria

Phytoplankton succession proceeds as described elsewhere (Margalef 1983; Reynolds 2006). At the onset of stratification in spring, colonization by opportunistic, small, fast-growing, unicellular flagellate algae are common. They include the green alga *Pedinomonas minor* and chrysophytes (*Chrysolykos planctonicus*, *Kephyrion* sp.), followed by small cryptophytes (mainly *Rhodomonas* spp.). They become scarce in the epilimnion but peak in the thermocline, which is located near the surface at the onset of stratification (Assemblage B). *Parvodinium umbonatum*, another dinoflagellate, is also abundant deeper in the metalimnion (Assemblage C, Fig. 8.9).

As summer advances, the epilimnion stretches deeper, and the metalimnion becomes thicker. In the upper part of the metalimnion a pronounced oxygen peak develops, while in its lowest part oxygen is depleted. Thus, at the end of summer, an oxic hypolimnetic stratum is absent. The epilimnion is virtually uninhabited, but the metalimnion is populated by large-sized green algal unicells or colonies covered by mucilage (Assemblage *D*), though species composition varies annually. In the 1980s, the colonial *Crucigenia rectangularis* bloomed in summers, whereas at the end of 1990s, the colonial *Kirchneriella* spp. as well as unicellular species with spines such as *Lagerheimia* and species wrapped with mucilage such as *Cosmarium* spp. (Desmids) were the most abundant. They were accompanied by other green algae (species of *Scenedesmus*, *Monoraphidium*, *Oocystis* and *Pseudoquadrigula*). Occasionally, the epilimnion may be slightly fertilized by inflow of water from the lake's slopes due to wind, and phytoplankton of Assemblage D may also develop in the upper waters (e.g. as in 1987, Fig. 8.9). Forced by meteorological events (storms producing some groundwater inflows), reversions to assemblage B, consisting of more opportunistic species (*Pedinomonas*, *Chlorella* and *Rhodomonas*), occur sporadically in summer. Later on, at the end of summer, *Peridinium cinctum* followed by *Ceratium hirundinella* becomes an important constituent of the phytoplankton (Assemblage E).

At the end of summer, the metalimnion becomes narrow and sinks towards the oxic–anoxic interface. In the steep thermocline, species of Assemblage D sharply peak: *Scenedesmus* spp. in the 1980s and *Cosmarium abbreviatum*, *Lagerheimia ciliata* and *Tetrahedron minimum* in 1998. By mid-autumn, instability favours regression to species from first successional stages (Assemblage B), e.g. *Pedinomonas* and *Rhodomonas*. In very dry autumns, groundwater inflow is low, and overturn is weakened, but night-time convection (cooling) is still strong. Then Desmids (*Cosmarium* spp.) can extend their occurrence, as in autumn 1998, and attain extremely dense concentrations in the mixed epilimnion. In tropical lakes, desmids growth has been consistently observed to be associated with atelomixis (alternated diurnal stratification and night overturn occurring daily, e.g. Barbosa and Padisák 2002). In addition, pennate diatoms (*Navicula* sp. in 1988, *Fragilaria ulna* in 1998) can flourish at this time of the year. However, when mixing becomes intense, mostly by late autumn, *Cyclotella* becomes dominant so that Assemblage A is re-established.

As shown in Fig. 8.9, the lower metalimnion is successively occupied by phytoplankton from the earlier epilimnetic–upper metalimnetic successional stages. This implies that winter–spring species can inhabit the lower part of the metalimnion in early summer. Then, in midsummer, epilimnetic early-summer plankton occupied this lower metalimnion. At the end of summer, within the compressed metalimnion, species of earlier successional stages than those dwelling in the enlarged epilimnion appear. This spatio-temporal distribution depends on nutrient availability; earlier species become replaced if nutrients are depleted in the upper layers (Fig. 8.6), but they can grow in nutrient-richer deeper waters maintained by stable density gradients if enough light is still available at these depths. Summarizing, first successional stages in the mixed column and in upper waters may later occupy the lower meta-upper hypolimnetic waters. This can also be seen at the level of congeneric species, being characteristic of different successional stages, which become spatio-temporally segregated. For instance, *Cosmarium botrytis—C. leave—C. abbreviatum—C. meneghinii* replace each other in this order their preponderance in the metalimnion, from early spring to late summer. In autumn, instead, *C. bioculatum* may become dominant. Nevertheless, many other genera (*Scenedesmus*, *Lagerheimia*, *Cryptomonas*) show similar diversification and segregation of congeneric species in the available niches.

At the oxic–anoxic interface there are well established dense populations of planktonic phototrophs stratified along a redox gradient. A three-layer model appears as characteristic for meromictic lakes (e.g. Vila et al. 1998): layer 1 is composed of eukaryotic algae and cyanobacteria, layer 2 is formed by purple sulphur bacteria (*Chromatiaceae*) and layer 3 is mainly composed by green sulphur bacteria (*Chlorobiaceae*). A multilayer is well established in Lake La Cruz through the stratification period, because the much deeper winter interface is light limited and there eukaryotic phytoplankton is almost non-existent. Nonetheless, persistent oxyclinal species of eukaryotic algae appear in low numbers. The first layer close to the oxygen extinction depths is formed by stratifying species, *Cryptomonas phaseolus* and other cryptophytes, with the presence of euglenophytes both autotrophic (*Euglena* and *Trachelomonas*) and heterotrophic (*Astasia* and *Menoidium*) (Assemblage F). However, the main autotrophic components of this layer are the *Synechococcus*-like picocyanobacterial species (Assemblage H, Fig. 8.9), which are very abundant in the lake, as previously mentioned. Phycoerythrin-containing *Synechococcus* may attain dense concentrations at the oxic–anoxic interface in summer forming the characteristic cyanobacterial upper microbial layer, which may be assimilated to the layer 1 described by Vila et al. (1998) for meromictic lakes. Cryptophytes also contain phycoerythrin and can thus harvest the green-enriched PAR in the upper hypolimnion. Below these organisms, depending on the gradient of PAR and the availability of electron donors for anoxygenic photosynthesis, the already mentioned purple and green sulphur bacteria develop underlying plates (Assemblage G, Fig. 8.9).

Planktonic autotrophs are by far the main primary producers in the lake, since the steep slopes of the sinkhole, harbouring the lake, limit the littoral zone to a narrow band. Macrophytes occur in small areas where falling rocks are like staggered terraces. The shore of the lake can be broadly described as an alternation of three types of areas: (1) rocky without vegetation, which occupies most of the lake shore; (2) *Scirpus lacustris* patches with sparse growths of *Chara vulgaris* and *Chara aspera* between rock; and (3) disperse but dense submersed *Myriophyllum spicatum* patches close to the shoreline, in which a small proportion of *Potamogeton pectinatus* and *Potamogeton lucens* can also be found (Cirujano 1995; Cirujano and Medina 2002).

8.4.3 Consumers: Zooplankton, Zoobenthos and Fish

Zooplankton of Lake La Cruz was studied simultaneously with phytoplankton. After early studies in the 1980s (Miracle and Vicente 1983), two interannual cycles were followed, respectively, during 1987–1988 (Esparcia et al. 1991, 2001; Esparcia 1993; Armengol-Díaz et al. 1993; Armengol 1997) and 1996–1998 (Mezquita et al. 2002; Boronat 2003). Further studies were focused on the vertical migration of zooplankton (Armengol and Miracle et al. 2000; Starkweather et al. 2005), and others consisted of seasonal samplings to compare zooplankton community structure among lakes of this region (Armengol and Miracle 1999; Boronat et al. 2001).

Diversity of microcrustacean zooplankton is quite poor in the pelagic zone of Lake La Cruz, and this community lacks calanoid copepods. It essentially consists of two species of cyclopoid copepods, *Tropocyclops prasinus* and *Cyclops divergens*, which differ in size, as well as three taxa of cladocerans: *Daphnia* gr. *longispina*, *Ceriodaphnia dubia* and *Diaphanosoma brachyurum*. Since 2000 an invader, *Cyclops vicinus*, also became an additional major component of the plankton. *C. vicinus* shows summer diapause, and dormant stages of copepodites (IV and V) can be found near the sediment surface, 10 m below oxygen disappearance. Nonetheless, they instantly recover mobility on placing them in surface lake water. With this behaviour, dormant stages can avoid invertebrate and fish predation. Very abundant in this lake are the planktonic *Chaoborus flavicans* larvae, which avoid visual fish predation by migrating upward at night from their abundant population in the deep anoxic hypolimnion, returning to their refuge at dawn (Starkweather et al. 2005). Thus, meromictic lakes favour the species that can use anoxic waters as a refuge from predation on a daily basis or during dormancy.

Zooplankton abundance is very low in winter and increases in spring and autumn. In summer, a paradoxical contrast between quite clear upper waters and densely populated deeper waters is characteristic. Copepods are permanent residents of the lake with several generations per year, including a long one from autumn to spring and two or more additional generations in summer, depending on the species. *T. prasinus* is the most abundant copepod, with abundance an order of magnitude higher than *C. divergens*. *T. prasinus* is an epilimnetic–upper metalimetic species, showing several peaks of naupliar larvae in May, July, September and December, in order of decreasing abundance. At the end of stratification period, because of the compression of the thermocline, this copepod may exhibit a slightly deeper distribution. *C. divergens*, in contrast, is a lower metalimnetic species, and its maximum, in the vertical profile, follows the deepening of the thermocline; at the end of summer, it is close to the oxycline. *C. divergens* has a longer life span and shows widely spread reproductive periods (at the most two generations in summer). Their nauplii peak in winter, June and at the end of August. Meanwhile, cladocerans show marked seasonal and interannual variability. Two of the three cladocerans species found in the lake, *Diaphanosoma brachyurum* and *Ceriodaphnia dubia*, are exclusively summer species, distributed every year within the epilimnion and the metalimnion. *D. brachyurum* is the most abundant, and it represents >50 % of the summer zooplankton biomass. *C. dubia* is much less abundant and occurs more briefly than *D. brachyurum*, with maxima in deeper waters. *D*.gr. *longispina*, which are not very abundant, appear irregularly in autumn and spring when *D. brachyurum* is absent. In short, the seasonal depth distribution of planktonic crustaceans in Lake La Cruz is quite similar to other Spanish karst lakes (Miracle et al. 1992).

Rotifers are the richest zooplankton group in Lake La Cruz. They present multilayered dense populations in the vertical profile. More than 20 strictly planktonic rotifer species have been usually found annually, 11 species being predominant. Pairs of congeneric species segregating in the vertical profile are usual. Four distinct groups of rotifer species can be recognized from their spatio-temporal distribution in the lake as follows:

1. Species with their maximum in late autumn, *Synchaeta pectinata*, or in spring, *Ascomorpha ecaudis*; they are distributed throughout the mixolimnion but tend to stay in upper waters.
2. Summer epilimnetic–upper metalimnetic species: *Hexarthra mira*, *Trichocerca similis*, *Ascomorpha saltans* and *Asplanchna girodi*.
3. Species that are widespread during mixing at low densities but during stratification develop maxima in the lower metalimnion and, as summer advances, move downwards with the thermocline, peaking in late summer at the oxic–anoxic interface: *Polyarthra dolichoptera*, *Anuraeopsis fissa* and *Keratella cf. quadrata*.
4. Species bound to the oxic–anoxic interface: *Filinia hofmanni* and *Anuraeopsis miracleae*; they are restricted to the oxic–anoxic interface and as summer advances move upwards with the oxycline.

Miracle et al. (1992) pointed out that if high-resolution vertical sampling is applied, a clear segregation can be found in the species peaking sharply at different depths of the metalimnion and the oxic–anoxic interface (Fig. 8.10). Dense rotifer

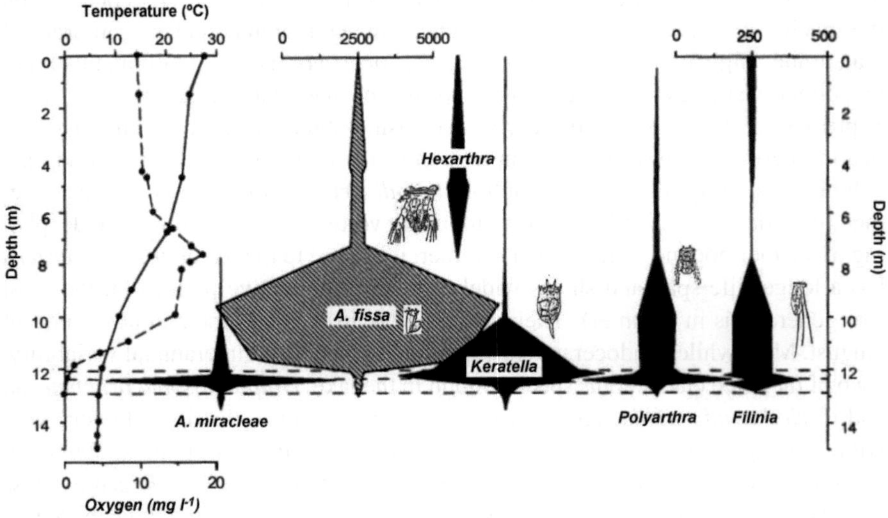

Fig. 8.10 Vertical distribution of rotifers in Laguna de la Cruz on July 21, 1988. Temperature (*solid line*) and oxygen (*dashed line*) profiles are also plotted. The width of the polygon at each depth represents the relative abundance of each taxon. For *Anuraeopsis fissa* (*striped kite*), use the 0–5000 individuals L^{-1} scale, and for the rest of rotifers (*Hexarthra mira*, *Keratella quadrata*, *Filinia hofmanni*, *Polyarthra* sp. and *Anuraeopsis miracleae*, *black kites*), use the 0–500 individuals L^{-1} scale. Note the vertical segregation of the two species of *Anuraeopsis* and the depth of the maxima of most rotifer species around the 1 mg L^{-1} oxygen concentration (Data from Dasí and Miracle, 1991; and Armengol and Miracle, 2000)

species populations, each dwelling at different strata of the oxic–anoxic interface, are characteristic of meromictic lakes. Concentrations of up to 30,000 in. L^{-1} of a single species have been registered in late summer peaks (Armengol-Díaz et al. 1993). The multilayered structure is well defined at the time when oxycline is deep and the above metalimnion is thick. However, it shows wide interannual variations according to mixing depth in winter and to the thermocline depth in summer. The years in which the metalimnion is shallower and thinner, strata at the oxic–anoxic interface overlap, causing suppression of the more specialized oxycline-bound species of group 4 and favouring the expansion of the more eurioic species of group 3 (Fig. 8.10).

The oxycline-bound species have been recorded from very few places in the world, dwelling in strata at low temperature and low oxygen contents. *A. miracleae* (Koste 1991) was described from the oxycline of Lake La Cruz. This species has been also found in the neighbouring sinkhole lake Lagunillo del Tejo and in Grosser Feichtaiersee (Jersabek and Koste 1993). The other oxyclinal species found in Lake La Cruz, *Filinia hofmanni* (Koste 1980), is a cold-stenotherm species that occurs in the hypolimnion of eutrophic and/or meromictic lakes (Hofmann 1982; Miracle and Alfonso 1993).

Studies on daily vertical migrations of zooplankton in Lake La Cruz (Armengol and Miracle 2000; Boronat 2003) indicate that the deep fine multilayer structure of the rotifer community at the oxic–anoxic interface is maintained throughout the day. Migrations of rotifers occupying these deep layers, with sharp density gradients, are much reduced. However, short distance active movements of rotifers in the cold and microaerophilic hypolimnetic layers are verified using traps that require ascending movements of at least 30 cm. On the other hand, in the epilimnion and upper metalimnion, there is a generalized migration of all microcrustaceans and rotifers following the ordinary diel migration cycle (dusk ascend, dawn descend). Rotifers, such *A. fissa*, that may inhabit these upper waters but also the oxic–anoxic interface only show diel migrations in the upper waters, maintaining unchanged the depth of their maxima in the oxic–anoxic interface. A special case is of the dipteran larva *Chaoborus flavicans* that migrates through the whole water column from the anoxic bottom layers to the lake surface water.

Microzooplankton (protists), mainly microaerophilic and anaerobic ciliate assemblages, also form conspicuous populations in Lake La Cruz, which are differentially distributed in the vertical profile. *Mesodinium* sp. in the upper waters and scuticociliates in lower depths are the dominant taxa. Meanwhile, scuticociliates of the genus *Pseudocohnilembus* are present over a wide range of dissolved oxygen concentration (DO), whereas *Sphenostomella* and *Cristigera* strictly follow the DO gradient. Mixotrophic ciliates, *Coleps* and *Pelagothrix* (*Prorodon*), together with *Spirostomum teres*, dominate the bottom part of the mixolimnion and the oxycline. *Spirostomum teres* is a very well-known ciliate from the oxic/anoxic boundary of stratified water bodies. This species has high grazing rates on picocyanobacteria and heterotrophic bacteria in Lake La Cruz. In the anaerobic monimolimnion, odontostomatids ciliates are also present.

The small size of the lake allows a quite homogenous, though diverse, assemblage of littoral microinvertebrates (Roca et al. 2000; Boronat 2003). Samples from littoral waters (hand tow net, mesh 75 μm) contained 60 rotifer species (including 15 euplanktonic), cyclopoid copepods (8 species, including 2 euplanktonic), cladocerans (12 species, including 3 euplanktonic), ostracods (2 species) and chironomid larvae. Nematodes, oligochaetes, turbellarians and gastrotrics are also abundant. With respect to other macroinvertebrates, Ephemeroptera larvae are the most frequent with low proportions of Odonata larvae and fewer numbers of Coleoptera and Collembola. Among all the animal taxa reported above, microcrustaceans are the only well-studied group. Main species are *Eucyclops macruroides*, *Macrocyclops albidus*, *Canthocamptus staphylinus*, *Acroperus angustatus* and *Cyclocypris ovum*. The last two species are the most abundant. *A. angustatus* is permanently found though with a decline in summer (Boronat 2003). *C. ovum* also remains permanently, showing low numbers in summer when only adults were found, whereas juveniles were observed only in early winter and spring (Mezquita et al. 2002).

The littoral zone is not much affected by stratification, except in autumn overturn, when water from the hypolimnion may break into the surface and carrying red *Lamprocystis purpurea* clumps that float along the shore facilitated by wind. Fertilization then occurs with nutrients coming from hypolimnion, and rotifers abundance increases. Later, at the end of autumn, ostracods and chydorids recover from their summer decline to reach their maxima in winter-early spring. Euplanktonic microcrustaceans are plentiful in littoral areas showing usually an earlier growth in the littoral before peaking in the pelagic in spring. On the other hand, deep benthos is very poor because of oxygen depletion in summer and, as said, the macroinvertebrate community is formed by a dense population of *Chaoborus flavicans* occupying exclusively lake's bottom waters.

This lake was naturally fishless, due to its mode of origin, as a closed and isolated sinkhole. However, fish are now relatively abundant, and they were probably introduced long ago. Main fish in the lake are *Barbus* (*Luciobarbus*) *guiraonis* and *Achondrostoma arcasii*. Even when stocking of fish was banned, anglers were releasing live fish used as bait, though angling has been recently forbidden. In addition, also *Gambusia* sp. inhabits the littoral areas, although its population seems to be controlled by the winter temperature. The effect of fish may be important in this small lake, and the interannual variability of fish fry production may explain the irregular presence of *Daphnia* gr. *longispina*.

8.5 Ecological Processes and Biogeochemical Markers

8.5.1 Photosynthetic Pigments

Photosynthetic pigments are among the most used biogeochemical markers to describe lacustrine processes (Wright et al. 2005; Romero-Viana 2007; Romero-

Viana et al. 2010). Meromictic lakes, as Lake La Cruz, usually show a complex mixture of photosynthetic pigments corresponding to the populations developing in the vertical profile. This is because the stratification and the high stability in the metalimnion and hypolimnion allow for the development and concentration of highly abundant populations of photosynthetic microorganisms. The study of photosynthetic pigments in aquatic ecosystems is relevant for the monitoring of the different populations of planktonic photosynthetic microorganisms as well as for more functional studies of ecological aspects such as the state of senescence or stress of these populations.

In Lake La Cruz, the concentrations of photosynthetic pigments in water were regularly determined by HPLC (Picazo et al. 2013); and they show high seasonal variability annually. The concentration of chlorophyll-*a* varied in the lake by two orders of magnitude, from 1 to near 100 μg L^{-1}, depending on the period and depth (Fig. 8.11). The high differences in concentration along the vertical profile are due to the seasonal development of a deep chlorophyll maximum (DCM). This DCM develops along with the stratification from spring to late autumn, around the oxycline at ca. 1 %–0.1 % of PAR. Chlorophyll-*a* derivatives (primarily chlorophyllide and phaeophytin) may represent between about 3 % in spring (April) and up to 60 % in late stratification (October) of the total chlorophyll-*a* concentration, providing information about the senescence of phytoplankton populations. Concentrations of chlorophyll-*b* and chlorophyll-*c* are relatively low but enough for monitoring the relative abundance of Chlorophyceae and Bacillariophyceae, respectively. In the anoxic layers, bacteriochlorophyll-a (from PSB) and bacteriochlorophyll-*d* from GSB) also vary widely (Fig. 8.11).

Different carotenoids were also measured in the lake waters using HPLC analyses. The more relevant carotenoids found were markers for the presence of cyanobacteria (zeaxanthin and myxoxanthophyll), all due to the picoplanktonic populations of *Synechococcus* sp., and also of Chlorophyceae (lutein, violaxanthin and antheraxanthin), Cryptophyceae (alloxanthin) and Bacillariophyceae (fucoxanthin and diatoxanthin). Other ubiquitous carotenoids, such as β-carotene, were also regularly observed in the lake. Among the bacterial carotenoids, okenone was the most abundant. This carotenoid is characteristic of some purple sulphur bacteria (PSB), such as *Lamprocystis purpurea* being the dominant species in Lake La Cruz. The signature carotenoids of green sulphur bacteria (GSB), isorenieratene and chlorobactene, were also widely found in the samples from anoxic waters. The dominance of picocyanobacteria (*Synechococcus* sp.) is linked to the presence of biliproteins during all the year in the entire vertical profile, with the most abundant biliproteins present in these populations of Lake La Cruz being phycoerythrin (PE) and phycocyanin (PC). The highest concentrations of phycocyanin were observed in winter in the mixolimnion and in epilimnetic waters during stratification period. Concentration of phycoerythrin increases seasonally with a maximum at the DCM, where these PE-rich picocyanobacteria take advantage of this pigment for the better harvesting of the low, green-enriched, light available in the lower metalimnion.

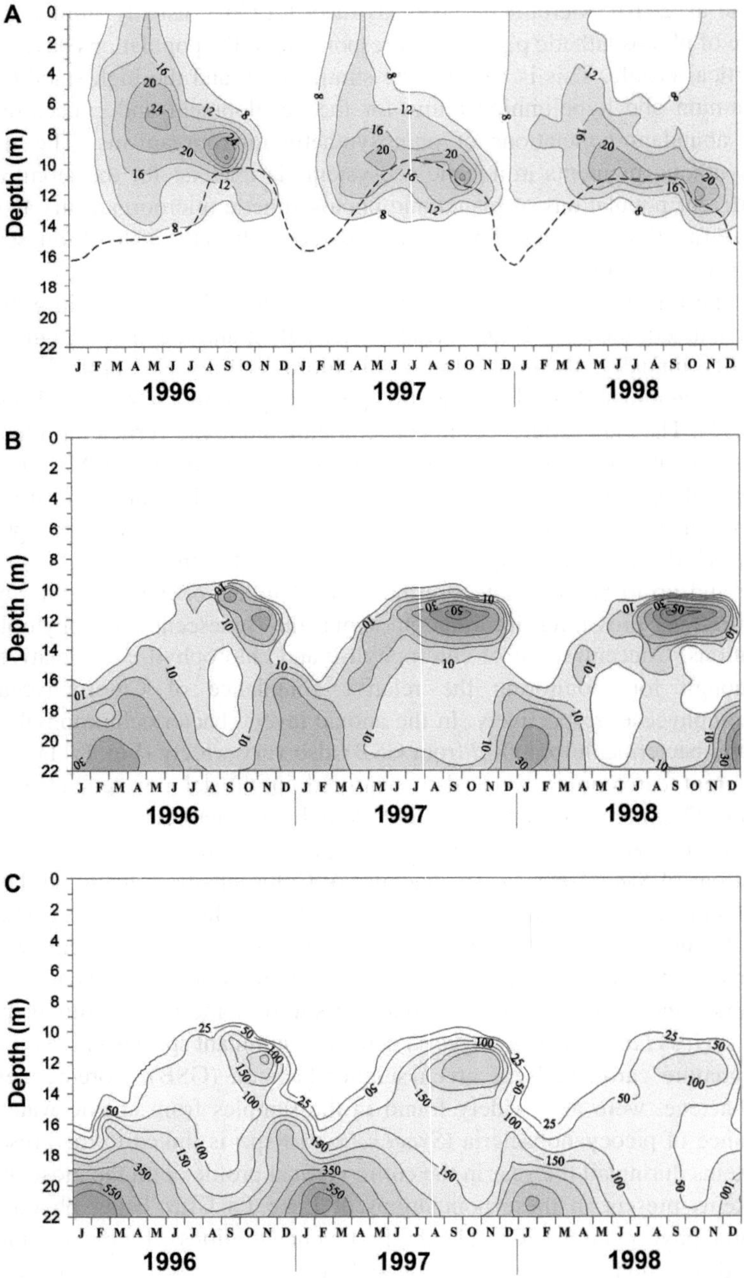

Fig. 8.11 Isopleths of the concentrations (mg m^{-3}) of (**a**) chlorophyll-*a*, (**b**) bacteriochlorophyll-*a* and (**c**) bacteriochlorophyll-*d* in the years 1996, 1997 and 1998 in Lake La Cruz. The *dashed line* in chart A represents the oxic–anoxic interface

Sedimentation processes of phototrophic plankton in Lake La Cruz, traced by photosynthetic pigments, mostly occur in some discrete pulses (Romero et al. 2006). A first pulse, mainly resulting in massive settling of oxygenic phototrophs from epilimnetic and metalimnetic waters, is associated with the annual massive calcium precipitation event that occurs in summer. Downwelling fluxes of chlorine derivatives and carotenoids from algae and picocyanobacteria reach the highest level just after the summer "whiting" (Fig. 8.12). By contrast, the massive calcium precipitation has almost no effect on sinking of sulphur bacterial populations. In turn, fluxes of bacteriochlorin derivatives and bacterial carotenoids from purple and green sulphur bacteria towards the sediment reach an annual maximum in late autumn after fading of the thermocline (Fig. 8.12). The collapse of thermal stratification causes exposure to oxygen to phototrophic sulphur bacteria, and thus a mass mortality of these populations occurs.

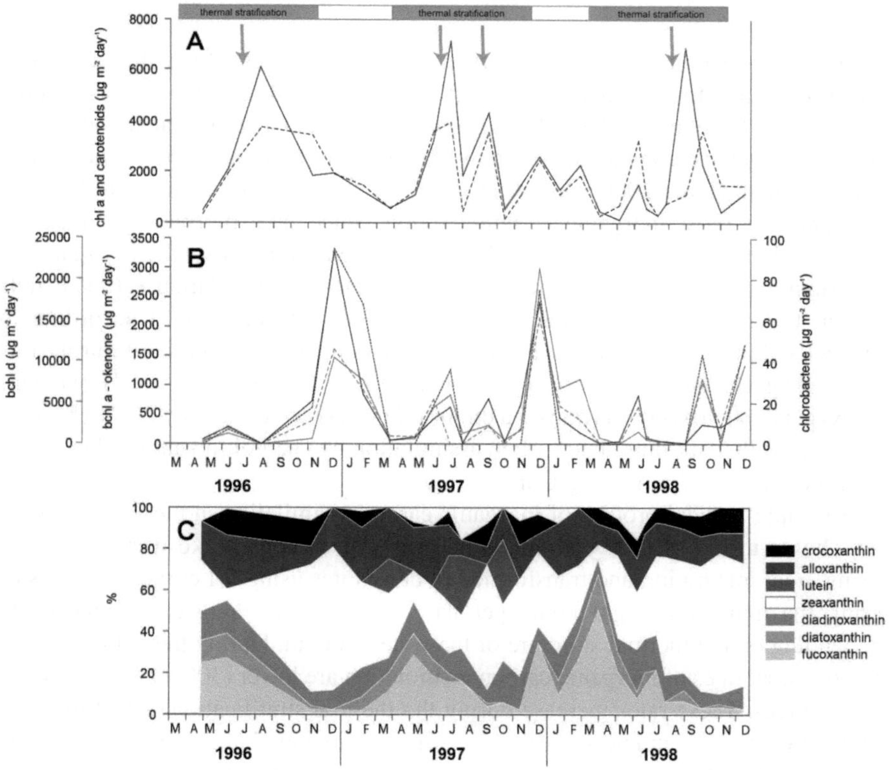

Fig. 8.12 (**a**) Fluxes of total chlorophyll-a (*solid line*) and carotenoids (*dashed line*) collected by sediment traps (µg dw m^{-2} day^{-1}). Whiting events are indicated by *arrows* and the thermal stratification periods by the *upper shaded area*. (**b**) Fluxes of bacteriochlorophyll-a plus derivatives (*grey solid line*) and the carotenoid okenone (*grey dashed line*) from the PSB populations, and bacteriochlorophyll d derivatives (*black solid line*) and the carotenoid chlorobactene (*black dashed line*) from the GSB populations. (**c**) Percentage contribution of taxa-specific carotenoids from the different algal groups to settled carotenoids recovered in the sediment traps

Although total fluxes of algal and picocyanobacterial pigments show a maximum during summer calcium precipitation (whiting), the relative contributions of taxa-specific carotenoids track the relative composition of the phototrophic community and the succession of planktonic populations (Fig. 8.12). Zeaxanthin, a specific carotenoid of the (pico)cyanobacterial populations, was the most abundant in the settled material during summer and autumn, whereas carotenoids derived from diatoms and chrysophytes such as diatoxanthin and fucoxanthin reached higher fluxes during winter and spring. Relative contributions of diadinoxanthin, a common pigment from dinoflagellates, and those derived from cryptophyceae, such as alloxanthin and crocoxanthin, were similar through the year suggesting more stable algal populations.

8.5.2 Primary Production and Bacterial Production

Primary production is a very important indicator for describing the ecosystem functioning and for the understanding of biogeochemical processes (Wetzel 2001; Dokulil and Kaiblinger 2009; Marra 2009). In Lake La Cruz, the calculation of the annual inorganic carbon assimilation by planktonic primary producers measured by the ^{14}C method for 2006 was of 184 g C m^{-2} yr^{-1}. This annual productivity accounts for most inorganic carbon fixation, since lake morphology does not allow a significant contribution of littoral and benthic photosynthetic organisms to primary production. This annual inorganic carbon assimilation estimate falls within the range of mesotrophic lakes (100–300 g C m^{-2} yr^{-1}) according to Wetzel's (2001) classification. From the total inorganic carbon fixed during this year, oxygenic photosynthesis represented 89.7 % (165 g C m^{-2} y^{-1}), whereas anoxygenic photosynthesis played a minor role by contributing only 2.8 % (5 g C m $^{-2}$ y^{-1}), while dark inorganic carbon assimilation by chemolithoautotrophic bacteria accounted for 7.5 % (14 g C m^{-2} y^{-1}).

Example vertical profiles of inorganic carbon assimilation in the water column are shown in Fig. 8.13. Depth-integrated rates of carbon uptake were calculated using areal integration and transformed to daily rates using P/I curves and considering the number of light hours per day. They varied along the annual cycle depending on the thermal structure of the water column. During the colder mixing period, areal rates of inorganic carbon assimilation are lower (300 mg C m^{-2} day^{-1} on average). These rates increase during the thermal stratification period to about 800 mg C m^{-2} d^{-1} in late stratification, when the maximal areal rates of inorganic carbon assimilation are obtained. Oxygenic photosynthesis was always the dominant inorganic carbon assimilation pathway. In winter, during the mixing period, it accounted for almost 90 % of total C fixation and occurred exclusively in the mixolimnion, namely, from surface to 16 m depth. The average rates measured throughout the mixolimnion were 3 mg C m^{-3} h^{-1} on average, with a recurrent subsurface maximum detected between 2 and 4 m. During the early stratification, the relative contribution of oxygenic photosynthesis was even higher (93 %). More

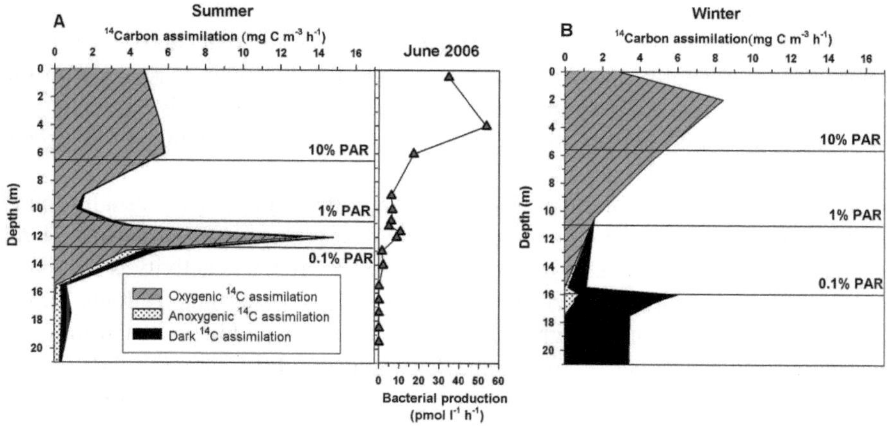

Fig. 8.13 Vertical profiles (cumulative area) of inorganic carbon assimilation rate (mg C m^{-3} h^{-1}) by oxygenic photosynthesis, anoxygenic photosynthesis and dark carbon fixation in the vertical profile of Lake La Cruz for the (**a**) summer stratification period, including the vertical profile of bacterial production (pmol L^{-1} h^{-1}, thymidine DNA incorporation) and (**b**) for the winter period

than half of it (53 %) was contributed by the epilimnion, the metalimnion contributed about 40 % of total due mainly to the incipient formation of a deep chlorophyll maximum (DCM) dominated by phycoerythrin-rich picocyanobacteria. At this stage, this population peaks around 12 m, where maximum rates are detected (up to 10 mg C m^{-3} h^{-1}). At the end of stratification, in early fall, the contribution of oxygenic photosynthesis to total carbon fixation in the lake dropped to around 74 %, though the abundance of picocyanobacteria in the DCM increased causing the highest rates (up to 40 mg C m^{-3} h^{-1}measured in 2006). During stratification photosynthetic rates were relatively constant throughout the epilimnion (~5 mg C m $^{-3}$ h^{-1}), representing nearly half (48 %) of the total oxygenic inorganic carbon assimilation during this period considering the lake hypsography.

Studies on selective inhibition of eukaryotic photosynthesis reveal that in winter, after mixing, up to about 80 % of the oxygenic carbon assimilation in the mixolimnion was due to picocyanobacteria. Also, during late stratification, ~80 % of the inorganic carbon fixation in the epilimnion and metalimnion can be attributed to picocyanobacteria. However, at the oxycline where the DCM establishes, inorganic carbon fixation of eukaryotic algae (52 %) balances that of picocyanobacteria (48 %). Their codominance coincided with the *Cryptomonas* sp. maximum near the chemocline.

The relative contribution of anoxygenic photosynthesis as carbon fixation pathway in stratified lakes depends of several factors, such as the sulphur and light availability (Van Gemerden and Mas 2004; Vila et al. 1998; Camacho 2009). In Lake La Cruz, anoxygenic photosynthesis relatively increased in its importance from mixing to stratification such that the areal rates were the lowest in early winter just after thermal mixing, while the maxima were measured in late autumn at the end of the stratification. During the mixing period, ~80 % of anoxygenic photosynthesis occurred in the monimolimnion below 16 m, while a smaller portion was detected in the permanent chemocline, just below the oxic–anoxic interface. During thermal stratification, anoxygenic carbon fixation mainly occurred in the upper and lower parts of the hypolimnion and metalimnion, respectively. High rates were also measured during stratification just below the maximum of oxygenic photosynthetic activity.

Dark assimilation of inorganic carbon (chemolithoautotrophy) is usually found in the chemocline and hypolimnion of stratified lakes (Camacho and Vicente 1998; Camacho et al. 2001; Casamayor 2012). In Lake La Cruz, this activity is relatively low during the mixing and early stratification periods, being mainly found in the anoxic monimolimnion, though ca. 10 % is detected within the mixolimnion, particularly in the oxycline. When the lake is thermally stratified, dark carbon assimilation occurs mainly (73 %) in the hypolimnion while the rest, about 27 %, in the microaerophilic metalimnion. A significant increase of the relative importance of dark inorganic carbon assimilation occurred in the lake at the end of the stratification, representing about 20 % of the total inorganic carbon assimilated in this period in Lake La Cruz.

Planktonic bacterial production is an important parameter for estimating the heterotrophic activity in a lake's pelagial (McDonough et al. 1986). Vertical profiles of bacterial production in Lake La Cruz during the stratification period, measured as incorporation of [^{3}H] thymidine, show the highest values in the aerobic epilimnion. In comparison, bacterial production decreases along the metalimnetic gradient, but close to oxycline and near the DCM, it increases slightly coupled with the highest rates of oxygenic photosynthesis occurring there. These results demonstrate the linkage of the carbon released by oxygenic phototrophs with bacterial production.

8.5.3 Food Web Structure

Food web in Lake La Cruz is considered in three compartments: (1) the pelagic oxic waters, (2) the littoral zone and (3) the bottom layers from the top of the oxycline. Within the pelagic and littoral oxic zones, and based on studies performed in 1998 in Lake La Cruz, interactions between phyto- and zooplankton appear to be similar to those described in the PEG model (Sommer et al. 1986). Phytoplankton peaks are followed by those of the zooplankton; e.g. a first important phytoplankton peak in May (*Cyclotella*) is followed by a zooplankton (*Diaphanosoma*) peak in early summer. Both groups decline markedly in midsummer, this being reinforced after the "whiting" due to enhanced phytoplankton settling. After autumn overturn an increase of phytoplankton is noted, which is not followed by zooplankton biomass increase because rotifers, which are small, dominate, whereas bigger zooplankters even reduce their abundance. In 1998, mean annual plankton biomass in the water column from the surface to the oxic–anoxic interface (12 m at midsummer) was about 2 g C m^{-2}, of which about 0.4 g C m^{-2} was zooplankton (rotifers and microcrustaceans). In the oxic waters, the classical grazer food web can be traced from phytoplankton to herbivorous zooplankton to carnivorous zooplankton, including *Chaoborus* (top planktonic predator), and finally to fish (juveniles of benthivorous species). Marked seasonality of the food web functioning occurs because of zooplankton species replacement and feeding changes during individual development (i.e. nauplius cyclopoid larva are herbivorous, but adults are omnivorous). Instars I and II of *Chaoborus* larvae would prey on rotifers, but the instars III and IV feed on relatively larger prey, the microcrustaceans. Fish juveniles appearing in spring also feed on microcrustaceans. Fish have been mainly introduced to the lake from the nearby mountain streams. The main species, *Achondrostoma arcasii* and *Luciobarbus guiraonis*, are benthivorous in their older stages and probably rely on the abundant littoral chironomids and other littoral insect larvae, as well as on detritus from macrophytes and other sources. However, in this small lake, they are also pelagic and may feed to some extent on *Chaoborus* larvae. Mosquito fish (*Gambusia* sp.) are restricted to littoral areas, though they are mostly planktivorous. Although the littoral zone is very restricted, nutrient recycling in this area may be relatively important because the lake is small. Fish and zooplankton make diel horizontal displacements between the littoral and the pelagic environments, as can be concluded from observations in a nearby sinkhole Lake El Tejo (Armengol et al. 2012). In short, the food web in the oxic waters is based mainly on organic matter production by phytoplankton and, to a lesser extent, by macrophytes. However, it is also linked to the terrestrial system by the incorporating both terrestrial organic matter (pollen, detritus) and losses by migration of flying insects.

The food web of the oxic layers represents only a part of the energy transfer from primary production. Ungrazed settling primary production is very important in this type of lakes with partial water recirculation. Microbial utilization of primary production occurs mainly very deep, at the oxic anoxic interfaces and the anoxic waters, mainly through anaerobic processes. When organic matter enters the permanent chemocline, it is trapped into the monimolimnion and does not recirculate upwards. Thus, a substantial part of the planktonic primary production of the lake is lost to the sediments, and an important part is also consumed by heterotrophic bacteria within the anoxic waters by anaerobic respiratory processes deriving from a detrital electron flux. These detritus particles, and especially the heterotrophic bacterioplankton, consuming this organic matter, are also the main food for anaerobic ciliates that develop conspicuous populations in the anoxic hypolimnion during stratification.

In contrast with the upper lake layers, which are nutrient depleted during most of the year and seemingly quite devoid of biomass in the upper half of the water column, in the deep bottom waters around the chemocline(s), very dense algal and bacterial populations develop. Stratifying algae, picocyanobacteria, and photosynthetic bacteria develop sharp and dense populations, altogether with heterotrophic and chemolithotrophic bacteria, at the lowest part of the metalimnion and in the oxic–anoxic interface. Zooplankton, both metazoan and microaerophilic protozoans, is concentrated in the upper part of this interface. Phototrophic sulphur bacteria, though forming dense populations in the anoxic waters, do not appear, however, as an important food source for zooplankton. Part of the food web is based on the existence of deep chlorophyll maximum and the considerable part of primary production owing to algae and bacteria that enhances growth of heterotrophic bacteria. Also, chemolithotrophic bacteria at the chemocline play an important role.

A food web based on the microbial loop has been postulated for meromictic lakes (Culver and Brunskill 1969). However, in the stagnant water of Lake La Cruz, the distance between the mean resident depths of primary producers and decomposers (trophogenic and tropholytic zones) is quite large. Because during the thermal stratification period an important part of the primary production occurs in the metalimnion and in the upper hypolimnion, most of it cannot enter the food web of the oxic waters, and the only consumers are organisms specialized to cope with the restricting conditions of the oxycline and the anoxic hypolimnion. These include several metazooplankton species, especially rotifers (Fig. 8.10), and microaerophilic and anaerobic ciliated protozoa. In Lake La Cruz, the coupling of this deepwater microbial loop with the classical grazer food web in the upper waters is very weak, because of the strong density gradients in the water column, the environmental restrictions (e.g. microaerobic or anoxic conditions) for most consumers to graze at these parts of the water column, and the lack of an efficient biological transport of nutrients to upper layers by migrating zooplankton. The energy transfer from the high standing stock of rotifers (and cyclopoids, partly preying on rotifers) to the upper-water food web is very poor. In fact, an inverse vertical distribution in the multilayer structure of trophic levels is evident. With this structure, predation is prevented because lower layers (anoxic and cold water, with low redox potential, hazardous gases and metal concentrations) are less suited to survival of predators, thus allowing the development of their more tolerant preys. In addition, the extreme case of *Chaoborus*, which spends the day in the monimolimnion to avoid fish predation in the layers above, is also remarkable. But *Chaoborus* is a planktonic animal, which feeds in the surface water layers, thus skipping the concentrated food items in the oxic–anoxic interface (Starkweather et al. 2005). In brief, and for the whole trophic functioning of the lake, most carbon fixed by the primary producers, both in the oxic and anoxic layers, sinks to the sediment or is consumed by detrital particle-attached or, mainly, osmotrophic anaerobic heterotrophic bacteria.

8.5.4 Calcium Carbonate Precipitation

Calcium carbonate precipitation is a well-known summer phenomenon in hard water lakes commonly named “cloudy” or “whiting” (Bathurst 1971). In Lake La Cruz, it occurs as a sudden and short-term event giving the water a milky colour. It lasts (Fig. 8.14) about 1 week in the surface waters. However, a second event also occurs in some years at the summer end. Such events cause a reduction of light penetration due to increased turbidity, caused by co-precipitation of soluble phosphorus with calcium carbonate and self-flocculation of phytoplankton and bacteria

Fig. 8.14 15-day time-span sequence pictures of Lake La Cruz showing the evolution (from *top* to the *bottom*) of a whiting event (tumultuous calcium carbonate precipitation) in summer period (modified from Boronat 2003)

due to their aggregation with crystals causing the removal of these organisms from the water column (Rossknecht 1980; Koschel et al. 1983).

Calcium carbonate precipitation can be favoured by high temperatures and a high pH promoted by intense photosynthetic activity. During spring and early summer, the lake water is oversaturated with calcium carbonate, though whiting suddenly occurs usually around the end of July. Such an event can be related to the nucleation of phytoplankton, especially picocyanobacteria, whose photosynthetic activity further increases pH around the cell thus enhancing calcium carbonate precipitation. Very small calcite crystals can attach to phytoplankton cells that act as nucleation sites for crystal growth, thus promoting an unrestrained precipitation, starting and extending through the whole epilimnion. As an example, a bloom of *Crucigenia rectangularis* was observed in Lake La Cruz during July in 1987 and 1988 when whiting started (Dasí and Miracle 1991), and its photosynthetic activity was probably linked to the increase of pH, leading to a decrease in calcium carbonate solubility, and its precipitation.

The mineralogical composition of calcite crystals based on X-ray diffraction studies shows that it is dominated by magnesium calcite (or by aragonite during the second precipitation event when it occurs at the end of summer). Varved sediments (Fig. 8.15) register the annual occurrence of these whiting events (Müller 1971;

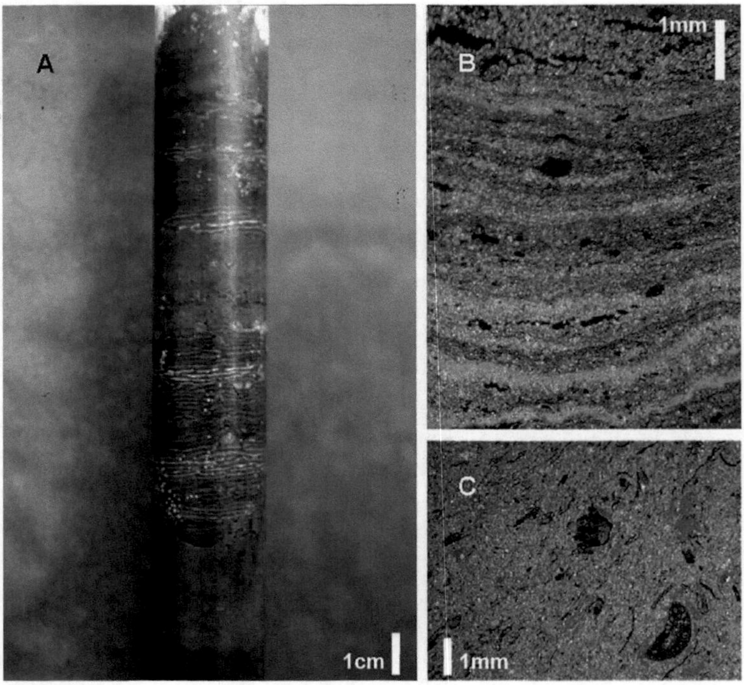

Fig. 8.15 (**a**) Photograph of a sediment core recovered from Lake La Cruz, (**b**) microphotographs of recent varves (1976–1988) and (**c**) detail of a detrital layer formed by equigranular of euhedral dolomite rhombohedrons, showing charcoal particles, charophyte remains and siliciclastic grains

Galat and Jacobsen 1985; Vicente and Miracle 1988; Miracle et al. 2000). Scanning electron microscopy, used for counting and the evaluation of crystal's volume, shows that the particulate $CaCO_3$ in Lake la Cruz appears as needle-shaped crystals with maximal dimensions of about $0.2 \times 1 \times 8$ μm with a maximum abundance in the epilimnetic waters of $7\text{–}8 \times 10^6$ crystals mL^{-1} that correspond to a volume of crystals of 4×10^6 μm^3 mL^{-1} (Rodrigo et al. 1993).

The Mg/Ca ratio in lake water determines which carbonate minerals will precipitate (Bathurst 1971; Kelts and Hsü 1978; Miracle et al. 2000). With these molar ratios below 2 Ca-calcite and Mg-calcite are the dominant precipitating minerals, but with ratios above 3–4 aragonite dominates (Müller et al. 1972). However, the relative effect of these ratios might depends on temperature and other conditions. According to Folk (1974), the presence of magnesium influences the shape of the precipitating crystals by inserting a selective "poisoning" effect on the side faces of calcite. Due to this inhibition, lateral growth of the crystals is blocked, and they proceed only along the longitudinal axis, which determines the needle-like shape of the crystals.

In Lake La Cruz, the "whiting event" occurs commonly when the lake is thermally stratified with a thermocline starting at 5–6 m and pH values of 8.8–8.9. Whiting strongly decreases water transparency, and Secchi disc depth drops from >6 m to about 0.5 m. In other lakes showing whiting events, e.g. Otisco Lake, the Secchi disc depth decreases 50 % (Weidemann et al. 1985), 65 % in Lake Michigan (Vanderploeg et al. 1987) and up to 76 % in Fayetteville Green Lake (Brunskill 1969). The strong decrease in light penetration affects the vertical distribution of photosynthetic organisms, and vertical movements are performed mainly by some species, either actively or by buoyancy changes. Because of these migrations and selective phytoplankton removal by settling, the mean depth of the deep chlorophyll maximum in Lake La Cruz moves upward from 12.5 m with clear waters to 8.5 m during whiting. Vertical migration of several phytoplankton species such as *Peridinium inconspicuum* and also the vacuolated green phototrophic bacteria *Chlorobium clathratiforme* has been reported in the lake (Dasí and Miracle 1991). Around 1 month later, once crystals sink into the monimolimnion, and also because of the sinking of most phytoplankton, Secchi disc depth increases to maximal values of about 12 m.

Soon after the epilimnetic precipitation, crystals slowly sink, and 10–15 days later, the maximum concentration of crystals is found at 10–12 m. Finally, part of the precipitate reaches the sediment forming a white 0.2 mm varve line of calcite that can easily be seen in the black sapropel, the latter being the dark-coloured sediments that are rich in organic matter (Fig. 8.15). Redissolving processes of crystals during sinking diminishes their numbers and size. However, since the dissolution rate of crystals is slow, most of them reach the sediment. The downward movement of calcium carbonate crystals causes a decalcification of epilimnetic waters and a calcium enrichment of the monimolimnion, as in other lakes; and this contributes to lake stability by increased density of the bottom waters (Wetzel 2001). Thus, the Ca/Mg ratio in an epilimnion dominated by magnesium is ca. 1:4, whereas this ratio is balanced (around 1:1) in the monimolimnion by the transport of calcium from surface waters (Rodrigo et al. 1993; Miracle et al. 2000).

8.6 Paleolimnology, Climate Change and the Origin of the Meromixis

8.6.1 Origin of the Meromixis as Revealed by the Sediment Record

The topmost 43 cm the sediment of Lake La Cruz shows a laminated sequence (Fig. 8.15), consisting of pairs of alternating light and brownish laminae. Sediment traps, which were examined monthly for 3 years (Romero et al. 2006), demonstrated the annual occurrence of laminations. The white layers composed of low-magnesium calcite and occasionally by aragonite, are formed by the seasonal pulses of calcite deposition resulting from summer whiting (Miracle et al. 2000; Romero-Viana et al. 2009, 2011; Valero-Garcés et al. 2014). Contrastingly, the dark laminae are formed by organic-rich silts and occasional clasts, which are mainly deposited after the annual water column overturn (Fig. 8.15). A total of 417 calcite laminations were counted in three cross-matched sedimentary sequences. The onset of laminated sediments in Lake La Cruz was then confidently dated at 1589 A.D. (Romero-Viana et al. 2010). The onset of laminations marks the beginning of a dynamic process towards meromixis. The lower values of the xanthophyll to other carotenoid ratio found in the bottom part of the laminated sequence suggest (Leavitt and Carpenter 1990) a differential degradation of organic compounds and therefore a not yet permanently stable anoxic bottom water layer until about 1660 A.D. (Romero-Viana et al. 2010). Indeed, calcite-laminated sediment also occurs in several holomictic lakes having optimal conditions for the preservation of annual laminations, such as long and uninterrupted periods of hypolimnetic anoxia (i.e. Gruber et al. 2000; Peck et al. 2002; De Vicente et al. 2006). These results may partly support the climate forcing hypothesis behind the meromictic origin in Lake La Cruz as firstly proposed by Julià et al. (1998).

Historical records of very cold episodes began in the fifteenth century, mainly in Europe, although the situation was extreme in the second half of the sixteenth century, with global extension of such cold episodes (Font-Tullot 1988). Both cold winters (extended ice-cover) and warm-dry summers (Ambrosetti and Barbanti 2005; Jankowski et al. 2006) could lead to meromixis by different ways, as proposed by Schmidt et al. (2002) for lake Längsee. For Lake La Cruz, extreme cold winters at the end of the sixteenth century likely resulted in prolonged anoxia in winter, promoted by the ice cover on the lake lasting for longer periods, thus enhancing the accumulation of dissolved ions in bottom waters. CO_2 oversaturation under strong hypolimnetic anoxia could cause calcite dissolution, and therefore the accumulation of dissolved calcium and bicarbonate ions could enhance the chemical stratification in the earlier stages of the development of meromixis (Wüest et al. 1992). Moreover, other factors are considered to have a synergic effect such as the increase of water level registered during the onset of laminations evidenced by ostracods records (Mezquita and Miracle 1997; Julià et al. 1998). However, the drastic drop in the temperatures during the end of the seventeenth century (Luterbacher et al. 2001, known as Maunder Minimum (1675–1700 A.D.) probably reinforced the process that was also linked to biological productivity. The location of the lake in a circular depression, sheltered from the wind, its morphometry and the lack of any significant subterranean inflow directly to the bottom of the basin, which could destabilize the monimolimnion, favour the maintenance of the meromictic condition (Rodrigo et al. 2001).

8.6.2 Lake La Cruz as Climate Tracer

Annually laminated sediments have become important paleoclimate archives because they provide (1) the highest time resolution in proxy data series and (2) an inherent and continuous time scale (Ojala et al. 2012). Based on our extensive knowledge of the biological community as well as the understanding of the sedimentation processes in Lake La Cruz, we evaluated the pathways of climate input signal into the sediment (Romero-Viana et al. 2008), as it was later done in nearby karstic lakes (Barreiro-Lostres et al. 2014). The calibration with meteorological variables over the period 1950–1988 showed that winter rainfall from December to March (DJFM) is the best predictor for the calcite laminae thickness explaining >52 % ($r^2 = 0.5251$) of thickness variability (Romero-Viana et al. 2008, 2011) (Figs. 8.16 and 8.17). Interestingly, the thickness of calcite laminae in Lake La Cruz over the calibration time period is not very well correlated with any summer temperature variables, although temperature increase has been commonly argued as main triggering factor for calcite precipitation (Brunskill 1969). On the other hand, the photosynthetic activity of phytoplankton increasing pH could act as triggering the process, though we have no conclusive results on this sense but instead some evidences.

Although temperature is undoubtedly a trigger for massive calcite precipitation, in Lake La Cruz, the amount of calcite crystals settled, and then the thickness of the annual white laminae mainly depends on calcium concentrations in water. The seasonal evolution of epilimnetic calcium concentrations in Lake La Cruz suggests a link between laminated record and climate variability (Fig. 8.16). Epilimnetic calcium concentration decreases substantially after the summer whiting process. However, the water column Ca concentrations are restored after winter rainfall when the aquifer increases the discharge and subsequently water level rises. In addition, autumn overturn could erode the upper limit of monimolimnion, where calcium concentrations are high and steady over the year (Rodrigo et al. 2001), thus providing additional calcium to the upper mixed layer. However, the increase in

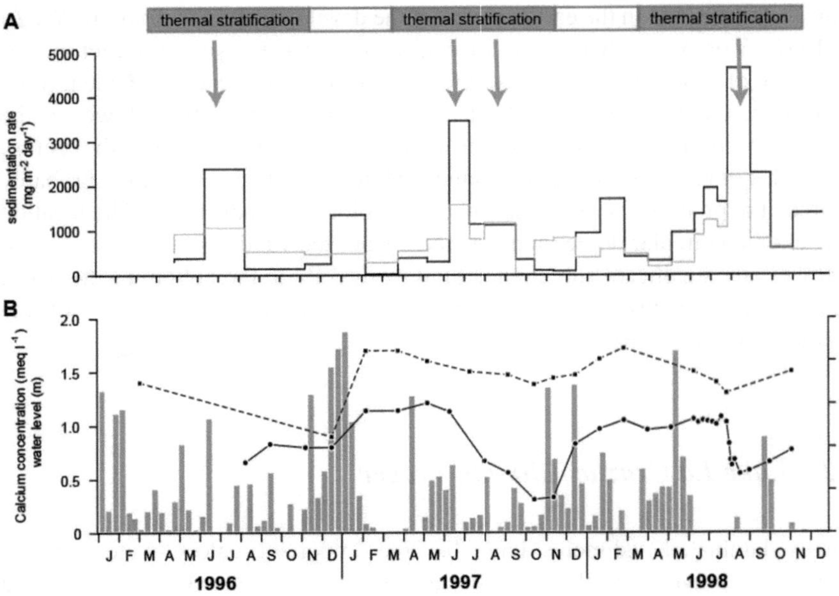

Fig. 8.16 (**a**) Sedimentation rates of mineral (*black line*) and organic matter (*grey line*) in Lake La Cruz. (**b**) Epilimnetic calcium concentration (*solid black line*) and water level fluctuations (*dashed black line*) over a reference level and accumulated ten days rainfall (*bars*) for the period in which the sedimentation patterns were studied. *Arrows* indicate the occurrence of the whiting episodes

mixolimnetic calcium concentration is mainly controlled by the aquifer discharge (Fig. 8.16).

Based on the calibration function developed ($y = 2.3 \times -157$; $r^2 = 0.52$) (Fig. 8.17), we reconstructed annual winter rainfall (Romero-Viana et al. 2010) since the onset of calcite laminations in Lake La Cruz. We observed the alternation of time periods of negative and positive anomalies (rainfall deviations from the means) in the region over the last four centuries (Fig. 8.17). One of the most intense negative anomalies was observed between 1630 and 1640, but then a positive trend started in 1660 reaching the highest positive values in 1685. The eighteenth century was also relatively wet but with values in 1720, 1740 and 1760. The second half of the eighteenth century stands out with high interannual variability and extreme values. This period, 1760–1800 (the Malda anomaly), was characterized by major climatic fluctuations, with a rapid succession of droughts and floods (Barriendos and Llasat 2003). However, the total annual precipitation in this period was similar to average values, as in other Mediterranean regions, e.g. Italy (Camuffo et al. 2000) and the Balkans (Xoplaki et al. 2001). In addition, the end of the eighteenth century was also relatively wet, but at the beginning of the nineteenth century, from 1800 to 1820, a persistent negative rainfall anomaly, with much lower precipitation, occurred during a minimum solar activity period, the Dalton Minimum (1790–1820). By contrast, around the middle and the end of the nineteenth century (1840–1850 and 1890), the inferred rainfall values were high, related to a positive

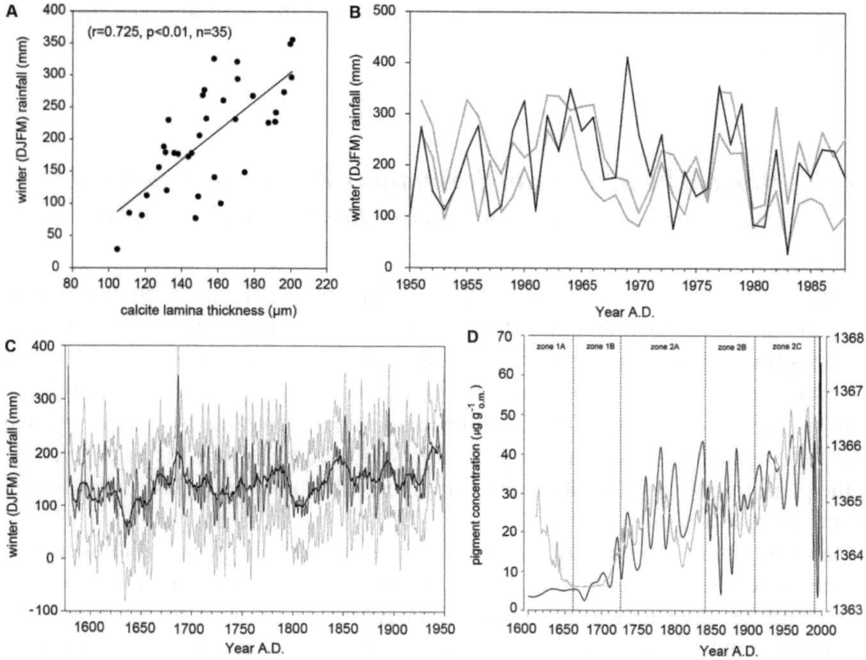

Fig. 8.17 (**a**) Calibration function of the thickness of calcite laminae. (**b**) Winter (DJFM) rainfall instrumental values (*black solid line*) and inferred range of winter rainfall from calcite laminae thickness of Lake La Cruz sediments (*grey lines*).(**c**) Inferred winter DJFM rainfall values (mm) (*black line*); *grey lines* corresponds to the standard deviation. The *thicker black line* shows the 10 years moving average. (**d**) Pigment concentrations (zeaxanthin plus total chlorophyll a derivatives) (*black line*) and solar irradiance variations arising from the 11-year activity cycle (Lean 2000) (*grey line*)

anomaly. Finally, the beginning of the twentieth century was drier than the mid-twentieth century.

At the interannual decadal time scale, the analysis of periodicities in the reconstructed winter rainfall time series on Lake La Cruz shows that regional winter rainfall may be governed by nonstationary components. This is in accordance with previous studies that report nonstationary phases in the European/North Atlantic climate system and also in areas of Mediterranean climate (Pauling et al. 2006; Touchan et al. 2005; Casty et al. 2005). Nevertheless, the signals of 0.12 and 0.25 year^{-1} periodicities observed in this regional rainfall reconstruction were similar to those observed for North Atlantic Oscillation (NAO) index series (Hurrell 1995). In fact, positive anomalies of regional rainfall are driven by the southwards position of the storms track during negative phases of the NAO (Romero-Viana et al. 2009). Nevertheless, the decoupled signal between the Luterbacher et al. (2001) NAO's reconstruction and the Cuenca regional time series during two long periods, e.g. as 1610–1680 and 1800–1850, point out that atmospheric circulation patterns other than NAO could also have conditioned the rainfall variability in the Iberian Peninsula (Romero-Viana et al. 2010).

8.6.3 Biogeochemical Transformations Within the Sediments and Trophic Evolution During the Meromictic State

Meromixis may have an overriding effect on lake biogeochemical processes since the permanent chemocline segregates the lake's water column into two main layers: the mixolimnion and anoxic monimolimnion. However, what was the response of the biotic component to the onset and development of meromictic condition? Only long-term perspectives could add information about community trajectory over time, and a paleolimnological approach allowed us to assess the historical changes in the phototrophic community and the trophic evolution over the last four centuries of this meromictic lake.

Lake sediments contain proxies for productivity, among which photosynthetic pigments, representing only a fraction of the preserved organic matter, are the most direct and accurate indicators. These biomarkers can be used to reconstruct the historical trends of primary production and the responses of the phototrophic community to environmental changes (Leavitt and Hodgson 2001). The analysis of acetone extracts of the lake sediments revealed a wide variety of algal and bacterial pigments (Romero-Viana et al. 2010). The stratigraphic profiles of pigment concentration in the laminated sediments (Fig. 8.18) and the biological signals preserved in lake sediment over the last four centuries of meromictic condition show two main features: (1) a stable composition of the phototrophic signature through the sedimentary sequence, which shows the conservative biological community following physical stable conditions, and (2) a long-term trend of increasing concentrations of photosynthetic pigments. This fact shows a progressive eutrophication of the lake, because of both natural and anthropogenic causes.

The pigment signature is contributed mainly by zeaxanthin (about 50 % of carotenoid concentrations through the laminated sequence), a specific carotenoid of cyanobacteria, which seems to have remained stable from 1600 A.D. to the second half of the twentieth century. Although sedimentary pigment signature could not be immediately interpreted as due to the phytoplankton assemblage, the high contribution of zeaxanthin through the sedimentary sequence and the high densities of *Synechococcus sp* in the lake (Camacho et al. 2003a, c) suggest that cyanobacteria significantly contributed the primary productivity of the lake ecosystem both before and during the periods of the increasing trophy. Among the cyanobacterial populations, *Anabaena sp.* and *Microcystis sp.* are very rare in the lake (Dasí and Miracle 1991). The absence of changes in photosynthetic pigments of cyanobacterial populations, as zeaxanthin, echinenone and myxoxanthophyll, suggests a biodiversity stability among the cyanobacterial group and the constant relative dominance of picocyanobacteria during the meromictic period. In spite of

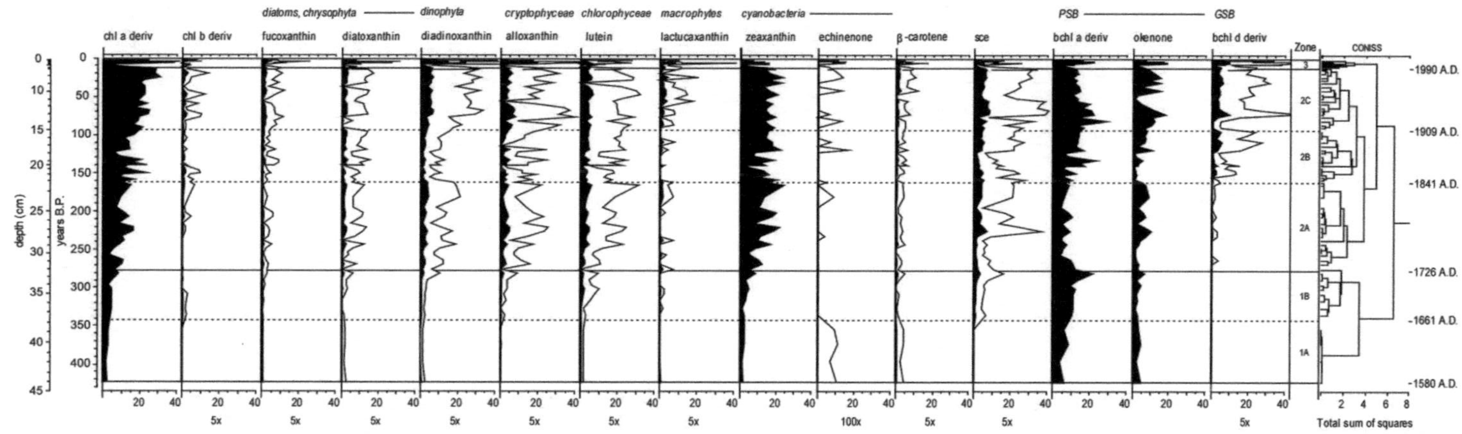

Fig. 8.18 Stratigraphic profiles of photosynthetic pigments in Lake La Cruz sediments. All compounds expressed as μg g^{-1} organic matter. Specificity of biomarkers is shown in the *upper part* of the chart, as follows: (1) Chlorophyceae, chlorophyll b and lutein; (2) cyanobacteria, zeaxanthin and echinenone; (3) Cryptophyceae, alloxanthin; (4) Chrysophyta and Bacillariophyta (diatoms), fucoxanthin, diadinoxanthin and diatoxanthin; (5) Macrophytes, lactucaxanthin; (6) grazing activity, steryl chlorin esters (sce); (7) GSB, bacteriochlorophyll d; and (8) PSB, bacteriochlorophyll a and okenone

the compositional stability of the phototrophic community during the last two decades, the pigment signature has shifted. Now there is a significant increase of biomarkers of dinophytes and cryptophytes. Even though 20 years is a very short-time window to infer factors operating behind the change, we speculate that the persistent positive anomalies of seasonal temperatures, especially during autumn, over the last two decades (Romero-Viana et al. 2008) could expand the duration of thermal stratification periods and delay the fall overturn, thus increasing the growing period for these groups that mainly thrive during stratification (Fig. 8.9).

The pigment signature indicates that, as at present (Rodrigo et al. 2000), the populations of phototrophic bacteria, both green, sulphur (GSB) and purple sulphur bacteria (PSB), were present in the lake during the last centuries. The changes in their relative dominance could be related to differences in optimal conditions for development of each bacterial type (Camacho 2009), such as the differences in the light quality and quantity at the oxycline depth (Parkin and Brock 1980; Vila and Abella 1994), as well as in sulphide availability (Van Gemerden and Mas 2004). At the beginning of the laminated sequence, from around 1660 until 1711 A.D., exclusively purple bacterial derivatives were present. The derivatives from GSB were detected only afterwards, and the GSB and PSB populations coexisted thereafter. Nevertheless, the low concentrations of bacteriochlorophyll *d* derivatives until 1878 A.D. suggest that the GSB growth was quite limited, probably because sulphide concentration even in anoxic waters is very low (Rodrigo et al. 2000). Though not well studied, methane bubbles have been commonly observed, both visually and by acoustic methods, escaping from the sediments and the monimolimnion of the lake (Starkweather et al. 2005). Thus, methanogenesis is a main respiratory pathway in the sediments and the monimolimnion of Lake La Cruz. Consequently, the production of sulphide occurs mainly by the degradation of –SH groups from proteins and not from sulphate acting as electron acceptor. This gives a much lower sulphide yield per unit of organic matter consumed. Both GSB and PSB use sulphide as main electron donor, but PSB can store intracellular sulphur granules that can also be used as alternative electron donor for photosynthesis; thus, they are less dependent on sulphide concentration in the water. Therefore, GSB started to develop conspicuous populations in the lake when the progressive increase of oxygenic productivity could provide enough sulphide derived from proteolysis of organic matter recycled in anoxic layers.

A satisfactory explanation of the changes occurring in primary production is difficult to give in lacustrine ecosystems with complex interactions. Nevertheless, in Lake La Cruz, the temporal pattern over the last three centuries of zeaxanthin plus total chlorine-*a* derivative concentrations (around 45 % of the total pigment concentration) has shown a significant relation with the solar activity (Lean 2000) (Fig. 8.17), as highlighted by similar other studies that provide indirect evidences of solar variability on lake primary production (Bradbury et al. 2002). However, the effects on lake production derived from other factors cannot be ruled out, especially the catchment-related factors. In fact, an earlier paleolimnological study in Lake La Cruz made at lower resolution (Julià et al. 1998) has shown that the pollen sequence coinciding with the initiation of laminated sequence indicates a decrease in the extent of forests and reclamation of land for agricultural purposes. Consequently, increased soil erosion and associated nutrient input to the lake could also have contributed to the eutrophication process.

8.7 Concluding Remarks

As a short conclusion of this chapter, Lake La Cruz, where meromixis originated in the seventeenth century, is a representative typical example of the doline-type karstic meromictic lakes in the world. The strong density gradients created by both the higher concentration of dissolved salt content in the monimolimnion, mainly calcium (bi)carbonate, and by the thermal gradient in the summer period, determine the chemical gradients, though the former is responsible for the maintenance of the meromictic condition. The strong stratification creates different niches for a variety of planktonic organisms that tend to concentrate themselves along the vertical profile where conditions are most suitable for their development. Eukaryotic algae, autotrophic prokaryotes such as *Synechococcus*-like picocyanobacteria, phototrophic sulphur bacteria and chemolithotrophic bacteria, develop dense populations at certain layers where they fix inorganic carbon at high rates. Specific photosynthetic pigments, phycoerythrin, certain bacteriochlorophylls and carotenoids, allow phototrophic plankton to collect the light available. This, together with the relatively higher nutrient availability at these layers, enhances the development of a deep-chlorophyll maximum mainly formed by picocyanobacteria and some eukaryotic microalgae overlying dense populations of sulphur bacteria. The zooplankton often traces the location of the most productive layers to increase the food intake. With its activity, the planktonic biota also shapes the ecological features of the lake. The organisms strongly modify the chemical conditions to such an extent that the meromixis can be considered to be partly biogenic origin. The biota can mediate the associated biogeochemical processes involving solutes in deep layers. Primary production by picocyanobacteria, eukaryotic algae and sulphur bacteria provides the conditions and materials for the settling of organic matter and inorganic materials, mostly calcium carbonate, which on degradation releases solutes and increase salt concentration in deep waters. The fine sedimentation patterns and the occurrence of "whiting" events confers Lake La Cruz sediments an extraordinary advantage as palaeoecological descriptors. The different proxies used, both biogeochemical and biological, allow the reconstruction of past ecological conditions within this area of the Mediterranean Basin. Moreover, the higher iron content of deeper waters in Lake La Cruz and the microbial processes mainly occurring at the interfaces make the lake an excellent analogue for the study of early life conditions in the iron-rich oceans of the primitive Earth (Walter et al. 2014). The lake offers some insights into the biogeochemical processes that led to the current biosphere.

References

Ambrosetti W, Barbanti L (2005) Evolution towards meromixis of Lake Iseo (Northern Italy) as revealed by its stability trend. J Limnol 64:1–11

Armengol X (1997) Caracterización estructural del zooplancton de las lagunas cársticas de Cuenca, con especial atención a su distribución vertical. PhD Dissertation, Universitat de València

Armengol X, Miracle MR (1999) Zooplankton communities in doline lakes and pools, in relation to some bathymetric parameters and physical and chemical variables. J Plankton Res 21:2245–2261

Armengol X, Miracle MR (2000) Diel vertical movements of zooplankton in Lake La Cruz (Cuenca, Spain). J Plankton Res 22:1683–1703

Armengol-Díaz J, Esparcia A, Vicente E et al (1993) Vertical distribution of planktonic rotifers in a karstic meromictic lake. Hydrobiologia 255(256):381–388

Armengol X, Wurtsbaugh WA, Camacho A et al (2012) Pseudo-diel vertical migration in zooplankton: a whole-lake 15N tracer experiment. J Plankton Res 34:976–986

Ávila A, Burrel JL, Domingo A et al (1984) Limnology of the Lago Grande de Estanya (Huesca) (Limnologia del Lago Grande de Estanya (Huesca)). Oecologia Aquat 7:3–24

Barbosa FA, Padisák J (2002) The forgotten lake stratification pattern: atelomixis, and its ecological importance. Verh Int Ver Theor Angew Limnol 28:1385–1395

Barreiro-Lostres F, Moreno A, Giralt S et al (2014) Climate, palaeohydrology and land use change in the Central Iberian Range over the last 1.6 kyr: the La Parra Lake record. Holocene 24:1177–1192

Barriendos M, Llasat MC (2003) The case of the Maldá anomaly in the Western Mediterranean basin (AD 1760–1800): an example of a strong climatic variability. Climatic Change 61:191–216

Bathurst RG (1971) Carbonate sediments and their diagenesis. Developments in sedimentology, 12. Elsevier, Amsterdam

Bertilsson S, Jones JB (2003) Supply of dissolved organic matter to aquatic ecosystems: autochthonous sources. In: SEG F, RL S (eds) Aquatic ecosystems: interactivity of dissolved organic matter, Aquatic Ecology Series, 1st edn. Academic, San Diego, pp 26–70

Botella y de Hornos F (1875) La ciudad encantada. Las hoces, salegas y torcas de la provincia de Cuenca. Anales Soc Esp Hist Nat 4:233–239

Boronat L (2003) Distribución de los microcrustáceos en Lagunas de Castilla—La Mancha. Ciclos estacionales y migración vertical en lagunas cársticas estratificadas. Dissertation, Universitat de València

Boronat L, Miracle MR, Armengol X (2001) Cladoceran assemblages in a mineralization gradient. Hydrobiologia 442:75–88

Bradbury P, Cumming B, Laird K (2002) A 1500-year record of climatic and environmental change in Elk Lake, Minnesota III: measures of past primary productivity. J Paleolimnol 27:321–340

Brunskill GJ (1969) Fayetteville Green Lake, New York. II: precipitation and sedimentation of calcite in a meromictic lake with laminated sediments. Limnol Oceanogr 14:830–847

Camacho A (2006) On the occurrence and ecological features of deep chlorophyll maxima (DCM) in Spanish stratified lakes. Limnetica 25:453–478

Camacho A (2009) Sulfur bacteria. In: Likens GE (ed) Encyclopedia of inland waters. Elsevier Academic Press, Oxford, New York, pp 261–278

Camacho A, Vicente E (1998) Carbon photoassimilation by sharply stratified phototrophic communities at the chemocline of Lake Arcas (Spain). FEMS Microbiol Ecol 25:11–22

Camacho A, Erez J, Chicote A et al (2001) Microbial microstratification, inorganic carbon photoassimilation and dark carbon fixation at the chemocline of the meromictic Lake Cadagno (Switzerland) and its relevance to the food web. Aquat Sci 63:91–106

Camacho A, Picazo A, Miracle MR et al (2003a) Spatial distribution and temporal dynamics of picocyanobacteria in a meromictic karstic lake. Algological Stud 109:171–184

Camacho A, Wurtsbaugh WA, Miracle MR et al (2003b) Nitrogen limitation of phytoplankton in a Spanish karst lake with a deep chlorophyll maximum: a nutrient enrichment bioassay approach. J Plankton Res 25:397–404

Camacho A, Miracle MR, Vicente E (2003c) Which factors determine the abundance and distribution of picocyanobacteria in inland waters? A comparison among different types of lakes and ponds. Arch Hydrobiol 157:321–338

Campbell P, Torgersen T (1980) Maintenance of iron meromixis by iron redeposition in a rapidly flushed monimolimnion. Can J Fish Aquat Sci 37:1303–1313

Camuffo D, Secco C, Brimblecombe P et al (2000) Sea storms in the Adriatic Sea and the Western Mediterranean during the last millennium. Climatic Change 46:209–223

Canfield DE, Farquhar J, Zerkle AL (2010) High isotope fractionations during sulphate reduction in a low-sulphate euxinic ocean analog. Geology 38:415–418

Casamayor EO, Llirós M, Picazo A et al (2012) Contribution of deep dark fixation processes to overall CO_2 incorporation and large vertical changes of microbial populations in stratified karstic lakes. Aquat Sci 74:61–75

Casty C, Handorf D, Raible CC et al (2005) Recurrent climate winter regimes in reconstructed and modelled 500 hPa geopotential height fields over the North Atlantic/European sector 1659–1990. Clim Dynam 24:809–822

Cirujano S (1995) Flora y vegetación de las lagunas y humedales de la provincia de Cuenca. JCCLM—Real Jardín Botánico de Madrid (CSIC), Madrid

Cirujano S, Medina L (2002) Plantas acuáticas de las lagunas y humedales de Castilla-La Mancha. JCCLM—Real Jardín Botánico de Madrid (CSIC), Madrid

Culver DA, Brunskill GJ (1969) Fayetteville Green Lake. V. Studies of primary production and zooplankton in a meromictic marl lake. Limnol Oceanogr 14:862–873

Dasí MJ, Miracle MR (1991) Distribución vertical y variación estacional del fitoplancton de una laguna cárstica meromíctica, la Laguna de la Cruz (Cuenca, España). Limnetica 7:37–59

De Vicente I, Cattaneo K, Cruz-Pizarro L et al (2006) Sedimentary phosphate fractions related to calcite precipitation in a eutrophic hardwater lake (Lake Alserio, Northern Italy). J Paleolimnol 35:55–64

Dokulil MT, Kaiblinger C (2009) Phytoplankton productivity. In: Likens GE (ed) Encyclopedia of inland waters, vol 1. Elsevier, Oxford, pp 210–218

Eraso AE, López-Azevedo V, López M et al (1979) Estudio de Las Torcas de Palancares y Cañada del Hoyo en el karst de la Serranía de Cuenca. Kobie 9:7–79

Esparcia A (1993) Distribuciones de las poblaciones de rotíferos en la oxiclina de La Laguna de La Cruz. Adaptaciones metabólicas a la microaerofilia en *Brachionus plicatilis*. PhD Dissertation, Universitat de València

Esparcia A, Armengol J, Vicente E et al (1991) Vertical distribution of *Anuraeopsis* species as related to oxygen depletion in two stratified lakes. Verh Int Ver Theor Angew Limnol 24:2745–2749

Esparcia A, Armengol J, Miracle MR (2001) Relación de la distribución de los rotíferos con la de los principales factores físicos y químicos en la laguna de la Cruz. Limnetica 20:305–320

Folk RL (1974) The natural history of crystalline calcium carbonate: effect of magnesium content and salinity. J Sediment Res 44(1):50–53

Font-Tullot I (1988) Historia del clima de España. Cambios climáticos y sus causas. Instituto Nacional de Meteorología, Madrid

Galat DL, Jacobsen RL (1985) Recurrent aragonite precipitation in saline-alkaline Pyramid Lake, Nevada. Arch Hydrobiol 105:137–159

Gruber N, Wehrli B, Wüest A (2000) The role of biogeochemical cycling for the formation and preservation of varved sediments in Soppensee (Switzerland). J Paleolimnol 24:277–291

Hofmann W (1982) On the coexistence of two pelagic *Filinia* species (Rotatoria) in Lake Plussee. I. Dynamics of abundance and dispersion. Arch Hydrobiol 95:125–137

Hurrell JW (1995) Decadal trends in North Atlantic Oscillation and relationship to regional temperature and precipitation. Science 269:676–679

Hutchinson GE (1957) A treatise on Limnology, Geography, physics and chemistry, vol I. Willey, New York

Imboden DM, Wüest A (1995) Mixing mechanisms in lakes. In: Leman A, Imboden DM, Gat JR (eds) Physics and chemistry of lakes, 1st edn. Springer, Berlin, pp 83–138

Jankowski T, Livingstone DM, Bührer H et al (2006) Consequences of the 2003 European heat wave for lake temperature profiles, thermal stability, and hypolimnetic oxygen depletion: implications for a warmer world. Limnol Oceanogr 51:815–819

Jersabek CD, Koste W (1993) Additional notes on taxonomy and ecology of *Anuraeopsis miracleae* Koste, 1991 (Rotatoria: Monogononta) from an Austrian alpine lake. Hydrobiologia 264:55–60

Julià R, Burjachs F, Dasí MJ et al (1998) Meromixis origin and recent trophic evolution in the Spanish mountain lake La Cruz. Aquat Sci 60:279–299

Kelts K, Hsü KJ (1978) Freshwater carbonate sedimentation. In: Lerman A (ed) Lakes. Geology, Chemistry, Physics, 1st edn. Springer, New York, pp 295–323

Kiss KT, Acs E, Szabó KE et al (2007) Morphological observations on *Cyclotella distinguend*a Hustedt and *C. delicatula* Hustedt from the core sample of a meromictic karstic lake of Spain (Lake La Cruz) with aspects of their ecology. Diatom Res 22:287–308

Koschel R, Benndorf J, Proft G et al (1983) Calcite precipitation as a natural control mechanism of eutrophication. Arch Hydrobiol 98:380–408

Koste W (1980) Uber zwei Plankton-Radertiertaxa *Filinia australiensis* n sp and *Filinia hofmanni* n sp, mit Bermerkun- gen zur Taxonomie der *Longiseta terminalis* Gruppe . Genus *Filinia* Bory de St. Vincent, 1824, Familie Filiniidae Bartos 1959 (Ubeordnung Monogononta). Arch Hydrobiol 90:230–256

Koste W (1991) *Anuraeopsis miraclei*, a new planktonic rotifer species in karstic lakes of Spain. Hydrobiologia 209:169–173

Lean J (2000) Evolution of the Sun's spectral irradiance since the Maunder Minimum. Geophys Res Lett 27:2425–2428

Leavitt PR, Carpenter SR (1990) Aphotic pigment degradation in the hypolimnion: implications for sedimentation studies and paleolimnology. Limnol Oceanogr 35:520–534

Leavitt PR, Hodgson DA (2001) Practical methods for analysis of sedimentary pigment. In: Smol JP, Birks HJB, Last WM (eds) Developments in paleoenvironmental research, tracking environmental changes using lake sediments, terrestrial, algal and siliceous indicators, vol 3. Kluwer, Dordrecht, pp 295–325

Lehours AC, Evans P, Bardot C et al (2007) Phylogenetic diversity of archaea and bacteria in the anoxic zone of a meromictic lake (Lake Pavin, France). Appl Environ Microb 73:2016–2019

Luterbacher J, Xoplaki E, Dietrich D et al (2001) Extending North Atlantic oscillation reconstructions back to 1500. Atmos Sci Lett 2:114–124

Margalef R (1983) Limnología. Omega, Barcelona

Marra J (2009) Net and gross productivity: weighing in with 14C. Aquat Microb Ecol 56:123–131

McDonough RJ, Sanders RW, Porter KG et al (1986) Depth distribution of bacterial production in a stratified lake with an anoxic hypolimnion. Appl Environ Microb 52:992–1000

Mezquita F, Miracle MR (1997) Chydorid assemblages in the sedimentary sequence of Lake La Cruz (Spain) subject to water level changes. In: Brancelj A, De Meester L, Spaak P (eds) Cladocera: the biology of model organisms. Springer, Netherlands, pp 277–285

Mezquita F, Boronat MD, Miracle MR (2002) The life history of *Cyclocypris ovum* (Ostracoda) in a permanent karstic lake. Arch Hydrobiol 155:687–704

Miracle MR, Vicente E (1983) Vertical distribution and rotifer concentrations in the chemocline of meromictic lakes. Hydrobiologia 104:259–267

Miracle MR, Vicente E, Pedrós-Alió C (1992) Biological studies of Spanish meromictic and stratified karstic lakes. Limnetica 8:59–77

Miracle MR, Alfonso T (1993) Rotifer vertical distributions in a meromictic basin of Lake Banyoles (Spain). Hydrobiologia 255/256:371–380

Miracle MR, Dasí MJ, Vicente E (1998) Forced phytoplankton vertical migrations due to lake water "whiting". Verh Int Ver Theor Angew Limnol 26:1749–1754
Miracle MR, Camacho A, Julià R et al (2000) Sinking processes and their effect on the sedimentary record in the meromictic Lake La Cruz (Spain). Verh Int Ver Theor Angew Limnol 27:1209–1213
Modamio X, Pérez V, Samarra F (1988) Limnologia del Lago de Montcortes (ciclo 1978–79), (Pallars-Jussa, Lleida). Oecologia Aquat 9:9–17
Mopper K, Schultz CA (1993) Fluorescence as a possible tool for studying the nature and water column distribution of DOC components. Mar Chem 41:229–238
Morata SM, Camacho A, Miracle MR et al (2003) Asociaciones fitoplanctónicas y su periodicidad en un lago marcadamente estratificado. Limnetica 22:35–52
Müller G (1971) Aragonite inorganic precipitation in a freshwater lake. Nature 229:18–18
Müller G, Irion G, Foerstner U (1972) Formation and diagenesis of inorganic Ca-Mg carbonates in the lacustrine environment. Naturwissenschaften 59:158–164
Ojala AEK, Francus P, Zolitschka B et al (2012) Characteristics of sedimentary varve chronologies—a review. Quat Sci Rev 43:45–60
Orem WH, Burnett WC, Landing WM et al (1991) Jellyfish Lake, Palau: early diagenesis of organic matter in sediments of an anoxic marine lake. Limnol Oceanogr 36:526–543
Parkin TB, Brock TD (1980) The effects of light quality on the growth of phototrophic bacteria in lakes. Arch Microbiol 125:19–27
Pauling A, Luterbacher J, Casty C et al (2006) Five hundred years of gridded high-resolution precipitation reconstructions over Europe and the connection to large-scale circulation. Clim Dynam 26:387–405
Peck JA, Khosbayar P, Fowell SJ et al (2002) Mid to Late Holocene climate change in north central Mongolia as recorded in the sediments of Lake Telmen. Palaeogeogr Palaeoecol 183:135–153
Picazo A, Rochera C, Vicente E et al (2013) Spectrophotometric methods for the determination of photosynthetic pigments in stratified lakes: a critical analysis based on comparisons with HPLC determinations in a model lake. Limnetica 32:139–158
Planas D (1973) Composición, ciclo y productividad del fitoplancton del lago de Banyoles. Oecologia Aquat 1:3–106
Reynolds CS (2006) The ecology of phytoplankton. Cambridge University Press, Cambridge
Roca JR, Mezquita F, Rueda J et al (2000) Endorheic versus karstic lakes: patterns of ostracod distributions and lake typology in a Mediterranean landscape (Castilla-La Mancha, Spain). Mar Freshw Res 51:311–319
Rodrigo MA (1997) Limnología comparada de las lagunas de dos sistemas cársticos de Cuenca: Bacterias fotosintéticas de la Laguna de La Cruz y la Laguna de Arcas-2. PhD Dissertation, Universitat de València
Rodrigo MA, Vicente E, Miracle MR (1993) Short-term calcite precipitation in the karstic meromictic Lake La Cruz (Cuenca, Spain). Verh Int Ver Theor Angew Limnol 25:711–719
Rodrigo MA, Vicente E, Miracle MR (2000) The role of light and concentration gradients in the vertical stratification and seasonal development of phototrophic bacteria in a meromictic lake. Arch Hydrobiol 148:533–548
Rodrigo MA, Miracle MR, Vicente E (2001) The meromictic Lake La Cruz (Central Spain). Patterns of stratification. Aquat Sci 63:406–416
Rodriguez-Puebla C, Encinas AH, Nieto S et al (1998) Spatial and temporal patterns of annual precipitation variability over the Iberian Peninsula. Int J Climatol 18:299–316
Rojo C, Miracle MR (1987) Poblaciones fitoplanctónicas de La Laguna de La Cruz (Cuenca), una laguna cárstica meromíctica. Act. VI Simp Nat Bot Criptogámica, pp 119–135
Romero L, Camacho A, Vicente E et al (2006) Sedimentation patterns of photosynthetic bacteria based on pigment markers in meromictic Lake La Cruz (Spain): paleolimnological implications. J Paleolimnol 35:167–177

Romero-Viana L (2007) Paleolimnological study in two karstic lakes: climate signal in varved sediment and phototrophic organisms variability. Dissertation, Universitat de València

Romero-Viana L, Julià R, Camacho A et al (2008) Climate signal in varve thickness: Lake La Cruz (Spain), a case study. J Paleolimnol 40:703–714

Romero-Viana L, Miracle MR, López-Blanco C et al (2009) Sedimentary multiproxy response to hydroclimatic variability in Lagunillo del Tejo (Spain). In: Buczkó K, Korponai J, Padisák J, Starratt SW (eds) Palaeolimnological Proxies as Tools of Environmental Reconstruction in Fresh Water. Developments in Hydrobiology. Springer, The Netherlands, pp 231–245

Romero-Viana L, Keely BJ, Camacho A et al (2010) Primary production in Lake La Cruz (Spain) over the last four centuries: reconstruction based on sedimentary signal of photosynthetic pigments. J Paleolimnol 43:771–786

Romero-Viana L, Julià R, Schimmel M et al (2011) Reconstruction of annual winter rainfall since AD 1579 in central-eastern Spain based on calcite laminated sediment from Lake La Cruz. Climatic Change 107:343–361

Rossknecht H (1980) Phosphatelimination durch autochthone Calcitfällung im Bodensee-Obersee. Arch Hydrobiol 88:328–344

Schmidt R, Psenner R, Müller J et al (2002) Impact of late glacial climate variations on stratification and trophic state of the meromictic lake Längsee (Austria): validation of a conceptual model by multi proxy studies. J Limnol 61:49–60

Sendra MD (2009) Distribución del fitoplancton en la laguna de la cruz y su relación con los nutrientes. Dissertation, Universitat de València

Sommer U, Gliwicz ZM, Lampert W et al (1986) The PGE-model of seasonal succession of planktonic events in fresh waters. Arch Hydrobiol 106:433–471

Starkweather PL, Miracle MR, Vicente E et al (2005) Diel distribution and migration of *Chaoborus flavicans*: a field analysis using broadband Acoustic Doppler Current Profiler (ADCP) techniques. Verh Int Ver Theor Angew Limnol 29:672–677

Touchan R, Xoplaki E, Funkhouser G et al (2005) Reconstructions of spring/summer precipitation for the Eastern Mediterranean from tree-ring widths and its connection to large-scale atmospheric circulation. Clim Dynam 25:75–98

Valero-Garcés B, Morellón M, Moreno A et al (2014) Lacustrine carbonates of Iberian Karst Lakes: sources, processes and depositional environments. Sediment Geol 299:1–29

Van Gemerden H, Mas J (2004) Ecology of phototrophic sulphur bacteria. In: Blankenship RE, Madigan MT, Bauer CE (eds) Advances in photosynthesis and respiration, anoxygenic photosynthetic bacteria, Part I, vol 2. Springer, Dordrecht, pp 49–85

Vanderploeg HA, Eadie BJ, Liebig JR et al (1987) Contribution of calcite to the particle-size spectrum of Lake Michigan seston and its interactions with the plankton. Can J Fish Aquat Sci 44:1898–1914

Vicente E, Miracle MR (1988) Physicochemical and microbial stratification in a meromictic karstic lake of Spain. Verh Int Ver Theor Angew Limnol 23:522–529

Vila X, Abella CA (1994) Effects of light quality on the physiology and the ecology of planktonic green sulphur bacteria in lakes. Photosynth Res 41:53–65

Vila X, Abella CA, Figueras JB et al (1998) Vertical models of phototrophic bacterial distribution in the metalimnetic microbial communities of several freshwater North-American kettle lakes. FEMS Microbiol Ecol 25:287–299

Walker KF, Likens GE (1975) Meromixis and a reconsidered typology of lake circulation patterns. Verh Int Ver Theor Angew Limnol 19:442–458

Walter XA (2011). Anaerobic iron cycling in a neoarchean ocean analogue. Dissertation, Université de Neuchâtel

Walter XA, Picazo A, Miracle MR et al (2014) Phototrophic Fe(II)-oxidation in the chemocline of a ferruginous meromictic lake. Front Microbiol 5:713. doi:10.3389/fmicb.2014.00713

Weber KA, Picardal FW, Roden EE (2001) Microbially catalyzed nitrate-dependent oxidation of biogenic solid-phase Fe (II) compounds. Environ Sci Technol 35:1644–1650

Weidemann AD, Bannister TT, Effler SW, Johnson DL (1985) Particulate and optical properties during $CaCO_3$ precipitation in Otisco Lake. Limnol Oceanogr 30:1078–1083
Wetzel RG (2001) Limnology: lake and river ecosystems. Academic, San Diego
Wright SW, Jeffrey SW, Mantoura RFC (eds) (2005) Phytoplankton pigments in oceanography: guidelines to modern methods. UNESCO, Paris
Wüest A, Aeschbach-Hertig W, Baur H et al (1992) Density structure and tritium-helium age of deep hypolimnetic water in the northern basin of Lake Lugano. Aquat Sci 54:205–218
Xoplaki E, Maheras P, Luterbacher J (2001) Variability of climate in meridional Balkans during the periods 1675–1715 and 1780–1830 and its impact on human life. Climatic Change 48:581–615

Chapter 9
Meromictic Pit Lakes: Case Studies from Spain, Germany and Canada and General Aspects of Management and Modelling

Martin Schultze, Bertram Boehrer, Katrin Wendt-Potthoff, Javier Sánchez-España, and Devin Castendyk

9.1 Introduction

Pit lakes are the product of open-pit, surface mining excavations that are deep enough to intersect one water table. Meromixis has been reported in pit lakes that have developed in former rock quarries (e.g. Hrdinka 2007; Molenda 2014) and in abandoned kaolinite mines (Hrdinka 2007), sulphur mines (e.g. Zurek 2006), lignite mines (e.g. Schultze et al. 2013b; Solski and Jedrczak 1991), coal mines (e.g. Denimal et al. 2005) and metal mines (e.g. Miller et al. 1996; Sánchez-España et al. 2009). The number of pit lakes is growing worldwide as former pit mining operations close and flood, and new open-pit mines are excavated (Castendyk and Eary 2009). Parallel to this trend, the global number of meromictic pit lakes is also on the rise.

In the past, it was even believed that almost all pit lakes would become meromictic because of the limited surface area and considerable depth characteristic of pit lake basins expressed by high relative depth (e.g. Lyons et al. 1994; Miller et al. 1996). However, the growing body of field observations shows there are considerably more holomictic than meromictic pit lakes (e.g. Kumar et al. 2013; Schultze et al. 2013b). Furthermore, Jöhnk (2001) and Castendyk and

M. Schultze (✉) • B. Boehrer • K. Wendt-Potthoff
Department Lake Research, UFZ Helmholtz-Centre for Environmental Research, Brückstraße 3a, Magdeburg, Germany
e-mail: martin.schultze@ufz.de; bertram.boehrer@ufz.de

J. Sánchez-España
Instituto Geológico y Minero de España (IGME), Unidad de Mineralogía e Hidrogeoquímica Ambiental (UMHA), c/Calera 1, 28760 Tres Cantos, Madrid, Spain

D. Castendyk
Hatch, Water and Tailings Management, Denver, CO, USA

© Springer International Publishing AG 2017
R.D. Gulati et al. (eds.), *Ecology of Meromictic Lakes*, Ecological Studies 228,
DOI 10.1007/978-3-319-49143-1_9

Webster-Brown (2007a) demonstrated that the relative depth alone is not sufficient for causing the pit lakes to become meromictic. Other factors, such as the geochemical variability of water within the pit lake catchment (i.e. total dissolved solids, chemical composition, pH), play an important role in stratification. Hence, the meromixis of existing pit lakes has been mainly caused by crenogenic processes (e.g. Boehrer et al. 2014; Sánchez-España et al. 2009; von Rohden et al. 2009) and occasionally by biogenic processes (e.g. Flite 2006; Nixdorf and Boehrer 2015). In some cases, meromixis results from the treatment of mine drainage within initially holomictic pit lakes (Lu 2004; Park et al. 2006).

In the past, meromixis in pit lakes was seen as a welcome prerequisite for the deepwater disposal of mine wastes and mine drainage (e.g. Brassard et al. 1996; Stevens and Lawrence 1997). Although there are successful cases (e.g. Sect. 9.4), experience has shown that such disposal practices often degrade the water quality of pit lakes, in particular, acidic lakes. Consequently, many countries now require better management of meromictic pit lakes.

In this chapter, we first present a comprehensive study on the meromictic Cueva de la Mora (Spain). Secondly, we report on the disappearance and recovery of meromixis in Lake Goitsche (Germany). Our third example is of the Island Copper Mine Pit Lake (Canada) where meromixis was created intentionally in order to use the monimolimnion as a deposition and treatment site for harmful mine waste and acid rock drainage (ARD). The first two case studies of pit lakes are based on research by the authors, whereas the third is based on a literature review. We conclude the chapter with a discussion of the management options for pit lakes and of predictive modelling as an essential tool for pit lake planning.

9.2 Cueva de la Mora Pit Lake (Spain): An Example of Multilayer Stratification in Meromictic Lakes

The pit lake of Cueva de la Mora mine (Iberian Pyrite Belt mining district, SW Spain, 37°46′58″N, 6°49′17″W) exhibits a complex permanent stratification pattern with sharp chemical gradients. These gradients are related to a complicated network of galleries connected to the pit (Fig. 9.1b) and to its high microbial activity. This small lake has a surface area of 17,800 m^2 and a maximum depth of 40 m (Fig. 9.1a) and was formed in the remaining mine void of the former surface mining operations in the 1970s by the flooding of a former base-metal sulphide mine. It represents a mature stage of pit lake development with in around four decades of hydrological history. Multidisciplinary research carried out in this and similar pit lakes improves our knowledge of natural processes controlling the hydrological and limnological dynamics of pit lakes and helps refine approaches for pit lake water quality prediction and management.

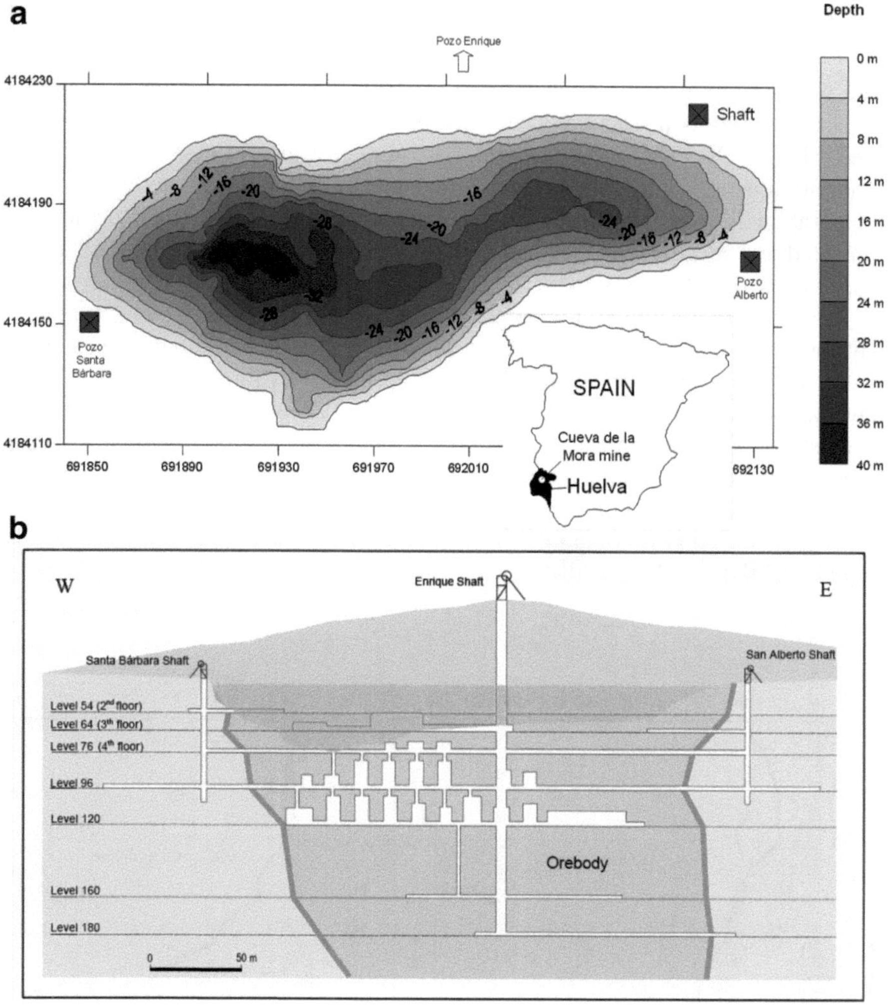

Fig. 9.1 Bathymetric map (**a**) and E–W cross section (**b**) of Cueva de la Mora pit lake. The cross section given in (**b**) shows the entire mine system, with location of the underground mine galleries, chambers and shafts, the open pit, the present lake and the ore body. Diagram is to scale (Sánchez-España et al. 2014b)

In recent years (2008–2014), a considerable effort has been made to understand the most important factors affecting the nature and development of Cueva de la Mora pit lake (Sánchez-España et al. 2009, 2011, 2013, 2014b; Wendt-Potthoff et al. 2012; Diez-Ercilla et al. 2014; Falagán et al. 2014). The most relevant findings of these investigations are summarized below.

9.2.1 Physical Limnology

The pit lake of Cueva de la Mora shows a 10 m-deep mixolimnion, a 2 m-thick transitional chemocline and a 30 m-thick monimolimnion, which is sharply stratified (Fig. 9.2). The lake bottom (18 °C) is warmer than the overlying monimolimnetic water (12–14 °C), but this density difference related to the water temperature is overcompensated by a much higher dissolved solid content, as indicated by specific conductance (12,000 μS cm^{-1} *vs.* 4000–5000 μS cm^{-1}, respectively; Fig. 9.2).

The most interesting feature of Cueva de la Mora pit lake is the "staircase" pattern displayed by profiles of specific conductance (κ_{25}), with homogenous water layers separated by sharp boundaries (Fig. 9.2). Four different conductive layers have been recognized with thicknesses varying from 2 to 5 m. The boundary zones show strong gradients in κ_{25}. This staircase pattern is stable throughout the year. The increase of κ_{25} parallels a corresponding increase in water density, which is determined by components, including temperature, dissolved substances (sulphate, metals), dissolved gases (mainly CO_2) and particulate matter (e.g. mineral colloids, organic debris).

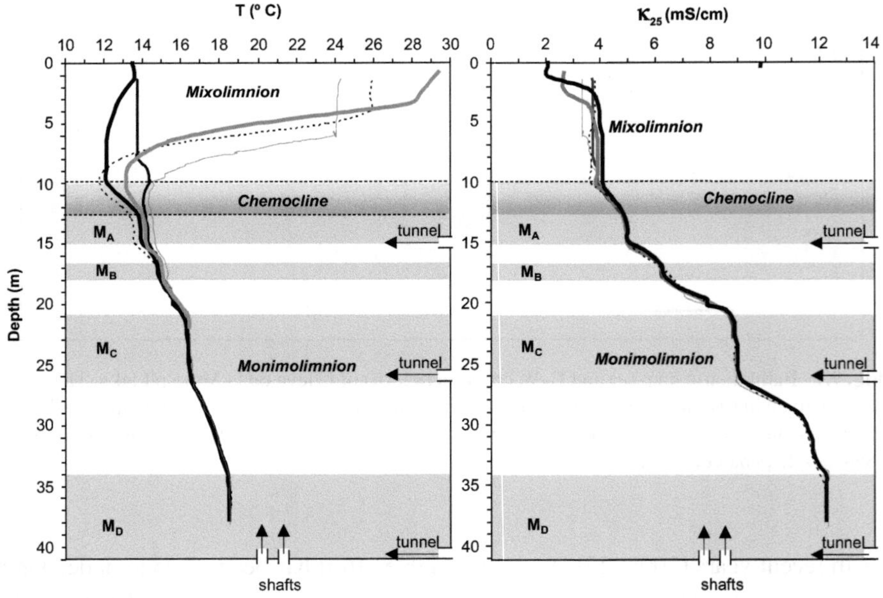

Fig. 9.2 Profiles of temperature (T, in °C) and specific conductance (κ_{25}, in mS cm^{-1}) taken in Cueva de la Mora pit lake between 2009 and 2011. Four layers (named M_A, M_B, M_C and M_D) with increasing temperature, conductance and density exist in the monimolimnion. The approximate position of mine tunnels and shafts intersecting the pit is indicated. *Dark-grey* areas denote homogeneous layers, and *light-grey* horizons represent transitional boundaries (modified from Sánchez-España et al. 2014b)

These staircase patterns are usually diagnostic of double-diffusive convection (DDC; see Chap. 2, Boehrer 2012) and have been observed in several aqueous environments (e.g. Kelley et al. 2003; Schmid et al. 2004). DDC occurs when both temperature and salinity increase downwards, such that the decrease in thermally induced density is compensated by the increase in salinity-induced density (Boehrer et al. 2009; Kelley et al. 2003; Turner 1973, 1978). DDC often results in upward flux of salinity and heat resulting in distinct layers, the thickness of which can be estimated quantitatively (e.g. Fedorov 1988). However, the observed four-layer structure of Cueva de la Mora's monimolimnion did not result from the DDC alone (Diez-Ercilla 2015).

The inflow of sulphate- and metal-laden groundwater entering the mine pit through intersecting galleries at different depths (Figs. 9.1 and 9.2) has apparently played a critical role in shaping the current stratification in Cueva de la Mora pit lake (Sánchez-España et al. 2014b). Among the four monimolimnetic layers recognized (M_A, M_B, M_C and M_D; Fig. 9.2), three are coincident with connections with former mine tunnels (M_A, M_C and M_D; Fig. 9.2) and seem to have resulted from the different density of the different inflowing waters. These layers show different tritium concentrations which have provided apparent ages of 1971–1972, 1974–1975 and >1975 for the sub-layers M_D, M_C and M_A, respectively (Sánchez-España et al. 2014b), suggesting that this structure was established during the initial flooding stages and has remained unaltered since then (Fig. 9.3). Thus, the original mine design seems to have been an important regulator of the physical structure of the resulting pit lake.

On the other hand, the development of layer M_A immediately below the chemocline is not related with any gallery and seems to be caused by DDC, which might be favoured by the progressive cooling of the mixolimnion at the end of the turnover period. The exchange of heat and dissolved solutes between the upper part of monimolimnion (layer M_A) and the lower part of mixolimnion results in the upward diffusion of both heat and dissolved solids (Fig. 9.2).

9.2.2 Hydrology

The relationship of the pit lake with the surrounding groundwater has been well established by the use of stable isotopes (Fig. 9.3). Vertical profiles of oxygen (δ^{18}O) and deuterium (δ^2H) isotopes obtained in different seasons have allowed inferring the hydrological dynamics of the lake and deducing residence times and evaporation rates of the different layers (Sánchez-España et al. 2014b). The upper mixolimnion shows marked seasonal isotope fluctuations, presenting highly evaporated water enriched with ^{18}O during the summer, and subsequent dilution by ^{18}O-depleted rain water during the winter (Fig. 9.3).

The monimolimnion shows neither seasonal nor interannual fluctuations (Fig. 9.3) suggesting that the deeper waters of this lake remain isolated and the influence of groundwater on the lake dynamics is negligible at present.

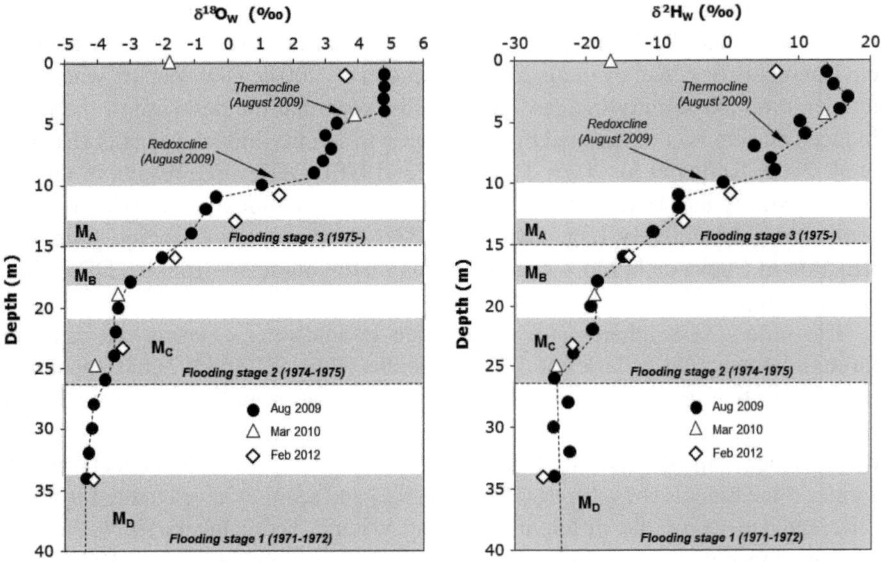

Fig. 9.3 Vertical profiles of $\delta^{18}O$ and $\delta^{2}H$ isotopic data in Cueva de la Mora pit lake in different seasons. The different layers and flooding stages (as deduced by tritium analyses) are indicated (Sánchez-España et al. 2014b)

9.2.3 Water Chemistry

The oxidative dissolution of pyrite and other metal sulphides (e.g. sphalerite, galena, chalcopyrite, arsenopyrite) in Cueva de la Mora mine has led to very acidic water with high metal and sulphate concentrations (Table 9.1). In addition to iron and sulphate (dominant ions), the pit lake has very high concentrations of base metals (Zn, Mn, Co, Ni, Cd) and metalloids (As), most of which can be toxic to aquatic microorganisms. The mixolimnion shows oxygen saturation during most of the year, and thus Fe is predominantly in oxidized form (Fe(III) $\approx$ 100 % Fe_{total} = 110–130 mg L^{-1}). The existing peaks of oxygen concentration (with values approaching 140 % O_2 sat.) are closely related to the corresponding peaks of chlorophyll-a concentration (Sánchez-España et al. 2009; Wendt-Potthoff et al. 2012; Diez-Ercilla et al. 2014). These deep chlorophyll maxima are critical in providing fresh organic carbon to a consortium of iron- and sulphate-reducing bacteria inhabiting the chemocline (Wendt-Potthoff et al. 2012; Diez-Ercilla et al. 2014; Falagán et al. 2014). The reducing environment of the monimolimnion is characterized by permanent anoxia and very high Fe concentrations which entirely correspond to ferrous iron (Fe(II) $\approx$ 100 %, Fe_{total} = 6 g L^{-1}).

The study of suspended particulate matter (SPM) retained in sediment traps indicates that the dissolved Fe(III) present in the mixolimnion precipitates as schwertmannite and hydronium jarosite. Schwertmannite is the first mineral phase formed after the oxidation of Fe(II), in agreement with previous studies.

Table 9.1 Typical values of pH and dissolved sulphate and trace metal concentration in the mixolimnion and four monimolimnetic layers of Cueva de la Mora lake (from Sánchez-España et al. 2009, 2013)

	pH	SO_4^{2-}	Fe_T	Mn	Zn	Al	Ni	Co	As
		mg L^{-1}	mg L^{-1}	mg L^{-1}	mg L^{-1}	mg L^{-1}	mg L^{-1}	mg L^{-1}	mg L^{-1}
Mixolimnion	2.2–3.1	2100	120	20	15	115	0.44	1.5	0.095
Layer M_A	3.0–3.5	4200	800	28	30	120	0.54	5.0	1.5
Layer M_B	3.3–4.0	5800	1800	50	60	200	0.64	5.2	4.8
Layer M_C	3.6–4.3	9800	3200	80	90	150	0.68	9.1	7.5
Layer M_D	4.1–4.5	13,100	6200	120	115	50	0.85	13.2	17.5

Schwertmannite is kinetically favoured with respect to other minerals, and its formation strongly buffers the lake pH around values of 2.6–3.1 [Eq. (9.1); Bigham and Nordstrom 2000].

$$8Fe^{3+} + SO_4{}^{2-} + 14H_2O \leftrightarrow Fe_8O_8(SO_4)(OH)_6 + 22H^+ \tag{9.1}$$

Schwertmannite is unstable and tends to be transformed to jarosite at lower pH, e.g. at pH values 2.2–2.5 (Sánchez-España et al. 2012a). Although the commonly described transformation of schwertmannite to goethite is limited here by the low pH and high sulphate concentration, the latter mineral is also present in the sediments at variable portion.

Geochemical modelling suggests a vertical cycling of iron across the water column. The mixolimnion is saturated with respect to jarosite, schwertmannite and goethite, while the anoxic and reducing monimolimnion is undersaturated with respect to all these phases, allowing the chemical redissolution and/or microbially mediated reduction of pelagic Fe(III) particles during their settling. As discussed below, reductive dissolution of Fe(III) particles occurs in the chemocline, being strongly catalysed by iron-reducing bacteria (Diez-Ercilla et al. 2014; Falagán et al. 2014).

These sequences of precipitation and dissolution may greatly influence the trace metal and nutrient dynamics of this pit lake. The reductive dissolution of Fe(III) precipitates is partly responsible for some of the observed increases in the concentration of trace elements such as arsenic, which is usually adsorbed on these compounds (Regenspurg et al. 2004; Sánchez-España et al. 2012a,b). Arsenic is partially mobilized into the aqueous phase by desorption and/or direct redissolution of the solid phases. The same applies to phosphate and organic carbon, both of which adsorb onto freshly precipitated ferric hydroxides at low pH (e.g. Cameron et al. 2006; Spiteri et al. 2008). Moderate concentrations of phosphate and organic carbon have been detected in these ferric precipitates (300–500 ppm P and 1.3 wt.% C_{org}). In the case of phosphate, its dissolved concentration in the deep monimolimnion could be counterbalanced by the release of adsorbed phosphate on one hand and the formation of diverse, metal-bearing, phosphate minerals on the other hand (e.g., strengite, plumbogummite, $MnHPO_4$, etc.).

The concentrations of most other elements in the bottom waters of Cueva de la Mora are much higher than those in the mixolimnion (Table 9.1). The most clear examples are those of sulphate (~2 g L^{-1} $SO_4{}^{2-}$ in the mixolimnion and >13 g L^{-1} $SO_4{}^{2-}$ in the bottom water) and As (0.07–0.12 mg L^{-1} in the mixolimnion and 17.5 mg L^{-1} in the bottom layer). Concentrations of Mn, Zn, Co and Ni also increase with depth, showing marked differences between the surface and the deepest part of the lake (Sánchez-España et al. 2009, 2013; Table 9.1).

Another important chemical feature distinguishing Cueva de la Mora pit lake from other lakes is the existence of a strong vertical pH gradient (2.2–3.1 in the mixolimnion, 4.1–4.5 in the lake bottom; Table 9.1). It results from different geochemical and biogeochemical processes. The pH of the oxygenated mixolimnion is strongly buffered by the hydrolysis and precipitation of Fe(III).

The absence of ferric iron in the monimolimnion triggers an increase of pH due to the occurrence of several proton-consuming biogeochemical reactions, including Fe(III) and sulphate reduction (Diez-Ercilla et al. 2014). Water/rock interactions taking place in wall rocks and in underground galleries are also partly responsible for this pH increase at depth, as the acid dissolution of aluminosilicates (feldspar, chlorite, muscovite) and minor carbonates present in volcanic and sedimentary host rocks contribute to neutralizing acidity.

Because of this vertical pH gradient, Al shows a decreasing concentration trend (Table 9.1). This implies undersaturation (dissolution) of Al-minerals in the mixolimnion and oversaturation (precipitation) in the monimolimnion. The most common solids formed by Al precipitation are hydroxysulphates, namely, hydrobasaluminite and felsöbanyaite (Sánchez-España et al. 2011). The formation of felsöbanyaite represents an additional and locally important buffering mechanism in the deep anoxic waters of this lake, which can be described by reaction (9.2):

$$4Al^{3+} + SO_4^{2-} + 14H_2O \leftrightarrow Al_4(SO_4)(OH)_{10} \cdot 4H_2O\ (s) + 10H^+ \quad (9.2)$$

In addition, hydrobasaluminite and felsöbanyaite are both metastable with respect to alunite, which can also form directly from dissolved Al by reaction (9.3):

$$3Al^{3+} + K^+ + 2SO_4^{2-} + 6H_2O \leftrightarrow KAl_3(SO_4)_2(OH)_6(s) + 6H^+ \quad (9.3)$$

Other common Al phases such as gibbsite or amorphous $Al(OH)_3$ (s) are undersaturated at all depths and do not seem to play a significant role in the Al geochemistry of Cueva de la Mora pit lake (Sánchez-España et al. 2011).

The precipitation of Al in the monimolimnion is also linked to the behaviour of certain elements which tend to be adsorbed to the mineral surfaces of the Al colloids (Sánchez-España et al. 2012b). Specifically, the Al precipitates are responsible for the removal of aqueous silica (SiO_2 (aq.)) and metals and metalloids like U, V, Se and Sb. A common feature of these elements is that they all form oxyanions which can be readily adsorbed onto the positively charged Al surfaces.

The water chemistry of Cueva de la Mora pit lake is also characterized by the presence of CO_2 in the deep waters (Fig. 9.4). The dissolution of trace carbonates (e.g. calcite, dolomite and ankerite) present in the massive sulphide deposits and in the host rocks consumes H^+ and releases CO_2 (aq.) [Eq. (9.4)]:

$$Ca_xMg_yCO_3 + H_2SO_4 \leftrightarrow xCa^{2+} + yMg^{2+} + SO_4^{2-} + CO_2(g) + H_2O \quad (9.4)$$

The decomposition of dead phytoplankton at depth also contributes to dissolved CO_2 (Eqs. 9.5 and 9.6).

The continuous release of carbon dioxide to this acidic lake [reactions (9.4), (9.5) and (9.6)] represents a potentially dangerous process. Although not as

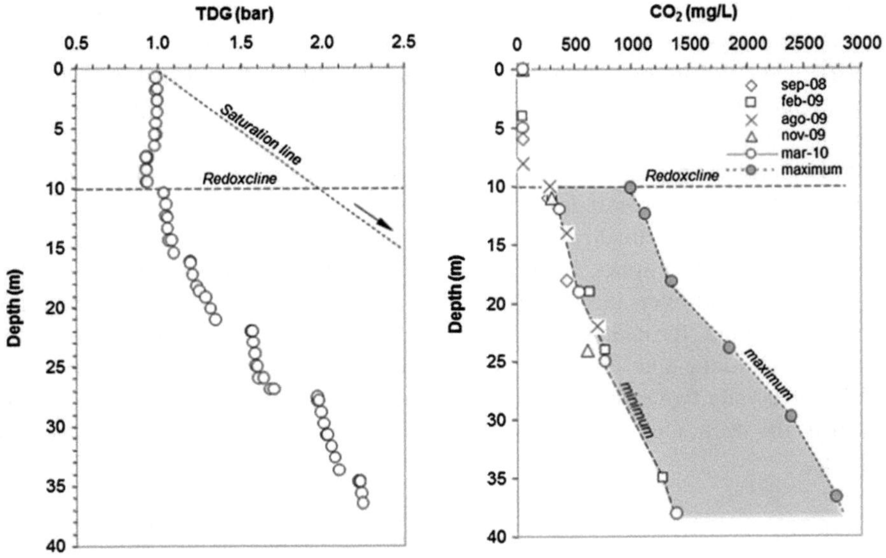

Fig. 9.4 Vertical profiles of total dissolved gas (TDG) pressure (*in bar*) in Cueva de la Mora pit lake, as measured in March 2012 (*left*) and carbon dioxide concentrations in this lake (*right*). The saturation line for TDG is based on absolute pressure. Minimum CO_2 concentrations were obtained on site in different seasons by UV–VIS spectrophotometry. They are underestimated due to partial degassing of water samples before analyses. Maximum CO_2 concentrations were calculated from TDG data assuming CO_2 is the only gas present. Due to the minor presence of other gases such as N_2, the calculated concentrations are overestimated. The actual CO_2 concentrations will therefore lie in between (*shaded area*)

dramatic as in the nearby Guadiana pit lake in Herrerías mine (Sánchez-España et al. 2014a), the high CO_2 concentration existing in the deep waters of Cueva de la Mora (estimated to be between 1600 and 2600 mg L^{-1} CO_2; Fig. 9.4) involves a total gas pressure of ca. 2.3 bar (including contributions from other gases), which represents around 50 % of the absolute (hydrostatic plus atmospheric) pressure (e.g. 4.6 bar at 36 m depth). The low pH and the high hydrostatic pressure favour the entrapment of CO_2 in the monimolimnion and preclude any natural decrease mechanism, so that gas pressure will probably continue to increase in this lake (Sect. 2.3). A degassing strategy could be necessary in the future to avoid any risk of a limnic eruption like those happened in Cameroon in Lake Monoun in 1984 and Lake Nyos in 1986 (Kling 1987). Such a degassing scheme is being tested in the adjacent Guadiana pit lake (Herrerías mine), where the existence of CO_2 (aq.) at concentrations of >5000 mg L^{-1} (equivalent to partial dissolved gas pressure of $pCO_2 = 3.5$ bar by CO_2 alone (Sánchez-España et al. 2014a, Boehrer et al. 2016) at depth strongly recommends the installation of degassing pipes similar to those used in the volcanic lakes Nyos and Monoun (Kling et al. 2005; Schmid et al. 2006).

9.2.4 Redoxcline Biogeochemistry

The most intense cycling of biogeochemically important elements (iron, sulphur, carbon) within the lake Cueva de la Mora occurs in the pelagic redoxcline. Here, a thick turbidity layer exists in the 10–20 m depth stratum (Diez-Ercilla et al. 2014; Fig. 9.5). This turbidity layer is singular in that it is chiefly comprised of bacterially induced metal sulphides, with a negligible presence of iron colloids. The primary production of planktonic microalgae provides suitable organic compounds for the growth of bacterial consortia, including iron oxidizers, iron reducers, sulphur oxidizers and sulphate reducers (Wendt-Potthoff et al. 2012; Diez-Ercilla et al. 2014; Falagán et al. 2014). The uppermost 2 m of the redoxcline appears to contain three sub-layers, or micro-niches, which are vertically arranged as follows: (1) an upper sub-layer of heterotrophic bacteria that can degrade algal exudates, (2) a middle sub-layer of Fe(III) reducers which use both dissolved and particulate (e.g. schwertmannite) Fe(III) as their terminal electron acceptor and (3) a bottom sub-layer of sulphate reducers that produce H_2S as a metabolic by-product. The reductive dissolution of schwertmannite in the upper part of the redoxcline is responsible for the absence of particulate Fe(III) in the turbidity layer and the release of previously adsorbed As, Cr, Pb and P. A few cm below the Fe(III)-reducing sub-layer, these released metals react with dissolved H_2S and form insoluble metal sulphides (e.g. As_3S_4, CuS, PbS), which sink in the water column. This sulphide precipitation is responsible for the removal of several metals (Cu, Pb, U, Th, As, Pb) and represents an efficient self-mitigation mechanism.

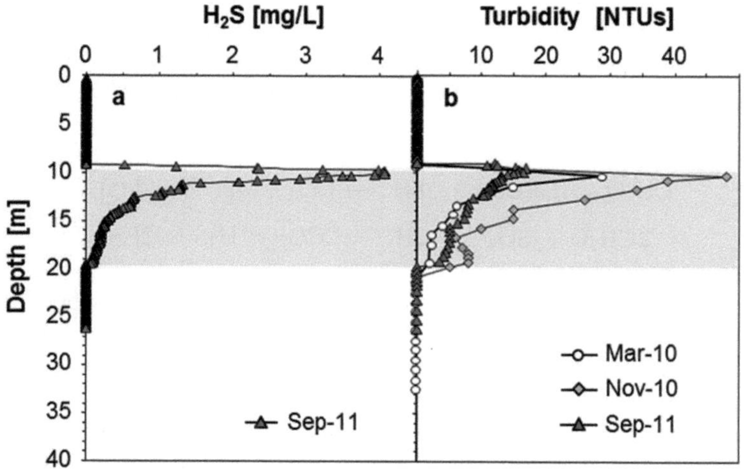

Fig. 9.5 (**a**) Vertical profiles of H_2S concentration and (**b**) turbidity (in nephelometric turbidity units—NTUs) in Cueva de la Mora pit lake, measured at three different times. The *shaded area* represents a biogenic sulphide precipitation layer (modified from Diez-Ercilla et al. 2014)

9.2.5 Sediment Biogeochemistry

Deep basin and shallow sediment cores have been collected by gravity coring. Chemical and X-ray diffractometric analyses of these sediments reveal the predominance of ferric compounds such as schwertmannite, jarosite and goethite in variable proportions, with lesser amounts of detrital minerals such as quartz, chlorite, muscovite or feldspar. Goethite is a product of jarosite and schwertmannite transformation, as these two minerals are metastable with respect to goethite (Bigham and Nordstrom 2000; Regenspurg et al. 2004). However, the low pH and high sulphate concentration of this pit lake make this transformation less feasible than in other acidic lakes (Sánchez-España et al. 2012a), so that this mineral is less common than in other pit lakes.

Vertical profiles of decreasing metal concentration suggest a progressive desorption of trace elements as the Fe(III) minerals are either transformed by mineralogical maturation (ageing), redissolved by chemical or bacterially catalysed reactions or both. However, biogeochemical cycling at the monimolimnetic sediment/water interface is less intense than in the pelagic redoxcline because of limited organic carbon input and the absence of light to sustain primary production (Wendt-Potthoff et al. 2012). The reduction of Fe(III) minerals (schwertmannite, jarosite) leads to the release of Fe(II), As and PO_4^{3-} back to the water column. These solutes slowly move upwards by chemical diffusion and may eventually reach the redoxcline at the top of the monimolimnion.

Chemical analyses of sediment pore waters yield significant gradients of pH and redox potential in the upper 10–15 cm of the sediment. These gradients likely reflect the occurrence of anaerobic microbial metabolisms, including Fe(III) and sulphate reduction (Wendt-Potthoff et al. 2012). These reactions are commonly catalysed by anaerobes that inhabit the water/sediment interface and couple the anaerobic oxidation of organic compounds (e.g. decomposing biomass settled from the photic zone) with the reduction of oxidized substances, mainly Fe(III) minerals and sulphate (Wetzel 2001; reactions (9.5) and (9.6) where CH_2O stands for any organic compound used by the bacteria):

$$CH_2O + 4Fe(OH)_3 + 8H^+ \rightarrow CO_2 + 4Fe^{2+} + 11H_2O \tag{9.5}$$

$$2CH_2O + SO_4^{2-} + 2H^+ \rightarrow 2CO_2 + H_2S + 2H_2O \tag{9.6}$$

A common feature of the reactions (9.5) and (9.6) is that they both consume protons and lead to an increase of pH in the pore waters (e.g. from 4.5 above the sediments to values around 5.5 at 4 cm below the water/sediment interface). This pH increase causes the precipitation of Al in the pore waters and the associated sorption of other elements. Another important implication of reactions (9.5) and (9.6) is that they both release CO_2. This microbial production of carbon dioxide also contributes to the elevated concentrations of dissolved CO_2 observed in the monimolimnion of Cueva de la Mora (Fig. 9.4).

Recent chemical (C/N molar ratio) and isotopic ($\delta^{13}C$, $\delta^{15}N$) analyses performed in benthic and suspended sediments in this lake confirm that the organic carbon input is dominated by algal debris (=sedimented phytoplankton) and detrital matter.

9.2.6 *Microbial Ecology*

Due to the very acidic nature of the oxic mixolimnion, higher organisms are absent in these waters, and, therefore, the ecology and food webs are restricted to micro-organisms. Bacterial counts in the water of up to 10^7 cells mL^{-1} may occur (Fig. 9.6), which is similar to findings of Tonolla et al. (2004) in the neutral sulphate-rich meromictic Lake Cadagno (Switzerland; see also Chap. 7). Bacterial counts reported earlier (Wendt-Potthoff et al. 2012) were an order of magnitude lower, reflecting a seasonal difference between the spring and late summer. The mixolimnion, redoxcline and monimolimnion represent three different habitats with respect to light, nutrients and electron acceptors. This separation into different habitats is illustrated by a series of microscopic images (see Fig. 9.6).

Phototrophic organisms in the mixolimnion comprise filamentous green algae and unicellular microalgae like diatoms and *Chlamydomonas*. Chlorophyll distribution and concentrations vary seasonally (Sánchez-España et al. 2009). Recently a chlorophyte, *Pseudococcomyxa* sp., was successfully isolated from the mixolimnion (Falagán et al. 2014).

Sulphur cycling within the water column, although well-known from neutral meromictic lakes with deeper saline waters (e.g. Kosolapov et al. 2003; Tonolla et al. 2004), is unusual in meromictic pit lakes (Rücker et al. 1999). Cueva de la

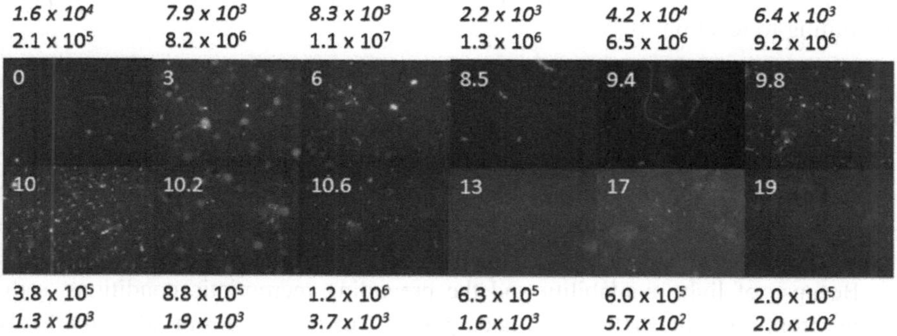

Fig. 9.6 Microscopic images of samples (September 2011) filtered onto 0.2 µm polycarbonate membranes. *White* numbers within the images indicate water depth (in meters). Numbers outside (both above and below) the images give total bacteria per mL (*normal font*) and colony counts of acidophilic aerobic heterotrophs (*in italics*)

Mora differs strikingly from other sulphidic meromictic lakes in that despite the sulphidic conditions at the chemocline, no "bacterial plate" of phototrophic H_2S-oxidizing microbes has been found to develop, as for Lake Cadagno in Switzerland (Tonolla et al. 2004) or as found in lakes Shira and Shunet in Siberia (Degermendzhy et al. 2010) (see for details Chaps. 5 and 7). Also, phototrophic sulphide oxidizers were not detected by molecular genetic methods (Falagán et al. 2014) nor reported so far from other acidic waters.

One of the most abundant cultivated prokaryotic groups in Cueva de la Mora was heterotrophic acidophiles. Their numbers reached 10^3–10^4 cells mL^{-1} at 13 m water depth in 2011 (Fig. 10.6) and were even higher in the chemocline in the following year (10^5–10^6 cells mL^{-1}, Falagán et al. 2014). Studies on microbial community composition using overlay media and molecular methods revealed that the majority of bacteria was related to iron cycling (Falagán et al. 2014). Cultivation-dependent counts using liquid dilution cultures were mostly below 10^3 mL^{-1} except for a distinct maximum of sulphide-oxidizing prokaryotes within the chemocline (1–8 $\times$ 10^3 cells mL^{-1}, Wendt-Potthoff et al. 2012; Wendt-Potthoff 2013).

The recurrent detection of H_2S at the chemocline (see above) indicates microbial sulphate reduction in this part of the water column; however, the sulphate reduction rate has not been quantified yet. Attempts to count sulphate-reducing bacteria applying "classical" media yielded no or very low counts (Wendt-Potthoff et al. 2012), but clones related to *Desulfomonile* were detected in September 2012 (Falagán et al. 2014). Moreover, if special acidic media with CO_2 and H_2 as carbon and energy sources (Meier et al. 2012) were used, up to 2 $\times$ 10^5 cells mL^{-1} of acidophilic sulphate reducers were cultivated (Wendt-Potthoff 2013). Assuming a typical per cell activity (Sahm et al. 1999), they would still reduce <1 pmol mL^{-1} d^{-1} of sulphate, which is below the detection limit determined for Cueva de la Mora (0.1 nmol mL^{-1} d^{-1}, M. Koschorreck, pers. comm.). Organisms growing under conditions of the applied medium belong to the genera *Thermodesulfobium* and *Desulfosporosinus* (Meier et al. 2012; Alazard et al. 2010). They reduce sulphate according to reaction (9.7):

$$SO_4{}^{2-} + 4H_2 + 2H^+ \rightarrow H_2S + 4H_2O \tag{9.7}$$

Thus, they do not depend on organic carbon sources. Molecular hydrogen is also more likely available in zones of high biological activity as in the chemocline or mixolimnetic sediment, from which *Desulfosporosinus*-related organisms were recently isolated (Falagán et al. 2014).

Because of light availability and the prevailing meromictic conditions, sediments from both the shallow and deep part of Cueva de la Mora differ markedly from each other. Where light reaches the bottom, a layer of filamentous algae grows and generates oxygen, but rates and seasonal variation of this process are not known yet. Generally, comparative biogeochemistry of mixolimnetic and monimolimnetic sediments in meromictic lakes has rarely been studied (e.g. Hongve 2003). For

Table 9.2 Range of microbial counts and process rates in mixolimnetic and monimolimnetic sediments

Microbial group or process	Mixolimnion	Monimolimnion
Total prokaryote cells mL^{-1}	0.9-2.2 × 10^{11}	7.5–9.7 × 10^{10}
Sulphate-reducer cells mL^{-1}	0.7–7.9 × 10^{4}	1.2–9.8 × 10^{4}
Fe(III)-reducer cells mL^{-1}	10^{3}–10^{5}	0.8–5.9 × 10^{4}
Acidophilic Fe(III)-reducer cells mL^{-1}	3×10^{3}–3 × 10^{5}	10^{2}–1.6 × 10^{3}
Sulphide-oxidizer cells mL^{-1}	0.1–2.1 × 10^{5}	0.2–2.4 × 10^{4}
Fe(II)-oxidizer cells mL^{-1}	5×10^{3}–5 × 10^{5}	2.1–2.4 × 10^{3}
CO_2 production nmol mL^{-1} d^{-1}	n.a.	170–230
Fe(III) reduction nmol mL^{-1} d^{-1}	302–627	97–218
Sulphate reduction nmol mL^{-1} d^{-1}	4.5–314	0.2–2.3
Methanogenesis nmol mL^{-1} d^{-1}	n.a.	b.d.
Percent reoxidation of sulphides	99	87

Values give the range of three sediment layers with a total depth of 9 cm (mixolimnion) and 8 cm (monimolimnion) (condensed from Wendt-Potthoff et al. 2012). *n.a.* not analysed, *b.d.* below detection limit

Cueva de la Mora, this has been attempted by Wendt-Potthoff et al. (2012) (see Table 9.2) and Falagán et al. (2014) who focussed on taxonomic composition of bacteria and archaea.

Fe(III) reduction rates and microbial counts indicate that microbial iron cycling is also important in the sediments, especially in the shallower parts of the lake. In monimolimnetic sediments, Fe(III) reduction was iron limited. Although methanogenic activity was not detected, low concentrations of methane were found in monimolimnetic sediments (20–65 μmol mL^{-1}, Wendt-Potthoff et al. 2012). This suggests that the rate of methanogenesis was either very low or the process was discontinuous in time. Methanogenic archaea seem to be rare in the sediments as well, as only one clone related to a methanogen was detected (Falagán et al. 2014). The extent of sulphide reoxidation (Table 9.2) was estimated from the amount of reduced sulphur, sulphate reduction rates and the age of the lake. The resulting values are close to previous estimates of 84 % and 97 % for productive and less productive lakes, respectively (Koschorreck and Tittel 2007; Holmer and Storkholm 2001). According to Kelly et al. (1995), periphytic filamentous algae alter the balance between sulphate reduction and sulphide oxidation by lifting the sulphate-reducing zone upwards. This facilitates reoxidation of the formed sulphides, which supports the 99 % of reoxidation of reduced sulphur estimated for mixolimnetic sediments.

While mixolimnetic sediments represent a biogeochemical hotspot of microbial activity in Cueva de la Mora, monimolimnetic sediments are limited by energy and organic carbon input (see above). This is important when estimating the potential for natural neutralization of the lake (Wendt-Potthoff et al. 2012). It also implies that comparative studies of mixolimnetic, chemocline and monimolimnetic sediments will give a more comprehensive understanding of whole-lake metabolisms and development.

9.3 Lake Goitsche (Germany): Disappearance and Reformation of Meromixis

The development of meromixis in lakes has rarely been investigated (e.g. Hemlock Lake, Fast and Tyler 1981; Mono Lake, see Chap. 11). The formation of pit lakes provides an opportunity to fill this gap if they are adequately monitored. Examples are the meromictic pit lakes at Berkeley Pit (USA, Gammons and Tucci 2013), Nuestra Señora del Carmen (Spain; Santofimia et al. 2012), Island Copper Mine (Canada; Fisher 2002), lakes Rassnitz and Wallendorf (Germany; Boehrer et al. 2014) and Kennecott Ridgeway Gold mine (USA; Flite 2006). There are only a few examples where the disappearance and reformation of meromixis has been observed in detail (e.g. Nuestra Señora del Carmen pit lake). Here, we present an example of Lake Goitsche (Germany, 51°36′45″N, 12°23′00″E) where both the disappearance and reformation of meromixis have been studied.

Lake Goitsche was formed in a former lignite mine about 100 km southwest of Berlin (Fig. 9.7). It has three subbasins due to the route of mining and dumping of overburden. Further, the lake bed has a number of local depressions (Fig. 9.7) due to geological conditions (Trettin et al. 2007). During lake remediation measures, mine dewatering was reduced in order to reduce costs (Laugwitz, pers. comm.). This allowed the intrusion of groundwater and the accumulation of surface runoff in small water bodies at the deepest sites of the pit. After the mine closure in 1991,

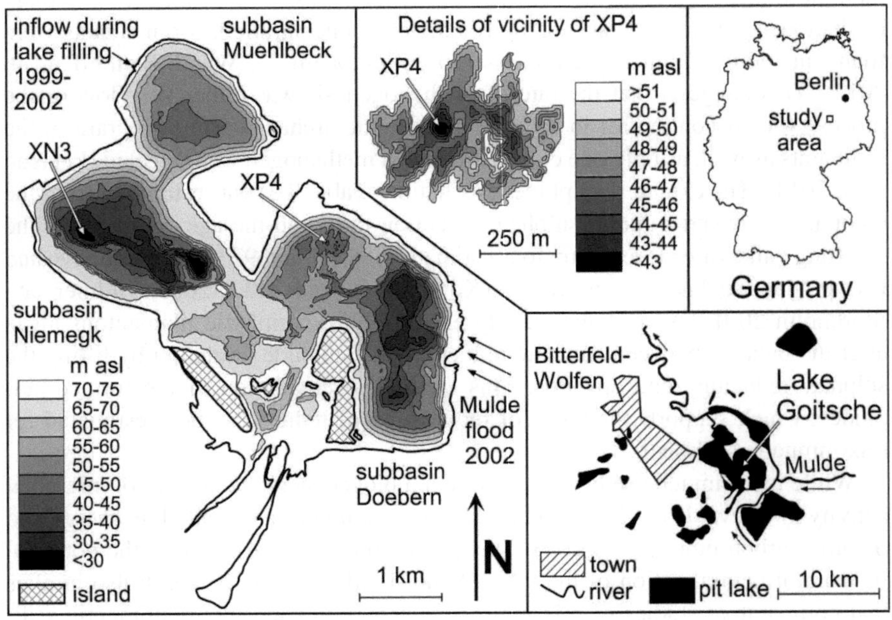

Fig. 9.7 Location and bathymetric map of Lake Goitsche

controlled filling with water from the nearby Mulde River began in 1999. In August 2002, a huge flood event occurred in Central Europe (Ulbrich et al. 2003) causing a dike breach and the uncontrolled flooding of over 90×10^6 m of water into the lake, causing a 7 m rise in the lake level, i.e. a rise from 71.5 m asl to 78.5 m asl, and an increase in lake volume from 165×10^6 m to 255×10^6 m. To preserve the shoreline, the lake was drained to 75 m asl during September to December 2002. This final water level represents a volume of 213×10^6 m^3. While the floodwater caused an almost complete overturn in subbasin Doebern and the deposition of a sediment layer ranging from about 0.5 m to over 5 m in thickness, only the epilimnion was flushed, and only a few cm of new sediment were deposited in the downstream subbasins Niemegk and Muehlbeck (Boehrer et al. 2005). For details of Lake Goitsche and its course of formation, see Schultze et al. (2002), Duffek and Langner (2002), Boehrer et al. (2003), Klemm et al. (2005), Dreher (2007), Trettin et al. (2007) and Schultze (2012).

Due to the density differences between acidic bottom waters in the mine void, groundwater seepage and river water, stable, chemically distinct layers formed and persisted in two of the local depressions (marked as XN3 and XP4 in Fig. 9.7; Boehrer et al. 2003). Here, we focus on site XP4 since disappearance and reformation of the monimolimnion happened only at this site, while the thin monimolimnion at XN3 disappeared for good after the flood. The changes in stratification (Fig. 9.8) were documented by regular measurements with a multiparameter probe (Ocean Seven, IDRONAUT, Milano, Italy).

The initial formation of the monimolimnion was probably caused by both: (i) the inflow of groundwater which had elevated concentrations of dissolved solids and (ii) the local morphology of the lake basin (Fig. 9.7). The inflow of groundwater was confirmed by radon measurements (A. Schmidt, pers. comm.) although groundwater resembling the ionic composition of the monimolimnion was not detected in the observation wells surrounding Lake Goitsche (Trettin et al. 2007).

Prior to the flood, the lower layer (monimolimnion) remained unaffected by the seasonal overturn of the upper layer (mixolimnion; Fig. 9.8). However, due to a high load of suspended solids (indicated by the amount of new sediment deposited after the flood; Boehrer et al. 2003), the floodwater entering Lake Goitsche was denser than the monimolimnion water and was able to enter and replace the monimolimnion partly. Electrical conductance measured immediately after the flood indicated that the monimolimnion did not disappear completely. Once the suspended solids had settled to the bottom, only a dilute monimolimnion remained (Fig. 9.8). This is also reflected by a decrease in the water density.

The complete removal of the monimolimnion happened during deep recirculation of lake Goitsche in the winter of 2002/2003. The density difference between the mixolimnion and the monimolimnion was not sufficient to withstand the mixing of the two layers. However, the monimolimnion re-established during summer stratification in 2003 from the groundwater flowing into the lake. The thermocline protected the evolving new chemocline against turbulent disturbance originating from wind-induced currents. The kinetic energy transferred into the

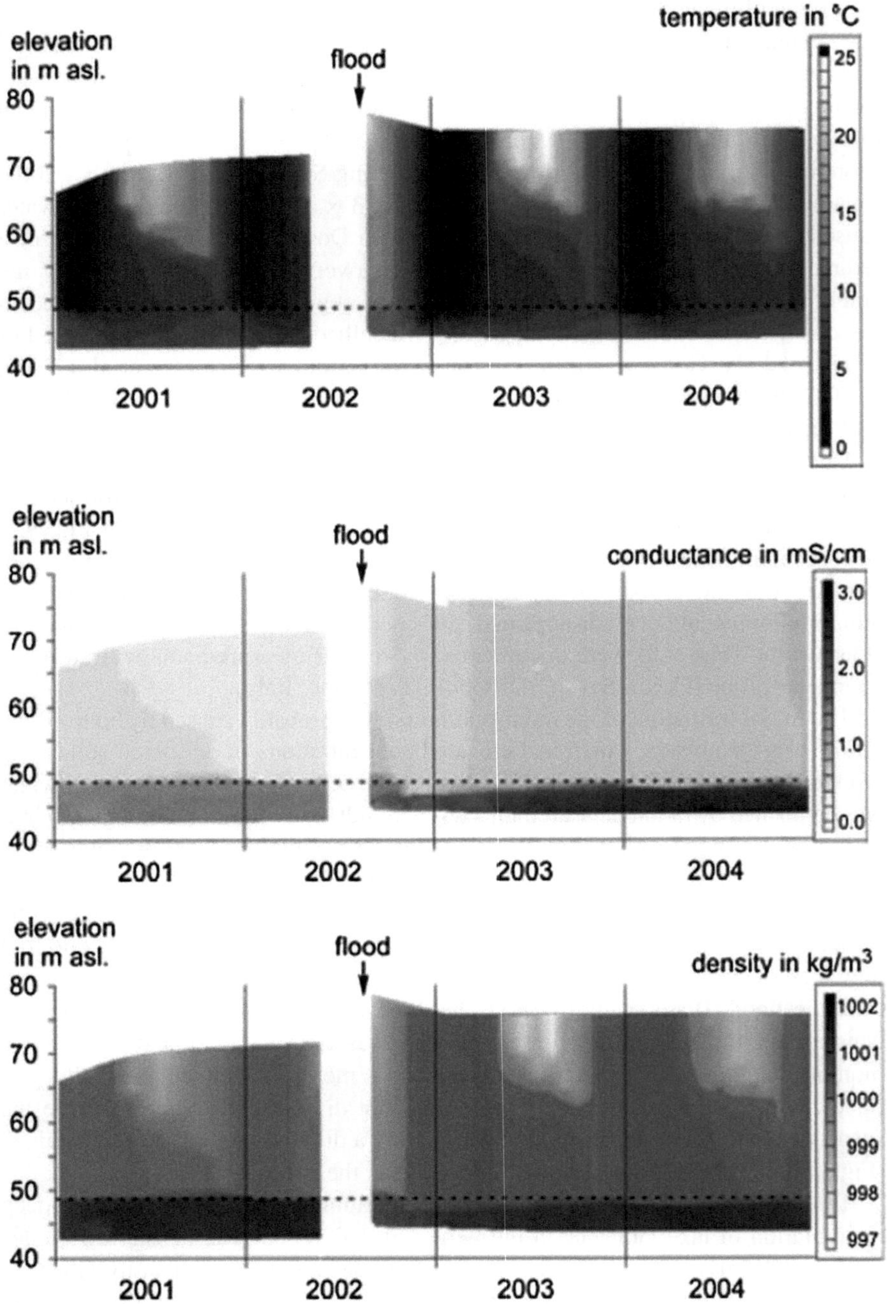

Fig. 9.8 Temperature, conductance and density at sampling site XP4. During and immediately after the flood, Lake Goitsche was not accessible (*blank areas*). Electrical conductance at 25 °C (κ_{25}) was calculated following ISO7888 (1985) using a temperature coefficient of $\alpha = 0.194$/K. Density was calculated from the water composition using the density calculator by Moreira et al. (2016); https://www.dropbox.com/s/tbxa3y06gzyybd4/density_calculator.zip)

water at the lake surface was mainly retained within the epilimnion, and only a limited portion reached the chemocline, which did not suffice to mix the hypolimnion with the entering groundwater completely. The protective effect of thermal and salinity stratification has been observed also in other meromictic pit lakes (e.g. Boehrer et al. 2014).

The resulting stability prevented recirculation during the winter of 2003/2004. During the summer of 2004, conductivity and density continued to rise in the monimolimnion due to groundwater inflow. Microbial decay and dissolution of suspended minerals in the monimolimnion may also have contributed to its stability. The chemocline rapidly recovered to the previous level at the upper edge of a smaller depression in the lake bed (see dashed lines in Fig. 9.8 and bathymetry around sampling site XP4 in Fig. 9.7).

9.4 Island Copper Mine Pit Lake: An Engineered Meromictic Lake

Most of the information presented in Sects. 9.4.1–9.4.3 originates from Fisher (2002), Poling et al. (2003), Fisher and Lawrence (2006) and Pelletier et al. (2009).

9.4.1 The Mine and Its Closure Plan

Island Copper Mine (Vancouver Island, Canada, 50°36′00″N, 127°28′30″W) was an open-pit surface mine. It was in operation from October 1971 to December 1995, and it produced 1.3×10^6 t copper, 31×10^3 t molybdenum, 340 t silver and 23 t gold (Wilton and Lawrence 1998). The tailings of the ore processing, which usually constitute one of the major environmental concerns related to a metal mine, were successfully deposited at the bottom of Rupert Inlet, a part of the Pacific Ocean adjacent to the mine (Poling et al. 2002). The waste rock was dumped around the mine as well as in Rupert Inlet (Fig. 9.9; Morin et al. 1995; Marcus 1997).

For mine closure, it was decided to create a meromictic pit lake using seawater with a freshwater cap for low-cost, passive treatment of ARD (Marcus 1997). Prior to pit flooding, a reactive waste rock dump (northwest dump) was pushed into the open pit, which released sulphide oxidation products as the pit was flooded. The waste rock surrounding the pit was also acid generating: it required long-term treatment after mine closure (Morin et al. 1995; Dagenais and Poling 1997).

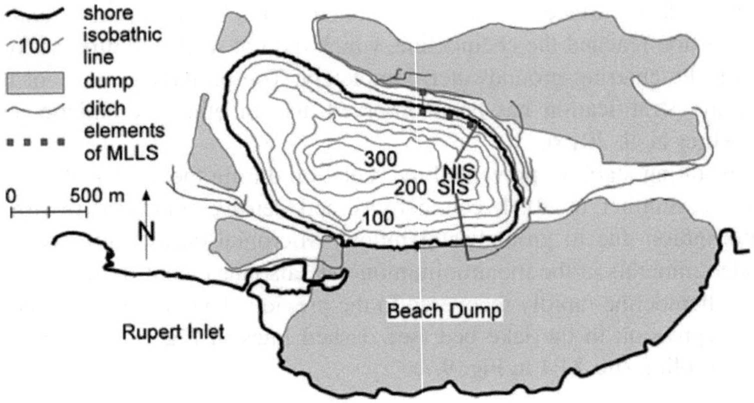

Fig. 9.9 Island Copper Mine Pit Lake (Canada) and its environment: *NIS* north injection system, *SIS* south injection system, *MLLS* middle-layer lifting system. Drawn from Fig. 2 in Dagenais and Poling (1997), Fig. 1.2 in Fisher (2002), Fig. 1 in Morin et al. (1995) and Fig. 17.2 in Pelletier et al. (2009)

9.4.2 Filling, Morphology and Water Balance of the Lake

In order to create a meromictic lake, seawater was diverted into the pit from Rupert Inlet through the former emergency tailings pond for 39 days in summer 1996 (Marcus 1997). Freshwater from the Marble River was then pumped on top of the seawater using the same infrastructure as used to bring water to the ore processing plant during the mining operation. With the addition of direct precipitation and local runoff, a stable meromictic lake was created (August 1996 to March 1998). Beginning in October 1996, ARD collected by a ditch system was additionally introduced into the lake at a depth of 220 m via the north injection system (NIS) and the south injection system (SIS) (Fig. 9.9). Both systems consisted of pipelines from surface to depth and diffusor systems at the lower end. Water flow was driven by gravity. The final nominal water level (2.5 m asl) was reached in December 1998.

The filling resulted in a meromictic water body consisting of three layers: a brackish layer at the surface (mixolimnion, about 5 m thick), an intermediate layer of diluted seawater which is gradually diluted over time (upper monimolimnion, about 210 m thick) and a bottom layer of nearly full-salinity seawater (lower monimolimnion, 135 m thick at its deepest point) (Fig. 9.10, Table 9.3). The pycnocline between mixolimnion and upper monimolimnion is about 2 m thick and referred to as the "upper pycnocline", whereas the one between the lower and the upper monimolimnion is about 6 m thick and referred to as the "lower pycnocline".

The Island Copper Mine Pit Lake receives substantially more water from direct and indirect precipitation than it loses by evaporation (Table 9.3). Excess water leaves (i.e. exfiltrates) as groundwater through the southern Beach Waste Rock Dump (i.e. Beach Dump; Fig. 9.9) before entering Rupert Inlet. A slurry wall along

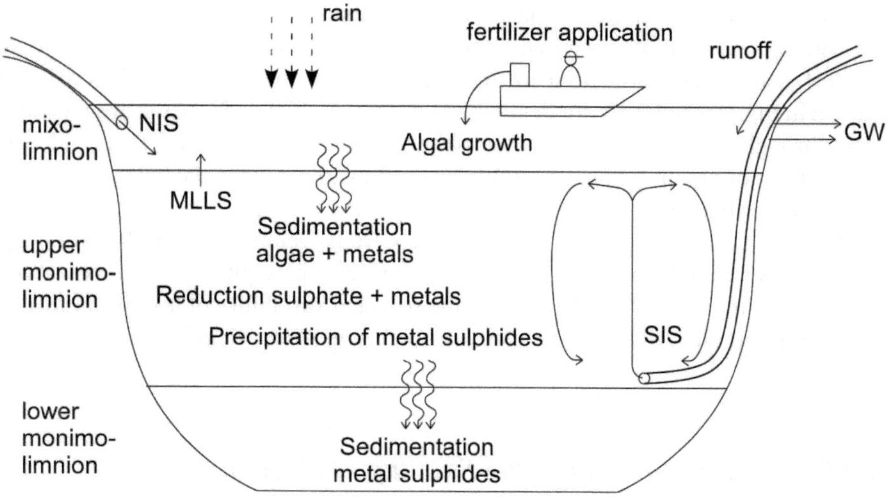

Fig. 9.10 Scheme of the Island Copper Mine Pit Lake (Canada) showing the main measures and factors influencing the stability of meromixis and the processes governing the behaviour of metals: *NIS* north injection system (current, formerly discharge at depth), *SIS* south injection system, *MLLS* middle-layer lifting system, *GW* groundwater (drawing not to scale). *Arrows* indicate the direction of fluxes of water and matter

Table 9.3 Morphometry and water balance of Island Copper Mine Pit Lake

Quantity	Unit	Value
Volume (total)	10^6 m^3	244
Volume mixolimnion	10^6 m^3	6.6
Volume upper monimolimnion	10^6 m^3	222
Volume lower monimolimnion	10^6 m^3	15
Surface area	km^2	1.735
Maximal depth	m	350
Thickness of mixolimnion	m	3.8
Thickness of upper monimolimnion	m	216
Thickness of lower monimolimnion	m	130
Mean depth	m	141
Relative depth	%	23.5
Precipitation (1880 mm)	10^6 m^3	3.34
Runoff	10^6 m^3	1.0
Inflow via NIS	10^6 m^3	2.30
Inflow via SIS	10^6 m^3	0.94
Evaporation	10^6 m^3	1.04
Outflow from mixolimnion to Beach Dump groundwater	10^6 m^3	6.54

Data on water balance are given as mean annual sum. Data source from Fig. 17.3 in Pelletier et al. (2009). Mean depth and relative depth calculated using equations by Wetzel (2001)

the southern perimeter of the pit served to extend the operational mine life by preventing seawater ingress into the open pit, and this wall now acts as a hydrologic control, setting the lower limit for the water level in the lake (i.e. 1.5 m asl; not shown in Fig. 9.9 and 9.10; for details see Poling et al. 2003). Since the groundwater flow through the Beach Dump is the only outflow of the lake and evaporation is < one-third of precipitation, the groundwater flow to the sea is the balancing factor for the water budget of the lake.

Most of the annual rain (1880 mm) falls from October to March, averaging 230 mm per month, while in the rest of the year, precipitation averages about 100 mm per month, or less. This drives the seasonal variability of lake water levels, and the runoff and ARD volumes are also higher in winter than in summer.

9.4.3 Factors Governing the Stability of Meromixis and the Behaviour of Metals

Intentionally, a permanently stratified waterbody was created by capping the seawater with less dense freshwater. This allowed for formation of stable anoxic conditions in the monimolimnion, which were needed for the passive treatment of ARD. Anoxic conditions favour metal reduction (e.g. iron, manganese), sulphate reduction which produces alkalinity, precipitation of metal sulphides and the long-term burial of the metal sulphides in lake sediments. Management of the lake was adapted over time based on monitoring and experiments to optimize and stabilize the system and to reduce long-term operational costs.

Initially, low-salinity ARD was injected through the NIS and SIS into high-salinity water at 220 m depth. This caused upward (i.e. buoyant) movement of the ARD, which in turn resulted in circulation within the upper monimolimnion and mixing of the ARD into this layer. Muggli et al. (2000) confirmed this process using dye (uranin) in tracer experiments. Further evidence of mixing came from the homogeneity of temperature, salinity and oxygen profiles within each layer and changes at the pycnoclines between the layers (Wilton and Lawrence 1998). The upper pycnocline was particularly important for sustaining a sufficiently large density contrast between the mixolimnion and upper monimolimnion. The dye experiments showed that the injected water did not reach the lake surface during periods of low injection rates in summer. However, when injection rates were high in winter, the injected dye spread across the entire area of the lake immediately below the upper pycnocline within 1 day (Muggli et al. 2000).

Detailed monitoring showed that the upward movement of the injected ARD resulted also in lateral differences in the thickness of the mixolimnion. Fisher and Lawrence (2000) reported that the mixolimnion was thinner above the injection systems compared with the rest of the lake. This phenomenon combined with the general rise of the upper pycnocline (discussed below) caused an increasing erosion of the upper pycnocline by wind-forced turbulence in the mixolimnion and a respective "sharpening" of the upper pycnocline (Fisher and Lawrence 2000). Furthermore, the activity of internal waves of higher frequency increased and contributed to the sharpening of the upper pycnocline, especially during periods of high injection rates and strong winds. This often coincided with storms in winter (Fisher and Lawrence 2000).

The injection of ARD also increased the volume of the upper monimolimnion, pushing the upper pycnocline closer to the lake surface (Fisher and Lawrence 2000). The risk arose that the mixolimnion would become too thin to survive as an effective cap, allowing metal-enriched water from the upper monimolimnion to reach the lake surface and flow into the groundwater and eventually into Rupert Inlet. Modelling initially predicted erosion of the upper pycnocline by wind-forced turbulence leading to a stable equilibrium thickness of the mixolimnion. Monitoring indicated that the erosion was smaller than predicted.

Therefore, the injection depth of the NIS was changed from 220 m depth to the lake surface to reduce the amount of water introduced into the upper monimolimnion. This measure was tested in 2000 and 2001 and was permanently implemented in August 2002. This change in management became possible due to water quality monitoring. ARD diverted to the NIS was found to be circum-neutral, with metal concentrations low enough for treatment involving fertilization within the mixolimnion (Table 9.4). In addition, it also provided alkalinity for algal growth, as well as slowed the decrease in concentration of alkalinity in the mixolimnion.

Changing the NIS discharge point to the lake surface slowed down the rise of the upper pycnocline considerably but did not stop its rise due to the continued injection of water via the SIS. To resolve this, the middle-layer lifting system (MLLS) was designed to pump water from the upper monimolimnion (from 15 m depth) into the mixolimnion (3 m depth). Based on the Venturi principle, the MLLS uses a negative head of water at the surface to drive suction of water from the upper monimolimnion into the mixolimnion. The pumping rate is self-regulating; the higher the rainfall, the greater the suction rate. The system is calibrated so that the pumping rate is always slightly greater than the rate of water injected at depth

Table 9.4 Water quality of the ARD diverted to the NIS and SIS compared with permitted concentrations for release of water into the environment around the Island Copper Mine Pit Lake

		NIS	SIS	Permitted release concentrations
pH		7.1–8.4	4.2–6.6.	6.5–11.5
SO_4^{2-}	ppm	400–780	1300–2000	–
Ca	ppm	140–230	280–410	–
Mg	ppm	20–35	60–100	–
Cu	ppm	0.01–0.04	0.03–3.3	0.05
Zn	ppm	0.5–3.9	4.2-–11.1	1.0
Cd	ppm	0.01–0.03	0.01–0.05	0.01
Mo	ppm	0.008–0.012	0.0005–0.001	0.5

Data from Pelletier et al. (2009)

through the SIS. The overall effect balances the amount of injected water by the amount of water transferred from the upper monimolimnion into the mixolimnion, thus stabilizing the depth of the upper pycnocline.

The inflow of low-salinity rain, runoff and ARD, combined with the outflow of the mixolimnetic water via groundwater flushes the mixolimnion (i.e. low residence time). Thus, the mixolimnion has a low density, and a strong density contrast has been maintained at the upper pycnocline. This was an essential component of the long-term stability of the entire system. Otherwise, the density of the mixolimnion would increase due to the introduction of saline monimolimnetic water via MLLS, which would facilitate overturn and aeration of the upper monimolimnion.

The dilution of the upper monimolimnion by the injected ARD from the SIS presents only a small risk to the stability of meromixis. Using a simple box model, we estimated that it will take nearly 400 years for the upper monimolimnion to be diluted to a salinity of 0.6 %, after which it will have a salinity similar to that of the mixolimnion. There is thus plenty of time to adapt the management strategy if the treatment of ARD is still necessary when the densities of the mixolimnion and the upper monimolimnion get that close that meromixis may get unstable.

Fertilization of the mixolimnion with liquid ammonium phosphate and urea ammonium nitrate did not influence the stability of meromixis, but it stimulated the growth of algae. This last was essential for the removal of metals by adsorption and sedimentation and for the neutralization of the ARD in the monimolimnia by providing organic matter for the reduction of iron and sulphate. From 1997, the fertilizers were added to the mixolimnion during spring or summer to stimulate plankton growth. Monitoring indicated that the produced biomass was not sufficient to cause rapid oxygen depletion in the monimolimnia. However, fertilization caused always a significant drop in metal concentrations in the mixolimnion. Moreover, the mixolimnetic metal concentrations rose if fertilization was interrupted. Based on detailed experimental studies, year-round fertilization was implemented in 2001. Recently, the weekly dose of fertilizer consisted of 300 mg/m^2 of nitrogen and 50 mg/m^2 of phosphorus. Silica was not added as the growth of diatoms was not desirable. The higher sinking rates of diatoms limit the amount of adsorbed metals per unit of algal biomass due to the short residence time in the mixolimnion. Algae other than diatoms have lower sinking rates and are therefore more suitable and efficient at metal removal.

The concentration of chlorophyll was about 20 µg L^{-1} after the implementation of year-round fertilization. In spring, the phytoplankton was dominated by green algae, while in autumn cyanobacteria dominated. The metal concentrations in the mixolimnion were around 0.02 mg L^{-1} zinc, 0.005 mg L^{-1} copper and 0.003 mg L^{-1} cadmium (all three well below the permitted values shown in Table 9.4). Clearly, the adsorption of metals onto the growing algae and then the sinking of dead algae worked well.

In 2004, the upper monimolimnion became anoxic. The precipitation of metal sulphides was inferred in 2005 by the detection of both dissolved hydrogen sulphide gas as well as by a rapid decrease of the metal content in the upper monimolimnion. Until the end of 2005, the removal of metals from the upper monimolimnion was dominated by adsorption onto organic matter and precipitating oxyhydroxides of iron and manganese. In the lower monimolimnion and the bottom sediment, sulphate reduction started earlier than in the upper monimolimnion.

9.5 Management Options for Meromictic Pit Lakes

Since pit lakes often have poor water quality, they need to be carefully managed. In particular, initial lake water is usually far from a hydrological, chemical and biological equilibrium. The most common water quality problems are associated with acidification due to pyrite oxidation and high concentrations of metals (Klapper and Schultze 1995; Miller et al. 1996), although neutral mine drainage and alkaline mine drainage may also have low water quality. Both environmental regulators and mining industry standards entail management of the pit lakes and their water quality (e.g. Jones and McCullough 2011).

There are several options to control the water quality of pit lakes during mining, during lake filling and after achieving equilibrium water levels (Skousen et al. 1998; Younger and Wolkersdorfer 2004; Gammons et al. 2009; Geller et al. 2013). The first step is planning of the pit lake management, ideally before the lake filling. In many countries, such planning is required when applying for mining permission. For this purpose, predictive modelling is of very great importance and feasibility (see Sect. 9.6). Planning also includes deciding whether meromixis is an acceptable option. This decision has to be based on predictive modelling results, availability of resources (i.e. technical and economical) for preventing or permitting meromixis and a discussion involving all the stakeholders, including the public and regulators (Swanson 2011; Vandenberg et al. 2015b). In general, meromixis is only acceptable if a long-term stability is expected and adverse impacts of meromixis on the quality of groundwater discharge can be eradicated.

Once a meromictic pit lake is formed, monitoring is among the essential management tasks. An early detection of undesired developments in the pit lake is essential in order to take counteractive measures. The validation of such measures and the improvement and validation of predictive modelling also require careful monitoring.

9.5.1 Sustaining Meromixis

Meromixis may require maintenance in order to stabilize stratification and prevent transfer of hazardous substances from the monimolimnion into the mixolimnion. Crusius et al. (2002), Castendyk and Webster-Brown (2007a) and Fraser et al. (2012) showed in their predictive modelling that the lake-filling process often is decisive for the final stratification type of a pit lake and the resulting water quality. Depending on the water properties and sequence of the introduction of water from the major water sources used for filling (e.g. rebounding groundwater, drainage from waste rock piles, river water), stable or temporary meromixis may be intentionally achieved (Crusius et al. 2002; Fraser et al. 2012). However, Boehrer et al. (2014) quantitatively demonstrated that the slow but steady erosion of monimolimnia can occur over time. For permanent stability, processes sustaining meromixis must be established. Continuous introduction of freshwater into the mixolimnion sustains the density difference to the monimolimnion (Santofimia et al. 2012; Boehrer et al. 1998, 2014).

Monimolimnia of pit lakes may accumulate gases, such as carbon dioxide (CO_2) (Murphy 1997; Sánchez-España et al. 2014a; see also Fig. 9.4; see also Chaps. 2 and 3). Accumulation of dissolved gases may destabilize stratification and, in extreme instances, enable a limnic eruption similar to that in the naturally meromictic lakes Nyos and Monoun (Cameroon; Kling 1987). To avoid such a situation, a controlled removal of CO_2 from the monimolimnion should be considered (Kling et al. 2005, Boehrer et al. 2016), whereby a vertical pipe conveys monimolimnetic water to the surface driven by bubbles. These bubbles form inside the pipe at a depth where absolute pressure (i.e. atmospheric plus hydrostatic pressure) is lower than the dissolved gas pressure. At the lake surface, the monimolimnetic water degasses in a fountain. The continuous operation of this type may remove the monimolimnion entirely. To maintain meromixis, degassed water can be redirected into the monimolimnion using devices similar to hypolimnetic aerators (Cooke et al. 2005). Dissolved CO_2 and H_2S increase water density, while dissolved CH_4 and N_2 decrease water density (Dietz et al. 2012). Accordingly, degassing influences water density.

9.5.2 *Monimolimnia as Sites for Waste Deposition and Water Treatment*

There are instances where monimolimnia of pit lakes have been successfully used for disposal of mine waste containing elements of concern, as arsenic, cadmium and lead (Brassard et al. 1996; see also Sect. 9.4 for Island Copper Mine). In Vollert Süd (Germany), organic substances were flocculated and removed from the mixolimnion, leading to a decrease in the water quality in the monimolimnion (Stottmeister et al. 2010). The South Mine pit lake in the Tennessee Copper Basin (USA) is used as disposal site for sludge from a treatment plant that is used to neutralize the North Potato Creek (Wyatt et al. 2006). Since the pit lake is meromictic, its iron-rich (600 mg L^{-1} Fe(II)) monimolimnion also provides a supplemental source of flocculant for the treatment plant. This monimolimnetic lake water is added to a mixing tank at the treatment plant together with lime (for neutralization) in order to achieve a better removal of metals by flocculation with iron hydroxide (Wyatt et al. 2006). An essential step of the overall treatment procedure is the redissolution of part of the iron contained in the deposited sludge under the reducing conditions in the monimolimnion. In other words, the monimolimnion is not only used as a safe disposal site but also as a geochemical reactor. The dissolution of iron is also the main process for the stabilization of meromixis as it keeps the salinity of the monimolimnion high enough to prevent mixing between the mixolimnion and monimolimnion.

In general, it is essential to manage the deposition of solid wastes so that they do not disturb the chemocline or contaminate the groundwater. This condition may be fulfilled by only disposing of the solid wastes with extremely low solubility under the biogeochemical conditions of the monimolimnion.

Monimolimnetic waters of concern should be precluded from discharging into the groundwater space. This requires careful consideration and modelling of local hydrological/hydrogeological conditions. Closed-basin pit lake may be a prerequisite for an engineered meromictic pit lake. Otherwise, direct pumping of lake water may be required in the future in order to maintain groundwater inflow. For example, the Berkeley Pit (USA) will be pumped to avoid groundwater discharge into a local aquifer until quality of the lake water is acceptable for infiltration into the aquifer (Castendyk et al. 2015a).

Finally, the reintroduction of deposited solid waste material into the mixolimnion by mixing processes must be evaluated a priori. A capping (a cover layer consisting of solid material on top of the waste material) may be considered to prevent re-suspension and upward migration of contaminants from the deposited waste material into the water column. Several concepts, material tests, laboratory and field studies as well as full-scale application of sediment capping with inert or reactive cover layers have been reported in literature (e.g. Jacobs and Förstner 2001; Zhang et al. 2016).

9.5.3 Controlled Removal of Meromixis

Under certain conditions, the removal of meromixis may become desirable or even necessary. This may occur: (1) if no technical option for stabilization of meromixis is available; (2) if stabilization of meromixis by active management is too expensive; and (3) if environmental requirements like the prevention of flow from monimolimnion into groundwater cannot be fulfilled.

Destratification of the lake may be achieved by blowing compressed air into a lake close to its bottom (Cooke et al. 2005). This method has been used to terminate meromixis and to force aerobic conversion of thiocyanate and ammonia in the Zone 2 Pit Lake (Canada; Pieters et al. 2015).

In order to prevent the contamination of the mixolimnion by substances enriched in the monimolimnion before destratification, the selective withdrawal of monimolimnetic water and its treatment may be an option. Such a treatment was implemented at the Berkeley Pit lake (maximal depth, 220 m, and volume, 150×10^6 m^3 in 2010, but still increasing) to recover copper from the monimolimnion (180 mg Cu L^{-1} in 2002; Gammons and Tucci 2013). Since the treated water was diverted into the mixolimnion, the density increased causing an unintended gradual deepening of the chemocline, which began in 2003 and was fully mixed in 2009 (Gammons and Tucci 2013).

The addition of oxygen to the monimolimnion using hypolimnetic aerators or deep oxygen injection systems (Cooke et al. 2005) to remove substances by oxidation can avert adverse effects in the mixolimnion. Moreover, if the monimolimnion contains ferrous iron (which is likely in mining environments), the subsequent precipitation of ferric iron and associated sorption of trace metals and metalloids may also be beneficial for the water quality. Adding flocculants (e.g. iron or aluminium) can also remove substances from the monimolimnion without impairing the quality of mixolimnetic water. If necessary, this can be combined with sediment capping (see Sect. 9.5.2).

9.5.4 General Aspects

In all cases, potential acidification accompanying hydrolysis and precipitation of metals has to be kept in mind. Neutralizing measures may be required in order to avoid incomplete precipitation of metal hydroxides or release of hydrogen sulphide (H_2S) to the atmosphere. Neutral conditions are also needed for microbial oxidation of ammonia which is often enriched in the monimolimnia (Jeschke et al. 2013). Acidification and neutralization with carbonates may increase the concentration of CO_2 in the monimolimnion. This may result in undesired destabilization of the stratification and uncontrolled mixing (see Sect. 9.5.1).

Landslides regularly occur along the side walls of open-pit mines and pit lakes (e.g. Zaruba and Mencl 1982, Rose and Hungr 2007; Schultze et al. 2013a). These landslides may cause unintended mixing of meromictic pit lakes. Gammons and Tucci (2013) reported that a landslide caused a holomictic phase in the Berkeley Pit lake. About 1×10^6 m^3 of solid material slid into the lake which was meromictic and had a volume of about 100×10^6 m^3 in October 1998. The holomictic phase was terminated by diverting less saline water to the lake surface in autumn 2000.

9.6 Modelling of Meromictic Pit Lakes

Modern regulations require companies to plan for site closure and pit lake remediation as part of the mine permit applications. Because pit lakes seem to have a higher probability of becoming meromictic than natural lakes, and monimolimnia tend to contain and accumulate hazardous substances, the potential for meromixis and the management of associated risks must be assessed in permit applications. Hydrodynamic models coupled with geochemical and groundwater models can provide the planners and stakeholders with an assessment of the future risks prior to mining. In addition, mining companies can use predictive models to demonstrate that future pit lakes will have adequate water quality to enable one or more favourable end uses, such as aquatic ecosystems, public recreation areas, reservoirs for irrigation or stock water, aquaculture facilities and research facilities. This section summarizes previous works on the limnological modelling of existing and future meromictic pit lakes and provides guidance for lake modellers.

9.6.1 Published Models of Meromictic Pit Lakes

Table 9.5 lists some published limnological and geochemical models of meromictic pit lakes. Except for Lake Waldsee (Germany), these predictions were done for management purposes and have not been fully verified. At the time of writing, Gahcho Kué mine (Canada), Martha Mine (New Zealand) and Aitik mine (Sweden) were still in operation, and two pit lakes existed at the Kori Kollo mine (Bolivia), but stratification data have not been published (Wagner et al. 2013). For Island Copper Mine Pit Lake (e.g. Fisher and Lawrence 2006), Lake Rassnitz and Lake Wallendorf (Boehrer et al. 2014), the predicted lake profiles showed strong agreement with the observed lake profiles. Long-range predictions for Brenda Mine and South Mine pit lakes had not been verified, but the short-term accuracy of these models was validated with data from the existing waterbody at the time of modelling (Stevens and Lawrence 1997; Hamblin et al. 1999; Colarusso et al. 2003). Moreira et al. (2011) included the chemical transformations in the density calculation: results from the model of Lake Waldsee are in good agreement with the measured data.

9.6.2 Available Models

Each of the studies in Table 9.5 utilized 1-D and 2-D hydrodynamic models. Small pit lakes with circular surface areas may be appropriately simulated with 1-D models provided that the lake is horizontally homogeneous in all respects. Imberger and Patterson (1981) provided a useful reference to the well-established, 1-D model *DYRESM*, which also formed the hydrodynamic background of the water quality and aquatic ecosystem models *DYCD-CORE* and *CAEDYM*. Outputs from *DYRESM* have been linked to geochemical models to generate water quality predictions (Balistrieri et al. 2006; Castendyk and Webster-Brown 2007b; Oldham et al. 2009). Dunbar (2013) provides a detailed description of *PitMod* which directly links a 1-D hydrodynamic model to the popular geochemical code *PHREEQC* (Parkhurst and Appelo 2000; Fraser et al. 2012).

Simple 1-D models have also been developed to explore the significance of specific processes that lead to meromixis. For example, Pieters and Lawrence (2014) developed a 1-D model in MATLAB to explore the potential for meromixis in pit lakes that experience ice cover. They conclude that lakes with significant

Table 9.5 Published modelling studies on meromictic pit lakes

Lake (status)	Model used	References
Island Copper Mine, BC, Canada (existing)	Box models and 1-D hydrodynamic model (DYRESM)	Wilton and Lawrence (1998) and Fisher (2002)
Brenda Mine, BC, Canada (existing)	1-D hydrodynamic model (DYRESM)	Stevens and Lawrence (1997) and Hamblin et al. (1999)
Lakes Wallendorf and Rassnitz, Germany (existing)	1-D hydrodynamic model, coupled offline with groundwater model	Boehrer et al. (1998, 2014)
Kori Kollo Mine, Bolivia (existing)	1-D hydrodynamic model, fully coupled with geochemical model (PitMod)	Crusius et al. (2002)
South Mine pit lake, TN, USA (existing)	2-D hydrodynamic model (CE-QUAL-W2)	Colarusso et al. (2003)
Lake Waldsee, Germany (existing)	1-D hydrodynamic model, fully coupled with geochemical model (DYCD-CORE)	Moreira et al. (2011)
Martha Mine, New Zealand (future)	1-D hydrodynamic model (DYRESM), separate modelling of geochemistry (PHREEQC) without feedback to hydrodynamics	Castendyk and Webster-Brown (2007a,b)
Aitik and Salmijärvi mines, Sweden (future)	1-D hydrodynamic model, fully coupled with geochemical model (PitMod)	Fraser et al. (2012)
Gahcho Kué mine, NWT Canada (future)	2-D hydrodynamic model (CE-QUAL-W2) linked to water balance model (GoldSim) and groundwater model (MODFLOW)	Vandenberg et al. (2015a) and Herrell et al. (2015)

depth, salinity and ice thickness have the potential for meromixis and that the addition of low-salinity runoff to the lake surface can increase stability over time (see Sect. 9.1 where we discussed relative depth as measure for the potential occurrence of meromixis). The Global Lake Ecological Observatory Network (GLEON) freely distributes *Lake Analyzer*, a set of tools that allows users to calculate common metrics for physical state of lakes, such as Lake Number, Schmidt stability and Wedderburn number (see www.gleon.org).

Pit lakes with large surface areas and complex shorelines are more likely to be subject to horizontal variability in heating, wind speed and water inputs and, therefore, are better represented by 2-D models. A detailed description of the freely distributed, widely used 2-D model *CE-QUAL-W2* is available from the web page of the Water Quality Research Group of the Portland State University (http://www.ce.pdx.edu/w2/). Both Vandenberg et al. (2015a) and Herrell et al. (2015) use *CE-QUAL-W2* to evaluate the potential development of meromictic pit lakes in the Northwest Territories of Canada as a result of diamond mining. Vandenberg et al. (2015c) and Prakash et al. (2015) present a custom-built, sediment diagenesis model for *CE-QUAL-W2* designed to evaluate future holomictic pit lakes in the Athabasca Oil Sands Mining District of Alberta, Canada.

Several 3-D hydrodynamic models are available, such as *ELCOM* managed by the Centre for Water Research at the University of Western Australia in Perth (http://www.cwr.uwa.edu.au/software1/models1.php?mdid=5; see also the discussion by Mooij et al. 2010). These models seem desirable from a scientific view. However, because the main drivers for modelling pit lakes are planning, regulation and management, 1-D and 2-D models that accurately represent the stability of meromixis can sufficiently achieve most modelling objectives and are simpler to design. To our knowledge, so far no published studies of meromictic pit lakes have used the 3-D models.

9.6.3 Modelling Objectives

Modellers need to answer the several following questions from physical limnology before they develop geochemical models:

- What temperature profiles and wind speeds produce minimal vertical stability within the mixolimnion?
- When in the annual cycle does minimum stability occur to allow for circulation of the mixolimnion, and when is the mixolimnion stratified?
- To what depth will mixing occur in the mixolimnion annually?
- What is the maximum depth of mixing and monimolimnion stability associated with extreme storm events (i.e. maximum wind speed) and extreme flood events (i.e. maximum surface water input) at 5-, 10-, 25-, 50- and 100-year recurrence intervals?
- What is the volume of each lake layer, and how likely are these volumes to change over time?
- What magnitude landslide would be required to induce whole-lake mixing, and what is the probability of such an event?
- How will changes in the volume and salinity of lake inputs over time affect lake stratification over time?
- What is the likelihood that an initially holomictic pit lake will evolve into a meromictic pit lake over time, and vice versa?
- Will stratification be strong enough to enable the monimolimnion to be used for the disposal of mine-impacted water and/or solid mine wastes?
- What is the optimal depth and longitudinal location for such disposal?
- Will the temperature, salinity and/or momentum of injected wastes generate buoyant plumes which drive circulation in the monimolimnion and destabilize the water column (see Sect. 9.4.3)?

Once the physical characteristics are defined, geochemical models can answer the following questions regarding the monimolimnion:

- What is the residence time, chemistry and toxicity of the predicted monimolimnion, and how will these characteristics change over time?

- What concentrations/partial pressures of O_2, CO_2, CH_4 and H_2S can be expected?
- Will sulphate reduction, hydrogen sulphide production and sulphide mineral precipitation occur?
- Will methanogenesis occur in the monimolimnion?
- Under what conditions could dissolved gases within the monimolimnion pose a hazard to the surface ecosystem and lake users (see Sect. 9.5.1)?

9.6.4 Model Development

Castendyk et al. (2015b) provide an overview of limnological concepts relevant to pit lakes, while Castendyk et al. (2015a) provide step-by-step instructions for the modelling of an existing holomictic pit lake using DYRESM. The same procedure may be followed to develop predictions of future, potentially meromictic pit lakes.

The first steps in modelling a future pit lake are to explicitly state the objectives of the model, to develop a conceptual model of critical lake processes and to select an appropriate 1-D, 2-D or 3-D program (Anderson and Woessner 1992). Next, the modeller obtains the most up-to-date spatial representation of the open pit at the conclusion of mining, meteorological data collected from a local meteorological station and a pit lake water balance for before, during and after the lake filling. The water balance must include the temperature, salinity and discharge rate of each input. Because initial conditions strongly influence the evolution of the pit lake, Castendyk and Webster-Brown (2007a) recommend modelling the water column during lake filling, rather than assuming a steady-state water column. This approach provides the mining companies with valuable information on the influence of management decisions during the lake-filling process.

Both DYRESM and CE-QUAL-W2 generate profiles of lake temperature, salinity and density over time from which the modeller can identify the frequency and depth of mixing events and calculate layer volumes. For models of existing lakes, the modeller typically adjusts the light extinction coefficient, wind-sheltering coefficient and/or light shading coefficient to obtain the best match between predicted and observed conditions. For future pit lakes, calibration is not straightforward. However, the modeller can run multiple scenarios to test the sensitivity of the model against individual parameters and can explore various scenarios for lake filling (e.g. Castendyk and Webster-Brown 2007a; Vandenberg et al. 2015a).

9.6.5 Conceptual Models

When developing a conceptual model for a meromictic pit lake, the relevant processes forming and stabilizing the density difference between monimolimnion and mixolimnion should be included. This determines the necessary effort of modelling. For example, meromixis in Lake Wallendorf and Lake Rassnitz was mainly caused by the high concentration of sodium chloride in the monimolimnion (Boehrer et al. 1998, 2014). In both cases, simulations did not require the consideration of biogeochemical reactions. However, in Lake Waldsee, such reactions had to be included because bicarbonate and ferrous iron from the microbial decay of organic matter caused density differences between the monimolimnion and mixolimnion (Dietz et al. 2012; Nixdorf and Boehrer 2015). Likewise, the import and export of dissolved and suspended solids, as well as the formation and settling of mineral precipitates, influence the water density and need to be included. Carbon dioxide may also require special attention since it increases the water density. However, bubble formation may cause mixing between monimolimnion and mixolimnion, or even a limnic eruption (Murphy 1997; Sánchez-España et al. 2014a).

To date, modelling of biological aspects in natural lakes has generally focused on eutrophication associated with nutrient loading and biological productivity (Jørgensen 2010; Mooij et al. 2010). These processes may also be important in pit lakes that receive labile carbon, phosphorus and nitrogen, such as oil sand mines, phosphate mines and hard-rock mines. Mining methods in hard rock typically involve the use of nitrogen-based explosives to break up wall rocks and liberate ore. Consequently, residual ammonia and nitrate salts are commonly flushed from pit walls during the first years of lake filling. Plankton influences the settling of organic matter, turbidity and the light extinction and the adsorption of trace metals onto organic surfaces.

Two approaches can be used to model the interplay between physical, geochemical and biological processes in long-range predictions. Stevens et al. (2002) proposed to run complex models for a few annual cycles to determine the important mechanisms in the pit lakes and then to parameterize these mechanisms in conceptual models for simulations of 50 years or longer. An alternative approach is to directly couple the biogeochemical models with hydrodynamic models. In order to minimize the modelling effort and the runtime of simulations, Dietz et al. (2012) provide a method to identify the most important substances responsible for the density difference between the monimolimnion and mixolimnion.

9.6.6 Model Applications

Models allow mine managers to compare different lake-filling strategies and management options (Castendyk and Webster-Brown 2007b; Castendyk and Webster-Brown 2010; Vandenberg et al. 2015a). Mine managers also utilize limnological models to design pit lake closure strategies that maximize the potential to passively treat specific contaminants of concern. For example, holomictic conditions favour the aerobic decomposition of organic contaminants in Alberta's oil sand pit lakes (Vandenberg et al. 2015c; Prakash et al. 2015), whereas meromictic conditions allow trace metal removal via sulphide precipitation in British

Columbia's Island Copper Mine Pit Lake (see Sect. 9.4). In both these locations, companies engineered pit lakes to produce the optimal physical limnology for the desired reactions.

9.6.7 Addressing Uncertainties

As shown in Table 9.5, a limited number of modelling studies of existing or future pit lakes have been published. Common reasons for this include intellectual property issues within consulting agencies, potential legal liabilities perceived by mining companies and their lawyers and resistance by scientific journals to publish unvalidated models. This status quo undermines the ability of modellers to learn from the successes and failures of previous modelling efforts, the development of more accurate predictions and public confidence in predicted results.

Models of existing pit lakes are far more accurate than predictive models of future pit lakes for the simple reason that observed data enables model calibration and the quantification of model accuracy. In the absence of lake data, limnological predictions should be considered "hypotheses" or testable scientific guesses of how we expect the lake will behave under a given set of boundary conditions. Important boundary conditions include (1) climate conditions which affect the energy balance over the lake surface (i.e. wind speed, wind direction and duration of ice cover), (2) climate conditions which affect the heat balance of the lake (i.e. incident long- and short-wave radiations, cloud cover, air temperature and latent heat of evaporation), (3) climate conditions that affect the pit lake water balance (i.e. direct rainfall, direct evaporation, watershed runoff and groundwater input/output), (4) the temperature, salinity and composition of lake inputs, (5) water turbidity and the light extinction coefficient which are related to sediment loading to the lake surface and phytoplankton productivity and (6) the bathymetry of the lake.

Changes to boundary conditions that develop between the conclusion of modelling and the filling of the pit lake can introduce significant errors in model predictions. For example, future climate change can directly affect the energy balance, heat balance and water balance of the lake, making climate change a significant knowledge gap in limnological modelling, especially for pit lakes that will not achieve steady-state lake levels (i.e. hydrologic equilibrium) for hundreds of years from now (Grimaldi 2009; Henny and Triyanto 2011).

Changes to mine plans typically occur over the lifetime of a mine because of new discoveries of adjacent resources, changes in the market value of the commodity being mined, corporate bankruptcy or pit wall failures, such as the 2010 landslide in the Bingham Canyon Copper Mine, Utah, USA (Pankow et al. 2013). To address these uncertainties, mining companies should regularly update limnological predictions as changes in the mine plans occur (Castendyk and Webster-Brown 2010). The impact of landslides on stratification needs to be simulated because pit wall collapses are common in post-mining pits.

In all modelling studies, the impact of potential extreme scenarios needs to be explored, such as the impact of very strong storm events (Boehrer et al. 1998; Colarusso et al. 2003). Short-term and long-term trends in the water balance over time need to be considered in predictions associated with long-lasting droughts and floods (Santofimia et al. 2012), as well as regional trends in temperature, evaporation and precipitation associated with climate change (e.g. Grimaldi 2009). This may involve adjusting the water balance and meteorological inputs to reflect changes predicted by regional climate models.

Acknowledgements We thank Olaf Büttner (UFZ, Magdeburg, Germany) for his help in preparing the bathymetric maps of Lake Goitsche and Marc Wen (ERM, Vancouver, Canada) for checking the correctness of the compiled data on Island Copper Mine Pit Lake and the interpretations. We acknowledge the permission of Elsevier B.V. for reprint of Figs. 9.1, 9.2 (modified) and 9.3 from Sánchez-España et al. (2014b) and the permission of Springer B.V. to use the modified reprint of Fig. 9.5 from Diez-Ercilla et al. (2014).

References

Alazard D, Joseph M, Battaglia-Brunet F, Cayol JL, Ollivier B (2010) *Desulfosporosinus acidiphilus* sp. nov.: a moderately acidophilic sulfate-reducing bacterium isolated from acid mining drainage sediments. Extremophiles 14:305–312

Anderson MP, Woessner WW (1992) Applied groundwater modeling; Simulation of flow and advective transport. Academic, San Diego, CA

Balistrieri LS, Tempel RN, Stillings LL, Shevenell LA (2006) Modeling spatial and temporal variations in temperature and salinity during stratification and overturn in Dexter Pit Lake, Tuscarora, Nevada, USA. Appl Geochem 21:1184–1203

Bigham JM, Nordstrom DK (2000) Iron and aluminum hydroxysulfates from acid sulfate waters. In: Alpers CN, Jambor JL, Nordstrom DK (eds) Sulfate minerals: crystallography, geochemistry, and environmental significance. Rev Mineral Geochem 40:351–403

Boehrer B (2012) Double-diffusive convection in lakes. In: Bengtsson L, Herschy RW, Fairbridge RW (eds) Encyclopedia of lakes and reservoirs, encyclopedia of earth sciences series. Springer, Dordrecht, pp 223–224

Boehrer B, Dietz S, von Rohden C, Kiwel U, Jöhnk KD, Naujoks S, Ilmberger J, Lessmann D (2009) Double-diffusive deep water circulation in an iron-meromictic lake. Geochem Geophys Geosyst 10:Q06006. doi:10.1029/2009GC002389

Boehrer B, Heidenreich H, Schimmele M, Schultze M (1998) Numerical prognosis for salinity profiles of future lakes in the opencast mine Merseburg-Ost. Int J Salt Lake Res 7:235–260

Boehrer B, Kiwel U, Rahn K, Schultze M (2014) Chemocline erosion and its conservation by freshwater introduction to meromictic salt lakes. Limnologica 44:81–89

Boehrer B, Yusta I, Magin K, Sanchez-Espana J (2016) Quantifying, assessing and removing the extreme gas load from meromictic Guadiana pit lake, southwest Spain. Sci Total Environ 563–564:468–477

Boehrer B, Schultze M, Liefold S, Behlau G, Rahn K, Frimel S, Kiwel U, Kuehn B, Brookland I, Büttner O (2003) Stratification of mining Lake Goitsche during flooding with river water. In: Tailings and mine waste '03. Swets & Zeitlinger, Lisse, pp 223–231

Boehrer B, Schultze M, Ockenfeld K, Geller W (2005) Path of the 2002 Mulde flood through Lake Goitsche, Germany. Verh Int Verein Limnol 29:369–372

Brassard P, Rosa F, Mudroch A (1996) Disposal of acid drainage generating mine tailings in lakes and man-made reservoirs. Environ Technol 17:1059–1069

Cameron D, Willett M, Hammer L (2006) Distribution of organic carbon in the Berkeley Pit Lake, Butte, Montana. Mine Water Environ 25:93–99

Castendyk DN, Eary LE (2009) The nature and global distribution of pit lakes. In: Castendyk DN, Eary LE (eds) Mine Pit Lakes: Characteristics, predictive modeling, and sustainability. Society for Mining, Metallurgy, and Exploration, Littleton, pp 1–11

Castendyk DN, Webster-Brown JG (2007a) Sensitivity analyses in pit lake prediction, Martha Mine, New Zealand 1: relationship between turnover and input water density. Chem Geol 244:42–55

Castendyk DN, Webster-Brown JG (2007b) Sensitivity analyses in pit lake prediction, Martha Mine, New Zealand 2: geochemistry, water-rock reactions, and surface adsorption. Chem Geol 244:56–73

Castendyk DN, Webster-Brown JG (2010) Effects of mine expansion on geochemical predictions of pit lake water quality: an example from Martha Mine, Waihi, New Zealand. J Geol Geophys 53:143–151

Castendyk DN, Balistrieri LS, Gammons C, Tucci N (2015b) Modeling and management of pit lake water chemistry 2: case studies. Appl Geochem 57:289–307

Castendyk DN, Eary LE, Balistrieri LS (2015a) Modeling and management of pit lake water chemistry 1: theory. Appl Geochem 57:267–288

Colarusso LA, Chermak JA, Priscu JC, Miller FK (2003) Modeling pit lake water column stability using Ce-Qual-W2. In: Tailings and mine waste '03. Swets & Zeitlinger, Lisse, pp 213–222

Cooke GD, Welch EB, Peterson SA, Nichols SA (2005) Restoration and management of lakes and reservoirs, 3rd edn. Taylor & Francis, Boca Raton

Crusius J, Dunbar D, McNee JJ (2002) Predictions of a pit-lake water column properties using a coupled mixing and geochemical speciation model. Trans Soc Min Metall Explor 312:49–56

Dagenais PJ, Poling GW (1997) An investigation into the geochemical history of a waste rock dump and its effect on water quality of the flooded open pit at Island Copper Mine, Port Hardy, British Columbia. In: Proceedings of the 4th international conference on acid rock drainage, Vancouver, BC, 31 May–6 June, 1997. Mine Environment Neutral Drainage Program, Ottawa, pp 1709–1725

Degermendzhy A, Zadereev E, Rogozin D, Prokopkin I, Barkhatov Y, Tolomeev A, Khromechek E, Janse J, Mooij W, Gulati RD (2010) Vertical stratification of physical, chemical and biological components in two saline lakes Shira and Shunet (South Siberia, Russia). Aquat Ecol 44:619–632

Denimal S, Bertrand C, Mudry J, Paquette Y, Hochart M, Steinmann M (2005) Evolution of the aqueous geochemistry of mine pit lakes—Blanzy-Montceau-les-Mines coal basin (Massif Central, France): origin of sulfate contents; effects of stratification on water quality. Appl Geochem 20:825–839

Dietz S, Lessmann D, Boehrer B (2012) Contribution of solutes to density stratification in a meromictic lake (Waldsee/Germany). Mine Water Environ 31:129–137

Diez-Ercilla M (2015) Estudio Hidrogeoquímico del Lago Minero Cueva de la Mora (IBP, Huelva)—Controles sobre la concentración de metales y modelo de estratificación. Dissertation, Universidad del Pais Vasco, Bilbao

Diez-Ercilla M, Sánchez-España J, Yusta I, Wendt-Potthoff K, Koschorreck M (2014) Formation of biogenic sulfides in the water column of an acidic pit lake: biogeochemical controls and effects on trace metal dynamics. Biogeochemistry 121(3):519–536

Dreher N (2007) Entwicklung des pelagischen Nahrungsnetzes in einem neu entstandenen Tagebausee. Dissertation, University Potsdam

Duffek A, Langner C (2002) P-retention in an acidic mining lake under the influence of flooding with river water. Verh Int Verein Limnol 28:1717–1720

Dunbar DS (2013) Modeling of pit lakes. In: Geller W, Schultze M, Kleinmann R, Wolkersdorfer C (eds) Acidic pit lakes. Springer, Heidelberg, pp 186–224

Falagán C, Sánchez-España J, Johnson B (2014) New insights into the biogeochemistry of extremely acidic environments revealed by a combined cultivation-based and culture-independent study of two stratified pit lakes. FEMS Microbiol Ecol 87:231–243

Fast AW, Tyler PA (1981) The re-establishment of meromixis in Hemlock Lake, Michigan, after artificial destratification. Int Rev Gesamten Hydrobiol 66:665–674

Fedorov KN (1988) Layer thicknesses and effective diffusivities in "diffusive" thermohaline convection in the ocean. In: Nihoul JCL, Jamart BM (eds) Small-scale turbulence and mixing in the ocean. Elsevier, Amsterdam, pp 471–479

Fisher TSR (2002) Limnology of meromictic Island Copper Mine pit lake. Dissertation, University of British Columbia

Fisher TSR, Lawrence GA (2000) Observations at the upper halocline of the Island Copper pit lake. In: Lawrence GA, Pieters R, Yonemitsu N (eds) Fifth international symposium on stratified flow, Vancouver, BC, 10-13 July, 2000. Department of Civil Engineering of University of British Columbia, Vancouver, pp 413–418

Fisher TSR, Lawrence GA (2006) Treatment of acid rock drainage in a meromictic mine pit lake. J Environ Eng 132:515–526

Flite O (2006) Onset and persistence of biogenic meromixis in a filling pit lake—a limnological perspective. Dissertation, Clemson University

Fraser CJD, Martin AJ, Dunbar D, Bourne J, Sjöblom AM, Josline CD, Lindström P (2012) Predictive pit lake modelling for Aitik Mine, Northern Sweden. In: Price WA, Hogan C, Tremblay G (eds) 9th international conference on acid rock drainage (ICARD 2012), Ottawa, Canada, 20-26 May 2012, volume 1. Golder Associates, Kanata, pp 789–799

Gammons CH, Tucci NJ (2013) The Berkeley Pit Lake, Butte, Montana. In: Geller W, Schultze M, Kleinmann R, Wolkersdorfer C (eds) Acidic pit lakes. Springer, Heidelberg, pp 362–376

Gammons CH, Harris LN, Castro JM, Cott PA, Hanna BW (2009) Creating lakes from open pit mines: processes and considerations—with emphasis on northern environments. Can Tech Rep Fish Aquat Sci 2826

Geller W, Schultze M, Wisotzky F (2013) Remediation and management of acidified pit lakes and outflowing waters. In: Geller W, Schultze M, Kleinmann R, Wolkersdorfer C (eds) Acidic pit lakes. Springer, Heidelberg, pp 225–264

Grimaldi R (2009) Climatologic characteristics. In: Castendyk DN, Eary LE (eds) Mine Pit Lakes: characteristics, predictive modeling and sustainability. Society for Mining, Metallurgy, and Exploration, Littleton, pp 19–31

Hamblin PF, Stevens CL, Lawrence GA (1999) Simulation of vertical transport in mining pit lake. J Hydraul Eng 125:1029–1038

Henny C, Triyanto (2011) Effects of climate change on the water quality of pit lakes in Banka Island, Indonesia. In: Haryani GS, Chrismada T, Fakhrudin M, Rosilla L, Sutapa I, Subehi L, Toruan RL, Maghfiroh M, Lukman (eds) Proceedings of the national symposium on ecohydrology, March 24, 2011, Jakarta, Indonesia. UNESCO, LIPI, APCE, Jakarta, pp 151–164

Herrell M, Vandenberg J, Faithful J (2015) Designing meromictic pit lakes as a mine closure mitigation strategy in northern Canada. In: Brown A, Bucknam C, Burgess J et al. (eds) Agreeing on solutions for more sustainable mine water management—Proceedings of the 10th ICARD & IMWA annual conference.—electronic document (paper 91); Santiago, Chile (GECAMIN), http://www.imwa.info/imwa-meetings/proceedings/293-proceedings-2015.html

Holmer M, Storkholm P (2001) Sulphate reduction and sulphur cycling in lake sediments: a review. Freshw Biol 46:431–451

Hongve D (2003) Chemical stratigraphy of recent sediments from a depth gradient in a meromictic lake, Nordbytjernet, SE Norway, in relation to variable external loading and sedimentary fluxes. J Paleolimnol 30:75–93

Hrdinka T (2007) Typology and potential utilization of anthropogenic lakes in mining pits in the Czech Republic. Limnol Rev 7:47–53

Imberger J, Patterson JC (1981) A dynamic reservoir simulation model: DYRESM 5. In: Fischer H (ed) Transport models for inland and coastal waters. Academic, New York, pp 310–361

Jacobs P, Förstner U (2001) Managing contaminated sediments. IV. Subaqueous storage and capping of dredged material. J Soil Sediment 1:205–212

Jeschke C, Falagán C, Knöller K, Schultze M, Koschorreck M (2013) No nitrification in lakes below pH 3. Environ Sci Technol 47:14018–14023

Jöhnk K (2001) 1D hydrodynamische Modelle in der Limnophysik: Turbulenz—Meromixis—Sauerstoff. Limnophysics report 1, Institute of Limnophysics, Konstanz

Jones H, McCullough CD (2011) Regulator guidance and legislation relevant to pit lakes. In: McCullough CD (ed) Mine Pit lakes: closure and management. Australian Centre for Geomechanics, Perth, pp 137–152

Jørgensen SE (2010) A review of recent developments in lake modelling. Ecol Model 221:689–692

Kelly CA, Amaral JA, Turner MA, Rudd JWM, Schindler DW, Stainton MP (1995) Disruption of sulfur cycling and acid neutralization in lakes at low pH. Biogeochemistry 28:115–130

Kelley DE, Fernando HJS, Gargett AE, Tanny J, Özsoy E (2003) The diffusive regime of double-diffusive convection. Prog Oceanogr 56:461–481

Klapper H, Schultze M (1995) Geogenically acidified mining lakes—living conditions and possibilities of restoration. Int Rev Gesamten Hydrobiol 80:639–653

Klemm W, Greif A, Broeckaert JAC, Siemens V, Junge FW, van der Veen A, Schultze M, Duffek A (2005) A study on arsenic and the heavy metals in the Mulde River system. Acta Hydrochim Hydrobiol 33:475–491

Kling GW (1987) Seasonal mixing and catastrophic degassing in tropical lakes, Cameroon, West Africa. Science 237:1022–1024

Kling GW, Evans WC, Tanyleke G, Kusakabe M, Ohba T, Yoshida Y, Hell JV (2005) Degassing lakes Nyos and Monoun: defusing certain disaster. Proc Natl Acad Sci USA 102:14185–14190

Koschorreck M, Tittel J (2007) Natural alkalinity generation in neutral lakes affected by acid mine drainage. J Environ Qual 36:1163–1171

Kosolapov DB, Rogozin DY, Gladchenko IA, Kopylov AI, EE Z (2003) Microbial sulfate reduction in a brackish meromictic steppe lake. Aquat Ecol 37:215–226

Kumar RN, McCullough CD, Lund MA (2013) Pit lakes in Australia. In: Geller W, Schultze M, Kleinmann R, Wolkersdorfer C (eds) Acidic pit lakes. Springer, Heidelberg, pp 342–362

Lu M (2004) Pit lakes from sulphide ore mining, geochemical characterization before treatment, after liming and sewage sludge treatment—cases studies at Rävlidmyran and Udden, Sweden. Dissertation, Luleå University of Technology

Lyons WB, Doyle GA, Petersen RC, Swanson EE (1994) The limnology of future pit lakes in Nevada: the importance of shape. In: Tailings and mine waste '94. Balkema, Rotterdam, pp 245–248

Marcus JJ (1997) Closing BHP's island copper mine. Eng Min J 1997:28–32

Meier J, Piva A, Fortin D (2012) Enrichment of sulfate-reducing bacteria and resulting mineral formation in media mimicking pore water metal ion concentrations and pH conditions of acidic pit lakes. FEMS Microbiol Ecol 79:69–84

Miller GC, Lyons WB, Davis A (1996) Understanding the water quality of pit lakes. Environ Sci Technol 30:118A–123A

Molenda T (2014) Impact of saline mine water: development of a meromictic reservoir in Poland. Mine Water Environ 33:327–334

Mooij WM, Trolle D, Jeppesen E et al. (2010) Challenges and opportunities for integrating lake ecosystem modelling approaches. Aquat Ecol 44:633–667

Moreira S, Boehrer B, Schultze M, Dietz S, Samper J (2011) Modeling geochemically caused permanent stratification in Lake Waldsee (Germany). Aquat Geochem 17:265–280

Moreira S, Schultze M, Rahn K, Boehrer B (2016) A practical approach to lake water density from electrical conductivity and temperature. Hydrol Earth Syst Sci 20:2975–2986

Morin KA, Hutt NM, Horne IA (1995) Prediction of future water chemistry from Island Copper Mine's on-land dumps. In: British Columbia Technical and Research Committee on Reclamation (ed) Proceedings of the 19th annual British Columbia mine reclamation symposium, Port Hardy, BC. BiTech Publishers, Richmond, p 224-233

Muggli DL, Pelletier CA, Poling GW, Schwamberger E (2000) Injection ARD plume behaviour in a pit lake utilizing in situ dye studies. In: Proceedings of the 5th international conference on acid rock drainage, Denver, CO, 21–24 May, 2000. Society for Mining, Metallurgy, and Exploration, Littleton, p 305-318

Murphy WM (1997) Are pit lakes susceptible to limnic eruptions. In: Tailings and mine waste '97. Balkema, Rotterdam, pp 543–547

Nixdorf E, Boehrer B (2015) Quantitative analysis of biogeochemically controlled density stratification in an iron-meromictic lake. Hydrol Earth Syst Sci 19:4505–4515

Oldham CE, Salmon SU, Hipsey MR, Ivey GN (2009) Modeling pit lake water quality: Coupling of lake stratification dynamics, lake ecology, aqueous geochemistry, and sediment diagenesis. In: Castendyk DN, Eary LE (eds) Mine Pit Lakes: Characteristics, predictive modeling, and sustainability. Society for Mining, Metallurgy, and Exploration, Littleton, p 127-136

Pankow KL, Moore JR, Hale JM, Koper KD, Kubacki T, Whidden KM, McCarter MK (2013) Massive landslide at Utah copper mine generates wealth of geophysical data. GSA Today 24 (1):4–9

Park BT, Wangerud KW, Fundingsland SD, Adzic ME, Lewis MN (2006) In situ chemical and biological treatment leading to successful water discharge from Anchor Hill Pit Lake, Gilt Edge Mine superfund site, South Dakota, U.S.A. In: Barnhisel RI (ed) Proceedings of 7th international conference on acid rock drainage (ICARD), ASMR, Lexington, pp 1065–1069

Parkhurst DL, Appelo CAJ (2000) User's guide to PHREEQC (version 2)—a computer program for speciation, batch-reaction, one-dimensional transport, and inverse geochemical calculations. U.S. Geological Survey. Water Resources Investigations Report 99-4259, Denver, CO, USA

Pelletier CA, Wen ME, Poling GW (2009) Flooding pit lakes with surface water. In: Castendyk DN, Eary LE (eds) Mine pit lakes. Society for Mining Metallurgy and Exploration, Littleton, pp 187–202

Pieters R, Lawrence GA (2014) Physical processes and meromixis in pit lakes subject to ice cover. Can J Civ Eng 41:69–578

Pieters R, Coedy W, Ashley KI, Lawrence GA (2015) Artificial circulation of a mine pit lake. Can J Civ Eng 42:33–43

Poling GW, Ellis DV, Murray JW, Parsons TR, Pelletier CA (2002) Underwater tailing placement at Island Copper Mine—a success story. Society for Mining, Metallurgy, and Exploration, Littleton

Poling GW, Pelletier CA, Muggli D, Wen M, Gerits J, Hanks C, Black K (2003) Field studies of semi-passive biogeochemical treatment of acid rock drainage at the Island Copper Mine pit lake. In: Farrell T, Taylor G (eds) Proceedings of the 6th international conference on acid rock drainage, 14–17 July, 2003, Cairns, QL. The Australian Institute of Mining and Metallurgy, Carlton Victoria, pp 549–558

Prakash S, Vandenberg JA, Buchak EM (2015) Sediment diagenesis module for CE-QUAL-W2 Part 2: numerical formulation. Environ Model Assess 20:249–258

Regenspurg S, Brand A, Peiffer S (2004) Formation and stability of schwertmannite in acidic mining lakes. Geochim Cosmochim Acta 68:1185–1197

Rose ND, Hungr O (2007) Forecasting potential rock slope failure in open pit mines using the inverse-velocity method. Int J Rock Mech Min Sci 44:308–320

Rücker J, Fyson A, Deneke R, Packroff G (1999) Meromiktische Seen mit saurem Mixolimnion—eine Besonderheit der Lausitzer Bergbaufolgelandschaft. In: Tagungsbericht DGL 1998. Eigenverlag der DGL, Tutzing, pp 239–243

Sahm K, MacGregor BJ, Jørgensen BB, Stahl DA (1999) Sulfate reduction and vertical distribution of sulfate-reducing bacteria quantified by rRNA slot-blot hybridization in a coastal marine sediment. Environ Microbiol 1:65–74

Sánchez-España J, López-Pamo E, Diez M, Santofimia E (2009) Physico-chemical gradients and meromictic stratification in Cueva de la Mora and other acidic pit lakes of the Iberian Pyrite Belt. Mine Water Environ 28:15–19

Sánchez-España J, Yusta I, Diez-Ercilla M (2011) Schwertmannite and hydro-basaluminite: A re-evaluation of their solubility and control on the iron and aluminum concentration in acidic pit lakes. Appl Geochem 26:1752–1774

Sánchez-España J, Yusta I, López GA (2012a) Schwertmannite to jarosite conversion in the water column of an acidic mine pit lake. Mineral Mag 76:2659–2682

Sánchez-España J, Diez-Ercilla M, Falagán C, Yusta I (2012b) An experimental and field study on P, Si, As, Cr, V and Se binding to Fe- and Al-hydroxysulfates under oxic and anoxic conditions in acidic pit lakes: chemical vs. microbial controls. Mineral Mag 75:2319

Sánchez-España J, Diez M, Santofimia E (2013) Mine pit lakes of the Iberian Pyrite Belt: Some basic limnological, hydrogeochemical and microbiological considerations. In: Geller W, Schultze M, Kleinmann B, Wolkersdorfen C (eds) Acidic pit lakes. Springer, Heidelberg, pp 315–342

Sánchez-España J, Boehrer B, Yusta I (2014a) Extreme carbon dioxide concentrations in acidic pit lakes provoked by water/rock interaction. Environ Sci Technol 48:4273–4281

Sánchez-España J, Diez-Ercilla M, Pérez Cerdán F, Yusta I, Boyce AJ (2014b) Hydrological investigation of a multi-stratified pit lake using radioactive and stable isotopes combined with hydrometric monitoring. J Hydrol 511:494–508

Santofimia E, López-Pamo E, Reyes J (2012) Changes in stratification and iron redox cycle of an acidic pit lake in relation with climatic factors and physical processes. J Geochem Explor 116-117:40–50

Schmid M, Lorke A, Dinkel C, Tanyileke G, Wüest A (2004) Double-diffusive convection in Lake Nyos, Cameroon. Deep-Sea Res Part I 51:1097–1111

Schmid M, Halbwachs M, Wüest A (2006) Simulation of CO_2 concentrations, temperature and stratification in Lake Nyos for different degassing scenarios. Geochem Geophys Geosyst 7(6): Q06019. doi:10.1029/2005GC001164

Schultze M (2012) Filling and remediation of pit lakes in former open cast lignite mines. Dissertation, Technical University Braunschweig

Schultze M, Boehrer B, Geller W (2013a) Morphology, age, and development of pit lakes. In: Geller W, Schultze M, Kleinmann R, Wolkersdorfer C (eds) Acidic pit lakes. Springer, Heidelberg, pp 11–17

Schultze M, Boehrer B, Kuehn B, Büttner O (2002) Neutralisation of acidic mining lakes with river water. Verh Int Ver Limnol 28:936–939

Schultze M, Hemm M, Geller W, Benthaus F-C (2013b) Pit lakes in Germany: Hydrography, water chemistry, and management. In: Geller W, Schultze M, Kleinmann R, Wolkersdorfer C (eds) Acidic pit lakes. Springer, Heidelberg, pp 265–291

Skousen J, Rose A, Geidel G, Foreman J, Evans R, Hellier W (1998) Handbook of technologies for avoidance and remediation of acid mine drainage. The National Mine Land Reclamation Centre, West Virginia University, Morgantown

Solski A, Jedrczak A (1991) Meromixis in acidotrophic reservoirs of "Anthropogenic Lake District". Pol Arch Hydrobiol 38:327–346

Spiteri C, Van Cappellen P, Regnier P (2008) Surface complexation effects on phosphate adsorption to ferric iron oxyhydroxides along pH and salinity gradients in estuaries and coastal aquifers. Geochim Cosmochim Acta 72(14):3431–3445

Stevens CL, Lawrence GA (1997) The effect of sub-aqueous disposal of mine tailings in standing waters. J Hydraul Res 35:147–159

Stevens C, Castendyk D, Fisher T (2002) The open-cast mine pit-lake: environmental fluid mechanics and long-term prediction. Mining 31:25–29

Stottmeister U, Kuschk P, Wiessner A (2010) Full-scale bioremediation and long-term monitoring of a phenolic wastewater disposal lake. Pure Appl Chem 82:161–173

Swanson SM (2011) What type of lake do we want? Stakeholder engagement in planning for beneficial end uses of pit lakes. In: McCullough CD (ed) Mine pit lakes: closure and management. Australian Centre for Geomechanics, Perth, pp 29–42

Tonolla M, Peduzzi S, Demarta A, Peduzzi R, Hahn D (2004) Phototropic sulfur and sulfate-reducing bacteria in the chemocline of meromictic Lake Cadagno, Switzerland. J Limnol 63:161–170

Trettin R, Gläser H-R, Schultze M, Strauch G (2007) Sulfur isotope studies to quantify sulfate components in water of flooded lignite open pits—Lake Goitsche, Germany. Appl Geochem 22:69–89

Turner JS (1973) Buoyancy effects in fluids. Cambridge University Press, Cambridge

Turner JS (1978) Double-diffusive intrusions into a density gradient. J Geophys Res 83:2887–2901

Ulbrich U, Brücher T, Fink AH, Leckebusch GC, Krüger A, Pinto JG (2003) The central European floods of August 2002: Part 1—rainfall periods and flood development. Weather 58:371–391

Vandenberg JA, Herrell M, Faithful JW, Snow AM, Lacrampe J, Bieber C, Dayyani S, Chisholm V (2015a) Multiple modeling approach for the aquatic effects assessment of a proposed northern diamond mine development. Mine Water Environ. doi:10.1007/s10230-015-0337-5

Vandenberg J, McCullough C, Castendyk D (2015b) Key issues in mine closure planning related to pit lakes. In: Brown A, Bucknam C, Burgess J et al (eds) Agreeing on solutions for more sustainable mine water management—Proceedings of the 10th ICARD & IMWA Annual Conference.—electronic document (paper 156); Santiago, Chile (GECAMIN)

Vandenberg JA, Prakash S, Buchak EM (2015c) Sediment diagenesis module for CE-QUAL-W2. Part 1: Conceptual formulation. Environ Model Assess 20:239–247

von Rohden C, Ilmberger J, Boehrer B (2009) Assessing groundwater coupling and vertical exchange in a meromictic mining lake with an SF_6-tracer experiment. J Hydrol 372:102–108

Wagner M, Lawrence J, Cederstav A (2013) Letter to Special Rapporteur on the rights of indigenous peoples at office of the high commissioner for human rights. http://www.google.de/url?sa=t&rct=j&q=&esrc=s&source=web&cd=5&ved=0CEQQFjAE&url=http%3A%2F%2Fearthjustice.org%2Fsites%2Fdefault%2Ffiles%2Ffiles%2FEJ-AIDA-SRRIP-Extractive-Industries.pdf&ei=MBu9U_DWL6bnygP0j4CABw&usg=AFQjCNH8tExYwG_-F8gv3I59JzuUtXRjLg&bvm=bv.70138588,d.bGE. Accessed 8 Jul 2014

Wendt-Potthoff K (2013) The role of iron cycling in the neutralization of extremely acidic waters. Habilitation thesis, University of Potsdam

Wendt-Potthoff K, Koschorreck M, Diez Ercilla M, Sánchez España J (2012) Microbial activity and biogeochemical cycling in a nutrient-rich meromictic acid pit lake. Limnologica 42:175–188

Wetzel RG (2001) Limnology: lake and river ecosystems. Academic, San Diego

Wilton MJ, Lawrence GA (1998) The evolution of the Island Copper Mine pit lake. In: British Columbia Technical and Research Committee on Reclamation (ed) Proceedings of the 22th annual british columbia mine reclamation symposium, Pentington, BC. BiTech Publishers, Richmond, pp 173–182

Wyatt G, Miller F, Chermak J (2006) Innovative water treatment plant utilizing the South Mine Pit at the Copper Basin mining site in Tennessee, USA. In: Barnhisel RI (ed) 7th International Conference on Acid Rock Drainage (ICARD), 26–30 March 2006, St. Louis, USA. American Society of Mining and Reclamation (ASMR), Lexington, pp 2529–2539

Younger P, Wolkersdorfer C (2004) Mining impacts on the fresh water environment: technical and managerial guidelines for catchment scale management. Mine Water Environ 23:S2–S80

Zaruba Q, Mencl V (1982) Landslides and their control. Elsevier, Amsterdam

Zhang C, Zhu M, Zeng G, Yu Z, Cui F, Yang Z, Shen L (2016) Active capping technology: a new environmental remediation of contaminated sediment. Environ Sci Pollut Res 23:4370–4386

Zurek R (2006) Chemical properties of water in a flooded opencast sulphur mine. Aquat Ecol 40:135–153

Chapter 10
Tropical Meromictic Lakes: Specifics of Meromixis and Case Studies of Lakes Tanganyika, Malawi, and Matano

Sergei Katsev, Piet Verburg, Marc Llirós, Elizabeth C. Minor, Brittany R. Kruger, and Jiying Li

S. Katsev (✉)
Large Lakes Observatory, University of Minnesota Duluth, 2205 E. 5th St, Duluth, MN 55812, USA

Department of Physics and Astronomy, University of Minnesota Duluth, 2205 E. 5th St, Duluth, MN 55812, USA
e-mail: skatsev@d.umn.edu

P. Verburg
National Institute of Water and Atmospheric Research, Gate 10 Silverdale Road, Hillcrest, Hamilton, New Zealand

M. Llirós
Department of Genetics and Microbiology, Universitat Autònoma de Barcelona, Campus Bellaterra, Bellaterra 08193, Spain

Institut des Sciences de la Vie, Université Catholique de Louvain, Place Croix du Sud 4/5, 1348 Louvain la Neuve, Belgium

E.C. Minor
Large Lakes Observatory, University of Minnesota Duluth, 2205 E. 5th St, Duluth, MN 55812, USA

Department of Chemistry and Biochemistry, University of Minnesota Duluth, 2205 E. 5th Street, Duluth, MN 55812, USA

B.R. Kruger
Large Lakes Observatory, University of Minnesota Duluth, 2205 E. 5th St, Duluth, MN 55812, USA

Division of Earth and Ecosystem Sciences, Desert Research Institute, 755 E Flamingo Rd, Las Vegas, NV 89119, USA

J. Li
Large Lakes Observatory, University of Minnesota Duluth, 2205 E. 5th St, Duluth, MN 55812, USA

© Springer International Publishing AG 2017

R.D. Gulati et al. (eds.), *Ecology of Meromictic Lakes*, Ecological Studies 228,
DOI 10.1007/978-3-319-49143-1_10

10.1 Specifics of Meromixis in Tropical Lakes

10.1.1 Physical Features

Temperature stratification—In tropical lakes, the specifics of meromixis are dictated by the weakness of seasonal fluctuations, high minimum temperatures, and high evaporation rates. Whereas temperate dimictic lakes overturn when their surface water temperatures cross the point of maximum density (at 4 °C), tropical lakes never reach that state. Their minimum temperatures, normally reached at the bottom of the water column, typically correspond to the annual minimum temperature reached by the surface waters (generally about 24 °C at sea level; Lewis 1996). Temperature differences between the top and bottom of the lake water columns are lowest (~1.5 °C) near the equator and reach 5–6 °C near the margin of the tropics (Lewis 1987). Mixing occurs at the time of the greatest surface cooling, which typically happens in the dry season due to enhanced evaporation. At other times of the year the water at the lake surface is warmer than the deep waters. Tropical lakes of moderate depth are therefore predominantly monomictic, whereas deeper lakes cannot be mixed completely by wind and become meromictic. This is different from a situation in temperate lakes where in the absence of a salinity gradient even deep lakes can be mixed thermally, and meromixis typically implies a strong gradient in salinity (Chap. 2). In tropical meromictic lakes, salinity gradients may not be necessarily as pronounced. Lakes without a stable salinity gradient may mix at depth irregularly, over intervals of many years. This is the case, for example, for Lake Lanao in the Philippines (Lewis 1973) and some Indonesian (e.g., Ruttner 1931) and African (Baxter et al. 1965) lakes. In principle, one can imagine a situation where a temperature gradient can be stable over long times without the stabilizing influence of salinity, e.g., if the heat being transferred downward from the surface waters is removed at depth to the sediments or consumed by cold groundwater inflows. This possibility (that may be termed "thermogenic meromixis"; Katsev et al. 2010), however, is rarely considered, and stable lakes without a salinity gradient have been traditionally assumed to be oligomictic, i.e., mixing at irregular intervals (Hutchinson 1975). Temperature differences in warm water correspond to greater density differences than in cold water. The same temperature differences in tropical lakes therefore produce stronger density stratification than in temperate lakes. For example, a temperature difference of 1°C around 25 °C produces a density difference of 0.26 kg m^{-3}, whereas the same difference around 10 °C produces only 0.09 kg m^{-3} (density differences disappear completely at the temperature of maximum density around 4 °C). Differences of a few degree Celsius between surface and deep waters in warm lakes thus can result in stable stratification (Fig. 10.1). The stratification stabilities in the tropical lakes, nevertheless, are generally lower than in seasonally stratified temperate lakes of comparable shape and size, because of considerably smaller temperature differences between surface and deep waters.

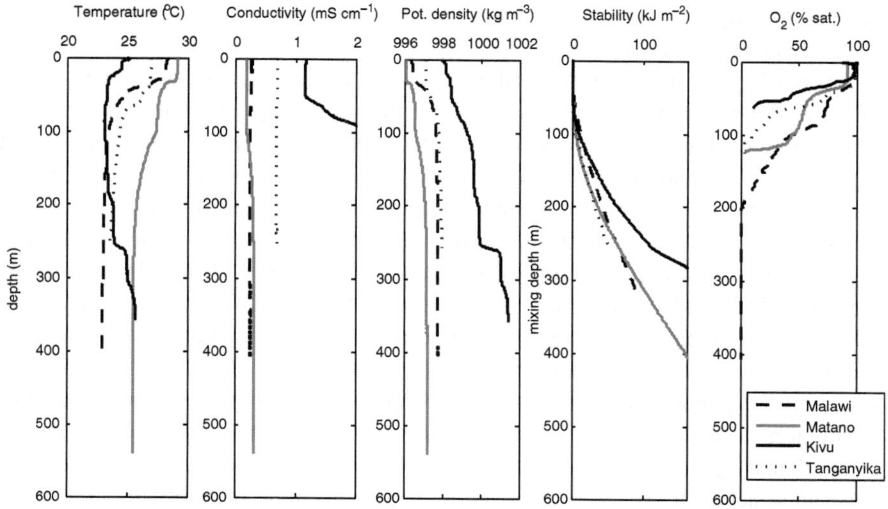

Fig. 10.1 Vertical profiles of temperature, conductivity, potential density, stability, and dissolved oxygen (% saturation) in several tropical meromictic lakes: the East African lakes Malawi (January 2012), Tanganyika (April 1997), and Kivu (October 2012) and Indonesian Lake Matano (January 2009). The stability is defined here as a function of depth as energy per unit area that is needed to mix the water column to that depth. The stability calculation is approximate, as the density calculation depends on a conversion from conductivity to salinity, which was corrected where data on total dissolved solids were available (e.g., Katsev et al. 2010; Wüest et al. 1996), but is not exactly known. Potential density, as opposed to actual in situ density, accounts for the compressibility of water at high pressures and is a better quantity for characterizing the stability of density stratification against buoyancy flows. The potential density for Lake Tanganyika was calculated from temperature only, as matching conductivity data were unavailable. The potential density calculation for Lake Kivu did not take into account the effects of nonionic substances, such as dissolved gases

Stability and mixing—Stability is commonly defined as the energy per square meter that is needed to mix the water column to a homogeneous state down to a specified depth (Hutchinson 1975). Figure 10.1 shows the stabilities (kJ m^{-2}) in several tropical meromictic lakes. The net result of the stronger density dependence on temperature combined with smaller actual temperature differences is a typically lower stability in tropical lakes. Typical wind energies, e.g., on the order of 5 kJ m^{-2}, are sufficient to mix a lake down to 40–80 m (Fig. 10.1), depending on the strength of stratification in a given season. The weakness of the Coriolis effect (deflection of currents due to the Earth's rotation), which normally limits current velocities, further enhances mixing in tropical lakes: by making wind more effective in generating mixing, it increases the depth of the mixed layer (Lewis 1996; Pollard et al. 1973). For example, wind-induced mixing during the (non-windy) stratified season reaches 30–40 m in large African lakes Malawi (Fig. 10.1 and unpublished data) and Tanganyika (Verburg et al. 2011), medium-sized Indonesian Lake Matano (Katsev et al. 2010), and intermediate-sized Lake Kivu (Katsev et al. 2014). Mixing during the windy dry season is deeper (e.g., ~55 m in Lake Kivu and ~80 m in lakes Tanganyika and Malawi).

Estimating the rates of vertical mixing, e.g., by eddy diffusion, below the mixolimnion is often difficult in stably stratified lakes. The constancy of temperature gradients presents a methodological challenge, as many commonly used methods rely on knowing the time derivatives of temperature profiles (Jassby and Powell 1975) and thus cannot be applied under the nearly steady-state conditions. Because such conditions are rare, mixing rates across weak but persistent density gradients in deep waters are poorly documented (Katsev et al. 2010). Tropical lakes are particularly underrepresented in the literature (Lewis 1982; Katsev et al. 2010). Direct measurements of the turbulence microstructure can yield realistic estimates for mixing rates due to turbulent diffusion (e.g., Sommer et al. 2013), and estimates based on the phenomenological correlations of mixing coefficients with stability parameters such as the Brunt-Väisäla frequency (which characterizes the strength of the density gradient) have been also suggested (Katsev et al. 2010). Neither of these approaches, however, is expected to be very accurate for estimating the long-term or whole lake-scale vertical transport of water, as they usually rely on the temperature measurements made at a particular moment of time, and may miss important episodic mixing or convection events (e.g., Katsev et al. 2014). For this purpose, methods based on balancing of the long-term vertical fluxes of substances or heat may be preferred.

Upwelling—In large meromictic tropical lakes, the exchange between the surface mixed layer and deep waters may be in large part facilitated, or even dominated, by coastal upwelling. This is particularly important in long lakes such as the African lakes Tanganyika and Malawi (both in Southern hemisphere), where the climatic gradient along the lake and the predominant wind directions generate a sustained upwelling system at the southern end of the lakes and a downwelling system at the northern end.

Heat budgets—Seasonal variations in irradiance (the total power of incident radiation per surface area) are low in the tropics, ranging typically between 200 and 260 W m^{-2} in the equatorial region (Lewis 1987). The annual heat budgets of tropical lakes are therefore comparatively low, on the order of a few hundreds of kW m^{-2}. Table 10.1 illustrates the components of heat budgets in two tropical

Table 10.1 Components of lake heat budgets (all in units of W m^{-2}). Heat gains are positive; heat losses are negative

	Lake Kivu[a,b]	Lake Tanganyika[c,d,e,f]
Solar (shortwave) radiation	229 ± 10	227
Longwave radiation	327 ± 5	379
Evaporation (latent heat)	−112 ± 12	−(123 − 184)
Longwave radiation loss from surface	−427 ± 5	−445
Loss to sensible heat	−7 ± 2.5	−(9.7 − 6.8)
Loss to precipitation and runoff	−1.4 ± 1	

References: [a]Aaberg (2013); [b]Katsev et al. (2014); [c]Verburg and Hecky (2009); [d]Verburg and Hecky (2003); [e]Verburg and Antenucci (2010); [f]Verburg et al. (2011)

meromictic lakes. The annual mean heat gain through net radiation (shortwave incoming + longwave incoming—longwave outgoing) is on the order of 130–160 W m^{-2} and is removed predominantly through evaporation (110–140 W m^{-2}). The net result of all heat gains and losses (sum of all components in Table 10.1) balances at nearly zero, within the uncertainties of estimation that are typically on the order of 5–10 % of the total budget (=several tens of kW m^{-2}). Meromictic conditions typically imply that only a small portion of the heat is transferred into the monimolimnion (through turbulent diffusion or deep-water exchange mechanisms), with most heat exchanges involving only the mixed layer. The monimolimnia thus remain effectively decoupled from short-term fluctuations in heat energy at the lake surface, and their heat content tracks longer-term trends, including those induced by climate changes.

Though tropical lakes experience relatively small variations in heat fluxes, the amount of heat exchange required to produce important changes in stability by convection is smaller there than in lakes at higher latitudes (Lewis 1996) (Fig. 10.1). This leads to a highly dynamic thickness of the mixed layer and a highly variable vertical position of the thermocline(s). Periods of cloudy weather can lead to a loss of a significant portion of the heat budget, erosion of the thermocline, and thickening of the mixed layer. Subsequent restoration of heat gain may lead to re-establishment of a thinner mixed layer marked by a secondary thermocline. Two thermoclines—a "permanent" one corresponding to the typical depth of maximum mixing and a seasonal one that forms during the stratified (wet) season—are common. Diurnal changes may establish yet another, tertiary thermocline (Katsev et al. 2010), typically in the upper 10–25 m, with deeper mixing occurring at night. Because of strong changes in water density with temperature, convection caused by nocturnal heat loss may be substantial. The thickness of the mixed layers during the stratified season in tropical lakes thus can vary by tens of meters (Lewis 1996). These changes, in turn, may strongly affect the positions of the main chemical gradients within the water column.

10.1.2 Geochemical and Ecological Features

Chemical stratification—Lack of a complete vertical turnover and the resultant separation of the water column into the surface mixed layer and the permanently isolated monimolimnion lead to strong differences in water chemistry (see also Chap. 3). Dissolved salts tend to accumulate in the monimolimnion (Fig. 10.1), enhancing the stability against mixing (though their stabilizing effect is not always as crucial in tropical lakes as it is in temperate meromictic lakes). Because oxygen can be supplied only from the atmosphere or by photosynthesis in the illuminated surface layer, consumption of oxygen by microbial respiration below the mixing depth results in a complete depletion of oxygen in the monimolimnion. The oxygen solubility decreases with increasing temperature; thus tropical lakes tend to have lower concentrations of oxygen in the surface waters than temperate lakes. As biochemical oxygen demand also normally increases with temperature, oxygen in tropical lakes tends to be consumed faster with depth. The exact position of the oxic-anoxic boundary below the mixing depth is determined by the rates and regimes of mixing (and deep-water renewal). In large lakes, like Lake Malawi and Lake Tanganyika (East Africa) where deep-water renewal by large-scale circulation takes place, oxygen may penetrate hundreds of meters below the wind-mixed layer (Verburg et al. 2003). In smaller lakes like Lake Matano (Indonesia), oxygen is exhausted within several meters of the permanent thermocline (Crowe et al. 2008a), whereas in strongly stratified small systems, like in the Kabuno Bay of Lake Kivu, the oxycline may be as narrow as several centimeters (Borges et al. 2011).

Anoxic conditions in the monimolimnion and the slow transport across the pycnocline density gradient favor the accumulation of reduced substances in the monimolimnion. Whereas respiration in the oxic surface waters is dominated by aerobic metabolisms, microbial respiration in the anoxic deep waters uses weaker electron acceptors such as sulfate, iron and manganese oxides, or CO_2 (methanogenesis). Reduced products such as hydrogen sulfide often accumulate. A sequential exhaustion of electron acceptors and toxicity of products to microbial populations leads to a vertical succession of chemical gradients and metabolic processes within the stratified water column (Canfield and Thamdrup 2009). These distinct layers serve as habitats for specially adapted microbial communities and delineate the vertical ranges of the geochemical reactions that those microbes catalyze.

The deep monimolimnetic waters also remain excluded from gas exchanges with the atmosphere. Gases such as CO_2, H_2S, or CH_4 often accumulate in concentrations far exceeding their concentrations in the mixolimnion, aided by the increases in gas solubility with hydrostatic pressure (Crowe et al. 2008a, b, 2011). Under certain conditions (e.g., where sublacustrine inputs of gases are aided by active volcanism), the increasing gas concentrations may pose a danger of a catastrophic "limnic eruption", an uncontrollable overturn of the lake accompanied by the release of large volumes of toxic gases to the atmosphere. In lakes Monoun and Nyos in Cameroon, limnic eruptions in 1982 claimed several thousand human lives (Kling et al. 1994; Sigurdsson et al. 1987). In Lake Kivu at the border of Rwanda and the Democratic Republic of the Congo (Descy et al. 2012a, b; Schmid et al. 2003, 2005), the CH_4/CO_2 levels in the monimolimnion are being monitored for the hazard of limnic eruption. The dissolved gases are also harnessed for energy generation, through the process of methane extraction. The permanent isolation and stagnation of deep waters may also serve to create environmental conditions that are similar to those that are thought to have dominated the early Earth, as recently investigated in both temperate lakes stratified by salinity (Lehours et al. 2007; Comeau et al. 2012; Klepac-Ceraj et al. 2012) and tropical lakes (Crowe et al. 2008a, b, 2014a, b).

Nutrients—The concentrations of nutrients, e.g., ammonium (NH_4^+) and phosphate (PO_4^{3-}), in the deep anoxic waters typically greatly exceed their levels in the mixolimnion, and the upward fluxes of these substances become important factors that affect biological productivity. Anoxia in tropical lakes stimulates the release of phosphorus (P) in the deep waters, as P bound to iron (oxyhydr) oxides is released upon their reduction (e.g., Katsev et al. 2006). The enhanced regeneration of phosphorus, together with the removal in the anoxic waters of reactive nitrogen (N) through denitrification, causes low N/P ratios in the deep waters. Mixing of these deep waters into the surface waters, e.g., by upwelling, may then create a nitrogen deficit, which is met through N fixation. This is often associated in tropical lakes with a proliferation of N-fixing filamentous cyanobacteria (Sarmento 2012; Hecky 2000).

Nutrients are critical regulators of biogeochemical processes, as the rates of supply of limiting nutrients directly affect the rates of photosynthetic and respiratory processes. Nutrient limitation in tropical aquatic environments tends to be stronger than in temperate lakes. The greater efficiency of nutrient recycling (both in the epilimnion and monimolimnion) also makes tropical lakes more sensitive to the external nutrient inputs. The degree of nutrient limitation can fluctuate seasonally, depending on factors such as rainfall (thus being susceptible to climatic drivers such as the Indian Ocean surface water temperatures; Tierney et al. 2013; Johnson and Odada 1996; Plisnier 2000; MacIntyre 2012) or the development of wind-induced temperature-density gradients that affect the nutrient fluxes to the photic zone (Hecky and Kling 1987; Spigel and Coulter 1996). The recycling rates of N and P in the photic zone may be a limiting factor for the growth rates of autotrophs (Lewis 2000), especially of ammonia-oxidizing Archaea with respect to cyanobacteria.

High temperatures favor high rates of N fixation in the euphotic zone but also high rates of denitrification in the deep anoxic waters, which reduces the N pools available to consumers (Lewis 2002, 2010). Accordingly, N often becomes a limiting element (Sarmento 2012), though P, inorganic carbon (C), and silicon (Si) can also become limiting. Limitation by inorganic C can only take place in lakes with high photosynthetic rates and large communities of autotrophic microbes (Talling 1976). Si limitation is less common and typically affects only diatoms. Si generated by weathering of rocks at high tropical temperatures is delivered to lakes in relatively large supply, primarily by rivers (Talling 1992). There is documented evidence, however, that in large lakes such as Lake Tanganyika, Si limitation becomes important (Verburg et al. 2003; Alleman et al. 2005; Descy and Sarmento 2008) and that Si/P ratios decrease with increasing lake size.

Biological productivity—Primary production in tropical lakes is about twice as high on a given nutrient base as in temperate lakes, mainly because of efficient nutrient and carbon recycling combined with higher temperatures and higher solar irradiance. In general, tropical lakes seem to be characterized by lower concentrations of life-supporting substances (nutrients, dissolved organic matter (DOM), etc.) but higher turnover rates. The temperature effect on biological production rates is often described with a Q_{10} factor of 2 (i.e., rates increase by a factor of 2 for a 10 °C temperature increase) for photosynthesis under nutrient-saturated conditions. The productivity, however, is often limited by nutrients. The efficiency of nutrient recycling is enhanced in tropical lakes by high temperatures and a highly dynamic mixed layer, as the vertical movements of the mixing depth facilitate the return of nutrients to the photic zone. These features complicate classifications of the tropical lakes based on their trophic status. For example, Lake Malawi could be classified as highly oligotrophic based on the concentrations of relevant chemicals but as mesotrophic to eutrophic based on primary production (e.g., Kruger 2014).

In tropical aquatic environments, high water temperatures (above 25 °C) and high incident light levels (of photosynthetically active radiation (PAR) that sustains primary production and of ultraviolet radiation (UVR) that stimulates a more efficient breakdown and recycling of organic material in the upper water column) result in the conditions of "endless summer" (Kilham and Kilham 1990). Such conditions support high metabolic rates, including the rates of bacterial production (BP) and bacterial respiration (BR), which in turn lead to higher rates of nutrient uptake. High respiration rates cause lower bacterial growth efficiencies (BGE) while increasing the rates of cell lysis and excretion of phytoplankton-derived DOC, which in turn also supports respiration (Zlotnik and Dubinsky 1989; Obernosterer and Herndl 1995; Myklestad 2000; Sarmento 2012; Merbt et al. 2012; Amado et al. 2013; Morana et al. 2014).

The supply of dissolved organic carbon (DOC) from both allochthonous and autochthonous sources is important for sustaining microbial growth (Sarmento and Gasol 2012; Pérez and Sommaruga 2006). Dependence of heterotrophic microbes on phytoplankton activity through DOC release has been shown in many aquatic ecosystems by positive correlations between particulate primary production (pPP) and bacterial production (BP) (Cole et al. 1988; Fouillar and Mostajir 2010). Given their efficient primary production and high nutrient recycling rates, tropical lakes experience higher heterotrophic microbial activity for given primary production levels (see Fig. 2 in Sarmento 2012). The dark monimolimnia in tropical stratified lakes, separated from the photosynthetically active surface waters, may be among some of the lowest-energy-flux environments known (Chapelle and Lovley 1990), but their metabolic potentials (and potential functional diversity) can be high (Rinke et al. 2013; Gies et al. 2014).

Bacterial production (BP) and respiration (BR) are typically positively correlated in both tropical and temperate freshwater lakes (Amado et al. 2013). The tropical lakes, however, show higher rates of both BP and BR, as well as of bacterial carbon demand (BCD, understood as BP+BR) (see Fig. 3 in Amado et al. 2013). The bacterial growth efficiency (BGE, understood as the ratio BP/BCD) in tropical lakes, on the other hand, is typically lower. This may be aided by several additional environmental conditions related to both bottom-up (nutrient availability and stoichiometry) and top-down (grazing pressure) processes, as well as physical processes (Amado et al. 2013). The BR/BP ratios in tropical lakes are estimated to be at least two times higher than in temperate lakes (Amado et al. 2013), while recent analyses (H. Sarmento, pers. comm.) point toward even higher ratios. Whereas high temperatures generally increase the rates of BP, BR, and bacterial specific growth rates (e.g., as demonstrated in oceanic samples; White et al. 1991; López-Urrutia and Morán 2007; Vázquez-Domínguez et al. 2007), they do not necessarily lead to higher bacterial growth efficiencies, as epilimnetic tropical waters often suffer from nutrient limitation (López-Urrutia and Morán 2007; Sarmento et al. 2010). Though currently available data on bacterial processes in temperate and tropical lakes (Amado et al. 2013) provide evidence for slightly higher values for BP at low latitudes, further studies are needed to fully understand the factors that regulate tropical microbial metabolisms.

In warm environments, small-sized phytoplankton (phototrophic picoplankton) may be more efficient than heterotrophic microbes in uptaking inorganic nutrients (Morris and Lewis 1992; Moutin et al. 2002). A tighter coupling between phytoplankton and heterotrophic microbes is therefore expected, resulting in tropical food web networks being significantly different (i.e., higher dissolved organic matter releases and remineralization rates) from those in temperate freshwater environments (Sarmento 2012). Under similar trophic conditions (e.g., similar Chlorophyll *a* concentrations), abundances of heterotrophic microbes in tropical lakes are lower than in temperate lakes (Sarmento 2012).

Biological stratification—The physical separation of mixolimnion and monimolimnion and the resultant chemical stratification result in the vertical layering of biological communities, often characterized by clear and sharp transitions. The differences are especially strong in the microbial fractions. Surface waters typically harbor less microbial diversity than deep waters (Lehours et al. 2005; Comeau et al. 2012; Klepac-Ceraj et al. 2012; Wright et al. 2012). The presence of oxygen, nutrients, and light in the mixolimnion contributes to the development of microbial communities that include both autotrophic [cyanobacteria, algae, ammonia-oxidizing archaea (AOA)] and heterotrophic groups (*Alpha*-, *Beta*-, and *Gammaproteobacteria*, *Actinobacteria*, *Bacteroidetes*, *Verrucomicrobia*, or uncultured representatives of Archaea) (Llirós et al. 2010, 2012; Newton et al. 2011; Villanueva et al. 2014). The groups reported for the mixolimnetic waters are often similar in temperate and tropical lakes. The monimolimnetic bacterial and archaeal communities may be represented by: *Delta*- and *Epsilonproteobacteria*, *Chloroflexi*, archaeal methanogens (e.g., acetoclastic), and several uncultured and new candidate divisions affiliated with Archaea or Bacteria (Daffonchino et al. 2006; Berdjeb et al. 2011; Llirós et al. 2011; Hugoni et al. 2013; Yakimov et al. 2013). The microbes in the warm, stagnant, and anoxic monimolimnia depend on metabolic processes that in turn depend on terminal electron acceptors (TEAs) other than oxygen. These organisms may exhibit relic metabolic characteristics, representing windows to the environmental conditions of the early Earth (e.g., Crowe et al. 2008a, b). Recent evidence points also toward the differences in microbial cell sizes between temperate lakes and tropical lakes. Tropical aquatic environments may harbor smaller cells (with lower width-length ratios) than temperate lakes (Hernández-Avilés et al. 2012). This may be related to stronger nutrient limitation in the tropics: smaller cells have higher surface-to-volume ratios and may be therefore better poised for success in environments that have tight nutrient recycling and uptake coupling and low ambient nutrient concentrations.

To the best of the authors' knowledge, no holistic (i.e., organism- and system-based) studies have been conducted in tropical freshwater meromictic lakes on the coupling between the microbial and macrobial networks [e.g., see a review of food web studies in Sarmento (2012)]. This is in contrast to marine ecosystems, as recently reviewed (Fuhrman et al. 2015 and Worden et al. 2015 and references therein).

Microbial communities in redoxclines—The microbial communities that develop within sharp (often as narrow as few meters or even centimeters) redox gradients (redoxclines), such as at the oxic-anoxic transitions, are highly diverse (i.e., Camacho et al. 2001; Grote et al. 2008; Glaubitz et al. 2010; Casamayor et al. 2000; Wright et al. 2012). These communities share, as a common trait, an important transitory repertoire of TEAs and metabolic capacities. At the same time, chemical gradients present at the redoxclines show overlapping limits reflecting transition zones (see discussion in Canfield and Thamdrup 2009) where the reduction of nitrate, manganese, iron, and sulfate takes place. Planktonic microbial species can take advantage of such chemical gradients as can be observed in several temperate (e.g., Camacho and Vicente 1998; Tonolla et al. 1999; Rodrigo et al. 2001; Llirós et al. 2008; Schubert et al. 2011) or tropical lakes (e.g., Llirós et al. 2010; Buckles et al. 2013). Interestingly, several microbial groups have been found exclusively at the depths of transition redox gradients. Among the most interesting ones are anoxygenic phototrophic bacteria belonging to the green sulfur bacteria (GSB), which have been described in such distinct environments with varying relative abundance but with putative high contributions to microbial processes at those depths (see some examples in Crowe et al. 2008b and Llirós et al. 2015). These redoxcline microbial communities may represent those microbes that developed major capabilities and adaptations to cope with the TEA availability and sharp gradients.

The use of TEAs alternative to oxygen in microbial metabolisms results in the production of dissolved gases, including climate active gases (e.g., CO_2, N_2O, CH_4; Paulmier and Ruiz-Pino 2009). Marine oxygen minimum zones and transition zones can account for a large fraction (>50 %; Gruber and Sarmiento 1997; Codispoti et al. 2001; Hawley et al. 2014) of global marine N_2 production, limiting primary production in surface waters. The N fixation and loss processes in the vicinity of transition zones can be balanced and coupled (Deutsch et al. 2007). The N cycle can be also coupled to sulfur (S) transformations, including a so-called cryptic sulfur cycle with potential to drive inorganic carbon fixation (Canfield et al. 2010). The roles and contribution of yet-uncultured microbial taxa that may couple the C, N, and S biogeochemical cycles in lakes remain to be determined and deserve further investigations.

In comparison to communities in temperate waters, heterotrophic microbes in tropical lakes may strongly depend on the rates at which labile organic carbon derived from PP becomes available to fuel BP. The microbes that predominantly use autotrophic metabolisms actively uptake CO_2 from the atmosphere and transport it downward in the form of remineralized carbon.

Microbial communities in the illuminated surface waters are sensitive to the amounts and spectrum of sunlight [including UVR (280–400 nm) and PAR (400–700 nm)] (Herndl et al. 1993; Kaiser and Herndl 1997; Sommaruga et al.

1997, 1999). PAR is responsible for carbon fixation through photosynthesis, whereas UVR affects the rates of cellular and molecular damage that cause mutations, cell inactivation, or death. Due to their small size (García-Pichel 1994) and cell volume (Jeffrey et al. 1996), microbes are among the most sensitive organisms to photodamage. In particular, UVR affects the in situ single-cell microbial activity such as DNA or protein synthesis and oxygen consumption (Bailey et al. 1983; Herndl et al. 1993; Müller-Niklas et al. 1995; Sommaruga et al. 1997). Microbes differ in their sensitivity to natural levels of incident solar radiation (Herndl 1997; Alonso-Sáez et al. 2006; Ruiz-González et al. 2012; Merbt et al. 2012). For example, members of *Gammaproteobacteria* and *Cytophaga-Flavobacteria-Bacteroidetes* groups are highly resistant to UVR and PAR, whereas members of *Alphaproteobacteria* are more sensitive: increased solar radiation causes a reduction in their metabolic capabilities (Alonso-Sáez et al. 2006; Ruiz-González et al. 2012). Little is known about the sunlight exposure response of Archaea (but see Merbt et al. 2012). Microbes may exhibit different irradiance thresholds, acclimatization processes, or the use of light-driven proton translocation mechanisms (e.g., rhodopsins) acquired through horizontal gene transfer (see Spudich et al. 2000; Frigaard et al. 2006; Merbt et al. 2012).

Table 10.2 provides a (simplistic) summary of the differences between deep temperate and deep tropical lakes.

10.1.3 Responses to Change

Responses to climate warming—Studies in the past two decades documented a widespread response of nearly all large tropical meromictic lakes to climate warming (Vollmer et al. 2005; Verburg and Hecky 2009; Katsev et al. 2014). The warming rates differ among lakes (Kraemer et al. 2015b), but generally warming rates in the mixed layer keep pace with atmospheric warming. Lakes warm primarily through increased fluxes of the incoming longwave radiation (Table 10.1), though other parts of their heat budgets (e.g., evaporation rates that depend on air humidity) may also be affected (Verburg and Hecky 2009). The mixed-layer temperatures in East African lakes have been increasing in the twentieth century (with a hiatus around 1940–1975) at rates of about 0.06–0.20 °C per decade: 0.13 in Lake Tanganyika (Verburg and Hecky 2009), 0.06 ± 0.02 in Lake Malawi (at 100 m; Vollmer et al. 2005), and 0.12–0.2 in Lake Kivu (at 78–54 m; Katsev et al. 2014). These surface warming rates do not significantly differ from those in non-meromictic lakes, e.g., 0.1 °C per decade in Lake Victoria (Hecky et al. 1994) and 0.15 °C per decade in Lake Albert (Lehman 1998). They approximately match the regional atmospheric warming rates: 0.05 °C per decade around Lake Malawi (Vollmer et al. 2005), 0.13 °C around Lake Tanganyika (O'Reilly et al. 2003; Verburg and Hecky 2009), and 0.22 °C around Lake Kivu (Katsev et al. 2014). Warming between the 1970s and early 2000s has been faster than on average over the past century.

Table 10.2 Simplified comparison between deep temperate and deep tropical lakes

	Temperate	Tropical
Solar radiation	Moderate, seasonally variable	High year-round. Moderated by clouds during wet season
Precipitation	Throughout the year; over ice accumulates as snow	Highly seasonal (wet season)
Air temperature	Seasonably variable Wide range	Limited annual cycle Small range
Water temperature	Decreases with depth in summer; increases with depth in winter. Bottom waters near 4 °C	Decreases with depth within limited range. Bottom waters near average minimum air temperatures
Growing season	Mainly restricted to stratified periods	Throughout the year
Lake stratification	Seasonal; one or two mixing periods. Oxygen depletion may occur in eutrophic lakes	Year-round, with low stability and variable thermocline depths. Persistent oxygen depletion occurs irrespective of trophic status
Nutrient limitation	Impacted by nutrient availability from watershed	Impacted by nutrient regeneration from anoxic deep waters; tight coupling of production and remineralization
Productivity	Low to high	Typically high
Nutrient uptake	Low to moderate	High
Metabolic rates	Low to moderate	High
Productivity/biomass ratio	Low to moderate	High

As the monimolimnia of meromictic lakes are not in direct contact with the atmosphere and experience only the downward heat fluxes from the mixed layer above, they effectively average the epilimnetic heat fluxes and can be treated as decadal-scale heat monitors. As heat fluxes take time to propagate, the warming rates in the deep waters should typically be slower than in the surface waters. Although in some cases, e.g., Lake Kivu (Katsev et al. 2014), the monimolimnion heat budgets can be significantly affected by groundwater inflows or hydrothermal heat, in most cases warming in the deep waters is likely to reflect the long-term trends in the atmospheric processes and air-water exchanges. Deep-water temperatures have increased at the rates of 0.02–0.05 °C per decade in Lake Tanganyika at 1000–600 m (Verburg and Hecky 2009; Verburg et al. 2003) and 0.12 °C per decade in Lake Malawi at 300 m (Vollmer et al. 2005).

The net heat flux into the lake may vary depending on specific factors at each lake, but it is generally reported to vary around 0.2 W m^{-2}. For example, the net heat flux into the monimolimnion was estimated in Lake Kivu at 0.18 W m^{-2} (Katsev et al. 2014) and in Lake Matano at 0.2 W m^{-2} (Katsev et al. 2010). The net heat absorption in the (entire) water column of Lake Tanganyika was estimated at about 0.4 W m^{-2} (Verburg and Hecky 2009). For comparison, in the global ocean, the rates of heat uptake associated with global warming have been estimated at 0.14 ± 0.04 W m^{-2} (for the 0–700 m layer between 1961 and 2003), corresponding to an average heating rate of 0.21 ± 0.04 W m^{-2} for the Earth surface (IPCC Fourth Assessment Report 2007).

Effects of warming—Warming of the tropical lakes and the typically greater rates of warming near the surface than in deep waters have been linked to several physical and ecological consequences. The increased temperature differences between surface and deep waters lead to stronger density stratification that reduces vertical water exchange. Though surface water temperatures in tropical lakes may be increasing slower than in temperate lakes (Austin and Colman 2007), the nonlinear dependence of water density on temperature causes stronger stratification changes in tropical lakes, with African Rift lakes and the ancient Indonesian lakes being most susceptible (Kraemer et al. 2015a). Stronger stratification results in the reduced delivery of nutrients from monimolimnion to photic zone. For example, in Lake Tanganyika, increase in stability of the stratification between the 1970s and 2000s has resulted in restricted water exchange, reduction of the oxygen penetration depth, as well as reduced delivery of nutrients to the photic zone (Verburg and Hecky 2003). This was hypothesized to have decreased the productivity of the lake, backed up by several lines of evidence. There is evidence that phytoplankton composition in lakes Tanganyika and Kivu may have recently shifted because of warming as well as species introduction (Descy and Sarmento 2008).

Climate projections for East Africa call for warming rates on the order of several degree Celsius over the twenty-first century (Mitchell and Hulme 2000; Hulme et al. 2001). In lakes whose productivity depends upon the delivery of nutrients from their deep monimolimnia, such as Lake Tanganyika and Lake Malawi, this warming has been projected to have negative impacts on the biological production. Because of restricted circulation, oxygen penetration into the monimolimnion can be expected to decrease. The exact responses of a particular lake, however, may be complex and dependent on other factors, such as changes in the patterns of humidity and prevailing winds, as they also affect the intensity of vertical mixing. For example, while strong predictions about the adverse effects of warming have been made based on studies in Lake Tanganyika, Lake Malawi's response to climate warming has been much more muted.

In addition to climatic factors, the biological conditions in the lakes may be affected by the rapidly developing agriculture in their watersheds. In particular, both soil erosion and fertilizer runoff may promote eutrophication. There may be feedbacks between the erosion and runoff inputs and nutrient inputs from the deep waters. These will depend upon the net balance from the competing influences of increasing P remobilization in the anoxic monimolimnia (due to decreased oxygen penetration) vs decreasing exchange of surface and deep waters (due to stronger water column stratification).

10.2 Case Studies

10.2.1 Lake Tanganyika

Geography—Lake Tanganyika is the second deepest lake (1470 m) in the world and the second largest by area in Africa. Located between 3 and 9° S, it is 670 km long and about 50 km wide (Fig. 10.2). The depth of this African Rift lake reflects its age: it has existed more or less in its present form probably for the past 5 million years (Cohen et al. 1993). Surrounded by mountain escarpments, the lake has a steep sloping bottom, except near the shelves at the north and south ends (Fig. 10.2). Lake Tanganyika has three basins: the southern basin (1470 m), the northern basin (1310 m), and the central basin (885 m) offshore from the Kungwe mountains, which is separated from the other two basins by sills at about 600 m depth. The largest inflowing rivers are the Ruzizi River in the north and the Malagarasi River in the east. The outflow is the Lukuga River in the west, a tributary of the Lualaba River, which is a headstream of the Congo River (Fig. 10.2). The average lake level is 775 m a.s.l. with an average annual amplitude of 80 cm. The meteorology on and around Lake Tanganyika shows marked differences along the north–south axis, which influences the distribution of energy fluxes across the water surface. Most notably, the highest heat loss to the atmosphere by evaporation occurs in the south.

General limnology—Lake Tanganyika is meromictic (Gonfiantini et al. 1979): below about 100–200 m depth, depending on the basin and season, the lake is permanently anoxic and relatively rich in nutrients (Hecky et al. 1991; Edmond et al. 1993). Its trophic status (Hecky and Fee 1981) is typically described as oligotrophic, with low concentrations of nutrients in its trophogenic layer. Average chlorophyll *a* concentrations are about 1 mg L^{-1} (Salonen et al. 1999). Nutrient inputs from catchment are low relative to the lake volume, and recycling of nutrients within the lake is important for primary production.

The large size of the lake provides for a complex physical limnology, which is important for the functioning of the ecosystem. Seasonal upwelling is important for the recycling of nutrients and sustaining food webs (Coulter 1991a). Upwelling at the southern end of the lake and metalimnetic entrainment occur primarily in the relatively cool dry season (May–September), driven by strong trade winds from the southeast (Coulter 1988). Stimulated by evaporation, epilimnetic temperatures fall to within 1–2° of the deep-water temperatures, decreasing the density gradient and facilitating mixing (Imberger and Patterson 1990). Wind-driven surface currents may disperse the nutrients northward, and internal waves may periodically introduce nutrients from deep waters into the epilimnion at either end of the lake (Coulter and Spigel 1991; Spigel and Coulter 1996). Deep mixing, while bringing nutrients to the photic zone, also carries algae down below the photic zone, so primary production is not immediately enhanced. As the upwelled nutrients enhance production at the end of the cool season (Descy et al. 2005), the lake is most turbid from September to December. This period of low transparency occurs in the north basin about 1–2 months later than in the south. Shortly after the annual

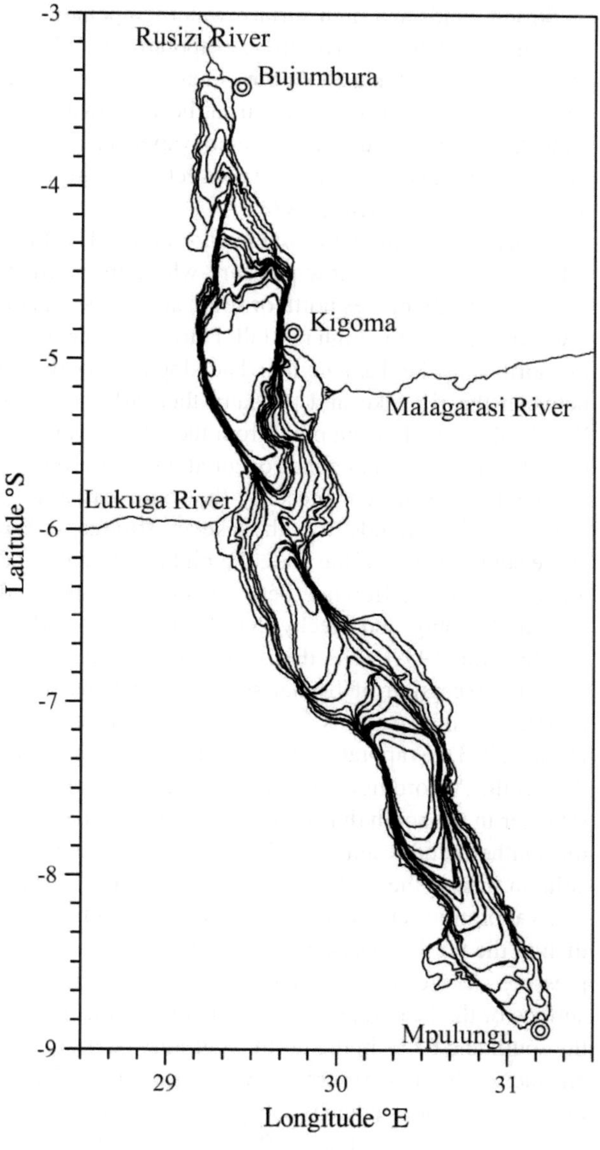

Fig. 10.2 Lake Tanganyika map. Reprinted from Verburg and Hecky (2009), with permission. Copyright 2014 by the Association for the Sciences of Limnology and Oceanography, Inc.

Secchi-depth low, transparency recovers and remains high until July of next year. During the long period of stability in the warm season (October–May), surface waters warm up, and a strong shallow thermocline around 50 m depth is generated more or less uniformly along the length of the lake, preventing vertical exchanges between epilimnion and metalimnion (Coulter 1991a, b, c, d; Hecky et al. 1991). At the onset of this season, stratification and higher light levels enhance algal photosynthesis.

Strong winds and high surface water temperatures, which are on average higher than air temperatures, drive strong evaporative heat losses, which account for 20–28 % of the lake's heat budget (Verburg et al. 2011). At nearly 2000 mm year^{-1} (Verburg and Antenucci 2010), evaporation has been estimated to account for 95 % of the water budget, i.e., 95 % of incoming water evaporates (Spigel and Coulter 1996). The high evaporative heat losses stimulate convective mixing and contribute to the formation of a deep mixed layer (Kling 1988).

Patterns in wind speed and evaporation—The Intertropical Convergence Zone (ITCZ) is a region of low pressure where the northern and southern trade winds meet. The ITCZ moves north or south across the equator with the seasons, bringing low wind speeds and high rainfall. During the dry season the position of the ITCZ is far north of Lake Tanganyika. The sloping coastline and temperature differences between the rift lake and the land then affect the local winds (Savijärvi 1997). Winds of thermal origin blow from the lake during the day, when the air is warmer over the land, and land winds occur at night (Verburg and Hecky 2003). These lake and land winds are enhanced by the south–east trade winds when they are in the same direction. Winds from the lake are stronger at Bujumbura and Kigoma, while at the southern end of the lake, the nighttime land winds are stronger. The daytime air temperature difference between the lake and land is greatest during the dry season (Verburg and Hecky 2003), and lake winds are generally strongest then (Beauchamp 1939). The diel wind patterns appear to be very stable and regular, with the exception of frontal storms during the wet season (Verburg and Hecky 2003), and diel patterns in evaporation follow the wind patterns (Verburg and Hecky 2003): evaporation is high at night in the south and during the daytime in the north. Accordingly, vertical mixing is stronger in the north in daytime and stronger in the south during the night. During the dry season, nocturnal heat loss at the southern end results in a deeper thermocline and better ventilation of the water column than at the northern end (Verburg et al. 2011; Verburg and Hecky 2003).

Evaporation rates reach higher maxima and are higher on average in the south than in the north, consistent with a greater cumulative wind stress and lower vapor pressure. As prevailing southerly winds carry air masses northward across the open surface of the lake, relative humidity during the dry season was on average 55 % in the south and 67 % in the north. Humidity peaks at night, especially at Bujumbura, enhancing the diel pattern in evaporation there (Verburg and Hecky 2003). Solar radiation was on average 9 % higher, and net radiation was 7 % higher at Mpulungu (at the southern end) than at Bujumbura (at the northern end). As annual extraterrestrial solar radiation levels differ by only 1 %, the difference is best explained by a higher cloud cover over Bujumbura.

Although the heat loss by evaporation decreases with increasing humidity and increases with temperature, the effect of wind speed dominates. For example, due to lower wind speeds, the lake surface in the southern basin did not cool down below the air temperature during the dry season of 1996 (Verburg and Hecky 2003), in contrast to the 1993–1995 seasons. In the southern basin, evaporative cooling is strongest in April–June, followed by upwelling, while lower evaporation rates in October–November allow heat gain. The heat content of the upper mixed layer is sustained throughout the remainder of the wet season under even lower evaporation rates. In the Kigoma basin, lateral inflow from the southern basin plays an important part in the cooling and deepening of the mixed layer (Verburg et al. 2011).

The range of surface water temperatures is greater in the south (about 3 °C between January and July), reflecting higher dry season evaporation, upwelling of cooler water by seasonal wind stress, and higher rates of wet season warming (Verburg and Hecky 2003). The more effective exchange with the nutrient rich deep waters that takes place there is important to the productivity of the entire lake.

Heat content and horizontal circulation—Interestingly, the timing of highest evaporation rates in Lake Tanganyika does not coincide with the greatest decreases in the water column heat content (Verburg et al. 2011). The sensible heat and latent heat fluxes (Verburg and Antenucci 2010) greatly differ between the north and south ends of the lake, and these differences also do not agree with the corresponding changes in heat content. The net surface heat fluxes for the two ends of the lake have opposite signs, which can only be explained by a net export of heat from the northern basin to the southern basin. The likely mechanism involves a northward movement of cool upwelled water from the south that is being replaced in the south by the warmer water from the north (Verburg et al. 2011; Fig. 10.3). In such an interbasin exchange, the upwelling at the southern end of the lake can explain the influx of cold water at about 50 m depth in the north basin (Figs. 10.4 and 10.5). A shallower return flow from north to south also agrees with observations (Van Well and Chapman 1976) off Kigoma that currents at 5 m depth were directed southward throughout most of the year, opposite to the general wind direction. The large net export of heat from north to south, driven by SE trade winds and upwelling in the south, upsets the local relationships between the heat exchanges with the atmosphere and the water column heat content. The effect of heat transport with the water masses is most acute in the south where warm water arrives at shallow depths and shuts off the cooler deeper water from the surface, terminating the period of deep mixing. The warm surficial return flow causes the heat content in the south to increase rapidly to a stable level early in the warm season. In the north, the deep mixing period ends later, because the cool water arrives there at the lower boundary of the mixed layer. These differences between basins affect the mixing rates and the dispersal of nutrients throughout the lake after local upwelling events. In particular, the advection of nutrient-rich water from south to north may explain occurrences of algal blooms observed around November in the northern part of the lake, which seem to be more frequent and persistent there (Dubois 1958; Hecky and Kling 1981).

At the end of the warm season in April, wind speeds already pick up substantially in the south before upwelling brings cool water to the surface. These southerly winds force the warm surface water northward. This deepens the mixed layer in the north, as the transported water, being similar in temperature, arrives near the surface and depresses the thermocline. Although the process of the mixed-layer deepening is completed in both basins at the same time (July), in the north it starts 1 month earlier, despite the surface temperatures there reaching their minimum 1 month later than in the south. As described above, at the end of the dry season

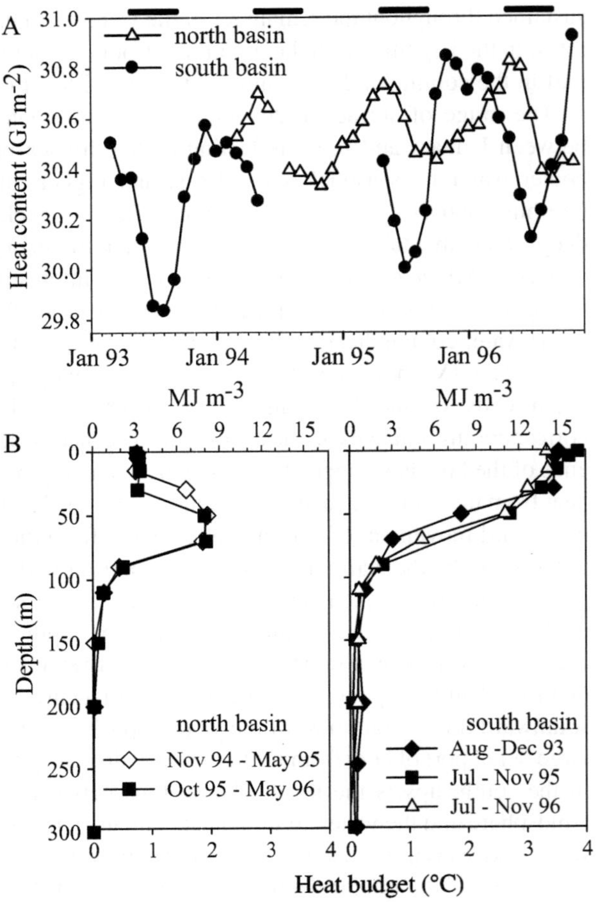

Fig. 10.3 Heat content variations in Lake Tanganyika. (**a**) Monthly mean heat content down to 300 m depth. The bars indicate dry season periods. (**b**) Annual heat uptake between the day of annual minimum heat content and the next day of maximum heat content across the water column in the north and south basins. One degree Celsius change is equivalent to a heat flux of 4.1868 J cm^{-3}. Reprinted from Verburg et al. (2011), with permission. Copyright 2014 by the Association for the Sciences of Limnology and Oceanography, Inc.

(around October), a large amount of cool upwelled water from the south arrives to the northern basin at or slightly above the thermocline depth. The heat content in the 50–70 m depth interval drops, at the time when the surface already starts to warm and the mixed layer becomes shallower.

Nutrients—The inputs of nutrients from deeper waters to the epilimnion are especially important for the cycles of phosphate and silica (Hecky et al. 1991). Exchanges of nitrogen are less important because of nitrogen losses to denitrification at the oxic-anoxic boundary. Denitrification lowers the N/P ratio of the deep waters and creates a strong N demand when nutrients are returned to the mixed layer (Hecky and Kling 1987). Apart from import through rivers and rainfall, nitrogen is thought to be mostly resupplied by fixation by cyanobacteria: *Anabaena*, the nitrogen-fixing cyanobacteria, is common in the lake (Descy et al. 2005). The relative contributions from these sources are open to debate, however. Nitrogen fixation has been estimated to account for as much as 97 % of total nitrogen input,

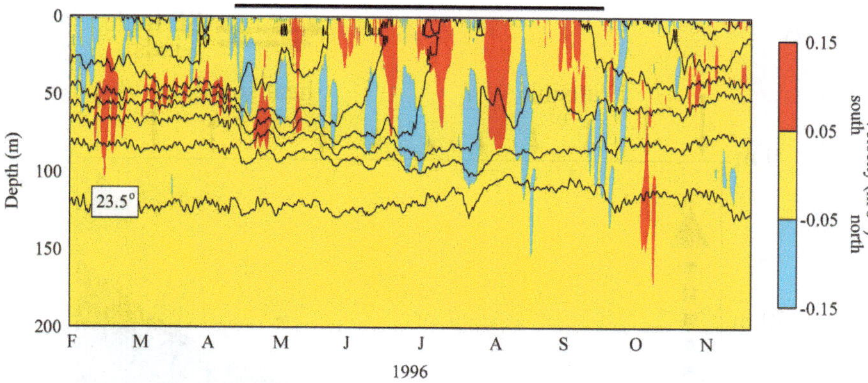

Fig. 10.4 North–south velocity (color contours) simulated at 300 km from the north end of the lake (6° S). Positive values indicate flow to the south. Black lines indicate temperature contours, with the contour interval 0.5 °C and the bottom isotherm 23.5 °C. The bar indicates the dry season. Reprinted from Verburg et al. (2011), with permission. Copyright 2014 by the Association for the Sciences of Limnology and Oceanography, Inc.

whereas about 90 % of phosphorus is supplied by vertical mixing (Hecky et al. 1991).

Deep waters were found to differ in their chemistry between the north and south basins (Edmond et al. 1993). In the deep water of the north basin, large excesses of PO_4 and NH_3 were present in 1975 relative to the south basin. This is probably explained by lower vertical exchange rates with the surface mixed layer in the north. Vertical mixing is deeper in the south because of higher wind speeds and the correspondingly lower surface temperatures during the dry season.

Biology—Fishery is primarily active offshore, targeting mainly pelagic fish species, centropomids and clupeids, and is an important source of protein to the human populations around the lake. The lake fauna is characterized by high diversity and endemicity (Muschick et al. 2014). Species richness is especially high in the littoral zones (Salzburger et al. 2005). In comparison (Hori 1983, 1991; Hori et al. 1993; Brichard 1989), the pelagic food web is a simple one (Dumont 1994; Lehman 1996), with few species and trophic levels (Hecky 1984, 1991).

There are over 250 cichlid species and 50 non-cichlid fish species in Lake Tanganyika. Most fish species are endemic and many of the littoral fish species occur only in a small area of the lake (Verburg and Bills 2007). Many other faunal taxa are diversified into species flocks. There are 13 shrimp species (all endemic), more than 80 mollusk species (Michel et al. 2003), 7 endemic sponges, 8 crab species (Marijnissen et al. 2008), probably more than 100 endemic ostracod species (Schon et al. 2014), etc. Medusae of a species of jellyfish (*Limnocnida tanganyicae*) can occur in dense masses (Salonen et al. 2012). Many species remain to be described. There are 39 cyclopoid and one calanoid zooplankton species (Coulter 1991d). Cladocerans are absent from the zooplankton, which is unusual for most temperate lakes. It is thought that *Stolothrissa* is too efficient a predator for the relatively large cladocerans to survive in Lake Tanganyika. The average size of

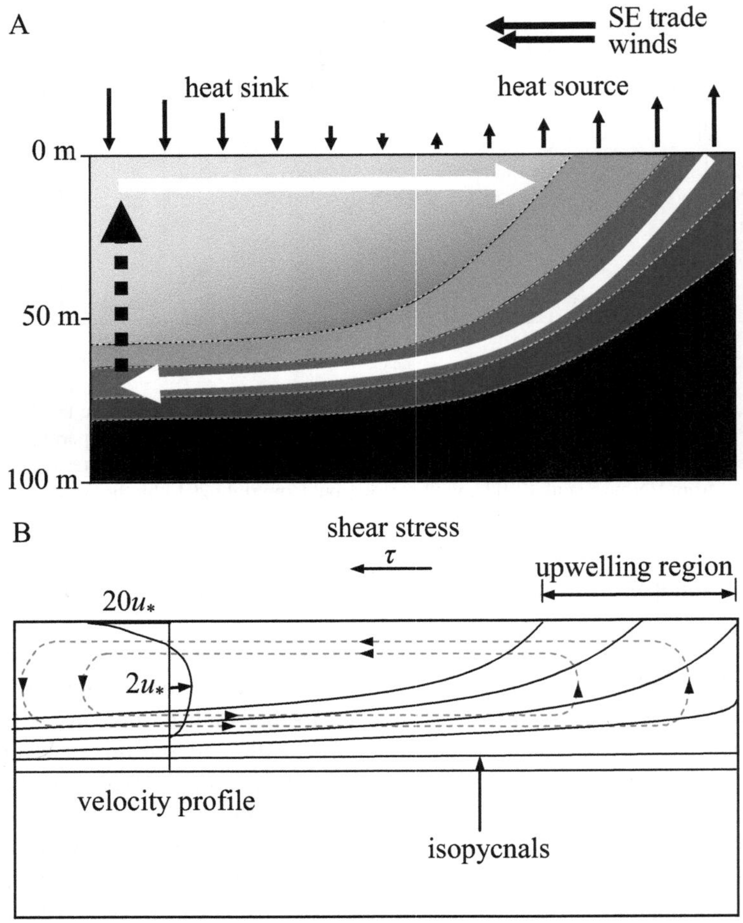

Fig. 10.5 (**a**) Diagram of the large-scale convective interbasin circulation in Lake Tanganyika, with the distribution of heat sinks and heat sources to the atmosphere indicated. Relatively cool water flows in the metalimnion in the downwind direction, while a return current of warmer water moves upwind to the south in the upper layer of the epilimnion. At the downwind end of the circulation, entrainment by vertical mixing (dashed arrow) enables transport across the thermocline. B ~ 1 or less. (**b**) Diagram of wind-driven circulation in lakes with mixed-layer deepening due to upwelling and a downwind current in the surface layer. B >> 1. Adapted from Monismith (1986). Reprinted from Verburg et al. (2011), with permission. Copyright 2014 by the Association for the Sciences of Limnology and Oceanography, Inc.

adult zooplankton in Lake Tanganyika is much smaller than in most temperate lakes, probably as a result of high predation pressure. The major consumers in the pelagic food web are two clupeid species (herrings) and four Nile perch species (*Lates* spp.: Centropomidae). The three larger *Lates* species have become relatively rare since industrial fisheries started in 1952 (Coulter 1991b, c). Only *Lates stappersi* is still caught in great quantities, especially in the south. *Lates stappersi* is pelagic throughout its life cycle and so is *Stolothrissa tanganyicae*, the most abundant clupeid. The three larger *Lates* species and *Limnothrissa miodon*, the other clupeid species, have an inshore juvenile phase.

Climate change effects and long-term consequences—Evidence for the effects of climate warming over the last century is limited in the terrestrial tropics (Hecky et al. 1994; Mizuno 1998; Tyson et al. 1999; Verburg et al. 2003), but consequences of past and future warming may be very important for deep tropical lakes (Hecky 1995). In Lake Tanganyika, since 1913, the deep-water temperatures (c. 1000 m) have increased by 0.2 °C, while shallow water temperatures (upper 100 m) increased by nearly 1 °C. This differential heating has increased the density gradient, restricting vertical mixing, reducing nutrient fluxes to the photic zone, and decreasing primary production. By comparing the 1994–1997 and 2000 temperature profiles at the northern basin with historical data, Verburg et al. (2003) investigated the ecological effects of these changes in the vertical density structure. The water transparency (Secchi depth) has increased between 1913 and 1996; the underwater light attenuation correspondingly decreased by 23 % since 1975. The euphotic zone, defined as the depth at which light levels are 1 % of their surface values, deepened from 16 m in 1913, 33 m in 1975, to 42 m in 1996 (Verburg et al. 2003). The increased transparency is most likely explained by a decrease in phytoplankton biomass. Dissolved organic carbon (DOC) and its associated absorbance are low in Tanganyika and other tropical lakes (Ramlal 2002). The phytoplankton biomass in the northern basin is now smaller than in 1975 (Hecky and Kling 1981; Verburg et al. 2003; Vuorio et al. 2003). The decrease in diatom biomass, in particular, was evidenced by the epilimnetic accumulation of silica. As diatoms have incorporated less silica, its concentrations have tripled since 1975 (Verburg et al. 2003). Silica concentrations decreased in the bottom sediments, a further evidence for reduced productivity by diatoms (Tierney et al. 2010).

Oxygen concentrations and penetration depth into the water column have decreased in Lake Tanganyika. Concomitantly, H_2S has accumulated in the monimolimnion, as SO_4 reduction rates have increased (Verburg et al. 2003). As H_2S is now found closer to the surface than earlier in the century (Verburg et al. 2003), the habitable depth range of deep-dwelling species has decreased. Strengthening of the stratification affects the oxygen distributions in two opposing ways: less oxygen is transported downward, and less oxygen is consumed below the thermocline by microbial respiration (Keeling and Garcia 2002) as inputs of organic matter from primary production decrease. Oxygen levels decrease when the restriction of the downward oxygen fluxes is stronger than the decrease in oxygen consumption.

Climate change has not strongly affected the water level in Lake Tanganyika, which was at 774 m in both 1913 and 2000. In Lake Issyk-Kul in Kyrgyzstan, a 3 m drop in lake level may have enhanced surface water salinity and reduced stratification (Peeters et al. 2003), the opposite effect of what was observed in Lake Tanganyika. A change in level of only a few meters should not have a noticeable effect on salinity and stratification in Lake Tanganyika, though there is evidence that a ~10 m transgression in the mid-nineteenth century correlated with a temporary decline in primary production (McManus et al. 2015). At only one-tenth of

Issyk-Kul's salinity, Lake Tanganyika has always been relatively fresh (Haberyan and Hecky 1987). The lake level was lower by about 20 m around 1000 BP but never lower since then (A.S. Cohen pers. comm.). During the Medieval Warm Period (ca. AD 1000–1270), East Africa was very dry (Verschuren et al. 2000). With the lake level 20 m lower than at present, there could not have been an outflow, possibly for many years, and losses by evaporation must have been high. It is possible that much of the water at present in the hypolimnion of Lake Tanganyika left the surface during that period (Verburg and Hecky 2009; Craig et al. 1974).

Climate warming is expected to affect Lake Tanganyika differently than the much shallower non-meromictic Lake Victoria where increased nutrient loading by runoff over the past century has led to eutrophication (Hecky 1993). In Tanganyika, deep mixing dominates P loading to the euphotic zone (Hecky 1991). As a result, decreasing mixing rates are likely to reduce the productivity of Lake Tanganyika further. Projected air temperature changes for the Tanganyika region are about 4.5 °C during this century or about seven times the mean global change over the past century (Mitchell and Hulme 2000; Hulme et al. 2001). Such air temperatures are likely to continue to increase the stability of the lake's water column, reduce nutrient loading from deep waters, reduce the depth of oxygen penetration, and diminish the productivity of the lake. As most of the lake is >100 m deep, nutrients in the deep pelagic region are responsible for sustaining most of the fish biomass. Potential fish yields from Lake Tanganyika are therefore likely to decline, affecting the availability of food to populations around the lake.

10.2.2 Lake Malawi

Background—Lake Malawi (Nyasa), situated in the East African Rift system, is the fourth largest freshwater lake in the world by volume (7775 km^3), the eighth largest by surface area (29,600 km^2), and the second deepest (>700 m) lake in Africa. It is 560 km long and 75 km wide (Fig. 10.6) (Eccles 1974). With surface elevation at ca. 500 m a.s.l., the deepest (ca. 700 m) point of the lake lies 200 m below sea level (Beauchamp 1953). Along with lakes Tanganyika and Baikal, Lake Malawi is one of the oldest lakes in existence. It is generally believed that the Malawi Rift Basin began formation in the late Miocene, roughly ~8.6 million years ago (although earlier rifting events are also believed to have occurred), and deep-water lake conditions may have existed for the past ~4.5 million years (Cohen et al. 1993; Ebinger et al. 1993; Delvaux 1995; Contreras et al. 2000). Nearly continuous sedimentation over this period has resulted in a pile of organic-rich sediment over 4 km deep (Cohen et al. 1993), which contains valuable paleolimnological records of regional climate. The lake has been known as a hotspot of biodiversity, with nearly 1000 endemic species of fish. It is an important resource for local populations, providing water, hydropower, and fish protein (Bootsma and Hecky 2003).

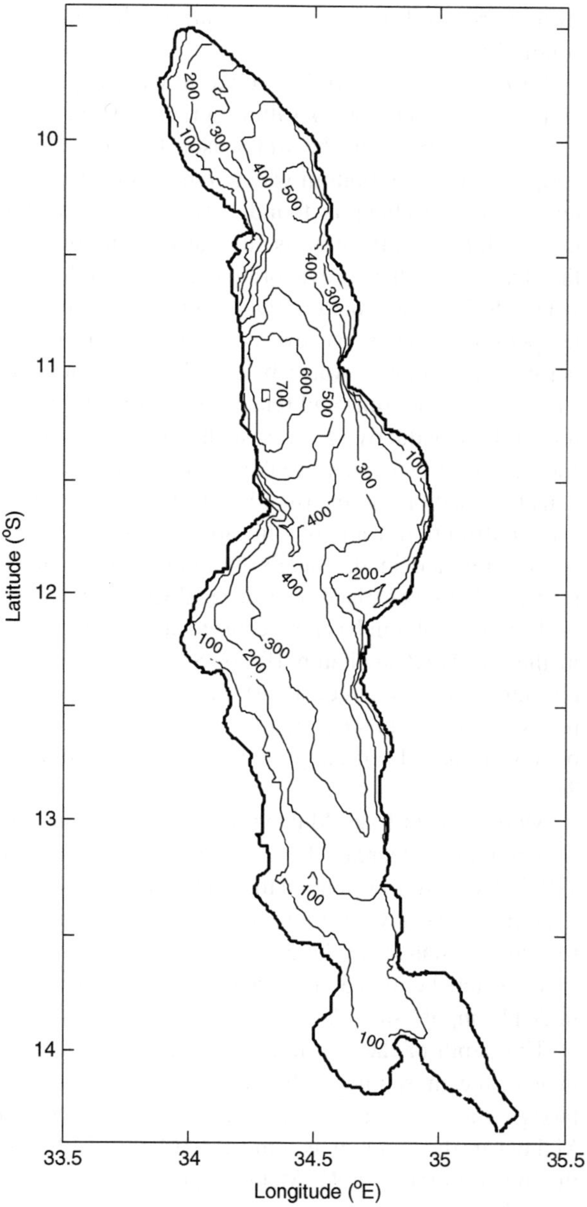

Fig. 10.6 Lake Malawi map

Although Lake Malawi experiences tropical climate, it is far enough south of the equator to experience significant seasonality. A rainy season occurs from November through March, at which time the Intertropical Convergence Zone (ITCZ) lies just south of the lake. Winds during this time are predominantly northerly. The dry season starts around April/May. During this season the ITCZ shifts northward, inducing a concurrent shift in wind direction to predominately southerly (Eccles 1974).

Stratification and mixing—Lake Malawi is permanently stratified by a temperature gradient and a weak salinity gradient (Fig. 10.1). Surface water temperatures vary from >28 °C in the wet season to about 24 °C in the dry season, whereas temperature in the bottom waters remains nearly constant at ~23 °C. The steepest temperature gradient is found between 50 and 100 m depth, forming a broad thermocline. Stratification is strongest during the wet season and weakens during the dry season due to evaporative cooling and mixing induced by dry southerly winds (Spigel and Coulter 1996). During austral winter (dry season), surface water temperatures can be as low as 23 °C: a mere 0.5 °C (or less) warmer than the permanently anoxic water below (Wüest et al. 1996). Salinity levels are low (0.211–0.236 per mil) and relative to temperature contribute to stratification insignificantly. Conductivity varies with depth non-monotonously, indicating that ionic species contribute only weakly to density stratification. A calculation of water density based on the measured concentrations of dissolved salts showed a significant contribution from nonionic species such as Si (Wüest et al. 1996). The salts do not accumulate in the lake's monimolimnion likely because of the deep-water renewal that takes place over decadal time scales (see below). A complete mixing of the water column has not occurred at least since the beginning of modern records in the 1930s (Beauchamp 1953). Sediment records do not indicate any obvious mixing events in the recent past, though based on the presence of layers enriched in redox-sensitive metals, significant excursions in the depth of the oxic-anoxic boundary have been suggested over a multimillennial time scale (Brown et al. 2000).

Meromixis in Lake Malawi results in permanent bottom water anoxia. Oxygen concentrations decrease below ~100 m and vanish below about 200 m (Halfman 1993). The bottom waters in the shallower (<200 m) southern region of the lake are hypoxic (<20 % O_2 saturation). Anoxic conditions promote the accumulation of reduced substances in the deep waters. Though free hydrogen sulfide was reported in the waters below 250 m (Eccles 1974), the concentrations are low (<5 μmol L^{-1} in 2011, unpublished data).

The depth of the thermocline/oxycline in Lake Malawi is significantly deeper than in the morphologically similar Lake Tanganyika (Beauchamp 1940). Even though the temperature difference between epi- and monimolimnion in Lake Malawi in the summer is almost 2 °C greater than in Tanganyika, the mixing in the metalimnion in Lake Malawi appears to be stronger, and oxygenated water is carried well below the depth of the strongest (seasonal) temperature gradient and the depth of seasonal mixing (60–80 m). This was suggested to be due to the stronger effect of winter trade winds, which cause stronger mixing in the dry season, as well as by more frequent switching of wind direction over Lake Malawi, which causes turbulence in the thermocline (Beauchamp 1940).

Deep-water renewal—Despite the prolonged and occasionally intense wind events, wind energy is generally insufficient to induce vertical mixing below about 100 m depth. Seasonal changes in stratification nevertheless have been reported to extend as deep as 250 m (Eccles 1974; Bootsma and Hecky 2003), at least locally. For example, Patterson and Kachinjika (1995) reported a mixed layer reaching 230 m at Nkhotakota in August/September 1992/1993. The "age" of the deeper waters estimated from the concentrations of CFCs, i.e., the effective time of the last contact with atmosphere, was estimated at 20–25 years (with a few caveats about the method of calculation; Vollmer et al. 2002), which suggests a relatively fast renewal of the deep waters relative to the hydrological residence time of 114 years. An earlier investigation using tritium isotopes suggested a much faster exchange, on the order of 5 years for the metalimnion-hypolimnion exchange and 4 years for the epilimnion-metalimnion exchange (Gonfiantini et al. 1979). Temperature time series indicate seasonal variations in the water temperature at depths down to 230 m, which indicates vertical movements of water extending to those depths (Beauchamp 1953). Plotting oxygen levels along the thermoclines also suggests that waters between 80 and 250 m receive inputs of oxygenated water (Beauchamp 1953). A critical part of the lake's hydrodynamics is therefore the mechanism that causes this deep circulation.

The predominant hypothesis for this mechanism invokes a basin-scale circulation. In the austral winter (dry season), temperatures that are cooler at the southern end of the lake than in the north cause colder surface waters in the south to become denser than in the mixolimnion, which makes them sink (Boehrer and Schultze 2008). The shallow (<100 m) southern basin of the lake undergoes complete mixing (Spigel and Coulter 1996), allowing for high rates of heat removal by evaporative cooling. This cooled water sinks along the lake's bottom slope, possibly reaching the deep basin north of Nkhata Bay (Eccles 1974). It is hypothesized that this influx of cooled water to the anoxic bottom waters is essential in maintaining the permanent thermocline at 250 m. The monimolimnion waters must be mixing internally to remain isothermal, as without an influx of cooled water one would expect the 250 m thermocline to gradually migrate deeper (Eccles 1974).

The direct evidence in support of this hypothesis of deep-water renewal is still scarce. Time series from moored instruments (unpublished data; Hamblin et al. 2003) and recent CTD profiles obtained in the southern part of the lake do not indicate mixing of surface waters below the diffuse thermocline at 200 m depth. A heat budget calculation is needed for its verification, which could verify the deficit of heat in the south and an excess of heat in the north. An alternative explanation involves the wind-driven coupled upwelling/downwelling systems that develop seasonally near the southern/northern ends of the lake. These systems are driven by seasonally strong winds, which become more effective at mixing during the time of weakest temperature stratification (dry season, July–September). Hamblin et al. (2003) estimated that the upwelling flows dominate the vertical mixing and the transport of nutrients, at least in the southern part of the lake.

Deep ventilation rates likely vary on a multi-decadal time scale in response to changes in climatic forcing, such as variations in temperature, precipitation, and wind speeds. The data are limited but a comparison of the distributions of water isotopes suggests significant changes over the last half of the twentieth century. The profiles of ^{18}O and ^{2}H measured in 2012 (Fig. 10.7) are significantly flatter than

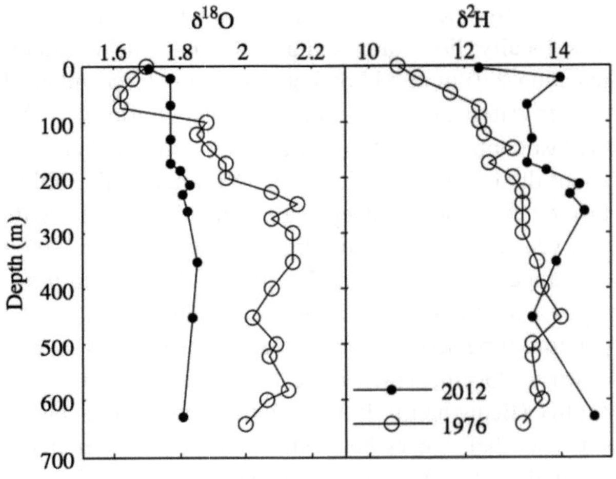

Fig. 10.7 ^{18}O and ^{2}H profiles in the water column of Lake Malawi. The 1976 data are from Gonfiantini et al. (1979)

those reported by Gonfiantini et al. (1979). The latter study has interpreted the non-monotonic variations in the isotopic profiles in the 1970s as being due to higher precipitation in the 1960s.

Smaller-scale mixing processes have been quantified in Lake Malawi to some degree (Eccles 1962, 1974) but are still poorly studied. Winds along the long fetch of Lake Malawi can generate longshore currents, allowing for lateral transport of sediment and nutrients along the shoreline (Johnson 1996). Such alongshore winds can also facilitate the offshore movement of water through Ekman transport, allowing for upwelling of the deeper waters (Bakun 1990). The intensity of vertical mixing by turbulent eddy diffusion has not been directly quantified: for example, the value of the eddy diffusivity K_z in the monimolimnion in the studies of both Hamblin et al. (2003) and Vollmer et al. (2002) was assumed at 10^{-5} m^2 s^{-1}.

Water balance and lake level fluctuations—The average monthly rainfall levels measured around Lake Malawi are comparable to those around such large temperate lakes as the North American Great Lakes. The evaporation levels, however, are much higher (>2× the levels in Lake Ontario; Spigel and Coulter 1996), due to high temperatures and strong winds along the long axis of the lake. Accordingly, the water inputs are balanced primarily by evaporation rather than by outflow (Eccles 1974). Water inputs are dominated by precipitation (67 %), with the remaining 23 % supplied by river inflows and storm runoff (Spigel and Coulter 1996). The dependence on evaporation is exacerbated by the nearly closed basin morphometry; the Shire River is the lone output from Lake Malawi and is typically only a few meters in depth (Eccles 1974). When this river was 3.5 m deep in 1974, it was estimated to account for only 20 % of the water loss from Lake Malawi, with evaporation accounting for the other 80 % (Eccles 1974). A subsequent water budget based on the data by Spigel and Coulter (1996) similarly indicated that 50 of the 52.7 km^3 water per year lost from the lake resulted from evaporation (~95 %), with river drainage making up the rest. Clearly, there is a delicate hydrologic balance in Lake Malawi, in which the rates of precipitation and evaporation are very important (Kruger 2014).

This delicate balance results in strong variations in the lake level. Historical records indicate that in the 1920s drier conditions resulted in the lake level dropping below the Shire River outflow. In the past, similar decreases in the lake level, lasting over longer time scales, likely contributed to the accumulation of salts and further stabilization of the lake stratification. The vast sediment record in Lake Malawi shows that both regional climate and the lake's water levels have fluctuated historically (Johnson 1996; Beadle 1981). Carbon isotopes measured in preserved plant leaf waxes (Castañeda et al. 2007) showed that Lake Malawi has shown low water levels during both wet and arid phases of the continent and that its fluctuations are typically out of phase with the Great Lakes to the north, likely a signal of its dependence on the position of the ITCZ (Johnson 1996; Finney et al. 1996). Records of calcite, ^{14}C, and diatoms suggest that the most recent major lake lowstand occurred between 6000 and 10,000 years ago, when the lake was ca. 100–150 m below its current level (Johnson 1996; Ricketts and Johnson 1996; Finney et al. 1996). Another, more drastic, lowstand occurred about 28,000–40,000 years ago, when water levels were thought to be 200–300 m below present levels (Finney et al. 1996; Johnson 1996).

Nutrient cycling—Nutrient distributions in the water column of Lake Malawi are largely shaped by its redox stratification (Fig. 10.8). Soluble reactive phosphorus levels increase with depth and are much higher in the anoxic deep waters below the oxic-anoxic boundary. The distributions of total oxidized nitrogen (nitrate and nitrite) exhibit a mid-water column maximum (e.g., Patterson and Kachinjika 1995), due to depletion at the surface resulting from primary production, depletion in the anoxic region due to denitrification, and mid-water column oxidation of ammonia. In the deep anoxic waters, the release of dissolved phosphorus upon the reduction of iron oxyhydroxides and removal of bioavailable nitrogen through denitrification lead to low nitrogen to phosphorus ratios (Patterson and Kachinjika 1995). Because of such deep-water nitrogen losses and because of possible

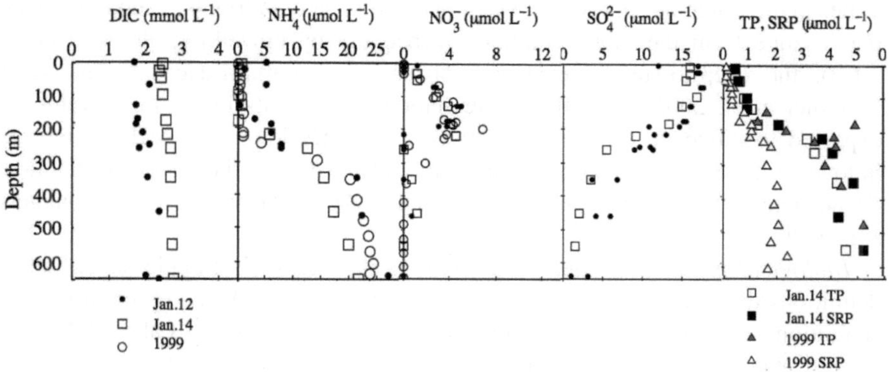

Fig. 10.8 Chemical distributions in the water column of Lake Malawi (from Li 2014). The data points marked as "1999" are from Bootsma and Hecky (1999). TP = total phosphorus; SRP = soluble reactive phosphorus

phosphorus-rich nutrient inputs from land-use changes, nitrogen fixation may be of regional importance in the lake, though its estimated importance to overall lake nitrogen levels has recently been revised downward, to 3–4 % of total N inputs to lake surface waters (Gondwe et al. 2008).

Estimates of the annual nitrogen, phosphorus, and silicon inputs to the surface mixed layer indicate that ca. 89 % of phosphorus and 88 % of silicon are delivered to the mixed layer from the anoxic waters below. Atmospheric deposition and river inflow account for the remaining inputs (Bootsma and Hecky 1999). For nitrogen inputs to the surface waters, vertical transport, river influx, and atmospheric deposition combined account for only about 21 %. The deficit may be balanced by high rates of biological nitrogen fixation, particularly by cyanobacteria (Bootsma and Hecky 1999 and sources therein), but recent nitrogen-fixation estimates (Gondwe et al. 2008) do not support this. Thus the nitrogen budget remains an area needing further research.

Sediment contributions to the water column nutrient inputs have been poorly quantified until recently. Anoxic conditions in the deep waters promote high efficiency of phosphorus regeneration, with sediment mass balance calculations suggesting that more than half of the deposited P is recycled back into the water column. Phosphorus efflux from sediments has been suggested to account for ~42 % of the total P inputs into the water column (Li 2014). In contrast to P, organic carbon deposited to the sediments is efficiently buried: the C burial efficiency (Katsev and Crowe 2015) in the anoxic sediments of Lake Malawi is high compared with that in oxygenated lakes, with over 60 % of the deposited carbon buried into the anoxic deep sediments (Li 2014; Li et al. 2012).

The lake-wide budgets of nitrogen, phosphorus, and silica were last estimated in the mid-1990s (Bootsma and Hecky 1999). Since then increases in population and agricultural activities have led to further losses of forest cover, altered river discharges, and significantly increased nutrient loadings (Hecky et al. 2003; Otu et al. 2011). A recent revision of nutrient budgets (Li 2014) suggested that phosphorus concentrations in the lake may be on the rise. The time scale on which the lake response may occur, however, is complicated by the different exchange time scales in the water column: while a significant portion of nutrients is supplied to the photic zone from the monimolimnion (especially by upwelling; Hamblin et al. 2003), the deep-to-surface exchange occurs over a multi-decadal time scale (Vollmer et al. 2002); thus the accumulation of nutrients in the deep waters is expected to influence the productivity over similarly long time scales. Riverine inflows in Lake Malawi contribute annually only ~1/600 of the volume of the lake (Beauchamp 1940), so their contributions to the whole-lake annual nutrient budgets are small.

Biological productivity and food webs—Lake Malawi's trophic status defies definitions that are based upon studies in temperate lakes. If the epilimnion concentrations of biologically relevant chemicals are considered, Lake Malawi is mesotrophic to eutrophic. Nutrient concentrations [total phosphorus, 0.1–0.5 μmol L^{-1} is highly oligotrophic. However, if primary production rates are used, Lake Malawi is mesotrophic to eutrophic. Nutrient concentrations [total phosphorus, 0.1–0.5 μmol L^{-1} (3–16 μg L^{-1}), and total nitrogen, 4.7–14.8 μmol L^{-1} (66–207 μg L^{-1})] in the lake's surface waters are lower than in most other aquatic systems, including oligotrophic Lake Superior, various Canadian lakes, and Lake Victoria (Guildford and Hecky 2000). Dissolved organic carbon levels (40–160 μmol L^{-1}, Ramlal et al. 2003) are in the range of values for open ocean surface water (Druffel et al. 1992; Carlson et al. 2010), and the lower values are similar to those in oligotrophic marine deep-water samples (Carlson et al. 2010). Particulate organic carbon (POC) levels in Lake Malawi are generally about 10 % of dissolved organic carbon (DOC) levels, with measured particulate carbon (primarily organic) ranging from 5 to 33 μmol L^{-1} (Ramlal et al. 2003). Chlorophyll *a* levels in the lake range from ca. 0.5 to 0.8 μg L^{-1}, considerably lower than average values in the oligotrophic Lake Superior of 0.95 μg L^{-1} (Guildford et al. 2000, 2007).

Despite the low concentrations of nutrients, organic carbon, and chlorophyll, Lake Malawi's estimated annual primary production rates are relatively high (mesotrophic to eutrophic, Bootsma and Hecky 2003), ranging from 123 to 518 g C $m^{-2}year^{-1}$ (Guildford et al. 2007; Hecky 2000; Patterson et al. 2000). More recent values are in the lower portion of this range. Phytoplankton growth in Lake Malawi appears moderately nutrient limited, with some light limitation occurring at times of deepest surface layer mixing in July and August (Guildford et al. 2000, 2007). There appears to be a tighter coupling of primary production, secondary production, and ultimately fish production, i.e., "greater trophic efficiency" in Lake Malawi than in temperate lakes (Guildford et al. 2000; Nixon 1988). Hypotheses attempting to explain this higher trophic efficiency include better nutritional value in Lake Malawi phytoplankton, which have fairly balanced N/P ratios relative to those in some temperate lakes (Guildford et al. 2000), to the better temperatures for year-round fish growth, and to food web structuring (Bootsma and Hecky 2003).

While Lake Malawi has only four prevalent zooplankton species, a calanoid copepod (*Tropodiaptomus cunningtoni)*, two cyclopoid copepods (*Mesocyclops aequatorialis aequatorialis* and *Thermocyclops neglectus*), and a cladoceran (*Diaphanosoma excisum*) (Lehman 1996; Irvine and Waya 1999), it is characterized by great diversity of fish species (~1000 fish species, most of which are endemic, Snoeks 2000). Despite the importance of Lake Malawi fisheries, there is only sparse data on the lower trophic levels, including initial primary production. There appear to be no published respiration values, and concurrent chlorophyll biomass and primary production measurements are spatially and temporally limited. A study in the southern part of the lake (Guildford et al. 2007) showed generally constant chlorophyll concentrations across seasons (seasonal averages from 0.7 in the wet stratified season of December–January to 1.0 μg L^{-1} in the dry mixed season of June–July). Integrated primary production rates were much more variable and not correlated to chlorophyll concentrations. The rates showed seasonal averages from 337 mg C m^{-2} day^{-1} in the dry stratified season (October–November) to 629 mg C m^{-2} day^{-1} in the wet stratified season (December to January). The main differences in primary production rates appeared to be the result of differing photosynthetic parameters, which can vary among algal species and with shade adaptation. There were differing slopes in photosynthesis-irradiation curves, which are estimates of quantum yields per mass chlorophyll (in other words, variations in the rate of biomass C produced per photon light per mass chlorophyll). The light-saturated photosynthetic rates per mass chlorophyll (i.e., the maximum measured rate, where the light dose is optimal) also varied with season.

In agreement with higher mixing rates, Lake Malawi is more productive than Lake Tanganyika. As nutrients are largely supplied from monimolimnion, a greater stability of stratification and a correspondingly reduced nutrient delivery to surface waters in Lake Tanganyika limit plankton development (Beauchamp 1940). In contrast to Tanganyika, the penetration of oxygen below the thermocline in Lake Malawi allows the existence of bottom-dwelling fauna, notably Chironomidae. Adult chironomids on emergence are swept into dense clouds, often carried hundreds of meters into the air (Beauchamp 1940). These clouds of "lake flies" are a remarkable feature of Lake Malawi, looking like bush fires over the surface of the lake.

Responses to change: warming, nutrient inputs, and erosion—As other East African Great Lakes, Lake Malawi is under considerable stress from both anthropogenic changes in its watershed and climate warming (Bootsma and Hecky 2003; Hecky et al. 2003; Otu et al. 2011). Historic water-temperature data revealed that the deep waters have warmed by ~0.7 °C between the 1940s and early 2000s (Vollmer et al. 2005). The temperature in the deep waters (>300 m) is currently higher than at any time since at least the 1940s. The warming rate in the deep waters over the past decade (unpublished data) is consistent with the rates of warming in Lake Malawi over the preceding decades. The temperature trends in the surface waters are not as clear, mainly due to greater seasonal variability. The associated changes in the stratification strength and nutrient inputs, however, did not seem to have had a significant effect on the depth of oxygen penetration into the Lake Malawi water column. Early reports put the oxycline depth at ~200 m in the summer and 220 m in winter (Beauchamp 1940), with some instances of oxygen penetration down to 250 m in winter (Beauchamp 1953). Measurements in 2012 seem to indicate slightly shallower oxygen penetration, but the magnitude of this perceived change is still within the variations that could be expected due to internal waves and seasonal cycles.

Intensification of agriculture in the Lake Malawi watershed has led to increased erosion and nutrient inputs: since the 1940s, the lake has been experiencing increases in sedimentation rates, concentrations of total sedimentary phosphorus, and abundances of eutrophic diatom taxa, at least at the shallower southern end of the lake (Hecky et al. 1999; Otu et al. 2011). Sedimentation rates have increased, most likely as a result of changing land use and shoreline erosion (Otu et al. 2011; Li 2014). Phosphorus inputs from the watershed into Lake Malawi likely have drastically increased in the last two decades, as sales of phosphorus fertilizers increased manyfold (Brown and Katsev 2012). As sediment burial removes a relatively small fraction of phosphorus from circulation, over decadal time scales, the increases in external P loading may be expected to affect the nutrient content in the epilimnion, though nutrient inputs from the monimolimnion are likely to remain dominant. The total phosphorus budget in Lake Malawi was previously considered as being close to balance (Bootsma and Hecky 1999). More recent estimates suggest an imbalance in P inputs, which may be slowly increasing the P concentrations in the lake (Li 2014), whereas the N budget remains essentially unchanged since the 1990s.

Capture fisheries are a major dietary component for the local human population. In the 1970s, they provided up to 70 % of the dietary protein for Malawians, but total protein in the diet and the proportion obtained from fish have decreased; fish catches have not increased as fast as the human population (Gondwe et al. 2011). To begin to address this deficit, fish aquaculture programs have been started. Their effects on regional nutrient budgets are an area of current research (Macuiane 2014).

10.2.3 Lake Matano

Background—Lake Matano on the Sulawesi island of Indonesia is a relatively small lake (161 km^2, about 20 km long and 8 km wide), but at 590 m depth it is the seventh deepest lake on the planet. It has recently attracted attention as a modern analogue for the conditions that prevailed in the ancient Archean ocean 3.5–2.5 billion years ago (Crowe et al. 2008a, b, 2014a, b). The lake's deep permanently anoxic waters contain high concentrations of dissolved ferrous (Fe^{2+}) iron (Fig. 10.9), a feature exceptionally rare today but characteristic of the deep ocean waters for a significant portion of the Earth's geologic history. Only a few other lakes on the planet are known to possess this quality [e.g., Lake Pavin in France and the Kabuno Bay of Lake Kivu (Lliros et al. 2015)]. In more typical instances of meromixis, anoxia leads to the accumulation of hydrogen sulfide, which is generated through the reduction of sulfate (SO_4^{2-}), and any available dissolved ferrous iron is removed through the precipitation of iron sulfides. Lake Matano's waters are also exceptionally transparent (Fig. 10.10), which allows for unique phototrophic microbial communities to exist within the illuminated chemocline at 100–110 m depth (Fig. 10.10) (Crowe et al. 2008b, 2011, 2014a).

The lake is located in a mountainous terrain: the highest point in the 436 km^2 catchment basin is at 1700 m a.s.l., while the lake surface is located at ~380 m, making the lower portion of the lake a cryptodepression (−210 m), i.e., below sea level. The lake is tectonic in origin and sits in a steep-sided graben, which controls the bathymetry. Located at about 2°S latitude, Lake Matano experiences minimal seasonal temperature fluctuations.

Stratification and mixing—The temperature difference between the surface and bottom water in Lake Matano is about 3.5°C and in the dry season can be as small as 2.5°C (Fig. 10.1) (Crowe et al. 2008a): the bottom waters remain at about 25.5° C, whereas the surface water temperatures typically vary from 27.5 °C in the dry

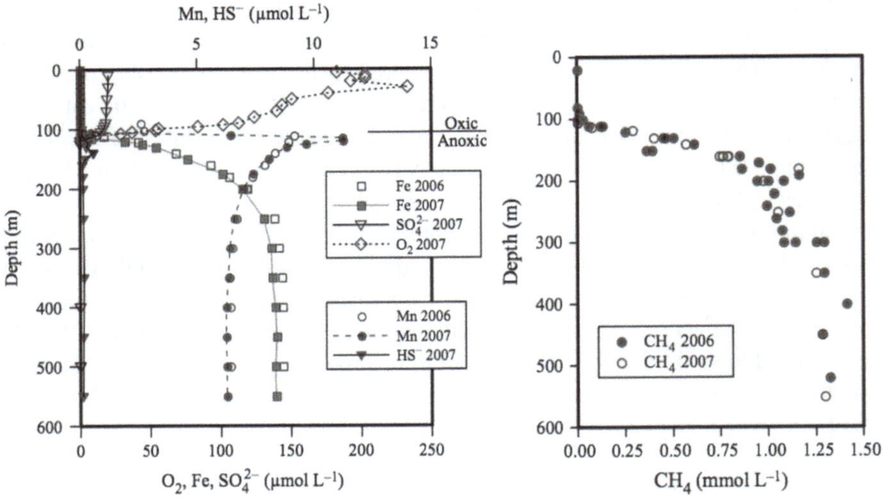

Fig. 10.9 Chemical distributions in the water column of Lake Matano (from Crowe et al. 2011; Copyright by John Wiley and Sons)

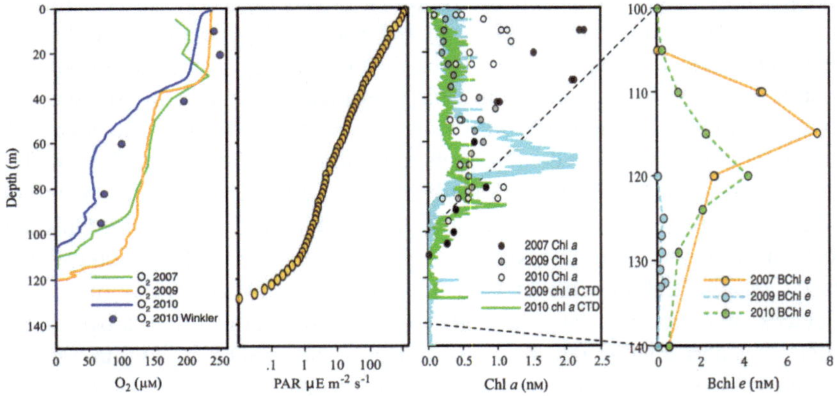

Fig. 10.10 Distributions of oxygen, photosynthetically active radiation (PAR), Chlorophyll *a*, and bacterial chlorophyll *e* (in the redoxcline) in the water column of Lake Matano. From Crowe et al. (2008a, b) (Copyright by the National Academy of Sciences) and Crowe et al. (2014a) (Copyright by John Wiley and Sons, reproduced with permission)

season to 30 °C in the wet season (Katsev et al. 2010). Despite the weak temperature gradient, a persistent pycnocline situated at ~100 m depth separates an oxic surface layer from anoxic bottom waters (Crowe et al. 2008a). Below 100 m depth, the salinity gradient is very weak: conductivity changes by less than 110 μS cm^{-1} over a 500 m depth interval. The temperature and conductivity gradients below the thermocline are virtually constant throughout the year and from year to year (Crowe et al. 2008a, 2014a), despite the apparently continuous downward heat fluxes from the epilimnion. These downward heat fluxes (~0.2 W m^{-2}) are expected to increase the bottom water temperatures only by about 0.06 °C per decade (Katsev et al. 2010). Though these rates have not been verified directly, such verification should be now possible, given nearly a decade of measurements. Based on the estimates of physical transport and chemical reaction rates in the lake, the lake has been suggested to have been meromictic for at least the past 130 years, and the renewal of the deep water was estimated to take place on time scales of 200–1000 years (Katsev et al. 2010).

Simulations revealed (Katsev et al. 2010) that depth variations in the physical transport rates are the dominant factor that shapes the vertical distributions of chemical substances in the water column of Lake Matano. Except in the chemocline region, transport rates outpace the reaction rates for most elements (except for Mn, Fig. 10.9; Jones et al. 2011), and vertical variations in their distributions reflect the variations in eddy diffusivity rather than geochemical sources or sinks (Katsev et al. 2010). The depth of the thermocline/redoxcline has fluctuated significantly from year to year. The oxic-anoxic boundary was at ~100 m depth in 2007, descended to 117 m in 2009, and ascended to ~105 m again in 2011 (Fig. 10.10; Crowe et al. 2014a) (after the effects of internal waves are averaged out). The exact mechanisms for these changes are poorly understood.

Biogeochemistry—Recent studies in Lake Matano have revealed an environment that is geochemically and microbiologically intriguing and possibly unique (Crowe et al. 2008a, b). The anoxic waters below 100 m depth are virtually devoid of sulfide (<0.6 μmol L^{-1}) but are rich in dissolved iron and methane (140 and 1400 μmol L^{-1}, respectively). The oxidation of ferrous iron at the redoxcline results in an intense redox cycling within a narrow depth interval near the oxic-anoxic boundary where Fe is recycled dozens of times (Jones et al. 2011). The precipitation of Fe oxides there serves as a strong sink for phosphate, which strongly limits the P availability to the mixed layer. An investigation into the iron mineralogy revealed that the primary Fe precipitate phase is a mineral called green rust (Zegeye et al. 2012), which is a mixed valence Fe(II)/Fe(III) mineral rather than the traditionally assumed Fe oxyhydroxides. An even more intense cycle of Mn with high Mn turnover rates was also reported (Jones et al. 2011). The strong immobilization of P below the base of the mixed layer leads to nutrient limitation. This results in an exceptional transparency of the Lake Matano water, which allows light (PAR) to penetrate down to the ~100 m deep redoxcline at the level of several μE (~0.1 % of incident radiation; Fig. 10.10; Crowe et al. 2014a). Oxidized nitrogen (nitrate and nitrite) concentrations in the mixed layer are very low (<100 nM), reaching only hundreds of nanomoles per liter at the redoxcline, and in deeper waters nitrate is completely depleted within the redoxcline (Crowe et al. 2011). Ammonium accumulates in the anoxic monimolimnion to about 250 μM, but its oxidation in the redoxcline limits the supply of reactive N to the photic zone. Whereas the upwelling strength in Lake Matano has not been quantified, the geochemical budgets of many elements agree relatively well with the suggested transport rates by vertical eddy diffusion (Katsev et al. 2010), suggesting that the role of upwelling in this relatively small lake is minor.

Whether the primary productivity in Lake Matano is limited by P, N, or some other factor (e.g., trace metals or Cr toxicity) is not exactly known. The rates of N fixation also have not been determined. In the absence of common electron acceptors (nitrate, Fe and Mn oxides, and sulfate), the organic matter reaching the anoxic monimolimnion is mineralized primarily by methanogenesis, which accounts for >50 % of authigenic organic carbon mineralization (Crowe et al. 2011). The dissolved inorganic carbon (DIC) concentrations reach 3.5 mM, while the DOM concentrations that are approximately uniform within the mixed layer at 1 mM quickly decrease below the redoxcline to <0.1 mM (Crowe et al. 2008a), suggesting that the DOC is used to fuel anaerobic respiration within and below the redoxcline.

Light penetration to 120 m depth allows photosynthesis by a low-light-adapted community of microorganisms that conduct anoxygenic photosynthesis (Crowe et al. 2008b, 2014a). A significant fraction of microbial primary production in the lake thus is associated with photoferrotrophic bacteria in the upper part of the anoxic water column. These organisms produce Fe oxides instead of oxygen, a pathway that has been postulated to have dominated the productivity of the ancient Archean ocean (Crowe et al. 2008b), and are exceptionally rare today because of the scarcity of ferruginous water columns with enough light to sustain photosynthesis. The pycnocline of Lake Matano is characterized by significant concentrations of bacterial Chl *e*, which underlies a layer enriched in the more common Chl *a* (Fig. 10.10). The redoxcline region of Lake Matano may also support several other rare biogeochemical pathways, such as anaerobic oxidation of methane with solid-phase ferric iron as an electron acceptor (Crowe et al. 2008a, b; Crowe et al. 2011).

10.2.4 Other Meromictic Tropical Lakes

The number of known tropical meromictic lakes for which at least some physical or geochemical information is available in the literature is in the dozens. They include, for example, lakes Nyos and Monoun in Cameroon (Kling 1987; Kling et al. 1989, 2005), small volcanic crater lakes such as Lake Sonachi in Kenya (MacIntyre and Melack 1982), Lake Nyahirya (Kizito and Nauwerck 1995) and several other lakes in Uganda (Rumes et al. 2005; Melack 1978), and lakes in the Basotu Lake District in Tanzania (Kilham and Cloke 1990). A recent inventory of meromictic lakes (Hall and Northcote 2012) identified 22 meromictic lakes in Africa, 10 in Mexico and Central America, and 13 in Indonesia, Australia, and New Zealand. Many more such lakes, however, are potentially undiscovered. Basic physicochemical data are missing, for example, for many deep tectonic and crater lakes in Indonesia. As examples in this chapter show, in tropics strong salinity gradients are not a prerequisite for the absence of mixing, and thus the great depths and lack of temperature seasonality in these lakes make meromixis likely. The 205 m deep Indonesian Lake Towuti (site of an international scientific drilling project; Russell and Bijaksana 2012), for example, appears to be meromictic at present but the mixing regime likely varied in the past in response to climate variations (Costa et al. 2015).

One of the most studied and interesting tropical meromictic lakes is Lake Kivu in East Africa, at the border of Rwanda and the Democratic Republic of the Congo. It is the highest of the East African Rift lakes. The specifics of meromixis in this lake are dictated by its very active geological setting, which includes multiple faults and Africa's two most active volcanoes in the immediate proximity to the lake (Zhang et al. 2014). The hydrothermal heat causes the temperature in the lake to strongly increase with depth (Fig. 10.1), reaching 26 °C near its bottom at 450 m depth, about 2 degrees above typical surface temperatures. Whereas meromixis is stabilized by a strong salinity gradient (Fig. 10.1), the destabilizing effect of this reverse temperature gradient results in a phenomenon of double diffusion, whereby local instabilities generate a characteristic "staircase" pattern in the vertical profiles of temperature and electrical conductivity (e.g., Sommer et al. 2013). Investigation of the water and substance transfer across these gradients presents an intriguing hydrodynamics problem, but complicates studies of the lake's mixing dynamics (Schmid and Wüest 2012). The isolated Kabuno Bay of the lake, separated from the main basin by a shallow (~7 m) sill, features its own salinity-controlled stratification with a sharp pycnocline at about 10 m depth (Borges et al. 2011; Katsev et al. 2014) and a ferruginous water column (Lliros et al. 2015). Due to concerns about the hazard of a catastrophic limnic eruption (Schmid et al. 2003), Lake Kivu has been the subject of multiple studies in the last decade (e.g., Lorke et al. 2004; Schmid et al. 2005; Pasche et al. 2009; Sommer et al. 2013; Katsev et al. 2014; Thiery et al. 2014; Zhang et al. 2014; Ross et al. 2015a, b), which characterized multiple aspects of its physical, chemical, and biological dynamics. Limnic eruption is a catastrophic overturn of the lake accompanied by a rapid ascent, depressurization, and ebullition of large quantities of gas accumulated under pressure in the deep monimolimnion. Eruptions of lakes Nyos (in 1986) and Monoun (in 1984) caused multiple human deaths in Cameroon (Kling et al. 1987). In Lake Kivu, most of the gas pressure in the monimolimnion comes from dissolved methane, which is produced from the significantly more soluble CO_2 by microorganisms within the lake (Pasche et al. 2011). In addition to hazard, the accumulation of methane also presents an economic opportunity, and efforts are underway to sustainably extract these gas reserves in Lake Kivu for electricity generation (Wüest et al. 2012).

The unique circumstances of the Lake Kivu meromixis make it difficult to compare its features below the surface mixed layer (0–55 m) with those in other tropical meromictic lakes. Given the space constraints in this chapter, providing only a brief overview of Lake Kivu cannot do justice to the complexities of this lake. We therefore refer an interested reader to recent publications (listed above, and references therein), as well as a recent book (Lake Kivu; Edited by J-P. Descy, F. Darchambeau, M. Schmid; Springer, Netherlands).

10.3 Summary and Conclusions

- The stratification in tropical meromictic lakes is characterized by relatively weak (few degree Celsius) temperature gradients that, in contrast to temperate lakes, persist throughout the year. In the absence of strong seasonality, stratification can be maintained even without significant gradients in the concentrations of dissolved salts. The salinity levels may be maintained low by slow basin-scale circulation and deep-water renewal. The thickness of the wind-mixed surface layer in tropical lakes is highly variable, increasing significantly during the dry season as evaporative cooling causes the temperature gradient to decrease. The water budgets are strongly affected by evaporation, which is also a major cooling mechanism.
- The depth of oxygen penetration below the permanent thermocline strongly depends on the productivity export from the photic zone and the strength of seasonal downwelling and can vary from essentially zero to hundreds of meters. The deep-water anoxia causes the removal of reactive nitrogen to N_2 by denitrification and an efficient recycling of phosphorus. The nutrient demands of the primary producers are met predominantly by the upward transport of P with upwelling, with nitrogen limitation being alleviated to some degree by N fixation. The recycling of nutrients in the water column of warm lakes is more efficient than in temperate lakes, enabling a significantly higher primary production on a given nutrient base.
- The large tropical lakes situated in the developing countries of East Africa are critical resources for the shoreline populations, supplying water, fish protein, and hydropower. These lakes, however, are under increasing stress from both climate warming and changing land use in their watershed. The surface water temperatures of tropical lakes are increasing at about the same rate as regional air temperatures. Responses of the biological productivity in these lakes depend on the degree to which these changes affect the strength of the upwelling, which is the primary source of nutrients to the surface waters. Strengthening of the stratification is expected to decrease the rate of primary production by limiting the vertical nutrient transport. The exact response, however, depends on the balance between the changes in the stratification strength and changes in the strength and direction of prevailing winds, as well as evaporation rates. It also depends upon the relative importance of enhanced P availability due to enhanced anoxia in deep waters, relative to predicted decreases in the exchange rate between surface and deep waters. Predicting such responses requires a combination of observations and geochemical budget and hydrodynamic modeling.

Acknowledgements S.K., E.C.M., B.K., and J.L. acknowledge funding from the University of Minnesota, Office of International Programs. E.C.M. and B.K. acknowledge support from the American Chemical Society Petroleum Research Fund (ACS-PRF #48595-AC2).

References

Aaberg AA (2013) Warming and stratification changes in Lake Kivu, East Africa. M.Sc. Thesis, University of Minnesota Duluth

Alleman LY, Cardinal D, Cocquyt C et al (2005) Silicon isotopic fractionation in Lake Tanganyika and its main tributaries. J Great Lakes Res 31:509–519

Alonso-Sáez L, Gasol JM, Lefort T et al (2006) Effect of natural sunlight on bacterial activity and differential sensitivity of natural bacterioplankton groups in northwestern Mediterranean coastal waters. Appl Environ Microbiol 72(9):5806–5813

Amado AM, Meirelles-Pereira F, Vidal LO et al (2013) Tropical freshwater ecosystems have lower bacterial growth efficiency than temperate ones. Front Microbiol 4:1–8

Austin JA, Colman SM (2007) Lake Superior summer water temperatures are increasing more rapidly than regional air temperatures: a positive ice-albedo feedback. Geophys Res Lett 34(6)

Bailey CA, Neihof RA, Tabor PS (1983) Inhibitory effect of solar radiation on amino acid uptake in Cheasapake Bay bacteria. *Appl Environ Microbiol* 46:44–49

Bakun A (1990) Global climate change and intensification of coastal ocean upwelling. Science 247(4939):198–201

Baxter RM, Prosser MV, Talling JF, Wood RB (1965) Stratification in tropical African lakes at moderate altitudes (1500 to 2000 m). Limnol Oceanogr 10(4):510–520

Beadle LC (1981) The inland waters of Africa. An introduction to tropical limnology. Longman, London, New York, p. 475

Beauchamp RSA (1939) Hydrology of Lake Tanganyika. Int Rev Ges Hydrobiol Hydrogr 39:316–353

Beauchamp RSA (1940) Chemistry and hydrography of Lakes Tanganyika and Nyasa. Nature 146:253–256

Beauchamp RSA (1953) Hydrological data from Lake Nyasa. J Ecol 226–239

Berdjeb L, Ghiglione JF, Domaizon I, Jacquet S (2011) A 2-year assessment of the main environmental factors driving the free-living bacterial community structure in Lake Bourget (France). Microb Ecol 61(4):941–954

Boehrer B, Schultze M (2008) Stratification of lakes. Rev Geophys 46(2):RG2005

Bootsma HA, Hecky RE (1999) Nutrient cycling in Lake Malawi/Nyasa. In: Bootsma HA, Hecky RE (eds) Water quality report: Lake Malawi/Nyasa Biodiversity Conservation Project. Southern African Development Community/Global Environmental Facility (SADC/GEF), Salima, Malawi, pp 215–241

Bootsma HA, Hecky RE (2003) A comparative introduction to the biology and limnology of the African Great Lakes. J Great Lakes Res 29(2):3–18

Borges AV, Abril G, Delille B et al (2011) Diffusive methane emissions to the atmosphere from Lake Kivu (Eastern Africa). J Geophys Res Biogeosci 116:G3

Brichard P (1989) Cichlids and all other fishes of Lake Tanganyika. TFH Publications, Neptune, NJ

Brown ET, Katsev S (2012) Southern East African climate recorded in laminated sediments of Lake Malawi. AGU Fall Meeting Abstracts 1:2113

Brown ET, Le Callonnec L, German CR (2000) Geochemical cycling of redox-sensitive metals in sediments from Lake Malawi: a diagnostic paleotracer for episodic changes in mixing depth. Geochim Cosmochim Acta 64(20):3515–3523

Buckles LK, Villanueva L, Weijers JWH et al (2013) Linking isoprenoidal GDGT membrane lipid distributions with gene abundances of ammonia-oxidizing Thaumarchaeotaand uncultured crenarchaeotal groups in the water column of a tropical lake (Lake Challa, East Africa). Environ Microbiol 15(9):2445–2462

Camacho A, Vicente E (1998) Carbon photoassimilation by sharply stratified phototrophic communities at the chemocline of Lake Arcas (Spain). FEMS Microbiol Ecol 25(1):11–12

Camacho A, Erez J, Chicote A et al (2001) Microbial microstratification, inorganic carbon photoassimilation and dark carbon fixation at the chemocline of the meromictic Lake Cadagno (Switzerland) and its relevance to the food web. Aquat Sci 63(1):91–106

Canfield DE, Thamdrup B (2009) Towards a consistent classification scheme for geochemical environments, or, why we wish the term "suboxic" would go away. Geobiology 7:385–392

Canfield DE, Stewart FJ, Thamdrup B et al (2010) A cryptic sulphur cycle in oxygen-minimum-zone waters off the Chilean coast. Science 330(6009):1375–1378

Carlson CA, Hansell DA, Nelson NB, Siegel DA, Smethie WM, Khatiwala S, Meyers MM, Halewood E (2010) Dissolved organic carbon export and subsequent remineralization in the mesopelagic and bathypelagic realms of the North Atlantic basin. Deep Sea Res II: Top Stud Oceanogr 57(16):1433–1445

Casamayor EO, Schäfer H, Bañeras L et al (2000) Identification of and spatio-temporal differences between microbial assemblages from two neighboring sulphurous lakes: comparison by microscopy and denaturing gradient gel electrophoresis. Appl Environ Microbiol 66 (2):499–508

Castañeda IS, Werne JP, Johnson TC (2007) Wet and arid phases in the southeast African tropics since the Last Glacial Maximum. Geology 35(9):823–826

Chapelle FH, Lovley DR (1990) Rates of microbial metabolism in deep coastal plain aquifers. Appl Environ Microbiol 56(6):1865–1874

Codispoti LA, Brandes JA, Christensen JP et al (2001) The oceanic fixed nitrogen and nitrous oxide budgets: moving targets as we enter the anthropocene? Sci Mar 65(S2):85–105

Cohen AS, Soreghan MJ, Scholz CA (1993) Estimating the age of formation of lakes: an example from Lake Tanganyika, East African Rift system. Geology 21(6):511–514

Cole JJ, Findlay S, Pace ML (1988) Bacterial production in fresh and saltwater ecosystems: a cross-system overview. Mar Ecol Prog Ser 43:1–10

Comeau AM, Harding T, Galand PE et al (2012) Vertical distribution of microbial communities in a perennially stratified Arctic lake with saline, anoxic bottom waters. Sci Rep 2:604

Contreras J, Anders MH, Scholz CH (2000) Growth of a normal fault system: observations from the Lake Malawi basin of the East African Rift. J Struct Geol 22(2):159–168

Costa KM, Russell JM, Vogel H, Bijaksana S (2015) Hydrological connectivity and mixing of Lake Towuti, Indonesia in response to paleoclimatic changes over the last 60,000 years. Palaeogeogr Palaeoclimatol Palaeoecol 417:467–475

Coulter GW (1988) Seasonal hydrodynamic cycles in Lake Tanganyika. Verh Int Ver Limnol 23:86–89

Coulter GW (1991a) Lake Tanganyika and its life. Oxford University Press, London

Coulter GW (1991b) Pelagic fish. In: Coulter GW (ed) Lake Tanganyika and its life. Oxford University Press, London, pp 111–138

Coulter GW (1991c) Fisheries. In: Coulter GW (ed) Lake Tanganyika and its life. Oxford University Press, London, pp 139–150

Coulter GW (1991d) Composition of the flora and fauna. In: Coulter GW (ed) Lake Tanganyika and its life. Oxford University Press, London, pp 200–274

Coulter GW, Spigel RH (1991) Hydrodynamics. In: Coulter GW (ed) Lake Tanganyika and its life. Oxford University Press, London, pp 49–75

Craig H, Dixon F, Craig V et al (1974) Lake Tanganyika geochemical and hydrographic study: 1973 expedition. Publ Scripps Inst Oceanogr 75(5):1–83

Crowe SA, O'Neill AH, Katsev S et al (2008a) The biogeochemistry of tropical lakes: A case study from Lake Matano. Indonesia Limnol Oceanogr 53(1):319–331

Crowe SA, Jones C, Katsev S et al (2008b) Photoferrotrophs thrive in an Archean Ocean analogue. Proc Natl Acad Sci 105(41):15938–15943

Crowe SA, Katsev S, Leslie K et al (2011) The methane cycle in ferruginous Lake Matano. Geobiology 9(1):61–78

Crowe SA, Maresca JA, Jones C et al (2014a) Deep-water anoxygenic photosythesis in a ferruginous chemocline. Geobiology 12(4):322–339

Crowe SA, Paris G, Katsev S et al (2014b) Sulphate was a trace constituent of Archean seawater. Science 346(6210):735–739

Daffonchio D, Borin S, Brusa T et al (2006) Stratified prokaryote network in the oxic-anoxic transition of a deep-sea halocline. Nature 440(7081):203–207

Delvaux D (1995) Age of Lake Malawi (Nyasa) and water level fluctuations. Mus R Afr Centr Tervuren (Belg) Dept Geol Min Rapp Ann 1996(99):108

Descy J-P, Sarmento H (2008) Microorganisms of the East African Great Lakes and their response to environmental changes. Freshw Rev 1:59–73

Descy J-P, Hardy M-A, Stenuite S et al (2005) Phytoplankton pigments and community composition in Lake Tanganyika. Freshw Biol 50:668–684

Descy J-P, Darchambeau F, Schmid M (2012a) Lake Kivu research: conclusions and perspectives. In: Descy J-P, Darchambeau F, Schmid M (eds) Lake Kivu: limnology and biogeochemistry of a tropical great lake. Springer, Dordrecht, pp 181–190

Descy J-P, Darchambeau F, Schmid M (2012b) Lake Kivu: past and present. In: Descy J-P, Darchambeau F, Schmid M (eds) Lake Kivu: limnology and biogeochemistry of a tropical great lake. Springer, Dordrecht, pp 1–169

Deutsch C, Sarmiento JL, Sigman DM et al (2007) Spatial coupling of nitrogen inputs and losses in the ocean. Nature 445(7124):163–167

Druffel ER, Williams PM, Bauer JE, Ertel JR (1992) Cycling of dissolved and particulate organic matter in the open ocean. J Geophys Res: Oceans 97(C10):15639–15659

Dubois JT (1958) Evolution de la temperature de l'oxygene dissous et de la transparence dans la baie Nord du lac Tanganika. Hydrobiologia 10:215–240

Dumont HJ (1994) Ancient lakes have simplified pelagic food webs. Arch Hydrobiol Beih Ergebn Limnol 44:223–234

Ebinger CJ, Deino AL, Tesha AL et al (1993) Tectonic controls on rift basin morphology: evolution of the Northern Malawi (Nyasa) Rift. J Geophys Res 98(B10):17821–17836

Eccles DH (1962) An internal wave in Lake Nyasa and its probable significance in the nutrient cycle. Nature 194:832–833

Eccles DH (1974) An outline of the physical limnology of Lake Malawi. Limnol Oceanogr 19 (5):730–742

Edmond JM, Stallard RF, Craig H et al (1993) Nutrient chemistry of the water column of Lake Tanganyika. Limnol Oceanogr 38(4):725–738

Finney BP, Scholz CA, Johnson TC et al (1996) Late Quaternary lake-level changes of Lake Malawi. In: Johnson TC, Odada EO (eds) The limnology, climatology and paleoclimatology of the East African lakes. Gordon and Breach, London, pp 495–508

Fouilland E, Mostajir B (2010) Revisited phytoplanktonic carbon dependency of heterotrophic bacteria in freshwaters, transitional, coastal and oceanic waters. FEMS Microbiol Ecol 73 (3):419–429

Frigaard NU, Martinez A, Mincer TJ, DeLong EF (2006) Proteorhodopsin lateral gene transfer between marine planktonic Bacteria and Archaea. Nature 439(7078):847–850

Fuhrman JA, Cram JA, Needham DM (2015) Marine microbial community dynamics and their ecological interpretation. Nat Rev Microbiol 13(3):133–146

Garcia-Pichel F (1994) A model for internal shelf-shading in planktonic organisms and its implications for the usefulness of ultraviolet sunscreens. Limnol Oceanogr 39:1704–1717

Gies EA, Konwar KM, Beatty JT, Hallam SJ (2014) Illuminating microbial dark matter in meromictic Sakinaw Lake. Appl Environ Microbiol 80:6807–6818

Glaubitz S, Labrenz M, Jost G, Jürgens K (2010) Diversity of active chemolithoautotrophic prokaryotes in the sulfidic zone of a Black Sea pelagic redoxcline as determined by rRNA-based stable isotope probing. FEMS Microbiol Ecol 74(1):32–41

Gondwe MJ, Guildford SJ, Hecky RE (2008) Planktonic nitrogen fixation in Lake Malawi/Nyasa. Hydrobiological 596:251–267

Gondwe MJ, Guildford SJ, Hecky RE (2011) Carbon, nitrogen and phosphorus loadings from tilapia fish cages in Lake Malawi and factors influencing their magnitude. J Great Lakes Res 37 (Suppl 1):93–101

Gonfiantini R, Zuppi GM, Eccles DH, Ferro W (1979) Isotope investigation of Lake Malawi. In: Isotopes in lake studies. International Atomic Energy Agency, Vienna, pp 195–207

Grote J, Jost G, Labrenz M, Herndl GJ, Jürgens K (2008) Epsilonproteobacteria represent the major portion of chemoautotrophic bacteria in sulfidic waters of pelagic redoxclines of the Baltic and Black Seas. Appl Environ Microbiol 74(24):7546–7551

Gruber N, Sarmiento JL (1997) Global patterns of marine nitrogen fixation and denitrification. Global Biogeochem Cycles 11(2):235–266

Guildford SJ, Bootsma HA, Fee EJ et al (2000) Phytoplankton nutrient status and mean water column irradiance in Lakes Malawi and Superior. Aquat Ecosyst Health Manag Soc 3:35–45

Guildford SJ, Bootsma HA, Taylor WS, Hecky RE (2007) High variability of phytoplankton photosynthesis in response to environmental forcing in oligotrophic Lake Malawi/Nyasa. J Great Lakes Res 33(1):170–185

Haberyan KA, Hecky RE (1987) The late Pleistocene and holocene stratigraphy and paleolimnology of lakes Kivu and Tanganyika. Palaeogeogr Palaeoclimatol Palaeoecol 61:169–197

Halfman JD (1993) Water column characteristics from modern CTD data, Lake Malawi, Africa. J Great Lakes Res 19(3):512–520

Hall KJ, Northcote TG (2012) Meromictic lakes. In: Bengtsson L, Herschy RW, Fairbridge RW (eds) Encyclopedia of lakes and reservoirs. Springer, Dordrecht, pp 519–524

Hamblin PF, Bootsma HA, Hecky RE (2003) Modeling nutrient upwelling in Lake Malawi/Nyasa. J Great Lakes Res 29:34–47

Hawley AK, Brewer HM, Norbeck AD, Paša-Tolić L, Hallam SJ (2014) Metaproteomics reveals differential modes of metabolic coupling among ubiquitous oxygen minimum zone microbes. Proc Nat Acad Sci 111(31):11395–11400

Hecky RE (1984) African lakes and their trophic efficiencies: a temporal perspective. In: Meyers DG, Strickler JR (eds) Trophic interactions within aquatic ecosystems. Westview Press, Washington, DC, pp 405–448

Hecky RE (1991) The pelagic ecosystem. In: Coulter GW (ed) Lake Tanganyika and its life. Oxford University Press, London, pp 90–110

Hecky RE (1993) The eutrophication of Lake Victoria. Verh Int Ver Limnol 25:39–48

Hecky RE (1995) African lakes and climate change. In: Watson RT, Zinyowera MC, Moss RH (eds) Hydrology and freshwater ecology, Impacts, adaptations and mitigation of climate change: scientific-technical analyses. Intergovernmental Panel on Climate Change. Cambridge University Press, Cambridge, 341 p

Hecky RE (2000) A biogeochemical comparison of Lakes Superior and Malawi and the limnological consequences of an endless summer. Aquat Ecosyst Health Manage 3(1):23–33

Hecky RE, Fee EJ (1981) Primary production and rates of algal growth in Lake Tanganyika. Limnol Oceanogr 26(3):532–547

Hecky RE, Kling HJ (1981) The phytoplankton and protozooplankton of the euphotic zone of Lake Tanganyika: species composition, biomass, chlorophyll content, and spatio-temporal distribution. Limnol Oceanogr 26(3):548–564

Hecky RE, Kling HJ (1987) Phytoplankton ecology of the great lakes in the rift valleys of Central Africa. Arch Hydrobiol Beih Ergebn Limnol 25:197–228

Hecky RE, Spigel RH, Coulter GW (1991) Nutrient regime. In: Coulter GW (ed) Lake Tanganyika and its life. Oxford University Press, London, pp 76–89

Hecky RE, Bugenyi FWB, Ochumba P et al (1994) Deoxygenation of the deep water of Lake Victoria, East Africa. Limnol Oceanogr 39(6):1476–1481

Hecky RE, Kling HJ, Johnson TC, Bootsma HA, Wilkinson P (1999) Algal and sedimentary evidence for recent changes in the water quality and limnology of Lake Malawi/Nyasa. Victoria 1:1–2

Hecky RE, Bootsma HA, Kingdon ML (2003) Impact of land use on sediment and nutrient yields to Lake Malawi/Nyasa (Africa). J Great Lakes Res 29:139–158

Hernández-Avilés JS, Bertoni R, Macek M, Callieri C (2012) Why bacteria are smaller in the epilimnion than in the hypolimnion? A hypothesis comparing temperate and tropical lakes. J Limnol 71(1):104–111
Herndl GJ (1997) Role of ultraviolet radiation on bacterioplankton activity. In: Häder DP (ed) The effects of ozone depletion on aquatic ecosystems. R.G. Landes Co., Austin, TX, pp 143–154
Herndl GJ, Müller-Niklas G, Frick J (1993) Major role of ultraviolet-B in controlling bacterioplankton growth in the surface layer of the ocean. Nature 361:717–719
Hori M (1983) Feeding ecology of thirteen species of Lamprologus (Teleostei; Cichlidae) coexisting at a rocky shore of Lake Tanganyika. Physiol Ecol Jpn 20:129–149
Hori M (1991) Feeding relationships among cichlid fishes in Lake Tanganyika: effects of intra- and interspecific variations of feeding behavior on their coexistence. Ecol Int Bull 19:89–101
Hori M, Gashagaza MM, Nshombo M, Kawanabe H (1993) Littoral fish communities in Lake Tanganyika: irreplaceable diversity supported by intricate interactions among species. Conserv Biol 7(3):657–666
Hugoni M, Taïb N, Debroas D et al (2013) Structure of the rare archaeal biosphere and seasonal dynamics of active ecotypes in surface coastal waters. Proc Natl Acad Sci 110(15):6004–6009
Hulme M, Doherty R, Ngara T et al (2001) African climate change: 1900-2100. Climate Res 17 (2):145–168
Hutchinson GE (1975) A treatise on limnology, Geography and physics of Lakes, vol 1, Part 1. Wiley, New York, London, p. 540
Imberger J, Patterson JC (1990) Physical limnology. Adv Appl Mech 27:303–475
Irvine K, Waya R (1999) Spatial and temporal patterns of zooplankton standing biomass and production in Lake Malawi. In: From limnology to fisheries: Lake Tanganyika and other large lakes. Springer, Netherlands, pp 191–205
Jassby A, Powell T (1975) Vertical patterns of eddy diffusion during stratification in Castle Lake, California. Limnol Oceanogr 20:530–543
Jeffrey WH, Pledger RJ, Aas P et al (1996) Diel and depth profiles of DNA photodamage in bacterioplankton exposed to ambient solar ultraviolet radiation. Mar Ecol Prog Ser 137:283–291
Johnson TC (1996) Sedimentary processes and signals of past climatic change in the large lakes of the East African Rift Valley. In: Johnson TC, Odada EO (eds) The limnology, climatology and paleoclimatology of the East African Lakes. Gordon and Breach, Amsterdam, pp 367–412
Johnson TC, Odada EO (1996) The limnology, climatology and paleoclimatology of the East African lakes. Gordon and Breach, Amsterdam
Jones C, Crowe SA, Sturm A et al (2011) Biogeochemistry of manganese in ferruginous Lake Matano, Indonesia. Biogeosciences 8(10):2977–2991
Kaiser E, Herndl GJ (1997) Rapid recovery of marine bacterioplankton activity after inhibition by UV radiation in coastal waters. Appl Environ Microbiol 63:4026–4031
Katsev S, Crowe SA (2015) Organic carbon burial efficiencies in sediments: the power law of mineralization revisited. Geology 43:607–610
Katsev S, Tsandev I, L'Heureux I, Rancourt DG (2006) Factors controlling long-term phosphorus efflux from lake sediments: exploratory reactive-transport modeling. Chem Geol 234 (1):127–147
Katsev S, Crowe SA, Mucci A et al (2010) Mixing and its effects on biogeochemistry in the persistently stratified, deep, tropical Lake Matano, Indonesia. Limnol Oceanogr 55(2):763–776
Katsev S, Aaberg AA, Crowe SA, Hecky RE (2014) Recent warming of Lake Kivu. PLoS One 9 (10):e109084
Keeling RF, Garcia HE (2002) The change in oceanic O2 inventory associated with recent global warming. Proc Natl Acad Sci 99(12):7848–7853
Kilham P, Cloke PL (1990) The evolution of saline lake waters: gradual and rapid biogeochemical pathways in the Basotu Lake District, Tanzania. In: Comín FA, Northcote TG (eds) Saline lakes. Springer, Dordrecht, pp 35–50

Kilham P, Kilham SS (1990) Endless summer: internal loading processes dominate nutrient cycling in tropical lakes. Freshw Biol 23:379–389

Kizito YS, Nauwerck A (1995) Temporal and vertical distribution of planktonic rotifers in a meromictic crater lake, Lake Nyahirya (Western Uganda). In: Ejsmont-Karabin J, Pontin RM (eds) Rotifera VII. Springer, Dordrecht, pp 303–312

Klepac-Ceraj V, Hayes CA, Gilhooly WP et al (2012) Microbial diversity under extreme euxinia: Mahoney Lake, Canada. Geobiology 10(3):223–235

Kling GW (1987) Seasonal mixing and catastrophic degassing in tropical lakes, Cameroon, West Africa. Science 237(4818):1022–1024

Kling GW (1988) Comparative transparancy, depth of mixing, and stability of stratification in lakes of Cameroon. West Afr Limnol Oceanogr 33(1):27–40

Kling GW et al (1987) The 1986 Lake Nyos gas disaster in Cameroon, West Africa. Science 236 (4798):169–175

Kling GW, Tuttle ML, Evans WC (1989) The evolution of thermal structure and water chemistry in Lake Nyos. J Volcanol Geotherm Res 39(2):151–165

Kling GW, Ewans WC, Tuttle ML, Tanyileke G (1994) Degassing of Lake Nyos. Nature 368:405–406

Kling GW, Evans WC, Tanyileke G et al (2005) Degassing Lakes Nyos and Monoun: defusing certain disaster. PNAS 102(40):14185–14190

Kraemer BM, Anneville O, Chandra S et al (2015a) Morphometry and average temperature affect lake stratification responses to climate change. Geophys Res Lett:GL064097. doi:10.1002/2015GL064097

Kraemer BM, Hook S, Huttula T et al (2015b) Century-long warming trends in the upper water column of Lake Tanganyika. PLoS One 10(7)

Kruger BR (2014) Sources, cycling, and fate of organic matter in large lakes: insights from stable isotope and radiocarbon analysis in Lakes Malawi and Superior. Doctoral dissertation, University of Minnesota

Lehman JT (1996) Pelagic food webs of the East African Great lakes. In: Johnson TC, Odada EO (eds) The limnology, climatology and paleoclimatology of the East African Lakes. Gordon and Breach Publishers, Amsterdam, pp 281–301

Lehman JT (ed) (1998) Environmental change and response in East African lakes. Kluwer Academic Publishers, Dordrecht

Lehours AC, Bardot C, Thenot A et al (2005) Anaerobic microbial communities in Lake Pavin, a unique meromictic lake in France. Appl Environ Microbiol 71(11):7389–7400

Lehours AC, Evans P, Bardot C et al (2007) Phylogenetic diversity of archaea and bacteria in the anoxic zone of a meromictic lake (Lake Pavin, France). Appl Environ Microbiol 73 (6):2016–2019

Lewis WM Jr (1973) The thermal regime of Lake Lanao (Philippines) and its theoretical implications for tropical lakes. Limnol Oceanogr 18(2):200–217

Lewis WM Jr (1982) Vertical eddy diffusivities in a large tropical lake (Lake Lanao, Philippines). Limnol Oceanogr 20:530–543

Lewis WM Jr (1987) Tropical limnology. Annu Rev Ecol Syst 18:159–184

Lewis WM Jr (1996) Tropical lakes: how latitude makes a difference. In: Schiemer F, Boland KT (eds) Perspectives in tropical limnology. SPB Academic Publishing, Amsterdam, pp 43–64

Lewis WM Jr (2000) Basis for the protection and management of tropical lakes. Lakes Reserv Res Manag 5:35–48

Lewis WM Jr (2002) Causes for the high frequency of nitrogen limitation in tropical lakes. Verh Int Ver Limnol. 28:210–213

Lewis WM Jr (2010) Biogeochemistry of tropical lakes. Verh Int Ver Limnol 30(10):1595–1603

Li J (2014) Sediment diagenesis in large Lakes Superior and Malawi, geochemical cycles and budgets and comparisons to marine sediments. University of Minnesota, Doctoral dissertation

Li J, Crowe SA, Miklesh D et al (2012) Carbon mineralization and oxygen dynamics in sediments with deep oxygen penetration, Lake Superior. Limnol Oceanogr 57(6):1634–1650

Llirós M, Casamayor EO, Borrego CM (2008) High archaeal richness in the water column of a freshwater sulphurous karstic lake along an interannual study. FEMS Microbiol Ecol 66 (2):331–342

Llirós M, Gich F, Plasencia A et al (2010) Vertical distribution of ammonia-oxidizing crenarchaeota and methanogens in the epipelagic waters of Lake Kivu (Rwanda-Democratic Republic of the Congo). Appl Environ Microbiol 76(20):6853–6863

Llirós M, Alonso-Sáez L, Gich F et al (2011) Active bacteria and archaea cells fixing bicarbonate in the dark along the water column of a stratified eutrophic lagoon. FEMS Microbiol Ecol 77 (2):370–384

Llirós M, Descy J-P, Libert X et al (2012) Microbial ecology of Lake Kivu. In: Descy J-P, Darchambeau F, Schmid M (eds) Lake Kivu: Limnology and biogeochemistry of a tropical great lake. Springer Netherlands, Dordrecht, pp 85–105

Llirós M, García-Armisen T, Darchambeau F et al (2015). Pelagic photoferrotrophy and iron cycling in a modern ferruginous basin. Sci Rep, 5.

López-Urrutia A, Morán XAG (2007) Resource limitation of bacterial production distorts the temperature dependence of oceanic carbon cycling. Ecology 88:817–822

Lorke A, Tietze K, Halbwachs M, Wüest A (2004) Response of Lake Kivu stratification to lava inflow and climate warming. Limnol Oceanogr 49(3):778–783

MacIntyre S (2012) Climatic variability, mixing dynamics, and ecological consequences in the African Great Lakes. In: Goldman CR, Kumagai M, Robarts RD (eds) Climate change and global warming of inland waters. Wiley-Blackwel, Chichester, pp 311–334

MacIntyre S, Melack JM (1982) Meromixis in an equatorial African soda lake. Limnol Oceanogr 27:595–609

Macuiane MA (2014) Cage aquaculture and environment in Lake Malawi: an assessment of water quality, food webs shifts, and development of a decision support tool for sustainable aquaculture. Doctoral dissertation, University of Minnesota

Marijnissen SAE, Michel E, Kamermans M et al (2008) Ecological correlates of species differences in the Lake Tanganyika crab radiation. Hydrobiologia 615:81–94

McManus J, Severmann S, Cohen AS et al (2015) The sedimentary response to a rapid change in lake level in Lake Tanganyika. Palaeogeogr Palaeoclimatol Palaeoecol 440:647–658

Melack JM (1978) Morphometric, physical and chemical features of the volcanic crater lakes of western Uganda. Arch Hydrobiol 84(4):430–453

Merbt SN, Stahl DA, Casamayor EO et al (2012) Differential photoinhibition of bacterial and archaeal ammonia oxidation. FEMS Microbiol Lett 327(1):41–46

Michel E, Todd JA, Cleary DFR et al (2003) Scales of endemism: challenges for conservation and incentives for evolutionary studies in a gastropod species flock from Lake Tanganyika. J Conchol (Special Publication) 3:1–18

Mitchell TD, Hulme M (2000) A country-by-country analysis of past and future warming rates. Tyndall Centre working paper no. 1., UEA, Norwich

Mizuno K (1998) Succession processes of alpine vegetation in response to glacial fluctuations of Tyndall Glacier, Mt. Kenya, Kenya. Arct Alp Res 30(4):340–348

Morana C, Sarmento H, Descy J-P et al (2014) Production of dissolved organic matter by phytoplankton and its uptake by heterotrophic prokaryotes in large tropical lakes. Limnol Oceanogr 59(4):1364–1375

Morris DP, Lewis WM Jr (1992) Nutrient limitation of bacterioplankton growth in Lake Dillon, Colorado. Limnol Oceanogr 37(6):1179–1192

Moutin T, Tron FT, van Wambeke F et al (2002) Does competition for nanomolar phosphate supply explain the predominance of the cyanobacterium *Synechococcus*? Limnol Oceanogr 47 (5):1562–1567

Myklestad SM (2000) Dissolved organic carbon from phytoplankton. In: Wangersky PJ (ed) Marine chemistry, vol 5. Springer, Berlin, pp 111–148

Müller-Niklas G, Heissenberger A, Puskaric S, Herndl GJ (1995) Ultraviolet-B radiation and bacterial metabolism in coastal waters. Aquat Microb Ecol 9:111–116

Muschick M, Nosil P, Roesti M, Dittmann MT, Harmon L, Salzburger W (2014) Testing the stages model in the adaptive radiation of cichlid fishes in East African Lake Tanganyika. Proc R Soc Lond B: Biol Sci 281(1795):20140605

Newton RJ, Jones SE, Eiler A et al (2011) A guide to the natural history of freshwater lake bacteria. Microbiol Mol Biol Rev 75(1):14–49

Nixon SW (1988) Physical energy inputs and the comparative ecology of lake and marine ecosystems. Limnol Oceanogr 33(4):1005–1025

O'Reilly CM et al (2003) Climate change decreases aquatic ecosystem productivity of Lake Tanganyika, East Africa. Nature 424:766–768

Obernosterer I, Herndl GJ (1995) Phytoplankton extracellular release and bacterial growth: dependence on the inorganic N:P ratio. Mar Ecol Prog Ser 116:247–257

Otu MK, Ramlal P, Wilkinson P et al (2011) Paleolimnological evidence of the effects of recent cultural eutrophication during the last 200 years in Lake Malawi, East Africa. J Great Lakes Res 37:61–74

Pasche N, Dinkel C, Müller B et al (2009) Physical and bio-geochemical limits to internal nutrient loading of meromictic Lake Kivu. Limnol Oceanogr 54(6):1863

Pasche N, Schmid M, Vazquez F et al (2011) Methane sources and sinks in Lake Kivu. J Geophys Res Biogeosci 116(G3).

Patterson G, Kachinjika O (1995) Limnology and phytoplankton ecology. In: Menz A (ed) The fishing potential and productivity of the pelagic zone of Lake Malawi/Niassa. Scientific report of the UK/SADC pelagic fish resource assessment project. Natural Resources Institute, Chatham, pp 1–67

Patterson G, Hecky RE, Fee EJ (2000) Effect of hydrological cycles on planktonic primary production in Lake Malawi/Niassa. Adv Ecol Res 31:421–430

Paulmier A, Ruiz-Pino D (2009) Oxygen minimum zones (OMZs) in the modern ocean. Prog Oceanogr 80(3–4):113–128

Peeters F, Finger D, Hofer M et al (2003) Deep-water renewal in Lake Issyk-Kul driven by differential cooling. Limnol Oceanogr 48(4):1419–1431

Pérez MT, Sommaruga R (2006) Differential effect of algal- and soil-derived dissolved organic matter on alpine lake bacterial community composition and activity. Limnol Oceanogr 51(6):2527–2537

Plisnier PD (2000) Recent climate and limnology changes in Lake Tanganyika. Verh Int Ver Theor Angew Limnol 27:2670–2673

Pollard RT, Rhines PB, Thompson RORY (1973) The deepening of the wind-mixed layer. Geophys. Fluid Dyn 3:381–404

Ramlal PS (2002) Sources, transport and sinks of organic matter in Lake Malawi and Lake Victoria, East Africa. PhD. Thesis, University of Waterloo

Ramlal PS, Hecky RE, Bootsma HA, Schiff SL, Kingdon MJ (2003) Sources and fluxes of organic carbon in Lake Malawi/Nyasa. J Great Lakes Res 29:107–120

Ricketts RD, Johnson TC (1996) Early Holocene changes in lake level and productivity in Lake Malawi as interpreted from oxygen and carbon isotopic measurements of authigenic carbonates. In: Johnson TC, Odada EO (eds) The limnology, climatology and paleoclimatology of the East African Rift Lakes. Gordon and Breach, Amsterdam, pp 475–493

Rinke C, Schwientek P, Sczyrba A, Ivanova NN, Anderson IJ, Cheng JF et al (2013) Insights into the phylogeny and coding potential of microbial dark matter. Nature 499:431–437

Rodrigo MA, Miracle MR, Vicente E (2001) The meromictic Lake La Cruz (Central Spain). Patterns of stratification. Aquat Sci Res Across Bound 63:406–416

Ross KA, Gashugi E, Gafasi A et al (2015a) Characterisation of the subaquatic groundwater discharge that maintains the permanent stratification within Lake Kivu; East Africa. PLoS One 10(3):e0121217

Ross KA, Schmid M, Ogorka S et al (2015b) The history of subaquatic volcanism recorded in the sediments of Lake Kivu; East Africa. J Paleolimnol 54(1):137–152

Ruiz-González C, Lefort T, Galí M et al (2012) Seasonal patterns in the sunlight sensitivity of bacterioplankton from Mediterranean surface coastal waters. FEMS Microbiol Ecol 79:661–674

Rumes B, Eggermont H, Verschuren D (2005) Representation of aquatic invertebrate communities in subfossil death assemblages sampled along a salinity gradient of western Uganda crater lakes. Hydrobiologia 542(1):297–314

Russell J, Bijaksana S (2012) The Towuti Drilling Project: paleoenvironments, biological evolution, and geomicrobiology of a tropical Pacific Lake. Sci Drill 14:68–71

Ruttner F (1931) Hydrographische und hydrochemische Beobachtungen auf Java, Su- matra, und Bali. Arch Hydrobiol Suppl 8:197–454

Salonen K, Sarvala J, Järvinen M, Langenberg V, Nuottajärvi M, Vuorio K, Chitamwebwa DBR (1999) Phytoplankton in Lake Tanganyika—vertical and horizontal distribution of in vivo fluorescence. In: From limnology to fisheries: Lake Tanganyika and other large lakes. Springer, Netherlands, pp 89–103

Salonen K, Hogmander P, Molsa H et al (2012) *Limnocnida tanganyicae* medusae (Cnidaria: Hydrozoa): a semiautonomous microcosm in the food web of Lake Tanganyika. Hydrobiologia 690:97–112

Salzburger W, Mack T, Verheyen E, Meyer A (2005) Out of Tanganyika: genesis, explosive speciation, key-innovations and phylogeography of the haplochromine cichlid fishes. BMC Evol Biol 5(1):1

Sarmento H (2012) New paradigms in tropical limnology: the importance of the microbial food web. Hydrobiologia 686(1):1–14

Sarmento H, Gasol JM (2012) Use of phytoplankton-derived dissolved organic carbon by different types of bacterioplankton. Environ Microbiol 14(9):2348–2360

Sarmento H, Montoya JMJ, Vázquez-Domínguez EE (2010) Warming effects on marine microbial food web processes: how far can we go when it comes to predictions? Philos Trans R Soc Lond B Biol Sci 365(1549):2137–2149

Savijärvi H (1997) Diurnal winds around Lake Tanganyika. Q J Roy Meteorol Soc 123:901–918

Schmid M, Wüest A (2012) Stratification, mixing and transport processes in Lake Kivu. In: Descy J-P, Darchambeau F, Schmid M (eds) Lake Kivu: Limnology and biogeochemistry of a tropical great lake. Springer, Dordrecht, pp 13–29

Schmid M, Tietze K, Halbwachs M et al (2003) How hazardous is the gas accumulation in Lake Kivu? Arguments for a risk assessment in light of the Nyiragongo volcano eruption of 2002. Acta Vulcanol 15(1–2):115–122

Schmid M, Halbwachs M, Wehrli B, Wüest A (2005) Weak mixing in Lake Kivu: new insights indicate increasing risk of uncontrolled gas eruption. Geochem Geophys Geosyst 6(7):1–11

Schon I, Poux C, Verheyen E, Martens K (2014) High cryptic diversity and persistent lineage segregation in endemic *Romecytheridea* (Crustacea, Ostracoda) from the ancient Lake Tanganyika (East Africa). Hydrobiologia 739:119–131

Schubert CJ, Vazquez F, Lösekann-Behrens T et al (2011) Evidence for anaerobic oxidation of methane in sediments of a freshwater system (Lago di Cadagno). FEMS Microbiol Ecol 76 (1):26–38

Sigurdsson H, Devine JD, Tchua FM et al (1987) Origin of the lethal gas burst from Lake Monoun, Cameroun. J Volcanol Geotherm Res 31(1–2):1–16

Snoeks J (2000) How well known is the ichthyodiversity of the large East African lakes? Adv Ecol Res 31:17–38

Sommaruga R, Obernosterer I, Herndl GJ, Psenner R (1997) Inhibitory effect of solar radiation on thymidine and leucine incorporation by freshwater and marine bacterioplankton. Appl Environ Microbiol 63(11):4178–4184

Sommaruga R, Sattler B, Oberleiter A et al (1999) An *in situ* enclosure experiment to test the solar UVB impact on plankton in a high-latitude mountain lake. II Effects on the microbial food web. J Plankton Res 21:859–876

Sommer T, Carpenter JR, Schmid M et al (2013) Interface structure and flux laws in a natural double-diffusive layering. J Geophys Res Oceans 118(11):6092–6106

Spigel RH, Coulter GW (1996) Comparison of hydrology and physical limnology of the East African great lakes: Tanganyika, Malawi, Victoria, Kivu and Turkana (with reference to some North American Great Lakes). In: Johnson TC, Odada EO (eds) The limnology, climatology and paleoclimatology of the East African. Gordon and Breach, Amsterdam, pp 103–140

Spudich JL, Yang CS, Jung KH, Spudich EN (2000) Retinylidene proteins: structures and functions from archaea to humans. Annu Rev Cell Dev Biol 16:365–392

Talling JF (1976) The depletion of carbon dioxide from lake water by phytoplankton. J Ecol 64 (1):79–121

Talling JF (1992) Ecological effects of wastewater. Chapman and Hall, London

Thiery W, Stepanenko VM, Fang X et al (2014) LakeMIP Kivu: evaluating the representation of a large, deep tropical lake by a set of one-dimensional lake models. Tellus A 66

Tierney JE, Mayes MT, Meyer N, Johnson C, Swarzenski PW, Cohen AS, Russell JM (2010) Late-twentieth-century warming in Lake Tanganyika unprecedented since AD 500. Nat Geosci 3(6):422–425

Tierney JE, Smerdon JE, Anchukaitis KJ, Seager R (2013) Multidecadal variability in East African hydroclimate controlled by the Indian Ocean. Nature 493(7432):389–392

Tonolla M, Demarta A, Peduzzi R, Hahn D (1999) In situ analysis of phototrophic sulphur bacteria in the chemocline of meromictic Lake Cadagno (Switzerland). Appl Environ Microbiol 65 (3):1325–1330

Tyson PD, Mason SJ, Jones MQW, Cooper GRJ (1999) Global warming and geothermal profiles: The surface rock-temperature response in South-Africa. Geophys Res Lett 25(14):2711–2713

Van Well P, Chapman DW (1976) Summary of limnological observations in Lake Tanganyika near Kigoma 1974-1975. United nations Food and Agriculture Organization URT/71/012/39, Rome.

Vázquez-Domínguez E, Vaqué D, Gasol JM (2007) Ocean warming enhances respiration and carbon demand of coastal microbial plankton. Glob Chang Biol 13:1327–1334

Verburg P, Antenucci JP (2010) Persistent unstable atmospheric boundary layer enhances sensible and latent heat loss in a tropical Great Lake—Lake Tanganyika. J Geophys Res 115:D11109. doi:10.1029/2009JD012839

Verburg P, Bills IR (2007) Two new sympatric species of *Neolamprologus* (Teleostei: Cichlidae) from Lake Tanganyika, East Africa. Zootaxa 1612:25–44

Verburg P, Hecky RE (2003) Wind patterns, evaporation and related physical variables in Lake Tanganyika. East Afr J Great Lakes Res 29(Suppl 2):48–61

Verburg P, Hecky RE (2009) The physics of the warming of Lake Tanganyika by climate change. Limnol Oceanogr 54:2418–2430

Verburg P, Hecky RE, Kling H (2003) Ecological consequences of a century of warming in Lake Tanganyika. Science 301:505–507

Verburg P, Antenucci JP, Hecky RE (2011) Differential cooling drives large-scale convective circulation in Lake Tanganyika. Limnol Oceanogr 56:910–926

Verschuren D, Laird KR, Cumming BF (2000) Rainfall and drought in equatorial east Africa during the past 1100 years. Nature 403:410–414

Villanueva L, Schouten S, Sinninghe Damsté JS (2014) Depth-related distribution of a key gene of the tetraether lipid biosynthetic pathway in marine Thaumarchaeota. Environ Microbiol. doi:10.1111/1462-2920.12508

Vollmer MK, Weiss RF, Bootsma HA (2002) Ventilation of Lake Malawi/Nyasa. Springer, Dordrecht, pp. 209–233

Vollmer MK, Bootsma HA, Hecky RE et al (2005) Deep-water warming trend in Lake Malawi, East Africa. Limnol Oceanogr 50(2):727–732

Vuorio K, Nuottajarvi M, Salonen K, Sarvala J (2003) Spatial distribution of phytoplankton and picocyanobacteria in Lake Tanganyika in March and April 1998. Aquat Ecosyst Health Manage 6:263–278

White PA, Kalff J, Rasmussen JB, Gasol JM (1991) The effect of temperature and algal biomass on bacterial production and specific growth rate in freshwater and marine habitats. Microb Ecol 21(1):99–118

Worden AZ, Follows MJ, Giovannoni SJ et al (2015) Rethinking the marine carbon cycle: factoring in the multifarious lifestyles of microbes. Science 347(6223):1257594–1257594

Wright JJ, Konwar KM, Hallam SJ (2012) Microbial ecology of expanding oxygen minimum zones. Nat Rev Microbiol 10(6):381–394

Wüest A, Piepke G, Johnson TC et al (1996) Combined effects of dissolved solids and temperature on the density stratification of Lake Malawi. Gordon and Breach, Amsterdam, pp. 183–202

Wüest A, Jarc L, Bürgmann H et al (2012) Methane formation and future extraction in Lake Kivu. In: Descy J-P, Darchambeau F, Schmid M (eds) Lake Kivu: limnology and biogeochemistry of a tropical great lake. Springer, Dordrecht, pp 165–180

Yakimov MM, La Cono V, Slepak VZ et al (2013) Microbial life in the Lake Medee, the largest deep-sea salt-saturated formation. Sci Rep 3:3554

Zegeye A, Bonneville S, Benning LG et al (2012) Green rust formation controls nutrient availability in a ferruginous water column. Geology 40(7):599–602

Zhang X, Scholz CA, Hecky RE et al (2014) Climatic control of the late quaternary turbidite sedimentology of Lake Kivu, East Africa: implications for deep mixing and geologic hazards. Geology 42(9):811–814

Zlotnik I, Dubinsky X (1989) The effect of light and temperature on DOC excretion by phyotplankton. Limnol Oceanogr 34(5):831–839

Chapter 11
Mono Lake: Plankton Dynamics over Three Decades of Meromixis or Monomixis

John M. Melack, Robert Jellison, Sally MacIntyre, and James T. Hollibaugh

11.1 Physical and Chemical Characteristics

11.1.1 Lake Morphometry, Hydrology, Geology of the Basin, and Climate

Mono Lake (38°N, 119°W) lies in a hydrologically closed basin on the western edge of the North American Great Basin below the eastern slope of the Sierra Nevada (Fig. 11.1). The lake's roughly elliptical basin and its islands were formed by tectonic and volcanic activity (Scholl et al. 1967; Gilbert et al. 1969). Paoha Island resulted from volcanic activity and uplift of lake sediments between ca. 1720 and 1850 (Patten et al. 1987). Situated at an elevation of 1943 m asl, the lake has a surface area of 160 km^2, mean depth of 17 m, and maximum depth of 45 m (Fig. 11.2). The catchment area of the lake ranges from 1642 to 2075 km^2 due to uncertainties in the flow paths of subsurface water from the east and south (Vorster 1985). Elevations around the lake reach 3960 m along the Sierra crest. Mesozoic granitic and Paleozoic metamorphic materials occur in the Sierra Nevada, while Miocene and Quaternary volcanic material is evident to the north, east, and south (Patten et al. 1987).

There is geological evidence that the lake is at least 500,000 years old (Lajoie 1968). In the late Pleistocene depths reached ca. 300 m (Benson et al. 1998), and

J.M. Melack (✉) • S. MacIntyre
Department of Ecology, Evolution and Marine Biology, University of California, Santa Barbara, CA, USA
e-mail: john.melack@lifesci.ucsb.edu

R. Jellison
Marine Science Institute, University of California, Santa Barbara, CA, USA

J.T. Hollibaugh
Marine Sciences, University of Georgia, Athens, GA, USA

© Springer International Publishing AG 2017
R.D. Gulati et al. (eds.), *Ecology of Meromictic Lakes*, Ecological Studies 228,
DOI 10.1007/978-3-319-49143-1_11

Fig. 11.1 NASA Landsat 7 satellite image of Mono Lake and its surroundings, December 16, 1999. Paoha Island is the large island near the center of the lake. The Sierra Nevada is partially snow covered to the west. From http://earthobservatory.nasa.gov/IOTD/view.php?id=8388

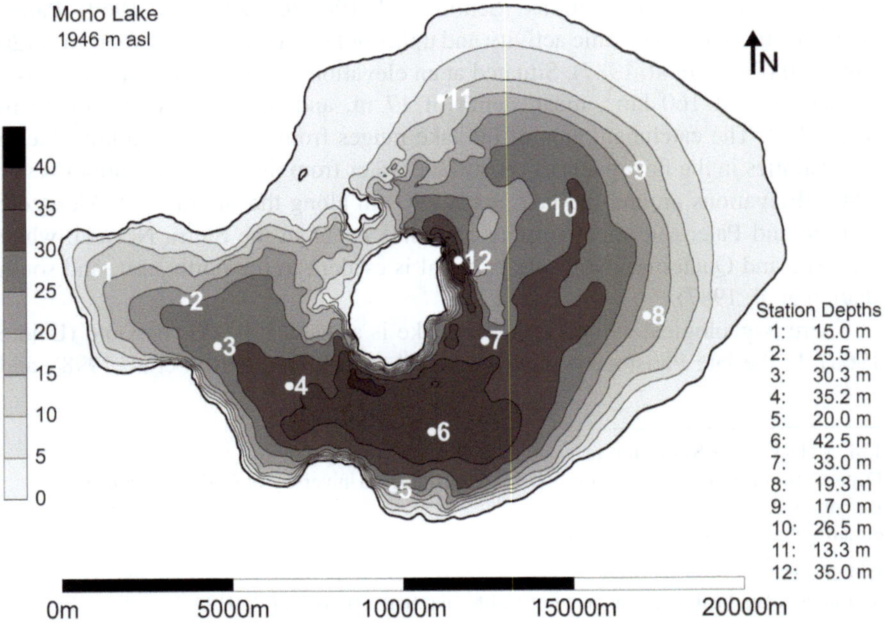

Fig. 11.2 Mono Lake bathymetry and location of sampling stations (*white dots*)

prolonged droughts in the last millennium resulted in low lake levels (Stine 1994). Diversions of freshwater streams that flowed into Mono Lake led to a 14-m decline in surface elevation in the 40-year period from 1941 to 1981 (Patten et al. 1987). During the Pleistocene glaciations, the lake sediments were composed largely of allochthonous material, while during the Holocene, autochthonous organic matter increased (Reed 1977). Finely laminated sediments provide a detailed record of organic matter accumulation and reveal a gradual doubling of accumulation rates during the period of increasing salinities starting in the late 1940s (Jellison et al. 1996).

Measurements of air temperature and precipitation near the lake are available since the early 1950s though snow and rain data from the Sierra Nevada date to the early 1920s. Most precipitation falls as snow and predominately on the western side of the basin as moisture from the Pacific Ocean ascends over the Sierra Nevada. Summer rains are associated with convective systems receiving moisture from the tropical Pacific, the Gulf of California, or occasionally the Gulf of Mexico (Hales 1974). Vorster's (1985) and Patten et al.'s (1987) descriptions of the water balance for Mono Lake note that the surface runoff originates mainly in the Sierra Nevada and is snowmelt occurring in May, June, and July. Groundwater enters the lake via springs and seepage (Lee 1969; Clark and Hudson 2001). Evaporation from the lake is estimated to be about 1 m per year (Vorster 1985).

Since 1990 meteorological data (wind speed and direction, air temperature, humidity, and precipitation) have been collected at a station located on the southern tip of Paoha Island and radiation (shortwave, longwave, and photosynthetically available radiation (PAR)) at a site located about 7 km southwest of the lake. PAR exhibits a regular sinusoidal curve dictated by the latitude of Mono Lake; maximum daily values typically range from about ~19 Einsteins m^{-2} at winter solstice to ~65 Einsteins m^{-2} in mid-June. Mean daily wind speeds generally vary from 1 to 13 $m\ s^{-1}$ over the year. Winds are predominately from the south southwest. Mean daily air temperatures range from around -11 °C in winter to a maximum of around 24 °C in summer. Mean daily relative humidity follows a general pattern of high values (mostly between 70 and 96 %) in January, decreasing to lows (mostly from 30 to 60 %) in April through September and then gradually to above 70 % through December.

11.1.2 Physical Processes: Density Structure, Internal Waves, and Boundary Mixing

Measurements and modeling of vertical stratification and mixing in Mono Lake are treated by Jellison and Melack (1993a), Jellison et al. (1993, 1998), MacIntyre et al. (1999), MacIntyre and Romero (2000), Melack and Jellison (1998), Romero and Melack (1996), Romero et al. (1998), Vidal et al. (2013), and other papers discussed below. Mono Lake alternates between being meromictic and monomictic depending on the quantity of freshwater introduced during snowmelt from the

Sierra Nevada and the proportion diverted to the City of Los Angeles (Figs. 11.3 and 11.4). Freshening with snowmelt water increases the density difference between the upper water column (mixolimnion) and lower water column (monimolimnion). As a monomictic lake, mixing from autumn to spring is caused by a combination of convection mediated by increases in density near the surface

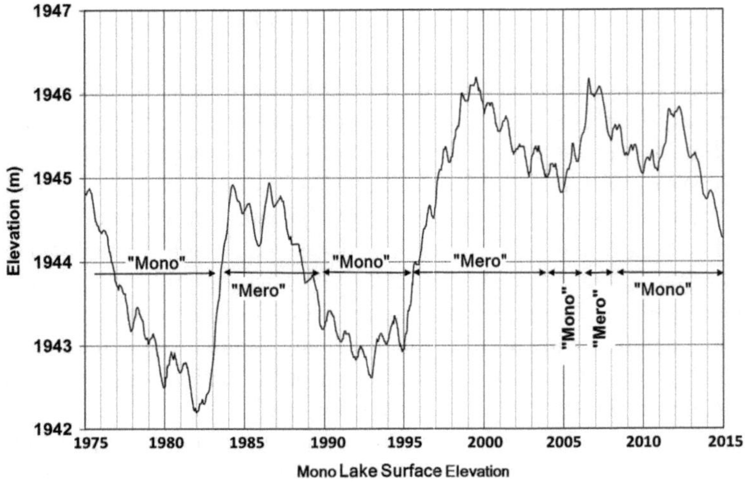

Fig. 11.3 Elevations of surface of Mono Lake with periods of monomixis and meromixis indicated

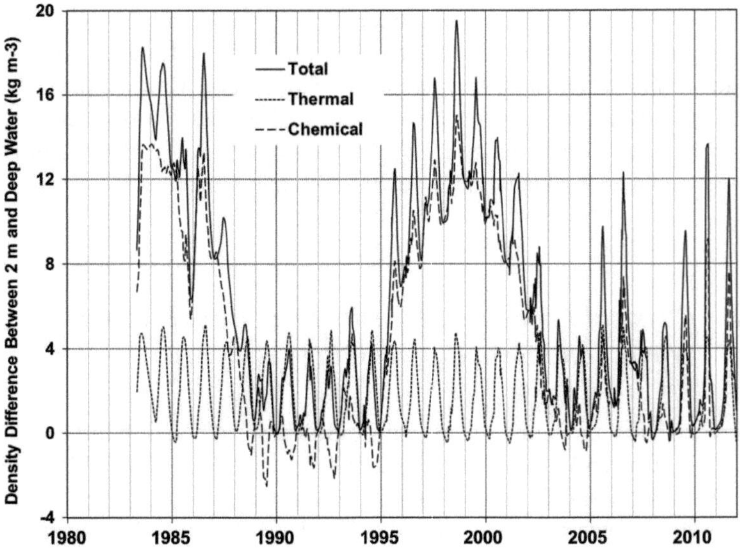

Fig. 11.4 Density difference (kg m^{-3}) between 2 and deep water (28 m, 1983–1990; 32 m, 1991–2011) at station 6 due to temperature and chemical stratification from 1991–2011

due to cooling and evaporative increase in concentration of salts and by wind-induced instabilities in the internal wave field. In the years when the lake is meromictic, the density gradient across the pycnocline is large enough that convection cannot penetrate to the bottom. The interannual variability in thermally induced convection is small; the interannual variability in convection induced by near-surface increases in salt concentration is somewhat greater but limited in range. When considerable freshening of the upper layer occurs, it can require several years for evaporation to increase the salinity of the upper water column and internal wave-induced freshening of the lower water column to weaken the density gradient across the main pycnocline sufficiently that complete mixing can occur. The depth of mixing in winter is a dominant factor in controlling nutrient supply to the upper water column and annual productivity. The interannual variations in vertical mixing in the context of plankton dynamics are summarized in Sect. 11.3.

During meromixis, Mono Lake is both thermally and chemically stratified (Fig. 11.4). The depth of thermal stratification varies seasonally, with the thermocline deeper in winter and often congruent with the chemocline. With warming, the thermocline shoals such that by early spring and through the summer, it lies above the chemocline. Thus, the lake is often three layered, may have an upper layer which is stratified to the surface, and may have an additional layer when, as is typical, a diurnal thermocline forms. The stability of stratification is high. For example, in spring 1998 as the thermocline was first rising in the water column due to spring heating, buoyancy frequencies across the thermocline at 10 m were 10 cycles per hour (cph) and 80 cph across the chemocline at 20 m. Late summer 2000, when solar heating had intensified the temperature stratification, buoyancy frequency across the thermocline at 10 m was 50 cph and 90 cph across the chemocline at 22 m. After the onset of convection in fall 1995, the thermocline and chemocline were colocated at 10 m; the buoyancy frequency across the pycnocline was 70 cph.

The layered density structure creates conditions favorable for the first vertical mode internal waves, in which the movements of the upper and lower portions of the pycnocline are in phase, and for higher-order vertical mode internal waves, which cause the movements of the upper and lower boundaries of the pycnoclines to be out of phase (Vidal et al. 2013; Fig. 11.5). MacIntyre et al. (2009) illustrated that at depths above the chemocline in spring, the upper water column initially behaved as a first vertical mode wave when a thermocline was present. As heat was mixed downward and the thermocline became more diffuse, higher vertical modes formed. Vidal et al.'s (2013) modeling illustrated the complexity of the internal wave field: as the thermocline upwelled, the chemocline downwelled. Vidal et al. (2013) further illustrated that the presence of Paoha Island effectively causes Mono Lake to act as two lakes connected to the south with the nonoverlapping thermoclines and chemoclines oscillating in each basin. Rotation affects the internal wave motions such that Kelvin waves form in the eastern and western basins. However, the lake is only sufficiently wide for rotation to occur at depths above the chemocline. Rotation also occurs when winds shift direction.

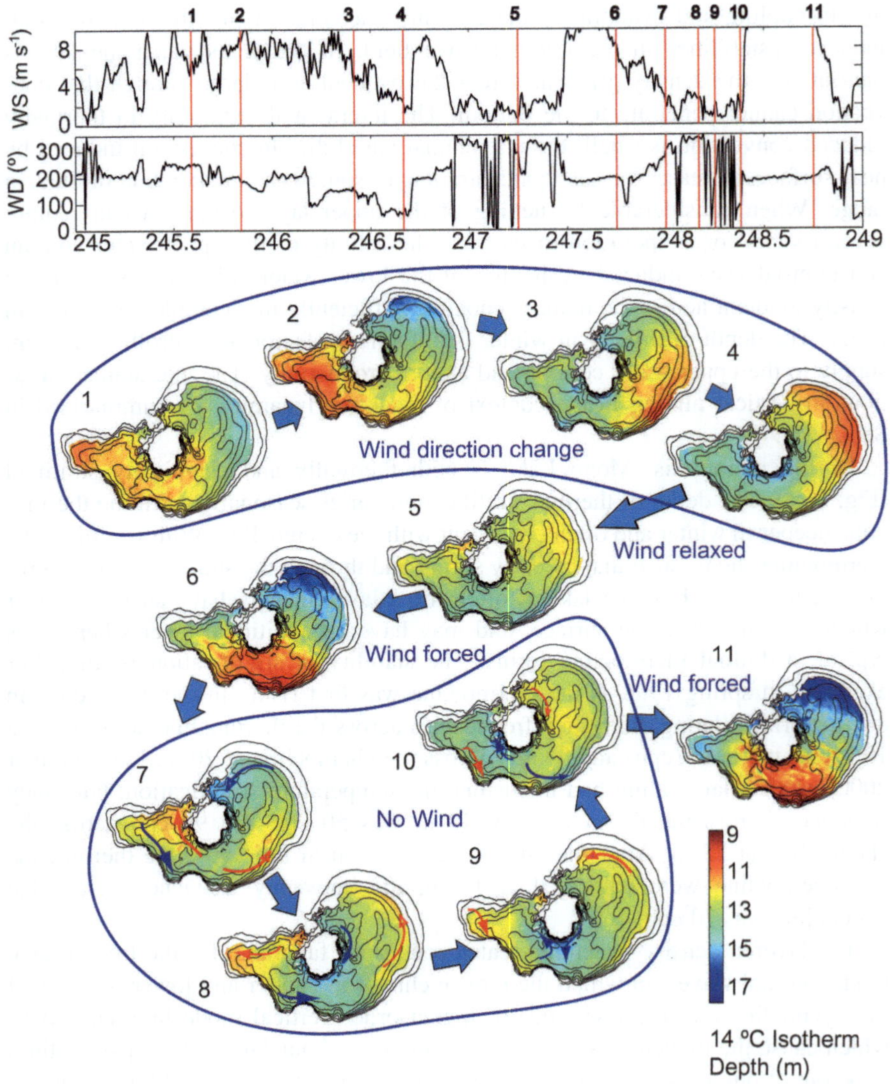

Fig. 11.5 Time series of wind speed and direction (*upper panels*) and spatial evolution of the internal wave field illustrated as the depth of the 14 °C isotherm (*lower figures*) in September 2000. *Red lines* in the upper panel correspond to 1–11 below. Downwelling occurred in response to southwesterlies (1) with the downwelling intensified as winds increased (2); the downwelled water rotated as winds shifted direction (3,4). Isotherms were level when winds relaxed (5); with southerly winds, upwelling occurred to the south and downwelling to the northeast and northwest (6). As winds relaxed, the rotation of the downwelled water indicated that Kelvin waves were present in both subbasins (7–10). Intensification of southerly winds once again caused upwelling to the south and downwelling at the other extremes of the lake (11). Redrawn from Vidal et al. (2013)

Knowledge of the internal wave field in lakes is important to the ecology. The vertical motions can expose phytoplankton to higher irradiance than in static layers enabling higher primary production near the base of the euphotic zone where chlorophyll maxima are often located (Evans et al. 2008). Breaking of nonlinear internal waves causes vertical fluxes (MacIntyre et al. 1999, 2009; MacIntyre and Jellison 2001), and the currents induced by internal waves can be important for horizontal exchange. While mixing due to convection causes entrainment of water from the upper pycnocline into the mixed layer, the breaking of internal waves enables fluxes from deeper in the water column, and they are an important mechanism enabling exchanges during stratification.

The degree of nonlinearity and extent of vertical mixing can be predicted from dimensionless indices that take into account stratification, lake morphometry, and the magnitude of winds. The Wedderburn, W, and Lake numbers, L_N, are the relevant indices, with L_N being an integral form of W (Imberger and Patterson 1990). When values of W or L_N exceed 10, little upwelling of the thermocline is expected, and internal waves which form will be linear. As values decrease below 10 but are >1, wind-induced tilting of the thermocline will occur with the extent of tilting and the degree of nonlinearity increasing with the decrease in W or L_N. Internal waves are expected to break nearshore. With W or L_N near 1, full upwelling will occur and mixing is predicted to occur along the horizontal extent of the thermocline. MacIntyre et al. (1999) were the first to demonstrate this process in the field using direct measurements of elevated turbulence in the thermocline from internal wave breaking near lateral boundaries. Turbulence, characterized by the coefficient of eddy diffusivity (K_z), was four orders of magnitude higher nearshore than at offshore sites. The winds in spring 1998 were strong enough to induce W near 1, and the abrupt deepening of the thermocline indicative of the intense mixing from internal bores occurred. The use of tracers, however, indicated that vertical mixing offshore was limited, which poses a challenge for growth of microbes and phytoplankton (MacIntyre et al. 2009). Thus, while the vertical fluxes can reduce nutrient limitation and support growth nearshore (MacIntyre and Jellison 2001), intrusions of well-mixed water are hypothesized to form and flow offshore and support growth offshore. Nonlinear waves are progressive, that is, they can flow offshore and induce transport. While most of the mixing occurs nearshore, the nonlinear second vertical mode waves that form as a result of the nearshore disturbances cause enhanced shear and dissipation at the top and bottom of the thermocline (S. MacIntyre, unpublished data). This mixing may also be important for supplying subsidies to the various guilds of bacteria found at different depths offshore in Mono Lake. Three-dimensional hydrodynamic modeling supports the observational data that most mixing would occur nearshore (Vidal et al. 2013). These several studies imply that the biological growth in offshore waters during the stratified period will be supported by subsidies mixed vertically nearshore and subsequently transported offshore.

A one-dimensional model, appropriate for multiyear simulations and modified to include boundary mixing (Romero and Melack 1996), was used to forecast the duration of meromixis under varying climatic conditions (Jellison et al. 1998).

Results indicated that meromixis would persist for decades, whereas in reality it lasted less than a decade. This discrepancy highlights the challenge of accurately capturing the vertical mixing induced by nonlinear internal waves when $W < 1$. That such events occurred several times during a 10-day study in April 1998 suggests that these events occur frequently during the winter with the common occurrence of high winds during the passage of frontal systems. One-dimensional models are frequently used to make predictions over long periods of time for climate-related studies. During conditions of weak stratification with high winds, such as in fall, winter, and spring in Mono Lake, internal wave motions are highly nonlinear and the implications for mixing as yet uncaptured in 1-D and 3-D hydrostatic models. In fact, winds are high enough over Mono Lake even during the more strongly stratified period for nonlinear waves to form. Thus, predictions as to whether Mono Lake will be meromictic or monomictic under future climate change scenarios will only become possible with improved incorporation of internal wave dynamics in hydrodynamic models.

11.1.3 Major Solutes and Geochemistry

Mono Lake is hypersaline with sodium the major cation (97 %) and carbonate (45 %), chloride (40 %), and sulfate (14 %) the major anions; both cations and anions are given as percentages of meq kg^{-1}, respectively (Oxburgh et al. 1991). The pH is about 10 (Mason 1967). The geochemistry is described by Domagalski et al. (1989) and Johannesson and Lyons (1994). From 1941 to 1981, the lake's salinity increased from about 50 g L^{-1} to about 100 g L^{-1} as a result of reduced inflows and evaporative concentration (Patten et al. 1987). The geochemical evolution of waters derived from incongruent weathering in the Sierra Nevada, followed by evaporative concentration and mineral precipitation, is described in the classic study by Garrels and MacKenzie (1967) and placed in a general context by Eugster and Hardie (1978).

The high salinity and particular chemical composition result in distinct physical properties of the lake water. Since the thermal capacity per gram is lower than pure water, less energy is required to heat a gram of water, and its viscosity is ca. 20 % higher than pure water (Mason 1967). Density increases as a monotonic function of temperature decreases, and the equation of state differs from that of seawater (Jellison and Melack 1993a; Jellison et al. 1999). The lake's epilimnion can have dissolved oxygen values above saturation during spring with abundant algae, while bottom waters are anoxic when the lake is stratified and attain sulfide concentrations of up to 3 mM during meromixis.

Mono Lake's alkaline, saline waters result in high solubilities of actinide elements (Simpson et al. 1980; Anderson et al. 1982). It has an unresolved radiocarbon budget (Broecker et al. 1988). Hydrothermally derived arsenic is high (~200 μM, Maest et al. 1992; Hollibaugh et al. 2005). Arsenic has an active geochemical cycle in the lake, with arsenate (As(V)) as the dominant species in the oxic epilimnion; arsenite (As(III)) as the dominant species in the anoxic, but not sulfidic, metalimnion; and a series of thioarsenic compounds as the dominant species in the sulfidic hypolimnion (Hollibaugh et al. 2005). Scenic tufa towers fringe the lake as a result of precipitation of calcite, high-magnesium calcite, and aragonite where sublacustrine springs flow into the lake (Dunn 1953; Cloud and Lajoie 1980).

11.1.4 Nutrients

Concentrations of soluble reactive phosphorus are high in Mono Lake (>400 μM). Inorganic nitrogen, primarily ammonium or free ammonia, varies seasonally and is often <5 μM in the epilimnion; nitrate is often below detection (Jellison et al. 1993; Jellison and Melack 2001). A positive response by phytoplankton to ammonium enrichments indicates that inorganic nitrogen can limit algal growth (Jellison et al. 1992; Jellison and Melack 2001). Interannual variations in nutrient levels in the context of meromixis are discussed in Sect. 11.3. Nitrogen fixation has been reported in nearshore sediments in association with aggregates of the filamentous green alga *Ctenocladus circinnatus* with anaerobic bacteria and *Oscillatoria* spp. implicated as the organisms responsible for fixing nitrogen (Oremland 1990; Herbst 1998). Nitrogen fixers have also been detected in the water column (Steward et al. 2004). Ammonia-oxidizing bacteria are present in the lake (Joye et al. 1999) leading to production of nitrite plus nitrate (Carini and Joye 2008). Ammonia-oxidizing Archaea have not been detected in the lake. Additional nitrate enters the lake as runoff, although the amounts are small (Jellison et al. 1993). Denitrification may reduce nitrate and remove fixed nitrogen, though these rates have not be determined. The two major sources of recycled ammonium are brine shrimp excretion and vertical mixing of ammonium-rich deep water (Jellison and Melack 1993a; Jellison et al. 1993; Romero et al. 1998; MacIntyre et al. 1999).

11.2 Biota

11.2.1 Phytoplankton and Benthic Algae

Few species of phytoplankton occur in the hypersaline waters of Mono Lake. A recently described unicellular green alga, *Picocystis* sp. strain ML, accounts for nearly 25 % of the primary production during the winter and more than 50 % at other times of the year (Roesler et al. 2002). *Picocystis* appears to be particularly adapted to low irradiance. Growth and oxygenic photosynthesis were observed under anoxic conditions at rates comparable with those measured under oxic conditions. A cyanobacterium with phylogenetic affinity to the *Cyanobium* lineage (Budinoff and Hollibaugh 2007) is unusual because it appears to be most abundant in anoxic, sulfidic, dark, bottom waters of the lake. It is also present, based on rDNA genes and apparently viable cells in sediment cores, with a prominent peak in abundance at 18 cm.

Cyanobacteria and bacillarophytes (mainly *Nitzschia* spp.) have also been identified (Mason 1967; Lovejoy and Dana 1977). Thirty taxa of benthic diatoms were identified from depths of 1, 5, and 10 m (Kociolek and Herbst 1992). *Navicula crucialis*, *Nitzschia frustulum*, *N. latens*, *N. reimerii* sp. nov., *N. monensis* sp. nov., and *Anomoeoneis sphaeophora* var. *minor* var. nov. were numerically dominant species. *Oscillatoria* spp. are widely distributed over the benthic environment and occur with *Ctenocladus circinnatus* aggregates and as part of microbial mats on rock and organic sediments (Herbst 1998).

The phytoplankton have large seasonal variations in abundance with maxima in early spring associated with elevated ammonium concentration and initiation of stratification in the upper water column, low values of phytoplankton from late spring to late summer as a result of grazing by *Artemia monica*, and increased abundances in autumn and winter as *Artemia* grazing declines (Melack 1983, 2002; Jellison and Melack 1993b; see Sect. 11.3).

11.2.2 Bacteria, Archaea, and Viruses

Strong redox gradients associated with vertical stratification are conducive to establishment of stratified microbial communities (Hollibaugh et al. 2001). Phylogenetic microbial diversity during meromixis was surveyed using denaturing gradient gel electrophoresis (DGGE) combined with cloning and sequencing (Humayoun et al. 2003). Most of the 16S rRNA gene sequences were in five major lineages: α- and γ-*Proteobacteria*, *Bacteroidetes*, high G+C Gram positive (*Actinobacteria*), and low G+C Gram positive (*Bacillus/Clostridium*). The others were represented by β- and δ-*Proteobacteria*, *Verrucomicrobiales*, *Planctomycetales*, and candidate divisions. 16S rRNA gene sequences retrieved from the mixolimnion and oxycline were dominated by three distinct ribotypes of *Actinobacteria* (49 and 63 % of the clones examined, respectively). The chemocline and monimolimnion were more diverse than epilimnetic communities and with broader phylogenetic representation, including a novel clade of *Bacteroidetes*. The population of sequences retrieved from the monimolimnion was dominated (52 %) by low G+C Gram-positive bacteria.

To evaluate spatial and temporal variations in biogeochemical processes, vertical distributions and seasonal variation in rates of aerobic and anaerobic methane oxidation and coupling between anaerobic methane oxidation and dissimilatory sulfate reduction have been determined. Rates of methane oxidation can be high in both monimolimnion and mixolimnion. Aerobic oxidation follows the oxycline/microaerophilic zone as it moves up and down through the water column in response to stratification and mixing (Carini et al. 2005). Aerobic oxidation is higher in the oxycline/microaerophilic zone than in the fully aerobic surface water, likely as a result of both methane supply and inhibition of methane oxidizers by high light. Rates are also high in the monimolimnion (Joye et al. 1999) due to the activity of novel sulfate reducers (Scholten et al. 2005). Oxidation of methane by sulfate reducers appears to compete with oxidation of more energy-yielding organic matter supplied by the spring increases in algal abundance and is lowest early after stratification, when there is a high flux of particulate organic matter to the monimolimnion, and increases later in the season as methane becomes the dominant electron donor for sulfate reducers. Vertically integrated rates are higher in the monimolimnion (Joye et al. 1999) because it has a thicker water column, but rates per unit volume are greatest in the oxycline.

Hydrogen sulfide consumption appears to be mediated by both aerobic and anaerobic processes, with H_2S oxidation linked to arsenate reduction playing a significant role in both H_2S distributions and arsenic speciation in the lake (Hoeft et al. 2004; Hollibaugh et al. 2005, 2006). There is likely a competition between arsenate respiring heterotrophs and arsenate-reducing sulfide oxidizers that is modulated by the supply of labile organic matter to sulfate reducers versus heterotrophs (Hollibaugh et al. 2005).

Arsenic is naturally concentrated in the lake (~200 μM) from geothermal sources in the lake's basin, and arsenate serves as an important alternative electron acceptor in Mono Lake. Arsenate reducers appear to fall into two guilds based on electron donors: heterotrophs using organic compounds like lactate and acetate and chemoautotrophs using sulfide (Hollibaugh et al. 2006) and probably other reduced sulfur compounds. Two isolates, δ-*Proteobacteria* (strain MLMS-1, Hoeft et al. 2004, and a *Desulfobulbus*), are members of the second guild. The *Desulfobulbus* is routinely detected in Mono Lake clone libraries and enrichments. Arsenic can also serve as an electron donor, and aerobic arsenite oxidation is well known (Oremland and Stolz 2003). A nonpigmented member of the *Ectothiorhodospiraceae* (strain MLHE-1) is capable of anaerobic arsenite oxidation using nitrate or ferric iron as electron acceptors (Oremland et al. 2002).

Sulfide oxidation via nitrate reduction proceeds rapidly when nitrate is available and appears to be a significant process that may also compete with arsenate reduction (Hollibaugh et al. 2006). Sulfide is important in the oxidation of arsenite (Fisher et al. 2008) and for the formation of thioarsenic compounds (Hollibaugh et al. 2005; Fisher et al. 2008; Edwardson et al. 2014), and both arsenite and thioarsenic can serve as an electron donor for anoxygenic photosynthesis (Budinoff and Hollibaugh 2008; Kulp et al. 2008; Edwardson et al. 2014).

Arsenate- and nitrate-reducing, sulfide-oxidizing bacteria and anoxygenic phototrophs related to MLHE-1 have been isolated from Mono Lake (Budinoff and Hollibaugh 2008; Kulp et al. 2008) as was another arsenite oxidizer (94 % similar to a *Bacillus* species) that is capable of oxidizing arsenite with selenate (Fisher and Hollibaugh 2008). Experiments with this organism suggest that it preferentially oxidizes thioarsenic complexes (Hollibaugh et al. 2005) formed when arsenite and sulfide are mixed in Mono Lake water.

Jiang et al. (2004) described the abundance and diversity of viruses, and Brum et al. (2005) reported spatial and temporal variability of viruses and viral infections of prokaryotes in Mono Lake. Bacterial abundance ranged from 0.3 to 4.4 $\times$ 10^7 cells mL^{-1} and was greatest in the hypolimnion during summer. Viral abundances were 10–100-folds greater than bacteria abundances. Viruses were estimated to be responsible for 3.7–16 % of the bacteria mortality.

11.2.3 Zooplankton and Zoobenthos

The brine shrimp (*Artemia monica*) is the dominant and often sole species of zooplankton in the pelagic waters (Lenz 1984). *Artemia* hatch in late winter and spring from cysts produced in previous years and that sink to the sediments. A second, summer generation is produced ovoviviparously. In autumn, *Artemia* population declines, because of predation by migratory grebes, decreasing water temperature and senescence. Seasonal and interannual variations are described further in Sect. 11.3.

Several protozoans and occasionally rotifers (*Brachionus plicatilis* and *Hexarthra jenkinae*) are present (Mason 1967; Jellison et al. 2001). In nearshore habitats, the alkali fly (*Ephydra hians*) is abundant seasonally (Herbst 1990). The pupae of the alkali fly were sufficiently abundant to provide a food source for local Paiute people (Williston 1883; Davis 1965) and are an important food for birds.

11.2.4 Birds

The brine shrimp and alkali fly larvae provide food for large numbers of birds including an important breeding colony of the California gull (*Larus californicus*) and migratory eared grebes (*Podiceps nigricollis*), Wilson's phalaropes (*Phalaropus tricolor)*, and red-necked phalaropes (*Phalaropus lobatus*) (Patten et al. 1987). No fish occur in the lake because of the high concentrations and particular dissolved solutes. The grebes depend on the abundant *Artemia* to molt and add reserves for their subsequent flight south (Cooper et al. 1984; Boyd and Jehl 1998; Jehl et al. 2002; Jehl 2007). A phenological shift toward an earlier peak in *Artemia* abundance is favoring spring breeding gulls (Wrege et al. 2006) over autumn populations of grebes.

11.3 Mixing Regimes and Zooplankton Dynamics

A series of publications has dealt with various aspects of the ecology of the lake in the context of the mixing regime (Melack 1983, 2002; Jellison et al. 1992; Jellison and Melack 1993a, b; Melack and Jellison 1998). Here, we update these findings and focus on changes in mixing and algal and *Artemia* abundance. Jellison and Rose

(2012) summarize major aspects of these varying conditions, and we have adapted their text here.

Monomixis and Declining Lake Levels, 1964 to 1982 The limnology of Mono Lake, including seasonal plankton dynamics, was first documented in the mid-1960s (Mason 1967). During this period the lake had declining water levels, increasing salinity, and a monomictic thermal regime with complete vertical mixing during the winter. No further limnological research was conducted until summer 1976 (Winkler 1977). Studies by Lenz (1984), Melack (1983), and Melack et al. (1985), beginning in 1979, focused on seasonality of mixing and plankton abundances. During the period from 1979 to 1981, Lenz (1984) documented a progressive increase in the ratio of peak summer to spring abundances of adult brine shrimp. The smaller spring generation resulted in greater food availability and higher ovoviviparous production by the first generations, leading to larger second generations.

In 1982, an intensive limnological program funded by Los Angeles Department of Water and Power was established to monitor changes in the limnological conditions in Mono Lake. Methods used for measuring different parameters are described in Jellison and Melack (1993a, b) and Melack and Jellison (1998). This monitoring program has continued to the present and has been complemented by studies funded by US National Science Foundation and other organizations. We summarize results from this work with a focus on meromixis, plankton variations, and related limnological conditions.

Meromixis, 1983 to 1987 In 1983, a large influx of stream water to the lake resulted in a condition of persistent chemical stratification (Figs. 11.3 and 11.4). In subsequent years evaporative concentration of the surface water led to a decrease in this gradient, and in November 1988 meromixis ended. These processes are documented in Jellison et al. (1992, 1993), Jellison and Melack (1993a, b), and Miller et al. (1993).

Following the onset of meromixis, ammonium concentrations in the mixolimnion fell to near zero during spring 1983 and remained below 5 μM until late summer 1988. Accompanying this decrease in mixolimnetic ammonium concentrations were low algal abundances from November through April though *Artemia* grazing was largely absent (Fig. 11.6). Ammonification of organic material and release from the anoxic sediments resulted in a gradual buildup of ammonium in the monimolimnion over the 6 years of meromixis to 600–700 μM. Under monomictic conditions in previous years, the summer ammonium accumulation beneath the thermocline reached 80–100 μM and was mixed into the upper water column during the autumn overturn.

The size of the first generation of adult *Artemia* in 1984 (~31,000 m^{-2}) was nearly ten times as large as observed in 1981 and 1982, while peak summer abundances of adults were much lower (Fig. 11.7). Subsequently, the two generations of *Artemia* were relatively constant in numbers during the meromictic period from 1984 to 1987. The size of the spring generation of adult *Artemia* varied from

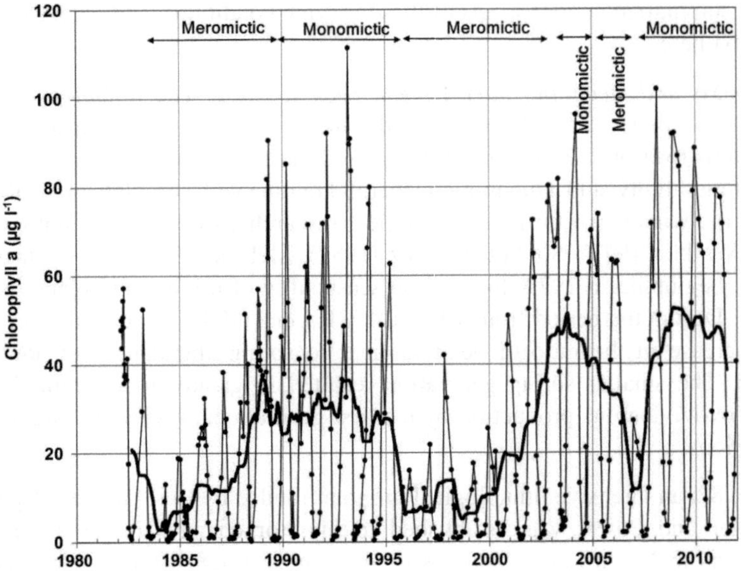

Fig. 11.6 Long-term, mixed-layer chlorophyll *a* in Mono Lake. *Bold lines* indicate seasonally filtered (365-day) running mean

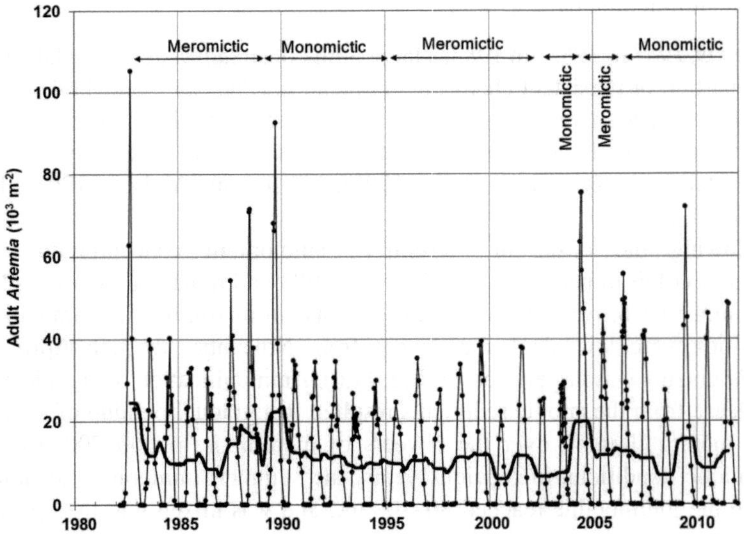

Fig. 11.7 Long-term, adult *Artemia* abundance in Mono Lake. *Bold lines* indicate seasonally filtered (365-day) running mean

23,000 to 31,000 m^{-2}, while the second generation of adult *Artemia* in late spring and summer varied from 33,000 to 54,000 m^{-2}. The relative abundances of the first and second generation are inversely correlated, and this is partially mediated by food availability as a large first-generation results in decreased algal levels for second-generation nauplii and vice versa. From 1984 to 1987, recruitment into the first-generation adults was nearly constant, and only a very small percentage (about 1–3 %) of the cysts were calculated as available to hatch (Dana et al. 1990). Fecundity had a significant correlation with ambient algal concentrations (r^2, 0.61).

Breakdown of Meromixis, 1988 and 1989 Although mixing of the full water column did not occur until November 1988, the successive deepening of the mixed layer during the period from 1986 to 1988 led to changes in the plankton. By spring 1988, the mixed layer included the upper 22 m of the lake, 83 % of the lake volume. In addition to restoring an annual mixing regime to much of the lake, the deepening of the mixed layer increased the nutrient supply to the mixolimnion by entraining deeper water with high ammonium concentrations. Mixolimnetic ammonium concentrations were fairly high during the spring (8–10 μM), and March algal populations were much larger than in 1987 (53 vs. 15 μg Chl *a* L^{-1}).

The peak abundance of spring adult *Artemia* in 1988 was twice as high as any previous year from 1979 to 1987. The pool of *Artemia* cysts available for hatching was potentially larger in 1988 since cyst production in 1987 was larger than in the four previous years (Dana et al. 1990), and significant lowering of the chemocline in the autumn and winter of 1987 allowed oxygenated water to reach cysts in sediments which had been anoxic since 1983. Naupliar survival may have been enhanced by food availability since chlorophyll *a* levels in the spring of 1988 were higher than the previous 4 years. During laboratory experiments, naupliar survival was higher in the ambient food treatment relative to the low food treatment (Dana et al. 1995).

Mono Lake returned to its previous condition of complete annual autumnal mixing from top to bottom with breakdown of meromixis in November 1988. Ammonium, which had accumulated to high levels (>600 μM) in the monimolimnion during meromixis, was dispersed throughout the water column raising its surface concentrations above the previously observed values to >50 μM. Dissolved oxygen concentration decreased by mixing with the anoxic water and is consumed by the biological and chemical oxygen demand previously created in the monimolimnion and fell to zero. *Artemia* populations crashed following deoxygenation. The whole water column in Mono Lake remained anoxic for a few months following the breakdown of meromixis. By mid-February 1989, dissolved oxygen concentrations had increased to 2–3 mg L^{-1} but were still below those observed in previous years (4–6 mg L^{-1}).

Epilimnetic chlorophyll concentrations in March and April 1989 were high (40–90 μg Chl *a* L^{-1}). Subsequently, the concentrations declined in late June to <0.5–2 μg Chl *a* L^{-1} due to brine shrimp (*Artemia*) grazing. In previous years (1982–1988), when the lake was meromictic, this chlorophyll *a* decline occurred up to 6 weeks earlier. Two effects of meromixis on the phytoplankton are evident:

decreased winter and spring concentrations and a shift in the timing of summer period of low chlorophyll.

The 1989 *Artemia* population had a small first generation of adults followed by a summer population an order of magnitude larger. A similar pattern was observed from 1980–1983. In contrast, during meromictic years, the first generation was larger and followed by a summer population of the same order of magnitude. The initiation of hatching occurred slightly later in the spring and coincided with the return of oxygenated conditions. First-generation numbers in 1989 were initially high (~30,000 ind. m^{-2}) in March and within the range observed from 1984 to 1988 but decreased by late spring to ~4000 ind. m^{-2}. High mortality may have been due to low temperatures, since March lake temperatures were 2–6 °C lower than the suspected lethal limit of ca. 5–6 °C for *Artemia* (Dana et al. 1995). Increased mortality of the animals may also have been associated with elevated concentrations of hydrogen sulfide, ammonium, and arsenite resulting from the breakdown of meromixis.

Spring chlorophyll *a* concentrations were high, 30–44 μg Chl *a* L^{-1}, but typical of pre-meromictic levels. This abundant food source led to large *Artemia* brood sizes and high ovigerity during the period of ovoviviparous reproduction and resulted in a large summer increase of *Artemia* (peak concentration , ~93,000 ind. m^{-2}). Negative feedback was apparent when the large summer population of *Artemia* grazed the phytoplankton to low levels (<0.5–2 μg Chl *a* L^{-1}) that led to decreased reproductive output by the females. Summer brood size and ovigerity were the lowest observed during 1983–1989.

Monomictic Conditions with Relatively Stable Lake Levels, 1990 to 1994 Mono Lake was monomictic from 1990 to 1994 (Fig. 11.4), and lake levels, which narrowly fluctuated between 1943 and 1943.3 m asl, were similar to those in the late 1970s. From 1990 to 1994, ammonium concentrations in the euphotic zone decreased to levels observed prior to meromixis in 1982. Ammonium was low (<2 μM) from March through April and increased to 8–15 μM in July. Ammonium concentrations declined slightly in late summer and then increased following the autumn turnover.

Summer population peaks of adult *Artemia* in 1990, 1991, and 1992 were ~35,000 m^{-2} despite the large differences among second-generation naupliar peaks. Thus, food availability and other environmental factors are more important in determining summer abundance than recruitment of second-generation nauplii. In 1993, when freshwater inflows were higher than usual and density stratification enhanced, the summer generation was slightly smaller (~27,000 m^{-2}). Summer abundance of adults increased slightly in 1994 when runoff was low and lake levels were declining.

Meromictic Conditions with Rising and Falling Lake Levels, 1995–2002 During 1995, above-normal runoff in the Mono Lake basin coupled with the absence of significant water diversions from the basin resulted in 1-m rise in water level and the onset of meromixis. However, due to holomixis during late 1994 and early

1995, nutrient and plankton dynamics during the first half of 1995 were similar to those observed during the previous 4 years. Effects of increased water column stability due to chemical stratification (Fig. 11.4) became evident later in the year with a shallower mixed layer, lower concentrations of ammonium in mixed layer and of chlorophyll a, and slightly smaller brood sizes.

Chemical stratification persisted and strengthened throughout 1996. Mixolimnetic salinity ranged from 78 to 81 g kg^{-1}, while monimolimnetic salinity was 89 to 90 g kg^{-1}. While ammonium concentrations were <5 μM in the mixolimnion throughout the year, concentrations in monimolimnion continued to increase. The spring epilimnetic chlorophyll a concentrations (5–23 μg Chl *a* L^{-1}) were similar to those observed in previous meromictic years but were much lower than the concentrations observed in March 1995 and previous monomictic years (1989–1994). Peak abundance of first-generation adults was observed about a month later than in previous years. The percent ovigery during June 1996 (42 %) was much lower than that observed from 1989 to 1994 (83–98 %). During the previous meromictic years (1984–1988), the female population was also slow to attain a high level of ovigery.

Chemical stratification continued to increase in 1997 as the water level rose 0.5 m, and the midsummer difference in density between 2 and 32 m attributable to chemical stratification increased from 10.4 kg m^{-3} in 1996 to 12.3 kg m^{-3} in 1997 (Fig. 11.4). Spring (February–April) chlorophyll *a* concentrations in epilimnion at 2 m (2–3 μg Chl *a* L^{-1}) were lower than those observed during 1996 (5–8 μg Chl *a* L^{-1}), and other meromictic years 1984–1989 (1.6–57 μg Chl *a* L^{-1}), and much lower than those observed during the spring months in the last period of monomixis (15–153 μg Chl *a* L^{-1}). In 1998 the surface elevation of the lake rose an additional 0.67 m, and the peak summer difference in density attributable to chemical stratification increased further (Fig. 11.4). Chlorophyll *a* concentrations at 2 m decreased from 14.3 μg Chl *a* L^{-1} in February to 0.3 μg Chl *a* L^{-1} in June, when the seasonal chlorophyll *a* concentration minimum was reached. Meromixis continued but weakened slightly in 1999, as the net change in surface elevation over the course of the year was negligible, 0.03 m. In 1999, the spring (February–April) epilimnetic chlorophyll *a* concentrations at 2 m (10–16 μg Chl *a* L^{-1}) were similar to those observed in 1998 but slightly higher than in the two previous years of meromixis (1996 and 1997).

In the years from 1996 to 1999, a single mid-July yearly peak of adult *Artemia* with little evidence of recruitment of second-generation *Artemia* into adults was observed. Peak midsummer adult abundance in 1997 was slightly lower than in 1996 but similar to 1995, while the peaks observed in 1998 and 1999 were slightly higher than in 1996. Maximum brood size in 1998 (50 eggs $brood^{-1}$) was within the range of maxima observed in 1995–1997 (62, 53, and 33 eggs $brood^{-1}$, respectively) but was significantly smaller than observed from 1987 to 1994 (81–156 eggs $brood^{-1}$).

In 2000, meromixis continued but weakened due to evaporative increase in salinity in the upper mixed layer accompanying a 0.2-m decline in surface elevation. Significant volumes of ammonium-rich monimolimnetic water were entrained, and only 16 % of the volume of the lake remained below the chemocline. Algal biomass varied in the mixolimnion from a midsummer minimum of 1.4 μg Chl *a* L^{-1} to a maximum in December maximum of 54.2 μg Chl *a* L^{-1}. Adult *Artemia* abundance was low (50 % of the long-term mean). Thus, while meromixis persisted in 2000, the combined effects of declining lake levels reduced proportion of the lake water below the chemocline and increased upward fluxes of ammonium due to the large buildup of monimolimnetic ammonium offset, to some degree, the effect of the absence of winter holomixis.

Chemical stratification weakened further in 2001 and 2002 accompanying a 0.24-m decline in water level each year as a result of evaporation and increased salinity in the mixolimnion. Colder than average mixolimnetic temperatures observed in February 2001 enhanced deep mixing. The midsummer difference in density attributable to chemical stratification declined to 8.9 kg m^{-3} in 2001 and to 5.5 kg m^{-3} in 2002. Ammonium concentrations in the monimolimnion continued to increase, and ammonium-rich monimolimnetic water was entrained upward. Chlorophyll *a* concentration was high during both spring (60–78 μg Chl *a* L^{-1}, February and March) and autumn 2002 (60–80 μg Chl *a* L^{-1}, November). In 2001 and 2002, the *Artemia* population had fairly rapid development of the first generation, a pulse of ovoviviparous reproduction in June, and peak of adult abundance in July or August, followed by a decline to low numbers by November. Analysis of seasonal *Artemia* dynamics and their food indicates small changes in algal biomass immediately following maturation of the first generation and leads to large changes of recruitment into the summer generation. In 2002, a larger spring hatch and spring adult generation lowered algal biomass and led to decreased recruitment into the summer adult population. This intergenerational compensatory interaction is a dominant feature of the seasonal and annual variation of adult abundance.

Response to Breakdown of Meromixis, 2003 and 2004 The meromixis initiated in 1995 nearly broke down early in 2003 prior to the onset of seasonal thermal stratification. This resulted in an upward pulse of ammonium, sulfide, and reduced sulfur compounds. Following a small rise in surface level and slight freshening of the mixed layer due to snowmelt runoff, decreased inflow and evaporative increase in concentration led to an inverse chemical gradient with slightly more saline mixolimnetic water overlying the colder monimolimnion. The combination of the inverse chemical gradient and autumn cooling led to holomixis in mid-November and the end of an 8-year period of meromixis (1995–2003). Chlorophyll *a* concentration was high throughout the winter and spring and autumn. While *Artemia* grazing and nutrient limitation normally result in low summer algal biomass, values in summer 2003 remained above 3 μg Chl *a* L^{-1} despite near-average *Artemia* abundance. The *Artemia* population had early development of a moderate first generation followed by recruitment balancing mortality through the summer.

The breakdown of meromixis in November 2003 mixed nutrient-rich bottom waters throughout the water column. Thus, 2004 began with high ammonia concentrations (10–29 μM), and a large algal bloom (105 μg Chl *a* L^{-1}) had developed by February, and dissolved oxygen concentrations in the lake had recovered following low values observed in November 2003 associated with the breakdown of meromixis. Hatching of overwintering *Artemia* cysts began in February as indicated by the presence of abundant first instar nauplii in late February. Record high (68,750 m^{-2}) naupliar numbers were observed in mid-March. A large hatch, abundant food, and warmer-than-average water temperatures led to the largest and earliest first generation of adult *Artemia* in Mono Lake observed during the period of study. This large first generation of adults depleted algal biomass and suppressed fecundity and recruitment into subsequent generations resulting in an early decline in adult abundance. *Artemia* grazing maintained low phytoplankton abundance throughout the summer.

Third Episode of Meromixis, 2005 to 2007 In early 2005, above-average snowmelt runoff led to a 0.55-m seasonal rise in surface water level of the lake. Late summer evaporative concentration and cooling of the upper mixed-layer decreased vertical stratification and almost initiated holomixis. Freshwater inputs late in 2005 increased salinity stratification enough to prevent winter holomixis and initiated a third period of meromixis. Deep mixing occurred in January–February 2006 resulting in significant upward flux of ammonium; hence, the reduction in spring chlorophyll caused by this third, brief episode of meromixis lessened. Spring chlorophyll *a* concentrations at 2 and 8 m averaged 58 μg Chl *a* L^{-1} compared with 98 μg Chl *a* L^{-1} in 2004. Abundant food and above-average water temperatures in 2005 led to a large first generation of *Artemia*. Although ovoviviparous reproduction was ca. 25 % above the long-term mean, the large first generation of adults depleted food availability and reduced recruitment into the second generation resulting in a rapid late summer decline in adults. While the near absence of adult *Artemia* in mid-October has only been observed in 2002, low (<5000 m^{-2}) mid-October abundances were also observed in 1986, 2000, 2003, and 2004.

A second year of above-average snowmelt runoff in 2007 resulted in a rise in surface elevation of 0.67 m and increased the salinity stratification and strengthening of meromixis. Growth and survivorship to adult *Artemia* were high. While a pulse of ovoviviparous reproduction by first-generation adults occurred in late May and early June, recruitment of these young ones into the adult population was low, and there was no midsummer July increase in adults. The abundance of adults declined through July, and by September adult population observed was the smallest at this time of year for the entire period of record.

Recent Period with Monomixis and Meromixis, 2008–2013 By early 2008 chemical stratification had been lost, and a monomictic regime of annual stratification and mixing occurred from 2008 to 2010 (Fig. 11.4). In 2008, a large pulse of ovoviviparous reproduction by first-generation adult *Artemia* occurred in late May and early June, but recruitment into the adult population was low, and there was no midsummer July increase in adults. The adults declined through July, and their numbers were very low (<200 m^{-2}) in mid-October. During 2009, the abundance of first-generation adults was high (~72,000 m^{-2}). Low phytoplankton abundance accompanying abundant *Artemia* resulted in below average summer ovoviviparous reproduction (58 %). Low ovoviviparous production and subsequent recruitment into the late summer adult population led to an early decline, and *Artemia* were virtually absent by mid-November.

In 2010, unusually cool spring temperatures resulted in slow seasonal warming and delayed maturation and reduced survival of *Artemia*. In June, 86 % of the adult females had empty egg sacs. Despite the slow start, the peak abundance of first-generation adults in July was ~46,000 m^{-2}, higher than all but five other years. As midsummer phytoplanktonic food was depleted by this large population, there was little recruitment into the late summer population, and August abundance was the lowest on record, and a very few *Artemia* were present in autumn and early winter. A multiyear trend of large spring generations of *Artemia* followed by a smaller-than-average late summer population of *Artemia* and autumn decline continued.

In 2011 above-average snowmelt runoff resulted in a 0.6-m rise in lake level by August and an 8 kg m^{-3} difference in salinity between the upper and lower waters. Although stratification weakened as mixing occurred between the November and December, the lake did not fully mix and remained weakly stratified through the winter. In 2012, lake level declined 0.6 m, and holomixis occurred as indicated by an isothermal water column in December, though salinity data are not available. Mono Lake declined 0.5 m and was monomictic in 2013, as indicated by salinity and temperature profiles (Kerby and House 2014).

Summary of Mean Annual Artemia Biomass The mean annual biomass of *Artemia* was estimated from instar-specific abundance and length-weight relationships for the period 1983–1999 and by direct weighing from 2000 to 2011 (Fig. 11.8). Dana et al. (1995) provide weight-length relations and statistics for these relations as a function of food supply and temperature. The mean annual biomass has varied from 5.3 to 17.6 g m^{-2}. The highest estimated mean annual biomass (17.6 g m^{-2}) occurred in 1989 just after the breakdown of meromixis during a period of elevated ammonium and phytoplankton abundance. The lowest annual estimate was in 1997 following 2 years of meromixis and increasing density stratification. Mean annual biomass was somewhat below the long-term mean during the first 3 years of the 1980s episode of meromixis, and then above the mean, the next 3 years as meromixis weakened and ended. The years with the lowest annual biomass of *Artemia* were 1997 (5.3 g m^{-2}) and 2002 (4.9 g m^{-2}), both during the extended period of meromixis, 1995–2003. However, mean annual *Artemia* biomass increased in 2003, as meromixis weakened, to 7.5 g m^{-2} and further to 11.0 g m^{-2} in 2004 following the breakdown of meromixis in late 2003. Mean annual *Artemia* biomass from 2005 through 2011 varied from 5.8 to 8.8 g m^{-2}.

Since 1979 Mono Lake has experienced four episodes of meromixis all of which transitioned back to monomixis after varying periods of time (2–8 years). These differences in mixing have multiple consequences for the lake's ecology, as described above. One notable feature is the relatively high abundance of *Artemia* at the onset of monomixis in 1989, 2004, 2009, and 2013. Further analysis and

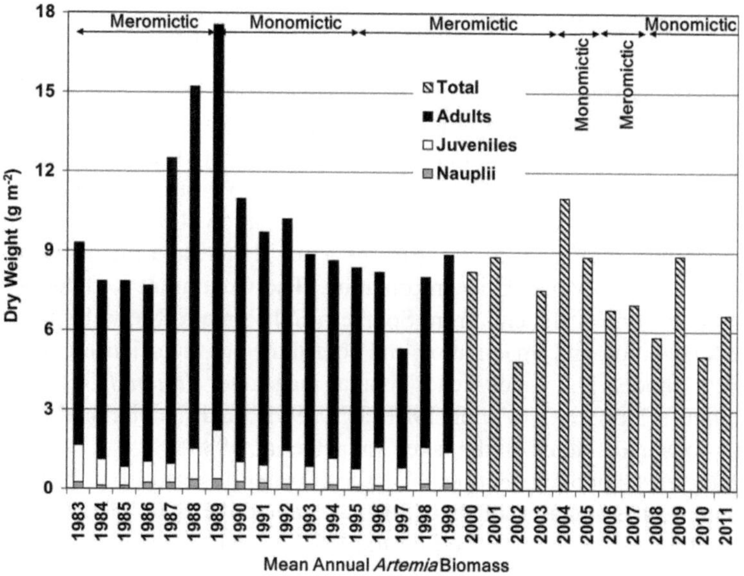

Fig. 11.8 Mean annual *Artemia* biomass, 1983–2011. Data for the period 1982–1999 estimated from instar-specific population data and previously derived weight-length relationships. From 2000 to 2011, *Artemia* biomass was measured directly by determining dry weights

modeling of the limnology of Mono Lake and population dynamics of *Artemia*, building on prior work, are needed to forecast conditions as climate and water diversions change. Field and laboratory studies (e.g., Dana et al. 1988, 1990, 1995) were used in linear-transfer and lag-Manly models of zooplankton cohort development that incorporated results from a multi-transfer model that included variance in development rate among individuals (Jellison et al. 1995). High relative standard errors (mean, 33 %), not atypical for zooplankton, were observed in life-stage abundances. Bruce et al. (2008) applied a coupled hydrodynamic and ecological model to distinguish the relative importance of vertical mixing versus nitrogen recycled by *Artemia* and microbes to primary production. The many terms in such models make validation and interpretation of the results difficult.

To develop a model capable of forecasting ecological changes and to empirically identify the key interacting physical and biological variables, i.e., variables that have acted as causal agents in the Mono Lake ecosystem, a promising approach has been developed by Sugihara and his colleagues based on nonlinear state space reconstruction (Hsieh et al. 2008; Sugihara et al. 2012). These methods can be used to objectively identify the interactions in complex systems and for prediction. The system behavior of Mono Lake is likely to be appropriate for state space reconstructions because three key variables (lake-wide monthly averages of *Artemia* abundances, chlorophyll *a*, and temperature at 2 m) show coherent structure.

11.4 Conservation Issues

In the 1970s expanding awareness of environmental issues and leadership by several individuals and organizations resulted in increased public concern about threats to Mono Lake. The Mono Lake Committee (www.monolake.org) was established in 1978 by David Gaines and associates to protect Mono Lake through litigation, legislation, cooperation, and public education. The Committee now has 16,000 members, publishes and distributes a newsletter, and maintains a website and compilation of Mono Lake information. The Committee's Outdoor Experiences program brings students from Southern California to Mono Lake to expand their understanding and imagination and contributes to the science curriculum of local high schools. The widely read, popular magazine, *National Geographic*, ran stories about Mono Lake (Young 1981; Jehl 1981), and the National Geographic Society sponsored studies on the lake (Melack et al. 1985).

With increased public attention, the issue of water diverted from the Mono Lake basin came to the attention of the US Congress and the State of California. As a result, studies were commissioned by the US National Research Council (Patten et al. 1987) and California Water Resources Center (Botkin et al. 1988). These analyses documented negative ecological consequences of continued water diversions and provide examples relevant to the use of environmental impact assessments to influence policy decisions (Wiens et al. 1993).

In 1994, following two decades of litigation and environmental controversy, the State Water Resources Control Board of California issued Decision 1631 to amend Los Angeles' water rights "to protect public trust resources at Mono Lake . . .." The decision restricts water diversions until the surface elevation of the lake reaches 1948 m and has resulted in increased inflows, the magnitude depending largely on amounts of precipitation in the watershed. Hart (1996) captured the politics, science, and personalities in the saga that led to the 1994 decision. The chronology of conditions in Mono Lake from 1994 to the present, summarized above, indicates how the lake has responded to the reduced inflow diversions in the context of climate variations and ecological dynamics.

Saline lakes throughout the world are vulnerable to diversions of inflows and sensitive to climatic variations (Melack et al. 1997; Williams 2002). The conservation and management of the large saline lakes in the Western Great Basin of the United States provide contrasting examples of the status of the lakes (Jellison and Sevon 2005). While Mono and Pyramid lakes have had their diversions reduced, Walker Lake, Nevada, continues to decline in level and increase in salinity. The level, salinity, and eutrophication of Great Salt Lake vary spatially and interannually (Wurtsbaugh et al. 2009). Though direct effects of changes in salinity can influence the biota (Melack et al. 1985; Dana and Lenz 1986; Dana et al. 1993; Herbst and Blinn 1998; Barnes and Wurtsbaugh 2015), indirect effects related to mixing dynamics and the occurrence of meromixis can lead to dramatic changes in ecological conditions, as demonstrated for Mono Lake.

References

Anderson RF, Bacon MP, Brewer PG (1982) Elevated concentrations of actinides in Mono Lake. Science 216:514–516

Barnes BD, Wurtsbaugh WA (2015) The effects of salinity on plankton and benthic communities in the Great Salt Lake, Utah, USA: a microcosm experiment. Can J Fish Aquat Sci 72:807–817

Benson LV, Lund SP, Burdett JW, Kashgarian M, Rose TP, Smoot JP, Schwartz M (1998) Correlation of late-Pleistocene lake-level oscillations in Mono Lake, California, with North American climate events. Quat Res 49:1–10

Botkin DB, Broecker WS, Everett LG, Shapiro J, Wiens JA (1988) The future of Mono Lake. California Water Resources Center, University of California, Riverside, Report 68

Boyd WS, Jehl JR (1998) Estimating the abundance of eared grebes on Mono Lake, California, by aerial photography. Colon Waterbirds 21:236–241

Broecker WS, Wanninkhof R, Stuiver M, Stine S, Herczeg A, Robinson S, Peng TH, Mathieu G (1988) The radiocarbon budget for Mono Lake: an unsolved mystery. Earth Planet Sci Lett 88:16–26

Bruce LC, Jellison R, Imberger J, Melack JM (2008) Effect of benthic boundary layer transport on the productivity of Mono Lake, California. Saline Systems 4:11 http://www.salinesystems.org/content/4/1/11

Brum JR, Steward GF, Jiang SC, Jellison R (2005) Spatial and temporal variability of prokaryotes, viruses, and viral infections of prokaryotes in an alkaline, hypersaline lake. Aquat Microb Ecol 41:247–260

Budinoff CR, Hollibaugh JT (2007) Ecophysiology of a Mono Lake cyanobacterium. Limnol Oceanogr 52:2484–2496

Budinoff CR, Hollibaugh JT (2008) Arsenite-dependent photoautotrophy by an *Ectothiorhodospira*-dominated consortium. ISME J 2:340–343

Carini SC, Joye SB (2008) Nitrification in Mono Lake, California (USA): activity and community composition during contrasting hydrological regimes. Limnol Oceanogr 53:2546–2557

Carini SC, Bano N, LeCleir G, Joye SB (2005) Patterns of aerobic methanotrophy and methanotroph community composition during seasonal stratification in a meromictic lake, Mono Lake, CA. Environ Microbiol 7:1127–1138

Clark JF, Hudson GB (2001) Quantifying the flux of hydrothermal fluids into Mono Lake by use of helium isotopes. Limnol Oceanogr 46:189–196

Cloud P, Lajoie KR (1980) Calcite-impregnated defluidization structures in the littoral sands of Mono Lake, California. Science 210:1009–1012

Cooper SD, Winkler DW, Lenz PH (1984) The effect of grebe predation on a brine shrimp population. J Anim Ecol 53:51–64

Dana GL, Lenz PH (1986) Effects of increasing salinity on an *Artemia* population from Mono Lake, California. Oecologia 68:428–436

Dana GL, Foley C, Starrett G, Perry W, Melack JM (1988) In situ hatching of *Artemia monica* cysts in hypersaline Mono Lake. Hydrobiologia 158:183–190

Dana GL, Jellison R, Melack JM (1990) *Artemia monica* egg production and recruitment in Mono Lake, California, USA. Hydrobiologia 197:233–243

Dana GL, Jellison R, Melack JM, Starrett G (1993) Relationships between *Artemia monica* life history characteristics and salinity. Hydrobiologia 263:129–143

Dana GL, Jellison R, Melack JM (1995) Effects of different natural regimes of temperature and food on survival, growth, and development of *Artemia*. J Plankton Res 17:2115–2128

Davis EL (1965) An ethnography of the Kuzedika Paiute of Mono Lake, Mono County, California. Anthropological Papers, vol 75. University of Utah, Salt Lake City, UT, pp 1–5

Domagalski JL, Orem WH, Eugster HP (1989) Organic geochemistry and brine composition in Great Salt, Mono and Walker lakes. Geochim Cosmochim Acta 53:2857–2872

Dunn RC (1953) The origin of the deposits of tufa in Mono Lake. J Sediment Petrol 23:18–23

Edwardson CF, Planer-Friedrich B, Hollibaugh JT (2014) Transformation of monothioarsenate by haloalkaliphilic, anoxygenic photosynthetic purple sulphur bacteria. FEMS Microbiol Ecol 90:858–868

Eugster HP, Hardie LA (1978) Saline lakes. In: Lerman A (ed) Lakes: chemistry, geology, physics. Springer, New York, pp 237–293

Evans MA, MacIntyre S, Kling GW (2008) Internal wave effects on primary productivity: experiments, theory and modeling. Limnol Oceanogr 53:339–353

Fisher JC, Hollibaugh JT (2008) Selenate-dependent anaerobic arsenite oxidation by a bacterium from Mono Lake, California. Appl Environ Microbiol 74:2588–2594

Fisher JC, Wallschläger D, Planer-Friedrich B, Hollibaugh JT (2008) Sulphur plays an unexpected role in the arsenic cycle of Mono Lake, CA. Environ Sci Tech 42:81–85

Garrels RM, MacKenzie FT (1967) Origin of the chemical composition of some springs and lakes. Adv. Chem, Series 67. American Chemical Society, Washington, DC, pp 222–242

Gilbert CM, Christensen MN, Al-Rawi Y, Lajoie KR (1969) Volcanism and structural history of Mono basin. Geol Soc Am Mem 116:275–329

Hales JE Jr (1974) Southwestern United States summer monsoon source- Gulf of Mexico or Pacific Ocean? J Appl Meteorol 13:331–342

Hart J (1996) Storm over Mono Lake: the Mono Lake battle and the California water future. Univ. of California Press, Berkeley

Herbst DB (1990) Distribution and abundance of the alkali fly (*Ephydra hians* Say) at Mono Lake (USA) in relation to physical habitat. Hydrobiologia 197:193–205

Herbst DB (1998) Salinity limits nitrogen fixation in sediments from Mono Lake, California. Int J Salt Lake Res 7:261–274

Herbst DB, Blinn DW (1998) Experimental mesocosm studies of salinity effects on the benthic algal community of a saline lake. J Phycol 34:772–778

Hoeft SE, Kulp TR, Stolz JF, Hollibaugh JT, Oremland RS (2004) Dissimilatory arsenate reduction with sulphide as the electron donor: experiments with Mono Lake water and isolation of Strain MLMS-1, a chemoautotrophic arsenate-respirer. Appl Environ Microbiol 70:2741–2747

Hollibaugh JT, Wong PS, Bano N, Pak SK, Prager EM, Orrego C (2001) Stratification of microbial assemblages in Mono Lake, California, and response to a mixing event. Hydrobiologia 466:45–60

Hollibaugh JT, Carini S, Gurleyuk H, Jellison R, Joye SB, LeCleir G, Vasquex L, Wallschlager D (2005) Arsenic speciation in Mono lake, California: Response to seasonal stratification and anoxia. Geochim Cosmochim Acta 69:1925–1937

Hollibaugh JT, Budinoff C, Hollibaugh RA, Ransom B, Bano N (2006) Sulphide oxidation coupled to arsenate reduction by a diverse microbial community in a soda lake. Appl Environ Microbiol 72:2043–2049

Hsieh C-h, Anderson C, Sugihara G (2008) Extending nonlinear analysis to short ecological time series. Am Nat 171:71–80

Humayoun S, Bano N, Hollibaugh JT (2003) Phylogenetic composition of the bacterioplankton from an alkaline, hypersaline lake, Mono Lake, California. Appl Environ Microbiol 69:1030–1042

Imberger J, Patterson JC (1990) Physical limnology. Adv Appl Mech 27:303–475

Jehl JR Jr (1981) Mono Lake: a vital way station for the Wilson's Phalarope. Natl Geogr 160:520–525

Jehl JR Jr (2007) Why do Eared Grebes leave hypersaline lakes in autumn? Waterbirds 30:112–115

Jehl JR Jr, Boyd WS, Paul DS, Anderson DW (2002) Massive collapse and rapid rebound: population dynamics of eared grebes (*Podiceps nigricollis*) during an ENSO event. Auk 119:1162–1166

Jellison R, Melack JM (1993a) Meromixis in hypersaline Mono Lake, California I. Vertical mixing and density stratification during the onset, persistence, and breakdown of meromixis. Limnol Oceanogr 38:1008–1019

Jellison R, Melack JM (1993b) Algal photosynthetic activity and its response to meromixis in hypersaline Mono Lake, California. Limnol Oceanogr 38:818–837

Jellison R, Melack JM (2001) Nitrogen limitation and particulate elemental ratios of seston in hypersaline Mono Lake, California, USA. Hydrobiologia 466:1–12

Jellison R, Rose K (2012) Mixing and plankton dynamics in Mono Lake, California. 2011 Annual report to Los Angeles. Department of Water and Power, Marine Science Institute, University of California, Santa Barbara

Jellison R, Sevon M (2005) Conservation and management of four large saline lakes in the Western Great Basin (USA). Verh Int Ver Limnol 29:137–139

Jellison R, Dana GL, Melack JM (1992) Ecosystem responses to changes in freshwater inflow to Mono Lake, California. Pages 107–118. In: Hall CA, Doyle-Jones V, Widawski B (eds) The history of water: Eastern Sierra Nevada, Owens Valley, White-Inyo Mountains, White Mountain Research Station Symposium, vol 4. Univ. of California, Los Angeles, pp 107–118

Jellison R, Miller LG, Melack JM, Dana GL (1993) Meromixis in hypersaline Mono Lake, California II. Nitrogen fluxes. Limnol Oceanogr 38:1020–1039

Jellison R, Dana GL, Melack JM (1995) Zooplankton cohort analysis using systems identification techniques. J Plankton Res 17:2093–2115

Jellison R, Anderson R, Melack JM, Heil D (1996) Organic matter accumulation in Mono Lake sediments during the past 170 years. Limnol Oceanogr 41:1539–1544

Jellison R, Romero J, Melack JM (1998) The onset of meromixis during restoration of Mono Lake, California: unintended consequences of reducing water diversions. Limnol Oceanogr 43:706–711

Jellison R, MacIntyre S, Millero F (1999) The density characteristics of Mono Lake brine. Int J Salt Lake Res 8:41–53

Jellison R, Adams H, Melack JM (2001) Re-appearance of rotifers in hypersaline Mono Lake, California, during a period of rising lake levels and decreasing salinity. Hydrobiologia 466:39–43

Jiang S, Steward GF, Jellison R, Chu W, Choi S (2004) Abundance, distribution, and diversity of viruses in alkaline, hypersaline Mono Lake, California. Microb Ecol 47:9–17

Johannesson KH, Lyons WB (1994) The rare earth element geochemistry of Mono Lake water and the importance of carbonate complexing. Limnol Oceanogr 39:1141–1154

Joye SB, Connell TL, Miller LG, Oremland RS, Jellison R (1999) Oxidation of ammonia and methane in an alkaline, saline lake. Limnol Oceanogr 44:178–188

Kerby M, House D (2014) Mono Lake limnology monitoring. 2013 annual report. Los Angeles Department of Water and Power, Bishop, California

Kociolek JP, Herbst DB (1992) Taxonomy and distribution of benthic diatoms from Mono Lake, California, U.S.A. Trans Am Microsc Soc 111:338–355

Kulp TR, Hoeft SE, Asao M, Madigan MT, Hollibaugh JT, Fisher JC, Stolz JF, Culbertson CW, Miller LG, Oremland RS (2008) Arsenic(III) fuels anoxygenic photosynthesis in hot spring biofilms from Mono Lake, California. Science 321:967–970

Lajoie KR (1968) Quaternary stratigraphy and geologic history of Mono basin, eastern California. Ph.D. thesis, University of California, Berkeley

Lee K (1969) Infrared exploration for shoreline springs: a contribution to the hydrogeology of Mono Basin. Ph.D. thesis, Stanford University

Lenz PH (1984) Life history analysis of an *Artemia* population in a changing environment. J Plankton Res 6:967–983

Lovejoy C, Dana GL (1977) Primary producer level. In: Winkler D (ed) An ecological study of Mono Lake, California. Inst Ecol Publ, vol 12. Univ. of California, Davis, pp 42–57

MacIntyre S, Jellison R (2001) Nutrient fluxes from upwelling and enhanced turbulence at the top of the pycnocline in Mono Lake, CA. Hydrobiologia 466:13–29

MacIntyre S, Romero JR (2000) Predicting upwelling, boundary mixing, and nutrient fluxes in lakes. Verh Int Ver Limnol 27:246–250
MacIntyre S, Flynn K, Jellison R, Romero J (1999) Boundary mixing and nutrient fluxes in Mono Lake, California. Limnol Oceanogr 44:225–242
MacIntyre S, Clark JF, Jellison R, Fram JP (2009) Turbulent mixing induced by nonlinear internal waves in Mono Lake, California. Limnol Oceanogr 54:2255–2272
Maest AS, Pasilis SP, Miller LG, Nordstrom DK (1992) Redox geochemistry of arsenic and iron in Mono Lake, California. In: Kharaka YK, Maest AS (eds) Water-rock interaction. A.A. Balkema, Rotterdam, pp 507–511
Mason DT (1967) Limnology of Mono Lake, California. Univ California Publ Zool, vol 83. University of California Press, Oakland, pp 1–110
Melack JM (1983) Large, deep salt lakes: a comparative limnological analysis. Hydrobiologia 105:223–230
Melack JM (2002) Ecological dynamics in saline lakes. Verh Int Ver Limnol 28:29–40
Melack JM, Jellison R (1998) Limnological conditions in Mono Lake: contrasting monomixis and meromixis in the 1990s. Hydrobiologia 384:21–39
Melack JM, Lenz PH, Cooper SD (1985) The ecology of Mono Lake. Nat Geogr Soc Res Rep 20:461–470
Melack JM, Dozier J, Goldman CR, Greenland D, Milner A, Naiman RJ (1997) Effects of climate change on inland waters of the Pacific coastal mountains and western Great Basin of North America. Hydrol Process 11:971–992
Miller LG, Jellison R, Oremland RS, Culbertson CW (1993) Meromixis in hypersaline Mono Lake, California III. Breakdown of stratification and biogeochemical response to overturn. Limnol Oceanogr 38:1040–1051
Oremland RS (1990) Nitrogen fixation dynamics of two diazotrophic communities in Mono Lake, California. Appl Environ Microbiol 56:614–622
Oremland RS, Stolz JF (2003) The ecology of arsenic. Science 300:939–944
Oremland RS, Hoeft SE, Santini JM, Bano N, Hollibaugh RA, Hollibaugh JT (2002) Anaerobic oxidation of arsenite in Mono Lake water and by a facultative chemoautotroph, strain MLHE-1. Appl Environ Microbiol 68:4795–4802
Oxburgh R, Broecker WS, Wannikhof RH (1991) The carbon budget of Mono Lake. Global Biogeochem Cycles 5:359–372
Patten DT, Conte FP, Cooper WE, Dracup J, Dreiss S, Harper K, Hunt GL, Kilham P, Klieforth HE, Melack JM, Temple SA (1987) The Mono Basin ecosystem—effects of changing lake level. National Academy Press, Washington, DC
Reed WE (1977) Biogeochemistry of Mono Lake, California. Geochim Cosmochim Acta 41:1231–1245
Roesler CS, Culbertson CW, Etheridge SM, Goericke R, Kiene RP, Miller LG, Oremland RS (2002) Distribution, production, and ecophysiology of *Picocycstis* strain ML in Mono Lake, CA. Limnol Oceanogr 47:440–452
Romero JR, Melack JM (1996) Sensitivity of vertical mixing to variations in runoff. Limnol Oceanogr 41:955–965
Romero JR, Jellison R, Melack JM (1998) Stratification, vertical mixing, and upward ammonium flux in hypersaline Mono Lake, California. Arch Hydrobiol 142:283–315
Scholl DW, von Huene R, St. Amand P, Ridlon JB (1967) Age and origin of topography beneath Mono Lake, a remnant Pleistocene lake, California. Geol Soc Am Bull 78:583–600
Scholten JCM, Joye SB, Hollibaugh JT, Murrell JC (2005) Molecular analysis of the sulphate reducing and methanogenic community in a meromictic lake (Mono Lake, California) by targeting 16SrRNA, Methyl CoM-, APS- and DSR- genes. Microb Ecol 50:29–39
Simpson HJ, Trier RM, Olson CR, Hammond DE, Ege E, Miller L, Melack JM (1980) Fallout plutonium mobility in an alkaline, saline lake. Science 207:1071–1072
Steward GF, Zehr JP, Jellison R, Montoya JP, Hollibaugh JT (2004) Vertical distribution of nitrogen-fixing phylotypes in a meromictic, hypersaline lake. Microb Ecol 47:30–40

Stine S (1994) Extreme and persistent droughts in California and Patagonia during Mediaeval time. Nature 369:546–549
Sugihara R, May R, Ye H, Ysieh C-h, Deyle E, Fogarty M, Munch S (2012) Detecting causality in complex ecosystems. Science 338:496–500
Vidal J, MacIntyre S, McPhee-Shaw EE, Shaw WJ, Monismith SG (2013) Temporal and spatial variability of the internal wave field in a lake with complex morphometry. Limnol Oceanogr 58:1557–1580
Vorster P (1985) A water balance forecast model for Mono Lake, California. Master's thesis, California State University, Hayward
Wiens JA, Pattern DT, Botkin DB (1993) Assessing ecological impact assessment: lessons from Mono Lake, California. Ecol Appl 3:595–609
Williams WD (2002) Environmental threats to salt lakes and likely status of inland saline ecosystems in 2050. Environ Conserv 29:154–167
Williston SW (1883) Dipterous larvae from the western alkaline lakes and their use as human food. Trans Conn Acad 6:87–90
Winkler DW (ed) (1977) An ecological study of Mono Lake, California. Inst Ecol Publ, vol 12. University of California, Davis
Wrege PH, Shuford WD, Winkler DW, Jellison R (2006) Annual variation in numbers of breeding California Gulls at Mono Lake, California: the importance of natal philopatry and local and regional conditions. Condor 108:82–96
Wurtsbaugh WA, Naftz DL, Johnson WP (2009) Eutrophication, nutrient fluxes and connectivity between the bays of Great Salt Lake, Utah (USA). In: Oren A, Palacios P, Wurtsbaugh W (eds) Saline lakes around the world: unique systems with unique values. Natural Resources and Environmental Issues, vol. XV. S.J. and Jessie E. Quinney Natural Resources Research Library, Logan, Utah, pp 153–160
Young G (1981) The troubled waters of Mono Lake. Natl Geogr 160:504–519

Chapter 12
Mexican Meromictic Lakes: What We Know So Far

Javier Alcocer

12.1 Introduction

According to Alcocer et al. (2000a), Mexican lakes show two main lake-mixing categories which fit the conclusions of Lewis (1996) regarding tropical lakes: (a) the relatively small and deep lakes are warm monomictic and (b) the comparatively large and shallow ones are warm polymictic. Most Mexican lakes are shallow and well mixed but it is expected that Mexican lakes with high relative depth may become meromictic as mentioned by Lewis (2000) for temperate lakes. However, it must be remembered that not only the temperature but also the salinity contributes to water density and thus salinity could have more influence than the temperature to induce meromixis. In this regard, in addition to the Mexican regions where deep lakes are found (e.g., tecto-volcanic regions), there are other two regions where meromictic lakes could be found: (a) coastal areas where freshwater and seawater interact (e.g., karstic coastlines) and (b) semiarid and arid areas where evaporation greatly surpasses precipitation (e.g., endorheic basins) and freshwater precipitates over saline water bodies. Up to date, there are reports of Mexican meromictic lakes located in four different parts of the country: two reports on inland (Coahuila and Guanajuato) and two on coastal lakes (Nayarit and Quintana Roo). These studies were carried out on a single or a few sampling dates. Following description shows what we know so far regarding Mexican meromictic lakes.

J. Alcocer (✉)
Proyecto de Investigación en Limnología Tropical, UNAM, FES Iztacala, Tlalnepantla de Baz, Mexico
e-mail: jalcocer@unam.mx

© Springer International Publishing AG 2017

R.D. Gulati et al. (eds.), *Ecology of Meromictic Lakes*, Ecological Studies 228,
DOI 10.1007/978-3-319-49143-1_12

12.2 Mexican Provinces

To understand and identify potential zones where meromictic lakes are located or could be found in Mexico, it is necessary to briefly introduce the geological and physiographic characteristics of the Mexican territory. Mexico's topography includes seven main mountain systems (Sierra Madre Oriental, Sierra Madre Occidental, Trans-Mexican Volcanic Belt, Sierra Madre de Oaxaca, Sierra Madre del Sur, Sierra Madre de Chiapas, and the Baja California Sierra), which turns the country in a mosaic of diverse physiographic settings and geology resulting in eleven provinces (Fig. 12.1) according to Ferrusquía-Villafranca (1993).

1. The *Baja California Peninsula* contains relatively flat valleys with thick alluvial deposits underneath, separated by mountain ranges. The mountainous formations display metamorphic and sedimentary rocks of Mesozoic age and volcanic rocks of Cenozoic age. The more recent Pleistocene alluvial basins are small.
2. The *Pacific Northwestern Coastal Plains* are composed of sandy sediments that extend in some areas several tens of kilometers into the continent. Mesozoic intrusive and metamorphic as well as Cenozoic volcanic rocks constitute the basement.
3. The *Sierra Madre Occidental* is a relatively high area, composed of a complex sequence of Cenozoic volcanic rocks covered by thin deposits of residual soils.

Fig. 12.1 An outline sketch of Mexico showing physiographic/geologic provinces and location of the meromictic lakes mentioned in the text. (1 = Baja California Peninsula, 2 = Pacific Northwestern Coastal Plains, 3 = Sierra Madre Occidental, 4 = Northern Alluvial Plains, 5 = Sierra Madre Oriental, 6 = Gulf Coast Plains, 7 = Central Plateau, 8 = Trans-Mexican volcanic belt, 9 = Sierra Madre del Sur, 10 = Sierra Madre de Chiapas, 11 = Yucatán Peninsula)

4. The *Northern Alluvial Plains* (Chihuahua–Coahuila plateaus) and ranges comprise the northern and central territories. The former consist of a rugged and arid highland covered by pyroclastic material, sierras of some 1000 m rise above mean plain level (i.e., 1000 m a.s.l.). The plateau declines in the south–north direction from more than 2000 m at the southern limit to 600 m along the Bravo (Grande) river. Although sierras delineate the plateau landscape, ample plains dominate the area. Deflation basins holding ephemeral saline—sometimes also alkaline—lakes are common (e.g., the Bolson or desert of Mapimí). These basins appeared in the Oligocene and Miocene by tectonic and volcanic processes and underwent later wind-driven erosion in an arid or semiarid climate. Watercourses originally flowing to the Bravo (Grande) river passed through southern endorheic basins, thus the central lakes filled rapidly with alluvial depositions to complete dryness. Now, only few terminal (i.e., where the water finds no way out of the lake but by evaporation) and playa lakes (i.e., desert basins with no outlet which periodically fill with water to form temporary or ephemeral lakes) remain.
5. The *Sierra Madre Oriental* is a relatively high area, constituted by anticline mountain ranges and syncline valleys, characterized by Mesozoic sedimentary rocks covered by thin deposits of residual soils.
6. The *Gulf Coast Plains* at the southern part enlarged from quaternary fluvial and lacustrine alluvial deposits from the Grijalva and Usumacinta rivers. Trade winds and a multitude of river deltas enhance the plains. Lake basins in the region are associated mostly with rivers, although solution basins originated on Jurassic salt domes. Lake Catemaco, south central Veracruz, in east central Mexico, relates to Tertiary and Quaternary volcanism as lava flows blocked an old river valley.
7. The *Central Plateau* is the world's largest tropical plateau. Volcanic mountains emerged by Mid-Tertiary and Plio-Pleistocene tectonic movement with lava flows and pyroclastic material accretion emitted through faults, to reach up to some 3000 m. The middle portion consists of uneven highlands ranging from 2000 m a.s.l to 2500 m a.s.l. Faulting and rising terrain modified drastically the Plateau's drainage system, thus creating deep lake basins. Volcanic cinder and sediments carried from the surrounding mountains filled up the lake's basins, and an active volcanism almost obliterated former lakes. A moderately dry temperate climate in this area includes isolated sections ranging from dry to desert. The low mean annual precipitation (about 400 mm), typical of Durango, Zacatecas, Aguascalientes, Guanajuato, and Queretaro, causes scarcity of water evident in arid basins, such as the Aguanaval, the middle Nazas, and Grande de Santiago, among others. The dry environment, active volcanism, and silting basins reduced the ancient large lakes to small, shallow, ephemeral lakes.
8. The *Trans-Mexican volcanic belt* dates back to Tertiary and Quaternary volcanism. The region still continues to rise by collision of the American continental plate and the Pacific Cocos and Rivera plates. Several closed humid mountain valleys surrounded by heights, some over 5000 m a.s.l. resulted from the presence of prominent strata volcanic structures and many small volcanoes.

The east–west Trans-Mexican Volcanic Belt actually became the southern limit of the Pleistocene glacial advance.

9. The *Sierra Madre del Sur* basement rocks date back to Precambrian, Paleozoic, and Mesozoic ages. Several lakes and alluvial plains appeared during the Pleistocene, and deposits of Quaternary basalt and limestone accumulated on top. Karst, some solution basins, and abrupt sierras characterize this area. High elevations surround small valleys connected through steep canyons.
10. The *Sierra Madre de Chiapas,* southeastern plains and ranges, differentiates the southeast region of Mexico. As a separated region from the Balsas River basin through the Tehuantepec Isthmus, it includes three main physiographic features: the Pacific coast sierras and mountains, the Central basin of Chiapas, and the Central Plateau of Chiapas. The two latter formations contain Cretaceous limestone. The Central basin of Chiapas contains many rivers, while the central plateau of Chiapas displays solution basins and karst topography.
11. The *Yucatán Peninsula*, at the southeastern part of Mexico, integrates a Tertiary and Quaternary flat limestone plain, without surface streams or rivers and lakes. Freshwater comes from rainwater that percolates to form a thin lens or layer floating above denser saline—marine—water. Groundwater flows through extensive networks of submerged cave systems that took place as the sea level rose underneath. This region is best described by its karstic topography. Highly soluble limestone, eroded before and after the Peninsula was submerged by the Caribbean Sea, led to a wide range of small sinkholes locally known as "cenotes."

12.3 Mexican Meromictic Lakes

Meromictic lakes have so far been found in four provinces (Table 12.1). These provinces could then be considered potential areas where more meromictic lakes could be discovered (Fig. 12.1).

Table 12.1 Mexican meromictic lakes and the physiographic-geologic provinces they were found

Meromictic lake	Province	Lake's origin
Cuatro Ciénegas, Coahuila	4 Northern Alluvial Plains	Man-made, salt harvesting
Rincón de Parangueo, Guanajuato	8 Trans-Mexican volcanic belt	Volcanic, phreatomagmatic maar
Isabela, Nayarit	8 Trans-Mexican volcanic belt	Volcanic, phreatomagmatic maar?
Dos Lagos, Chiapas	10 Sierra Madre de Chiapas	Solution, sinkholes, doline
Nohoch Hol, Quintana Roo	11 Yucatán Peninsula	Solution, sinkholes, doline

The numbers in the *middle column* (Province) refer to geological areas (see above) where these lakes are found

12.3.1 Cuatro Ciénegas, Coahuila

Cole and Minckley (1968) found two small artificial pools slightly north of and connected to Laguna Salada, 3.5 km south and 1.5 km west of Cuatro Ciénegas, central Coahuila, Mexico. The pools were described as rectangular, about 3.5 × 5.0 m, with vertical sides, and contained water 45–50 cm deep. They were man-made as part of a salt-harvesting operation. Climate in the area is arid, with scarce summer rains (annual precipitation $<$200 mm), and very high temperatures (44 °C) in summer (Minckley 1969). The area is composed of Mesozoic limestone of marine origin with abundant precipitated mineral salts (e.g., calcium sulfate).

12.3.2 Rincón de Parangueo, Guanajuato

Rincón de Parangueo is a phreatomagmatic maar (i.e., a volcanic lake whose basin was created by an explosion produced when magma vaporized groundwater) lake situated in the southern portion of the Guanajuato State (20°25′ N, 101°15′ W, 1700 m a.s.l.), not far from the northern boundary of Michoacán State (Alcocer et al. 2000b). It is a small crater lake (0.74 km^2), elliptical in shape (1050 m × 913 m). Its water depth reached around 50 m in the past, but at the time the meromixis was detected (1995), it was 7.5 m (see photograph in Alcocer, 2002). Mean annual precipitation in the area is about 650 mm (range, 600–800 mm) of which 90 % occurs from May to October. Mean annual temperature averages 19 °C, ranging from 14 °C in January to 23 °C in May (García 1973).

12.3.3 Isabela, Nayarit

Isabela is a crater lake on the volcanic Isabela Island (21°50′ N, 102°52′ W) 32 km off the Nayarit coast in the Mexican Pacific. The island is a complex of coalescent stratovolcanoes lying along the Cocos plate (Alcocer et al. 1998), and the mechanism of formation of this crater might have been similar to that of a maar (*sensu stricto*), with interaction between rising magma and water contained in a saturated zone of porous material (Housh et al. 2010). Is a small lake (area 2 km^2), almost circular in shape with a diameter of 270 m. The crater rim ranges between 19 m and 25 m above the water surface. The shoreline is very steep. The lake's maximum depth is 17.5 m according to Alcocer et al. (1998) and 24.5–25 m according to Romero-Viana et al. (2013) and Kienel et al. (2013). Alcocer et al. (1998) indicate the lake's surface is about 7 m a.s.l. at a maximum depth of 17.5 m. The lake occupies a cryptodepression ($Zc = -10.5$ m). Climate is tropical with a maximum monthly rainfall of 300–400 mm $month^{-1}$ from June to October; during the rest of the year, precipitation is almost negligible. The nearest weather station in San Blás

(70 km SE) reports a negative annual water budget with 1470 mm precipitation and evaporation of 2700 mm annually (Kienel et al. 2013).

12.3.4 Dos Lagos, Chiapas

The "Lagunas de Montebello" National Park contains more than 50 lakes (sinkholes) spread over an area of almost 65 km^2 between latitude of 16°04′–16°20′ N and longitude of 91°37′–91°47′ W. Alcocer et al. (unpublished data) from 18 lakes suggest Dos Lagos to be meromictic. The lake lies at 16°05′ N, 91°38′ W, and 1427 m a.s.l. Dos Lagos is a small doline (area 0.052 km^2), almost circular in shape with a perimeter of 990 m, and maximum length of 0.34 km; the maximum depth is 42 m and the average depth 25.2 m (Alcocer et al. 2016). Climate is humid temperate with abundant summer rains; the mean monthly temperature is 23.6 °C (range 20.9–25.6 °C) and the mean annual precipitation is 1862 mm (García 1973).

12.3.5 Nohoch Hol, Quintana Roo

Nohoch Hol is a "cenote" (i.e., a Mexican term for a sinkhole) 9 km northeast of Tulum (20°28′ N, 87°40′ W) 2 km away from the Caribbean coast of Quintana Roo. The karst area of the Yucatan Peninsula developed in Quaternary limestone, with a marine origin. The shape of Nohoch Hol resembles a typical sinkhole with vertical walls. The water surface is about one meter below ground level, with a circular shape, a surface diameter of 33 m, an area of 855 m^2, and a maximum depth of 17.8 m. The connectivity of the deeper basin generally appears to be poor, and groundwater exchange may be principally via pores, fractures, and fissures (Torres-Talamante et al. 2011). Climate in the area is tropical subhumid with a summer rainy season; the mean annual air temperature is 26 °C, and annual precipitation is >1500 mm (García 1973). Stoessell et al. (1993, 2002) mention four cenotes (Big Calica, Little Calica, Angelita, and Chemuyil) close to Nohoch Hol that seem to be meromictic.

12.4 Origin of Meromixis in Mexican lakes

The origin of meromixis in Mexican lakes could be assigned to one or a combination of the following processes (Table 12.2).

(a) Evaporation–precipitation water balance (E >> P) in semiarid/arid areas. During the dry season, enhanced evaporation of water causes a progressive increase in salinity. However, during the rainy season when the highest rainfall

Table 12.2 The probable origin of the meromixis in Mexican meromictic lakes

Meromictic lake	Meromixis origin
Cuatro Ciénegas, Coahuila	Semiarid/arid areas where E >> P Less saline water flows over a more saline water body (*ectogenic*)
Rincón de Parangueo, Guanajuato	Semiarid/arid areas where E >> P Less saline water flows over a more saline water body (*ectogenic*) Groundwater withdrawal? (*crenogenic*)
Isabela, Nayarit	Semiarid/arid areas where E >> P Less saline water flows over a more saline water body (*ectogenic*)
Dos Lagos, Chiapas	Morphometry Saline groundwater (*crenogenic*)
Nohoch Hol, Quintana Roo	Freshwater–marine water interaction Freshwater flows over a saline water body (*ectogenic*) Saline water–marine water flows below a freshwater body (*crenogenic*)

Meromixis type according to Stewart et al. (2009)

amounts are recorded in a brief period of time, freshwater input dilutes the upper water layer, resulting in meromixis.

(b) Freshwater–marine water interaction. The dissolution of the carbonate rock in the karstic areas, particularly in the Yucatan Peninsula, has led to several sinkholes. In some sinkholes, particularly those with poor hydraulic connection, underneath a lens of freshwater (meteoric/groundwater), one finds saline (seawater), where meromixis can take place.

(c) Lake morphometry. Some deep Mexican lakes have high relative depth and a natural predisposition to meromixis. Mexican deep lakes resulted from tectonic/volcanic processes (e.g., graben, maar), mostly along the Trans-Mexican volcanic belt, but also from carbonate dissolution (e.g., sinkholes) in karst areas (e.g., Yucatan Peninsula, Tamaulipas, Chiapas).

12.4.1 Cuatro Ciénegas, Coahuila

The artificial ponds for salt harvest reported by Cole and Minkley (1968) from Cuatro Ciénegas were connected to Laguna Salada. Through evaporation, the salt water of the ponds increases with time; the higher density of this water causes it to form a distinct layer beneath the inflowing, less saline water at the bottom of the pond. The ponds have become meromictic. This type of shallow, meromictic pond can be described as heliothermal water bodies.

12.4.2 Rincón de Parangueo, Guanajuato

The short rainy season—when most (80–90 %) of the annual rainfall occurs—seems to be a key factor in establishing meromixis in lakes located in the arid and semiarid areas of Mexico. Seasonal desiccation of lakes in these regions associated with high evaporation rates facilitates an increase in the concentration of lake water and salinity. Aranda-Gómez et al. (2009) state that the lake sediments of Rincón de Parangueo, Guanajuato, are rich in soluble salts (e.g., halite, sylvite, trona); these salts could easily dissolve during the wet season and get washed into the lake. Kienel et al. (2009) also found halite in Rincón de Parangueo.

Once a highly saline water layer has been developed and favored by increased evaporation in arid and semiarid climates, most of the annual precipitation delivered in a short period of time dilutes surface waters, thus producing meromixis. It seems that meromixis in Rincón de Parangueo could have been initiated after 1930–1940 with a drop of groundwater levels due to the expansion of irrigated agriculture in the region (Alcocer et al. 2000b). The drop of groundwater levels was followed by decrease in lake levels, desiccation of springs, and salinization of the natural waters (Kienel et al. 2009).

12.4.3 Isabela, Nayarit

With a negative annual water budget (i.e., evaporation 2700 mm year^{-1} and precipitation 1470 mm year^{-1}) and most annual precipitation occurring in a short period of time, changes in lake water salinity occur. Alcocer et al. (1998) suggested the relevance of the water budget in producing and maintaining meromixis as they explain for the three haloclines found in Lake Isabela. The uppermost halocline probably resulted from November rains; the second halocline may reflect the depth of mixing associated to the summer storms, while the deepest halocline may be a permanent feature. During the wet season, rainfall dilutes the upper water layer decreasing salinity. By the end of the dry season, the conductance measurements rise in the upper water layer because of evaporation (Romero-Viana et al. 2012).

Housh et al. (2010) state that the lake level oscillates with marine tides, indicating a connection with the sea through porous pyroclastic deposits. Romero-Viana et al. (2012) did not observe a tidal response in the lake level during 45 days of fieldwork and concluded that seawater intrusion could be ruled out as a factor in water level fluctuations. Moreover, Kienel et al. (2013) affirm that seawater infiltration may be possible but is currently too slow to be detected in conductance profiles.

Salinity in Lake Isabela exceeds that of seawater (35 g l^{-1}, 50 mS cm^{-1}), even in the upper layer ($\leq$68 mS cm^{-1}), and particularly from 4 m to the bottom, where salinity has remained high and almost constant for about 17 years (100–112 mS cm^{-1}, 1993–2009). Seawater intrusion probably had a role in the origin and filling of the lake. However, since then, evaporation played a role in explaining the hypersalinity of Lake Isabela. Most annual precipitation is seasonal; it dilutes the upper water layer and decreases salinity thus maintaining meromixis.

Alcocer et al. (1998) suggest another factor contributing to meromixis in Lake Isabela: the lake is well protected; walls surrounding the lake inhibit mixing by wind. In addition, its relative depth of the lake is reasonably high (6.5–9.3 %; 17.5–25 m maximum depth); thus the morphometry favors meromixis.

Alcocer et al. (1998) who measured salinity in Lake Isabela in November 1993, showed three pycnoclines, suggesting the lake to be permanently meromictic below 13 m. However, Romero-Viana et al. (2013), based on a salinity profile in May 2008 displaying an isohaline hypersaline water column, suggest that tropical cyclones could have an effect on the stratification in water column and could have caused the loss of the chemocline previously observed by Alcocer et al. (1998) at the 13 m depth.

Kienel et al. (2013) considered four profiles including the two discussed by Romero-Viana et al. (2013), and they concluded that Lake Isabela is chemically stratified on an annual basis. This conclusion differs from what has been observed in other meromictic lakes, where chemical stratification resulted from exceptionally large freshwater input and persisted for several years. By the end of the dry season, the water column in the lake becomes isohaline through evaporation, while the lake becomes meromictic through the freshwater input during the rainy season that dilutes the upper layer, i.e., the meromixis is ectogenic in origin. Although isohaline, Lake Isabela was thermally stratified (surface = 31 °C, 3 m deep = 25 °C).

The dissolved oxygen (DO) profiles in Kienel et al. (2013) suggest that the water column mixes annually to the bottom; and even though none of the three profiles reached anoxia (0 mg l^{-1}), the highest DO concentration measured at the subsurface was rather low (i.e., <1.5 mg l^{-1}), which resulted awkwardly. Alcocer et al. (1998) showed also a subsurface DO maximum of 7.5 mg l^{-1} -instead of <1.5 mg l^{-1}-, but differently to Kienel et al. (2013) with anoxia from 2.5 m to the bottom.

12.4.4 Dos Lagos, Chiapas

Alcocer et al. (2016) describe solution lakes (sinkholes) in Montebello that are among the deepest lakes in Mexico. The great depth of these lakes morphometrically promotes meromixis. Additional limnological information on these lakes (Alcocer et al. unpublished data) suggests one of them, Dos Lagos, to be meromictic. Besides its depth, the conductance profiles display two layer or strata (Fig. 12.2). The upper "fresher" stratum (<500 mS cm^{-1}) is fed mostly with rainwater and runoff. The lower, saltier stratum (>1000 mS cm^{-1}) receives saline groundwater. Groundwater's salinity originates from gypsum-rich salts of marine origin present in the region.

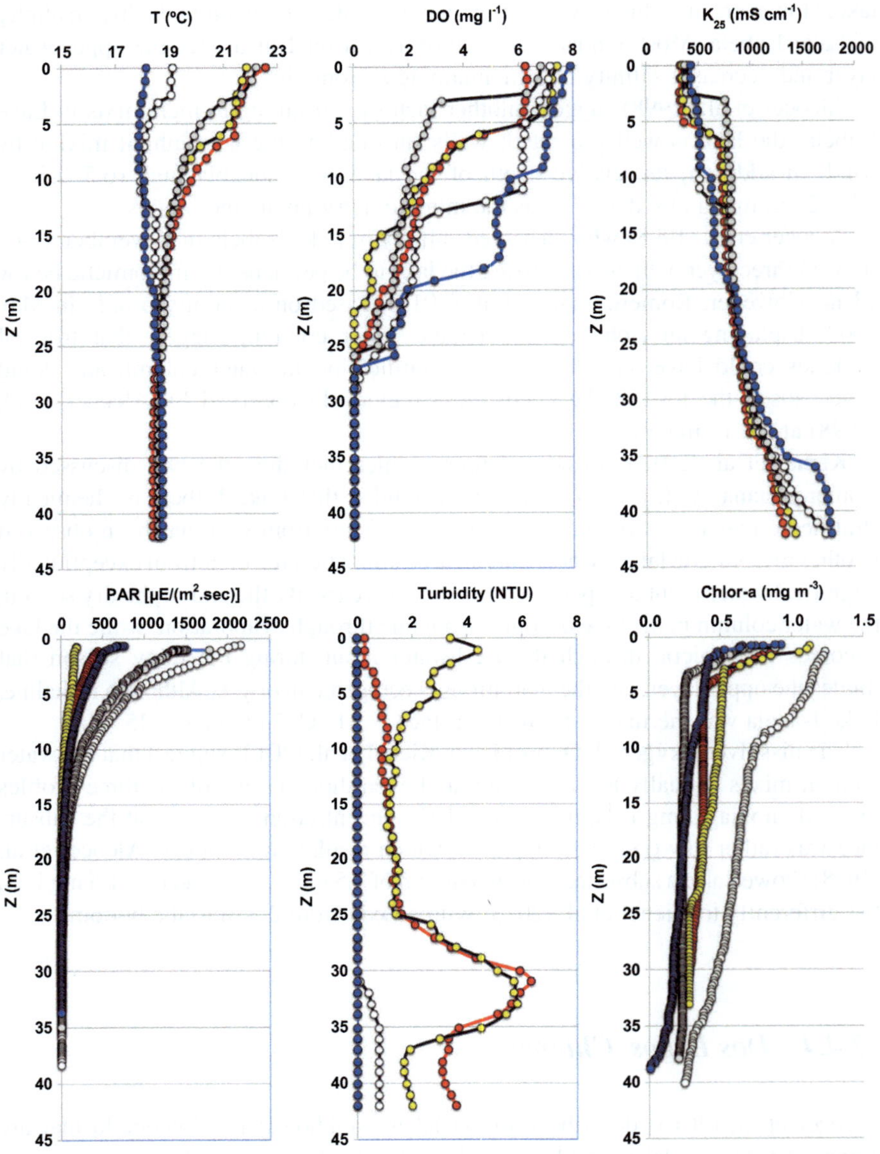

Fig. 12.2 Temperature (T), dissolved oxygen (DO), electrical conductivity (K25), photosynthetic active radiation (PAR), turbidity and chlorophyll "a" (Chl-a) profiles recorded (*red* = May 2013, *yellow* = June 2014, *white* = February 2015, *grey* = October 2015, and *blue* = March 2016) at Dos Lagos, Montebello, Chiapas (Alcocer et al. unpublished data)

12.4.5 Nohoch Hol, Quintana Roo

In recently formed cenotes, the water is well interconnected with the groundwater through fractures and dissolution features; its residence time is short. Rainwater infiltrates and flows underground toward the sea, while seawater flows inland, below the freshwater layers, propelled by tides and storms through tunnels and crevices. On the other side, older cenotes show slow flow and turnover through sedimentation and blocking of the water source and the siphon. Groundwater may still flow, but many of the pathways are blocked, and the exchange of ground and free-overlying water in the cenote is restricted. Two processes may restrict groundwater flow to and from the cenote: roof or wall collapse and sedimentation. These two different hydrodynamics in the cenotes led Schmitter-Soto et al. (2002) used the terms "lotic" and "lentic" to designate the younger and older cenotes, respectively.

Thus, lentic cenotes combine two meromixis-initiating processes: ectogenic (freshwater flows over a saline water body) and crenogenic (marine water flows below a freshwater body). It must, however, be emphasized that once the meromixis has developed, the ectogenic process is more active (freshwater entering through conduits) than the crenogenic one (salt water entering through pores, fractures). In addition to conductance, the temperature gradient also reflects the temperature of the source waters in this eastern Yucatan Peninsula area: the groundwater temperature is similar to the average annual air temperature (26 °C) and the saline water temperature matches the coastal shallow Caribbean surface water temperature (27 °C).

12.5 Physical and Chemical Characteristics of Mexican Meromictic Lakes

12.5.1 Water Transparency and Turbidity

Green (1986) reports Secchi disk transparency depth of 0.6 m in Rincón de Parangueo due to dense bloom of *Spirulina*. Nearly a decade later in 1995, Alcocer et al. (2002) also observed the *Spirulina* bloom and reported 0.8 m of Secchi disk depth.

Dos Lagos has transparent water; the thickness of the euphotic zone (i.e., $Z_{eu} = 1\ \%$ Surface PAR) ranges between 22 m and 23 m (Alcocer et al. 2013–2016 unpublished data). The lake displays a turbidity maximum in May 2013 and March 2016 at 32–33 m with 5.7 and 6.3 NTU, respectively. The turbidity maxima apparently are not related to deep phytoplankton accumulation as shown by the chlorophyll "a" profiles with surface maxima (Fig. 12.2).

Kienel et al. (2013) found a turbidity maximum at 3 m in the chemocline of Lake Isabela where a thin layer of photosynthetic sulfur bacteria occurs; the authors mention this to suggest a passive accumulation of phytoplankton at the density gradient.

Torres-Talamante et al. (2011) report that Nohoch Hol has a clear mixolimnion; >1 % PAR is reached 9 m, at the top of the chemocline, where turbidity increases abruptly and PAR attenuates. Turbidity is low along the mixolimnion (≈7.2 NTU) and the monimolimnion (≈6.2 NTU), but increases abruptly at the chemocline up to as much as 19.3 NTU. The turbidity maximum in Nohoch Hol is biogenic; associated with the presence of large number of bacterioplankton and phytoplankton at the chemocline.

Turbidity in Mexican meromictic lakes relates mainly to primary productivity (i.e., it is biogenic turbidity). Two cases could be distinguished: (a) turbidity in the mixolimnion associated to algal blooms such as in Rincón de Parangueo (*Spirulina*) and Lake Isabela (diatoms) and (b) turbidity in the chemolimnion associated to dense layers of bacteria and/or phytoplankton such as in Nohoch Hol (*Scenedesmus*). In the last case, the mixolimnion is transparent allowing light to reach the chemolimnion (9–15 m).

12.5.2 Temperature

Cole and Minckley (1968) found an anomalous thermal profile (i.e., inverse stratification) in an artificial pool, Cuatro Ciénegas with cooler water (24 °C) lying above warmer water (47 °C). The mean rate of temperature increase from surface to bottom was 0.5 °C cm^{-1}. The colder water had lower salinity (15 g l^{-1}) and the warmer was saltier (111 g l^{-1}). Rincón de Parangueo shows a similar inverse stratification with 19 °C in the mixolimnion (~0–2 m) with a sharp change to 23 °C below, in the monimolimnion, where it remained homothermal to the bottom (7.5 m). Salinity increase is the main explanation for this anomalous thermal profile (Alcocer et al. 2002). Later, in 1999, Armienta et al. (2008) found a similar thermal profile with (18.1 °C and 23.6 °C in the mixo and the monimolimnion, respectively).

In contrast to the anomalous temperature profiles found in Cuatro Ciénegas and Rincón de Parangueo, the temperature profile in Lake Isabela (Alcocer et al. 1998) shows a surface minimum (32.0 °C), which could be explained by both cooling through surface water evaporation and input of cooler fresher water (also suggested by its lower conductance), a subsurface (0.5–1.0 m) temperature maximum (33.03 °C), and an homogeneous water layer (26.71–26.91 °C) at the bottom (13.0–17.5 m), which is higher than the annual mean air temperature of 25.5 °C. Within the mixolimnion, there is a well-defined thermocline with an average gradient of 0.5 °C m^{-1} (Alcocer et al. 1998).

Kienel et al. (2013) visited Lake Isabela four times: December 2006, May 2008, December 2008, and February 2009. The lake was thermally stratified on all sampling dates. In May 2008, the temperature profiles were quite similar to those already described: well-developed thermocline and homothermy of bottom water.

In May 2008, the maximum temperature was at surface. Temperature profiles recorded by Filonov (unpublished data) verified that Isabela (a) remained thermally stratified throughout the year, (b) displayed subsurface temperature maxima by the end in the dry season (October and November 2006, October 2007), and (c) showed a surface maximum at the beginning of the rainy season (July 2006 and June 2008). Water temperatures recorded by Kienel et al. (2013) are in general somewhat lower ($\approx$25 to <31 °C) than those reported by Alcocer et al. (1998) (26.71–33.03 °C, a range that coincides with Filonov's unpublished data).

Dos Lagos temperature profiles (Alcocer et al. unpublished data) suggest a warm monomictic lake with winter circulation period and subtle thermoclines in May 2013, June 2014, and October 2015, while nearly homothermal profiles in February 2015 and March 2016 (Fig. 12.2). The thermoclines are thin (3–4 m width) and superficial (2–9 m), with an average temperature gradient of $0.45 \pm 0.16\,°C\,m^{-1}$. In spite of the homothermal profiles of February 2015 and March 2016, the dissolved oxygen and electric conductivity profiles (see below) suggest Dos Lagos lacks complete overturn with persistent anoxia below 22–30 m and a saltier/denser (>1000 $\mu S\,cm^{-1}$) layer below 28–34 m.

The most significant feature of the Nohoch Hol mixolimnion is the continuous temperature increase between the surface and the base of the layer (9-m depth). Descending through the chemolimnion, temperature has a marked maximum of 26.7 °C occurring at 10.0 m, near the top of the chemolimnion, which is 0.6 °C higher than at 9.0 m at the base of the mixolimnion and 0.7 °C higher than the monimolimnion. The temperature of the monimolimnion is around 26 °C. This profile exemplifies the complexity of the layering developed in meromictic lakes where temperature and conductivity play major roles in modifying the density gradient.

According to Torres-Talamante et al. (2011), the thermal maximum within the chemolimnion supports near-lentic conditions at $\geq$10 m water depth. Nohoch Hol is therefore an interesting case with vertical stratification. This study indicates that a lentic condition dominated by diffuse groundwater flowing in the chemolimnion and monimolimnion, while some degree of circulation occurs in the mixolimnion waters. In addition, the temperatures of the mixolimnion and the monimolimnion in Nohoch Hol reflect the temperatures of the source waters: groundwater and marine water, respectively.

A thermal anomaly *sensu* Stoessell et al. (2002) is present in Nohoch Hol where a temperature maximum occurs within the chemocline. The authors also report similar anomalous temperature profiles from other cenotes (i.e., Big Calica, Little Calica, Angelita, and Chemuyil) close to Nohoch Hol. Out of the four, only the Big Calica profile is quite similar to that of Nohoch Hol; the other profiles, although also with maxima within the chemocline, are not as well defined as Big Calica and Nohoch Hol.

Temperature profiles found in Mexican meromictic lakes are diverse. There are (a) thermal "anomalies" with colder waters (mixolimnion) above warmer ones (monimolimnion) in, for example, Cuatro Ciénegas and Rincón de Parangueo, (b) subsurface temperature maxima (within the chemolimnion) such as in Lake

Isabela and Nohoch Hol, and (c) double thermoclines like in Montebello. Temperature profiles are tightly coupled to the salinity profiles.

12.5.3 Salinity

Even shallow saline meromictic ponds as found by Cole and Minckley (1968) in Cuatro Ciénegas, can have steep salinity gradients. In this case, the colder mixolimnion with a salinity of 14.6 g l^{-1} is around 7.5 times lesser saline than the warmer monimolimnion that has a salinity of 111.3 g l^{-1}. The same is true for Rincón de Parangueo. According to Alcocer et al. (2002), salinity in the mixolimnion (0–2 m) is <3 g l^{-1}, while in the monimolimnion it reaches 53–56 g l^{-1} (70–80 mS cm^{-1}). Also, Armienta et al. (2008) measured conductance of 165 mS cm^{-1} on a surface sample from Rincón de Parangueo.

The first vertical profile of Lake Isabela from November 1993 (Alcocer et al. 1998) revealed the lake to be meromictic with three well-defined water masses of high conductance separate by sharp haloclines (0–0.5 m, 2.5–4.5 m and 12–13 m). The conductance ranges from 68 mS cm^{-1} at the surface to 112.5 mS cm^{-1} at the bottom. As mentioned before, Kienel et al. (2013) made four fieldtrips to Lake Isabela between December 2006 and February 2009. Three of these trips (December 2006 and 2008, and February 2009) have conductance profiles similar to the one previously described with a fresher superficial layer over a saltier one. However, the May 2008 profile shows the water column to be hypersaline (around 105 mS cm^{-1}) and entirely homogeneous. Conductance profiles recorded by Filonov (unpublished data) show comparable conductance profiles.

The profiles recorded in 1993 by Alcocer et al. (1998) and those from Kienel et al. (2013) differ in that instead of three haloclines in the former, there were just two in the latter. The highest salinity (105 mS cm^{-1}) layer in Kienel's profiles started around 5 m and extended to the bottom. In Alcocer's profiles, the saltier portion (4.5–17.5 m) is divided in two layers separated by a sharp halocline: the first one with around 100 mS cm^{-1} from 4.5 m to 12 m and the second one with 112.5 mS cm^{-1} from 13 m to 17.5 m.

As mentioned, Dos Lagos electrical conductivity profiles (Alcocer unpublished data) revealed the presence of two layers (Fig. 12.2). The conductance of the upper layer averages 662 ± 172 μS cm^{-1}, meanwhile in the lower layer, electric conductance averages 1143 ± 249 μS cm^{-1}. The difference between the conductances of both layers suggests a difference in the water source, fresher surface, and saltier groundwater.

The conductance profiles of Nohoch Hol reveal three distinct water layers (Torres-Talamante et al. 2011). The mixolimnion (0–9 m) is a freshwater layer with an average of 5.0 ± 0.8 mS cm^{-1}. The upper boundary (9.0 m) of the second layer is well defined as conductance inflection point at 6.7 mS cm^{-1}, whereas the lower boundary (15.5 m) is less distinct but corresponds to 54.6 mS cm^{-1}. The monimolimnion is saline with 55.1 ± 0.1 mS cm^{-1} and therefore similar to seawater conductance. Stoessell et al. (2002) comment on the presence of well-defined haloclines in four cenotes, with changes in salinity from <5 % of the seawater salinity in the upper layer up to >95 % of seawater salinity in the lower layer. The width of the halocline varies among the four cenotes from ~2 m (Little Calica) to ~10 m (Big Calica).

Salinity profiles in Mexican meromictic lakes mostly identify three water masses: a "fresher" upper mixolimnion, a "saltier" lower monimolimnion, and in between, the halocline. The halocline width varies from a few centimeters (Cuatro Ciénegas) to a few meters (Nohoch Hol) to up to tens of meters (Montebello). It seems the shallower the water body, the thinner the halocline. However, salinity profiles are dynamic in time as evidenced by Lake Isabela where Alcocer et al. (1998) found the presence of three haloclines, while Kienel et al. (2013) registered just one. The halocline salinity range could be subtle like in the lakes of Montebello ($\approx$200–350 $\mu S\ cm^{-1}$) and most likely maintained by its association to the thermocline and depth ($\approx$70–90 m), moderate like in Lake Isabela ($\approx$50–70 $g\ l^{-1}$) and Rincón de Parangueo ($\approx$3–56 $g\ l^{-1}$), or quite large such as Cuatro Ciénegas ($\approx$15–111 $g\ l^{-1}$).

12.5.4 Dissolved Oxygen

Dissolved oxygen (DO) in Lake Isabela was 6 $mg\ l^{-1}$ (nearly 100 % saturation) at the surface, 7.5 $mg\ l^{-1}$ (145 % saturation) at 0.5 m, and 0.0 $mg\ l^{-1}$ below 2.5 m (Alcocer et al. 1998). Kienel et al. (2013) show three DO profiles with all concentrations <1.5 $mg\ l^{-1}$ but >0 $mg\ l^{-1}$, even in the bottom hypersaline layer. It is easier to develop anoxic conditions on the bottom waters of a stratified, hypersaline lake with temperatures $\geq$25 °C and large inputs of autochthonous—such as phytoplankton blooms—(hypertrophic) and allochthonous organic matter. Both the above studies show subsurface DO maxima that presumably are associated with the accumulation of phototrophic microorganisms at the chemolimnion, with high rates of primary production.

The DO concentration of the upper layer of Dos Lagos reaches 6–8 $mg\ l^{-1}$ (94–106 %). The DO clinograde profiles show sharp oxyclines (Fig. 12.2) in varying number ranging from one and up to three (March 2016). Dos Lagos is anoxic below 22 m (June 2014) and 30 m (February 2015) (Alcocer et al. unpublished data).

Nohoch Hol shows a DO clinograde profile decreasing from 7.3 $mg\ l^{-1}$ (90 % DO saturation) at the surface to 1.0 $mg\ l^{-1}$ (10 % DO saturation) at the base of the mixolimnion. DO decreased abruptly in the chemolimnion down to anoxia and remained anoxic in the monimolimnion (Torres-Talamante et al. 2011). Comparably, cenotes Big Calica, Little Calica, Angelita, and Chemuyil have anaerobic haloclines as well as anaerobic underlying seawater, as evidenced by the occurrence of sulfides and sulfate depletion (Stoessell et al. 2002).

DO profiles of Mexican meromictic lakes display subsurface maxima coinciding with the halocline (e.g., as in Lake Isabela) or with the thermocline (e.g., lake Montebello) most likely related to the presence of dense layers of phytoplankton accumulated at the pycnocline. However, there is also the case of Nohoch Hol where even though the largest amount of photosynthetic organisms is accumulated at the halocline, no subsurface DO maximum develops. Similarly, Dos Lagos does not display subsurface DO maxima. Whereas the mixolimnion shows DO saturation/oversaturation, the monimolimnion remains anoxic.

12.5.5 pH

Rincón de Parangueo was homogeneous throughout the water column and had a pH of 9.8. Armienta et al. (2008) found similar pH (10.0–10.2) throughout the water column in 1999. The soda-alkaline nature and high primary production rates explain the elevated pH values.

The pH of Lake Isabela fluctuated from basic (9.3) in the surface oxygenated waters to slightly acidic (6.4) in the deep anoxic layers. Presumed high rates of primary production in the surface waters and anaerobic decomposition in the deep waters may explain this profile.

The pH in Dos Lagos fluctuated between 6.9 and 8.1, the surface slightly alkaline layer (7.6) to almost neutral (7.1) in the deep layer. Surface waters (<20 m) are more variable with 7.5 ± 0.33 than the deeper ones (>20 m) with 7.1 ± 0.13.

The pH in the Nohoch Hol water column decrease rapidly in the first 2 m from 7.5 to 7.0, but below this depth pH remains stable to the base of the mixolimnion. The pH shows a polymodal profile increasing from neutrality to slightly basic (7.4) at the top of the chemolimnion and from neutral to slightly acidic (6.7) at the base of chemolimnion; the pH remains close to neutrality along the monimolimnion. Stoessell et al. (2002) cite the anoxic monimolimnia of cenotes Big Calica, Little Calica, Angelita, and Chemuyil as having lower than expected pH.

The pH values in the mixolimnion of Mexican meromictic lakes are alkaline, varying from just above neutrality (7.5 in Nohoch Hol) to up to alkaline (<9 in Rincón de Parangueo and Lake Isabela). The monimolimnion pH remains basic in those lakes with high alkalinity (e.g., Rincón de Parangueo), but becomes acidic (6.4) in those with lower alkalinity (e.g., Nohoch Hol).

12.5.6 Nutrients

Alcocer et al. (1998) compared the aerobic (mixolimnion) to the anaerobic (monimolimnion) layers of Lake Isabela and found NO_2^--N, NO_3^--N, NH_4^+-N, and PO_4^{3-}-P were lower in the aerobic strata (0.4, 4.6, 2.6 and 40.8 mg l^{-1}, respectively) than in the anaerobic strata (0.5. 10.3, 5.2 and 96.9 mg l^{-1}, respectively). In contrast, SiO_2 was higher in the mixolimnion (133 mg l^{-1}) than in the monimolimnion (109 mg l^{-1}).

Kienel et al. (2013) found that total phosphorus (TP), silica (Si), and ammonia (NH_4^+) profiles of Lake Isabela had lower concentrations than what Kienel et al. (op cit.) labeled as rainwater body corresponding to the mixolimnion and higher in the hypersaline lake water monimolimnion. TP, Si, and NH_4^+ in the rainwater body are <1 mg l^{-1}, <<5 mg l^{-1}, and <<5 mg l^{-1}, respectively. In contrast, the concentration of these three nutrients in the hypersaline watersheds are 3–5 mg l^{-1}, 10–15 mg l^{-1}, and 15–20 mg l^{-1}, respectively. Alcocer et al. (1998) found a similar pattern. Frigate bird guano is the predominant source of phosphorus and nitrogen and an important source of carbon to the lake (Kienel et al. 2013).

According to Torres-Talamante et al. (2011), nutrient concentrations in the mixolimnion of Nohoch Hol are homogeneous. $N\text{-}NO_3^-$ is the nutrient with the highest concentration reaching 34.8 μM (0.49 mg l^{-1}), while soluble reactive phosphorous (SRP) is below detection limit. $N\text{-}NH_4^+$ and soluble reactive silica (SRSi) are low with 0.9–2.4 μM (0.01–0.03 mg l^{-1}) and 91.0–101.9 μM (2.56–2.86 mg l^{-1}), respectively. $N\text{-}NO_3^-$ concentrations are the highest in the mixolimnion, diminishing within the chemolimnion and reaching minimum in the monimolimnion. $N\text{-}NH_4^+$ reached the highest concentrations in the chemolimnion and diminished toward the monimolimnion, while the mixolimnion has the lowest concentrations. SRP, TP, and SRSi show two peaks, a small one at 10.0 m and a larger one at 13.0 m. Nutrients in the monimolimnion were generally low: $N\text{-}NO_3^- = 0.116$ μM (0.002 mg l^{-1}), SRP = 0.1 μM (0.003 mg l^{-1}), TP = 0.4 μM (0.01 mg l^{-1}), and SRSi = 160.0 μM (4.49 mg l^{-1}). The exception is $N\text{-}NH_4^+$ that is relatively high at 9.3 μM (0.13 mg l^{-1}).

Stoessell et al. (1993) estimated nutrients in the nearby cenotes Little Calica and Angelita. Nitrates and phosphates in the mixolimnion of Little Calica are 9.3 and 0.019 mg l^{-1} and 10.54 and 0.114 mg l^{-1} in Angelita, respectively. In the monimolimnion, nitrates are absent, while phosphates increased to 1.9 and 0.142 mg l^{-1}, respectively, in Little Calica and Angelita.

The trophic status in meromictic Mexican lakes ranges from oligotrophic unpolluted lakes in karst areas where available P is rather scarce (e.g., Nohoch Hol) to highly eutrophic soda-alkaline lakes (e.g., Rincón de Parangueo) or where a large source of nutrients such as frigate bird guano is available (e.g., Lake Isabela). Notably, soda-alkaline lakes are among the most productive ecosystems in the world. In most meromictic lakes, nutrients concentrations are larger in the monimolimnion than in the mixolimnion; this is true independently of their trophic status. Differently, Nohoch Hol shows higher concentrations at the chemolimnion

12.5.7 Chemical Composition

The chemical composition of the Mexican meromictic lakes is provided in Table 12.3. Rincón de Parangueo is a soda lake; the dominant salts are sodium chloride, sodium carbonate, and sodium bicarbonate. Although not mentioned in Armienta et al. (2008), it seems magnesium concentration is quite high. Lake Isabela is a sodium chloride lake. While there are no data available on the ionic composition of Nohoch Hol, it is quite probable that, based on their water source (i.e., rainwater and marine water), the monimolimnion resemble marine water and the mixolimnion would be dominated by calcium and magnesium carbonates/bicarbonates. As a proxy, we include the ionic composition of Little Calica and Angelita, a couple of cenotes nearby Nohoch Hol provided by Stoessell et al. (1993)

Table 12.3 Chemical composition of some Mexican meromictic lakes

Variable	RP[a] Mixo	LI[b] Mixo	LI[b] Moni	LC[c] Mixo	LC[c] Moni	CA[c] Mixo	CA[c] Moni
Tot. Alk	76,000	n.m.	n.m.	6.83	19.36	7.15	4.54
Ph. Alk	33,000	n.m.	n.m.	n.m.	n.m.	n.m.	n.m.
Na^+	4250	16,853	22,992	252.9	7586.4	160.2	9655.4
K^+	5070	450	827	8.8	263.5	4.6	344.5
Ca^{2+}	n.d.	509	749	119	388.4	127.1	781.5
Mg^{2+}	n.d.	2050	2820	47.4	930.9	31.6	1113.2
SO_4^{2-}	76	4082	5154	77	1354.4	4.2	3045
Cl^-	64,000	31,111	41,236	471.5	13,769	291.4	17,438
B	414	6.6	9.5	n.d.	n.d.	n.d.	n.d.
SiO_2	29	n.m.	n.m.	3.5	39.1	6.8	10.8
CO_3^{2-}	40,000	n.m.	n.m.	n.d.	n.d.	n.d.	n.d.
HCO_3^-	12,000	n.m.	n.m.	n.d.	n.d.	n.d.	n.d.
Br^-	n.m.	104	137	n.d.	n.d.	n.d.	n.d.
Sr^{2+}	n.m.	9.7	13.8	0.8	3.1	1.1	8.7
Fe	n.m.	n.m.	n.m.	0.28	0.50	0.28	0.50
TDS	n.m.	n.m.	n.m.	1410	24,800	1100	31,650

RP Rincón de Parnagueo, *LI* Lake Isabela, *LC* Little Calica cenote, *CA* Cenote Angelita. (Concentrations in mg l^{-1}; Alk = mg l^{-1} $CaCO_3$; TDS = ppm). *Alk* alkalinity, *Tot. Alk* Total Alk, *Ph. Alk* Phenolphthalein Alk, *TDS* Total Dissolved Solids, *n.d.* not determined, *n.m.* not mentioned. *Mixo* mixolimnion, *Moni* monimolimnion)

[a]Armienta et al. (2008) estimates due to analytical interferences

[b]Kienel et al. (2013)

[c]Stoessell et al. (1993)

12.6 Biota of Mexican Meromictic Lakes

The Cuatro Ciénegas was inhabited in the upper, fresher waters by three fish species (*Gambusia marshi*, *G. longispinis*, and an undescribed species of *Cyprinodon*), while the bottom and saltier layer contained a cyanobacterial mat (Cole and Minckley 1968).

Green (1986) visited Rincón de Parangueo in 1981 and found the lake with a dense layer of what he identified as the cyanobacteria *Spirulina* and some *Anabaena*. According to Komárek and Komárková-Legnerová (2002), the species found in Rincón de Parangueo are *Arthrospira maxima* (95 %) and *A.* cf. *indica* (5 %), respectively. The zooplankton is composed of the rotifer *Brachionus inermis* (99 %), *Hexarthra polyodonta* (1 %), and a large holotrich ciliate that fed upon *Spirulina*. The lake's surface was covered with alkali flies (*Ephydra hians*) reaching high densities (100–500 ind m^{-2}). No organisms were reported from the deeper water layers.

The planktonic microbial community of Lake Isabela comprised four groups (Alcocer et al. 1998): bacteria (photosynthetic sulfur bacteria and heterotrophic bacteria), phytoflagellates (mainly *Cryptomonas* spp.), heterotrophic nanoflagellates (HNF, mostly *Bodo* spp. and *Spumella* spp.), and ciliates (mainly oligotrichs *Strobilidium* spp. and hypotrichs, *Euplotes*). Bacteria were found throughout the water column. Heterotrophic bacteria increased in numbers from the mixolimnion (5.2×10^6 ind ml^{-1}) to the chemolimnion (6.9×10^6 ind ml^{-1}) and even more in the monimolimnion (9.7×10^6 ind ml^{-1}). Photosynthetic bacteria decreased by an order of magnitude from the mixolimnion (8.3×10^6 ind ml^{-1}) to the chemolimnion (8.1×10^5 ind ml^{-1}) and even further in the monimolimnion (4.7×10^5 ind ml^{-1}). Limnetic macroinvertebrates were an ostracod (quite probably an undescribed species of *Thalassocypris*) dwelling between the aerobic and anaerobic strata, and a corixid (*Trichocorixa reticulata*) inhabiting the upper, oxygenated part of the water column.

Romero-Viana et al. (2012) reported >80 lipid compounds in the sediment that could be attributed to three major groups of organisms: (1) algal populations (*n*-alkyl diols and derivatives, unsaturated alcohols, ketones, and long-chain alkanes); (2) a mixed community of ciliates, bacteria, and cyanobacteria (tetrahymanol derivatives, hopanols, phytene derivatives, diploptene, short-chain alkanes, and 4-methyl C_{28} sterol); and (3) photosynthetic sulfur bacteria (farnesene and carotenoids). Kienel et al. (2013) observed a layer of photosynthetic bacteria in chemocline of Lake Isabela. During the rainy season when the upper layer becomes fresher, algae (mostly diatoms) constitutes the main group; during the dry period, with increasing salinity, bacteria and ciliates dominate.

The bacterioplankton of Nohoch Hol is composed of bacillus, cocci, and filamentous bacteria; bacillus with >80 % dominates. According to Torres-Talamante et al. (2011), the bacterioplankton density increases with depth from 8.3×10^3 cells ml^{-1} at 2.0 m to 35.2×10^3 cells ml^{-1} at 10.0 m, where also phytoplankton maximum occurred. Within the mixolimnion, the bacterioplankton density increases slightly toward the top of the chemolimnion. Within the chemolimnion, the bacterioplankton display two maxima, one at 10.0 m with 35.2×10^3 cells ml^{-1} and another at 12.0 m with 23.1×10^3 cells ml^{-1}; both maxima match the $N\text{-}NH_4^+$ peaks at these two depths. In the monimolimnion, the bacterioplankton density attained values similar to those found in the mixolimnion with 12.9×10^3 cells ml^{-1}. Stoessell et al. (1993) report the probable occurrence of *Thiobacillus* sp. bacteria throughout the water column of cenotes Little Calica and Angelita, including the freshwater zone.

Phytoplankton was composed of eleven taxa in Nohoch Hol (Torres-Talamante et al. 2011): Chlorophyta, *Chlorella*, *Monoraphidium*, and *Scenedesmus*; Bacillariophyta, *Cyclotella*, two species of *Navicula*, and three species of *Nitzchia*; and Cyanophyta, *Chroococcus*, and *Phormidium*. Chlorophyta dominates (96 %) the phytoplankton abundance, with *Scenedesmus* comprising >93 % of the total phytoplankton density (21.9×10^3 cells ml^{-1}). *Scenedesmus* and *Cyclotella* dominate the mixolimnion; however, at 10 m depth in the upper part of the chemolimnion, *Scenedesmus* forms an almost monospecific (>99 %) dense layer reaching up to 20.5×10^3 cells ml^{-1}. One meter below this layer, *Cyclotella* dominates (90 %) but in much lower densities (0.058×10^3 cells ml^{-1}) than those mentioned for *Scenedesmus*. As expected, phytoplankton is absent in the anoxic monimolimnion.

An interesting case of the influence of weather conditions on the ecology of a meromictic lake is illustrated by Lake Isabela. The limnology of the lake is ruled by the influence of the strong seasonal variations in rainfall/evaporation (Romero-Viana et al. 2012, 2013). The lipid biomarkers of algal populations preserved in the sediments were related to variability of rainfall through salinity fluctuations controlled by seasonal rainfall. Wetter periods led to a fresher mixolimnion, which enhanced algal growth, while the relative abundance of the more halotolerant populations (Cyanobacteria, bacteria and ciliates) decreased. Kienel et al. (2013) mentioned that, in addition to the decrease in salinity during the wetter periods, the seabird guano is introduced through runoff from catchment deposits. The guano is the main source of nutrients (C, N, P) to the lake and fuels the high algal primary productivity.

Torres-Talamante et al. (2011) mention an interesting example of bacterial succession in the chemolimnion of Nohoch Hol, controlled by DO concentration, which also controls the nutrient cycling: "*Within the dysoxic zone at the top of the chemolimnion, colonies of Beggiatoa and/or Thiothrix, or similar, could actively process sulfur increasing the activity of the sulfur cycle. Deeper within the chemolimnion, Thiobacillus-like and other Rhodosperillaceae may be driving the nitrogen cycle. At the anoxic monimolimnion, the rod-shape bacteria similar to Allochromatium of the sulfur cycle must be dominant drivers of the nitrogen cycle, while bacteria like Desulfovibrio, Chromatium, and Thiocapsa (coccus) could drive the sulfur cycle.*" The chemolimnion stands for an ecotone characterized by high biodiversity and biological activity, playing an important role in the biogeochemistry of the waterbody.

12.7 Failed Meromictic Lakes

Sistema Zacatón is a unique karst area in the southern part of the state of Tamaulipas (23°00′ N, 98°09′ W, 196 m a.s.l.). This area has numerous sinkholes; Gary et al. (2008) hypothesized they have been formed by hypogenic karst processes influenced by volcanic activity. Three sinkholes of Sistema Zacatón are bathymetrically (very deep and small area) predisposed to meromixis: (a) El Zacatón which is the deepest underwater cave shaft in the world with 319 m deep and 113 m in diameter, (b) La Pilita is around 105 m deep and has a surface of ca. 20 m × 30 m, and (c) Caracol is 70 m deep and 20 m in diameter. Nonetheless, Gary (2009) found the three cenotes were very homogenous with respect to temperature, pH, dissolved oxygen, and specific conductance. The waters are hydrothermal (i.e., the water is heated geothermally, this is, groundwater heated through the proximity of a magmatic chamber); with a constant temperature ~30 °C and anoxic associated to geothermal convective currents which have resulted in a homogeneous water column.

Further, Gary et al. (2008, 2009) mention that Cenote Verde (also Poza Verde) is the only body of water of the Sistema Zacatón to have "*discernable thermoclines and chemoclines*" thus suggesting a meromictic lake. Cenote Verde is nearly circular, 150 m in diameter, and 45 m deep. However, the data provided in Gary (2009) show dissolved oxygen values ranging from 7 mg l^{-1} at the surface to 2 mg l^{-1} on the bottom and specific conductivity varying from 700 μS cm^{-1} at the surface to up to 750 μS cm^{-1} at the bottom, which questions the presence of meromixis.

12.8 Concluding Remarks

1. The climatic, geological, and physiographic characteristics of the Mexican territory resulted in a large diversity of inland aquatic bodies. So far, a few meromictic lakes have been reported from three and probably four different physiographic–geologic regions of Mexico covering wide ranges of latitude (≈16–26 °N), longitude (≈87–102 °W), and altitude (≈7–1700 m a.s.l.) range.
2. The meromictic lakes in Mexico originated from one or a combination of the following factors: (a) coastal lakes where the freshwater above and seawater below in karst zones interact, (b) lakes located in semiarid and arid areas where evaporation greatly exceeds precipitation and rainfalls are abrupt so that freshwater accumulates above the saline one, and (c) deep lakes with relatively small surface area (high relative depth) that lack complete circulation.
3. It seems the factors leading to meromixis have no effect in originating patterns on the lakes' physical, chemical, or biological characteristics. Mexican meromictic lakes display a mosaic of environmental and ecological features.
4. The meromictic lakes are small or large, shallow or deep, with either turbid mixolimnion or transparent mixo- and turbid chemolimnion; with diverse

temperature profiles from thermal inversions to subsurface maxima or double thermocline; with one to three haloclines showing from subtle to quite large salinity ranges, and thickness from few cm to up to tens of meters; the DO profiles are typical clinograde or with a subsurface maximum; pH is alkaline in the mixolimnion but varies from acidic to alkaline in the monimolimnion; nutrient (e.g., nitrogen, phosphorus) concentrations range from low (oligotrophic) to high (eutrophic/hypertrophic); dominant salts are sodium chloride, sodium carbonate/bicarbonate, and/or calcium sulfate.

5. Our understanding of Mexican meromictic lakes is limited. There are several aspects yet to be studied such as their general, microbial, and trophic ecology, biodiversity, as well as the potential effects of global climate change on these peculiar ecosystems.
6. Finally, holomictic lakes with irregular circulation tendencies are not truly meromictic lakes. Single or short sampling periods may suggest a lake to be meromictic; however, a longer study period could reveal irregular circulation patterns instead. Then, long sampling programs are required to identify the true Mexican meromictic lakes. It is known that "*truly meromictic lakes are found occasionally in the tropics, as they are at temperate latitudes*" (Lewis 2000).

References

Alcocer J (2002) Surface water—major challenges in Mexico. LakeLine Winter:28–31

Alcocer J, Lugo A, Sánchez MR et al (1998) Isabela crater-lake: a Mexican insular saline lake. Hydrobiologia 381:1–7

Alcocer J, Lugo A, Escobar E et al (2000a) Water column stratification and its implications in the tropical warm monomictic Lake Alchichica, Puebla, Mexico. Verh Int Verein Limnol 27:3166–3169

Alcocer J, Escobar E, Lugo A (2000b) Water use (and abuse) and its effects on the crater-lakes of Valle de Santiago, Mexico. Lakes Reserv Res Manage 5:145–149

Alcocer J, Lugo A, Oliva MG (2002) Los lagos cráter de Valle de Santiago, Guanajuato. In: Lanza G, JL G (eds) Lagos y presas de México. Centro de Ecología y Desarrollo, Mexico, pp 193–212

Alcocer J, Oseguera LA, Sánchez G et al (2016) Bathymetric and morphometric surveys of the Montebello Lakes, Chiapas. J Limnol 75(s1):56–65

Aranda-Gómez JJ, Chacón-Baca E, Charles-Polo M et al (2009) Collapse structures at the bottom of a recently desiccated maar lake: Rincón de Parangueo maar, Valle de Santiago, México. In: Abstracts of the IAVCEI—CVS—IAS 3IMC Conference, Malargüe, Argentina, 14–17 April 2009

Armienta MA, Vilaclara G, De la Cruz-Reyna S et al (2008) Water chemistry of lakes related to active and inactive Mexican volcanoes. J Volcanol Geotherm Res 178:249–258

Cole GA, Minckley WL (1968) Anomalous-thermal conditions in a hypersaline inland pond. J Arizona Acad Sci 5(2):105–107

Ferrusquía-Villafranca I (1993) Geology of Mexico—a Synopsis. In: Ramamoorthy TP, Bye R, Lot A et al (eds) Biological diversity of Mexico: origins and distribution. Oxford University Press, New York, pp 3–107

García E (1973) Modificación al sistema de clasificación climática de Köeppen. Para adaptarlo a las condiciones de la República Mexicana. Instituto de Geografía UNAM, Mexico

Gary MO (2009) Karst hydrogeology and speleogenesis of Sistema Zacatón, Tamaulipas, Mexico. Dissertation, The University of Texas at Austin

Gary MO, Fairfield N, Stone WC et al (2008) 3D mapping and characterization of Sistema Zacatón from DEPTHX (DEep Phreatic THermal eXplorer). In: Yuhr LB, Alexander EC Jr, Beck BF (eds) 11th Multidisciplinary Conference on Sinkholes and the Engineering and Environmental Impacts of Karst. ASCE, pp 202–212

Gary MO, Halihan T, Sharp JM Jr (2009) Detection of sub-travertine lakes using electric resistivity imaging, Sistema Zacatón, México. In: White WB (ed) Proceedings of the 15th International Congress of Speleology, Texas

Green J (1986) Associations of zooplankton in six crater lakes in Arizona, Mexico and New Mexico. J Zool 208:135–159

Housh TB, Aranda-Gómez JJ, Luhr JF (2010) Isla Isabel (Nayarit, México): quaternary alkalic basalts with mantle xenoliths erupted in the mouth of the Gulf of California. J Volcanol Geotherm Res 197:85–107

Kienel U, Wulf Bowen S, Byrne R et al (2009) First lacustrine varve chronologies from Mexico: impact of droughts, ENSO and human activity since AD 1840 as recorded in maar sediments from Valle de Santiago. J Paleolimnol 42:587–609

Kienel U, Plessen B, Schettler G et al (2013) Sensitivity of a hypersaline crater lake to the seasonality of rainfall, evaporation, and guano supply. Fundam Appl Limnol 183(2):135–152

Komárek J, Komárková-Legnerová J (2002) Contribution to the knowledge of planktic cyanoprokaryotes from central Mexico. Preslia 74:207–233

Lewis WM Jr (1996) Tropical lakes: how latitude makes a difference. In: Schiemer F, Boland KT (eds) Perspectives in tropical limnology. SPB Academic Publishing, Amsterdam, pp 43–64

Lewis WM Jr (2000) Basis for the protection and management of tropical lakes. Lakes Reserv Res Manage 5:35–48

Minckley WL (1969) Environments of the Bolsón of Cuatro Ciénegas, Coahuila, México, with special reference to the aquatic biota. University of Texas, El Paso, TX

Romero-Viana L, Kienel U, Sachse D (2012) Lipid biomarker signatures in a hypersaline lake on Isabel Island (Eastern Pacific) as a proxy for past rainfall anomaly (1942–2006 AD). Palaeogeogr Palaeoclimatol Palaeoecol 350–352:49–61

Romero-Viana L, Kienel U, Wilkes H et al (2013) Growth-dependent hydrogen isotopic fractionation of algal lipid biomarkers in hypersaline Isabel Lake (México). Geochim Cosmochim Acta 106:490–500

Schmitter-Soto JJ, Comín FA, Escobar-Briones E et al (2002) Hydrogeochemical and biological characteristics of cenotes in the Yucatan Peninsula (SE Mexico). Hydrobiologia 467:215–228

Stewart KM, Walker KF, Likens GE (2009) Meromictic lakes. In: Likens GE (ed) Encyclopedia of inland waters. Elsevier, Oxford, pp 177–190

Stoessell RK, Mooore YH, Coke JG (1993) The occurrence and effect of sulfate reduction and sulfide oxidation on coastal limestone dissolution in Yucatan cenotes. Ground Water 31 (4):566–575

Stoessell RK, Coke JG, Easley DH (2002) Localized thermal anomalies of coastal Yucatan sinkholes. Ground Water 40(4):416–424

Torres-Talamante O, Alcocer J, Beddows PA et al (2011) The role of chemolimnion in meromictic cenotes of the Yucatan Peninsula, Mexico. Hydrobiologia 677:107–127

Part III
Conclusions

Chapter 13
Conclusion: Ecology of Meromictic Lakes

Ramesh D. Gulati and Egor S. Zadereev

13.1 Introduction

Relatively little was known about meromictic lakes until 1935 when the term 'meromixis' was introduced by the Austrian limnologist Findenegg (1935) based on his observations on Caranthian lakes. While working on the sub-alkaline lakes in Carinthia, he found that some deep lakes did not completely mix at all; he called these lakes as *meromictic*. In such deep lakes, during the warming cycle, the wind-induced currents cannot override the large density differences between the surface and bottom waters. The latter contain relatively higher concentrations of dissolved salts so that density difference between the surface and bottom water is quite manifest. The stability of the vertical density differences may be further reinforced by the lake's morphometry, size, shape and depth in particular. Thus, the meromictic conditions in such lakes can persist for decades and even centuries.

Meromictic lakes have an upper mixed stratum called 'mixolimnion', which has features of a typical lake where mixing by wind occurs; this upper stratum is homogeneous in both its temperature and chemical composition. This mixolimnion can stratify seasonally into an upper epilimnion, and a lower hypolimnion, with a thin intervening layer, the thermocline. Thus, mixolimnion in such lakes may behave like holomictic lakes that mix annually at least once but often twice. Such lakes have below the mixolimnion a denser stratum the 'monimolimnion', which does not mix even under the wind influence. The monimolimnion remains estranged from the atmosphere and, therefore, it lacks a direct supply of oxygen;

R.D. Gulati (✉)
Netherlands Institute of Ecology (NIOO), 6708 PB Wageningen, The Netherlands
e-mail: r.gulati@nioo.knaw.nl

E.S. Zadereev
Institute of Biophysics SB RAS, Akademgorodok, 660036 Krasnoyarsk, Russia
e-mail: egor@ibp.ru

© Springer International Publishing AG 2017

R.D. Gulati et al. (eds.), *Ecology of Meromictic Lakes*, Ecological Studies 228,
DOI 10.1007/978-3-319-49143-1_13

also, it does not mix with the mixolimnion above it. There is an interceding boundary layer between the mixolimnion and monimolimnion usually called 'chemocline': the sharp chemical change caused by a vertical chemical gradient in the water column. This boundary between mixolimnion and monimolimnion is often also called pycnocline (to stress that this marked difference is based on water density) or halocline if this difference is caused by salinity increase alone in the water. Unlike the holomictic lakes, the meromictic lakes are formed if recirculation in the deeper layers is insufficient to uniformly mix the entire waterbody and remove any existing chemical gradients. In other words, the monimolimnion is per se excluded from recirculation, and it has a markedly different chemical milieu, which persists due to its higher density than that of mixolimnion. Such a density increase in the deeper layers (monimolimnion) creates the so-called meromixis. Also biogeochemical processes, e.g. decomposition of organic material, iron oxidation and calcite precipitation, contribute partly to meromixis.

The mixolimnion in the meromictic lakes displays seasonal stratification and circulation patterns that resemble the mixing process in holomictic lakes. The mixolimnion of meromictic lakes can be monomictic (one annual recirculation period during the cold season) as in some Siberian lakes (Chap. 5) or dimictic with a complete circulation of the lake in both spring and autumn (Wetzel 2001). These patterns of seasonal circulation are governed by the climate zone (Chap. 2). Many dissolved substances can be enriched by precipitation and flocculation in the mixolimnion once the solubility product is reached or gas pressure increases and exceeds the absolute pressure. This causes the composition of solutes to unmistakeably differ from typical water composition. The permanent density stratification in lakes limits the vertical transport of water and thus the solutes. In several lakes, double-diffusive convection (DDC) has been reported (Boehrer and Schultze 2008; Chap. 2). During the summer stratification in the mixolimnion, little is happening in the chemocline stratum with the monimolimnion below and mixolimnion above it. Groundwater may enter the monimolimnion and increase its volume so that the chemocline gradually rises (von Rohden et al. 2010). In addition, diffusion and turbulent transport of solutes from the monimolimnion can raise the density gradient locally and lead to an increase in the oxygen demand in the lowest part of the hypolimnion. The properties of the affected hypolimnetic water are transformed to become very similar to those of the monimolimnion. This hypolimnetic water eventually becomes a part of the monimolimnion. In general, meromixis is much more dynamic than it often appears. The lake will remain meromictic until different processes (Chap. 2) can balance mixing and diffusive effects that erode the stratification.

In this book, we started with the Introduction (Chap. 1) on terminologies and worldwide distribution of meromictic lakes. Thereafter, we examined first the physical aspects of meromixis (Chap. 2) and then devoted attention to biogeochemical processes and paleolimnological aspects of studies in meromictic lakes (Chap. 3). The biology and ecology of meromictic lakes are discussed in Chap. 4. Thereafter, we described several case studies of meromictic lakes from different geographical areas and with different mechanisms to sustain meromixis (Part II of book: Chaps. 5–12) followed by the Conclusions (Chap. 13) in Part III, which are based mainly on both chapters of authors' summary remarks and on our editorial excerpts.

13.2 General Properties of Meromictic Lakes

Physical Aspects of Meromixis (Chap. 2) Lakes turn meromictic, if mixing and deep-water recirculation are insufficient to mix the waterbody to eliminate any chemical gradients. The monimolimnion is excluded from the deep recirculation, and hence it develops a distinctly different chemical milieu. It persists through all the seasons due to its often markedly higher density than the mixolimnion above it. A number of purely mechanical processes is known to create either such a density upsurge of the deep waters or a density decrease of surface waters to create meromixis. These mechanical processes include inflows of salty water into lakes or freshwater into saline lakes. Also, there is a partial deep-water renewal by the evaporation or cooling in side bays and salty intrusions from ice in saline lakes in cold climate. In addition, geochemical processes, such as decomposition of organic material, iron oxidation and redissolution, temperature-dependent solubility of mirabilite and calcite precipitation, can contribute to meromixis (Chap. 2).

The monimolimnion does not have direct contact with the atmosphere. Hence, there is no direct supply of oxygen from the monimolimnion, which is generally anoxic. Moreover, the composition of solutes in the monimolimnion deviates unambiguously from the usual water composition. The permanent density stratification within the monimolimnion limits the vertical transport of both water and solutes. At the same time, the stratification in meromictic lakes should not be viewed as a static phenomenon. There are several discernible exchange processes (diffusion, transport of solutes) between the mixolimnion and monimolimnion, via the chemocline. Moreover, for meromictic lakes, it is possible to measure the age of monolimnetic water and calculate its renewal rate. Due to its continuous renewal, the age of monimolimnetic water can be younger than the duration of meromixis in a lake (e.g. Vollmer et al. 2002).

Lastly, the perpetually anoxic conditions above the monimolimnetic sediment of meromictic lakes provide better circumstances for conservation of the settling organic material and prevent its disturbance by bioturbation.

Biogeochemistry of Meromictic Lakes (Chap. 3) Meromictic lakes vary widely in their chemical composition: the concentrations of total dissolved substances (TDS) range widely, from very low (<20 mg L^{-1}) to very high (>300 g L^{-1}). Many meromictic lakes are saturated with certain compounds and metals, which include bicarbonates, carbon dioxide, dissolved iron and dissolved silica that invariably contribute to the density gradients. The pH of the meromictic lakes varies from acidic (<3) to alkaline (>10); the redox conditions range from well oxygenated and dominated by high concentrations of dissolved ferric iron (Eh ~ 600 mV) in acidic pit lakes to strongly reduced water (Eh < -100 mV).

Reduced chemical species, e.g. of ferrous iron, hydrogen sulphide and ammonia, vary very much in the monimolimnion depending on local conditions, lake's developmental history and geochemistry of the area where lake was formed.

The variety of chemical constituents in meromictic lakes implies that chemical conditions in such lakes will differ greatly. In freshwater lakes, bicarbonate, carbon dioxide, dissolved iron, manganese and silica and their biogeochemical cycling can be key contributors to the density gradients. Chemistry of hard-water meromictic lakes can be dominated by the presence of limestone in the lakes' catchment areas and by the biogenic precipitation in its mixolimnion (Laguna La Cruz, Chap. 8). Also, inputs of dissolved gypsum from the catchment (Lake Cadagno, Chap. 7) or even carbon dioxide from the volcanic sources underneath the lake floor (Lake Kivu, Chap. 10) may dominate. Many meromictic lakes are characterised by high levels of sulphate as well as high concentrations of H_2S in the monimolimnion (e.g. lakes Shira and Shunet, Chap. 5). For saline lakes, high concentrations of sulphate and sodium and precipitation of different compounds such as pyrite or calcite can assist the lake to endure its meromictic conditions.

Many meromictic lakes are rather rich in methane in the monimolimnion. Exceptionally, the concentrations of biogenic methane are high enough for commercial exploitation (e.g. Lake Kivu, Chap. 10). Sea water from underground seepage into meromictic lakes can significantly influence the biogeochemistry of such lakes (e.g. Lake Rogoznica in Croatia, Chap. 6). Island Copper Mine Pit Lake, Canada, has a strong influence of sea water (Chap. 9) capped with a layer of fresh to brackish sea water. Both alkaline and acidic lakes can be meromictic; soda meromictic lakes (such as Mono Lake, Chap. 11) are generally both alkaline and saline. Acidic meromictic lakes usually originate from surface mining and their acidification is caused by pyrite or other compounds (Chap. 9). Concentrations of metals in lakes that receive waters from metal mining areas may reach grams per litre of iron and several hundred milligrams per litre of other (heavy) metals.

Most biological processes that contribute to meromixis involve microbial decomposition of the settling particulate organic matter (POM) and release of the soluble compounds, such as bicarbonate, contributing significantly to water density (Boehrer and Schultze 2008). The reduction of manganese, iron and sulphate and fermentation and acetotrophic methanogenesis are important pathways of bicarbonate production. Such products generally increase the density of water. Conversely, methane as one of the reaction products of acetotrophic methanogenesis lowers the density of water (Dietz et al. 2012).

Eutrophication can cause meromixis through increased primary production in the mixolimnion of lakes and the subsequent greater decay of the produced biomass in the deeper lake parts. If high primary production rates in the mixolimnion and the correspondingly high uptake rates of carbon dioxide cause calcite to precipitate, the settling calcite may be dissolved again in the monimolimnion. This last is supported by both high levels of dissolved CO_2 and acidic pH due to decay of the POM in the monimolimnion (Chap. 3).

Not only the precipitation of calcite but also precipitation, redissolution and high concentrations of iron are also responsible for sustaining meromixis. Ferrous iron produced from reduction of ferric iron in monimolimnion may on diffusion into oxygenated mixolimnion be again oxidised and precipitate. Also, iron may be oxidised by the photo-ferrotrophic microorganisms in the anoxic lake parts. Settling ferric hydroxide particles are redissolved on being reduced by microorganism. Like iron, manganese precipitates on its oxidation, and, like iron, it contributes to stabilising meromixis.

Paleolimnological Aspects of Sediment Chemistry (Chap. 3) Lake sediment preserves records of the environmental changes, which enable us to reconstruct the past conditions in lakes. Hence, biomarkers are well-preserved chemical compounds that carry signatures of the water column or sediment processes at the time of deposition. The origin and breakdown dynamics of organic matter can be tracked by the records of ^{13}C and ^{15}N stable isotopes: other proxies are algal assemblages, microfossils, pollen, pigments, fish bones, etc. The interpretations for many of these proxies for both meromictic and non-meromictic lakes bear similarities. However, sediment chemistry can influence the interpretations of paleolimnological records (see Chap. 3).

Both the composition and vertical distribution of substances within the aquatic sediments are modified after they are deposited by various chemical, physical and biological processes (collectively called diagenesis). A sequence of redox reactions are fuelled by oxidation of the deposited organic matter and catalysed microbially. In non-meromictic lakes, the most intense redox cycling transpires within the sediment column, near the sediment oxic-anoxic boundary. Long-term anoxia in meromictic lakes effectively shifts these reactions into the water column, where oxic-anoxic boundary lies around the pycnocline depth. Anoxic conditions favour preservation of organic substances and thus prevent oxidation of organic matter. Thus, sediment in meromictic lakes acts particularly as an important repository or storehouse of paleoenvironmental information.

The absence of mixing by benthic macrofauna and of bioturbation due to the lack of oxygen further minimises post-depositional disturbances. Varved or annually layered sediment provides predominantly high-resolution record. Seasonal varves can be preserved, e.g. when particle settles fast enough to transmit the seasonal signal to the lake floor. Since both bottom currents and gas formation are minimal in the sediment, they help retain the laminations. East African Rift Valley lakes are good examples of paleolimnological reconstruction. These lakes contain valuable records of the past climatic conditions and changes herein. The absence of bioturbation during the meromictic (anoxic) phases of lake histories facilitates the preservation of evidence for high-resolution episodic events such as volcanic eruptions. The disappearance of laminations in sediment generally indicates shifts in the lake's hydrodynamic regime, caused usually by fluctuations in in deep-water oxygen levels. Well-preserved high-resolution signals in the anoxic sediment of such ancient lakes can enable us to assess the historical behaviour of high-frequency signals of events, such as El Niño/Southern Oscillation, and shifts in carbonate precipitation/deposition as measures of environmental changes. The enrichment by precipitation of calcium carbonates (primarily as calcite or aragonite), regulated by factors such as the dissolved CO_2 (DIC) levels, Ca^{2+} concentrations and pH, are frequently considered to indicate high biological or photosynthetic activity of the primary producers.

Evidence of past changes in a lake's mixing regime, including episodes of deep-water circulation and lake overturns, may be also preserved in sediment records. Specific for the meromictic lake fossil proxy are the remains of carotenoids (plant pigments) and bacteriochlorophylls and their derivatives (i.e. phaeopigments and DNA) that serve as indicators of anaerobic conditions in the photic zone of lake at some stage of its existence (Overmann et al. 1993). Variations in sediment concentrations of organic carbon have been often viewed as proxies for changes in primary productivity, or organic sedimentation rates. Organic carbon that is buried into the deep sediment can be considered as indicator of paleoproductivity. However, caution is needed because evaluating the relationship between organic carbon in sediment and paleoproductivity is not easy (Katsev and Crowe 2015).

Biology and Ecology of Meromictic Lakes (Chap. 4) It is well known that the water column of meromictic lakes is not entirely mixed, and this affects the lake water chemistry and lake biology, including the organism and their ecology. Moreover, meromictic lakes have some exclusive ecological and biological features. The chemocline—a habitat created at a depth with mixolimnion above and monimolimnion below—is characterised by complex community of bacteria, autotrophic and heterotrophic protists and metazooplankton, invariably dominated by rotifers. The rates of anoxygenic photosynthesis by purple sulphur bacteria (PSB) and green sulphur bacteria (GSB) in deep, anoxic layers are high. Such the lakes have partially coupled sulphur, carbon and nitrogen cycles.

Most bacterial and archaeal taxa, especially in the anoxic waters, have still not been identified, and little is, therefore, known about their specific role in these environments. Meromictic lakes need to be investigated for both their uncultivated diversity and about the metabolic potential of the bacterial communities. The chemocline communities represent the essential ingredients of microbial loop, formed by bacteria and protists, which probably link the production of organic matter in the anoxic waters with the classical grazer food web in the mixolimnion. The functioning of the food web in such lakes is virtually unknown. The classical grazer food web in meromictic lakes is often strongly truncated, because fish and other predators are invariably entirely absent, most probably because of high salinity levels and limited volume of inhabitable oxic water. Many saline meromictic lakes, e.g. the Siberian lakes (Chap. 5), also contain certain taxa of crustacean zooplankton, including calanoid copepods and cladocerans. With further increase of salinity, the zooplankton may be limited to just *Artemia* spp., e.g. Mono lake (Chap. 11) and in many hypersaline lakes.

Among the most inimical features of meromictic lakes are the uninterrupted stratification and the vast monimolimnion that is depleted of oxygen. Other inimitable features of meromictic lakes are the coupling of nutrients cycles and anoxic and oxic food-web components. Briefly, meromictic lakes can be considered as natural laboratories to study the complexity of the food webs and biological interactions in habitats spatially separated by vertical chemical and physical gradients.

13.3 Case Studies on Different Meromictic Lakes

Siberian Lakes Shira and Shunet (Chap. 5) Lakes Shira and Shunet (South Siberia, Russia) are both saline and ectogenic meromictic, i.e. meromixis is sustained by ice formation in winter because the ice lowers the density of water further. In spring, due to the melting of ice, a second halocline is formed in near-surface layers of mixolimnion that precludes the lakes from full circulation. Based on study period 1998–2014, the mixolimnion in both lakes, Shira and Shunet, was monomictic, circulating in autumn only. Because of their geographical proximity, similar chemical composition, stable stratification and absence of fish and other plankton predators, the lakes have quite similar species compositions of bacteria, phytoplankton and zooplankton. However, their food webs differ notably because of the differences in their food-web structure and the morphometry and salinity that control the stability of permanent stratification.

Lake Shira, which is relatively deeper and has a monimolimnion with a variable, has a bacterial community in the chemocline that does not significantly contribute to the food web. In the shallower Lake Shunet with a shallow chemocline located at constant depth ca. 5 m, the extremely dense bacterial community, a population of *Cryptomonas* sp. and several species of ciliates occupy this stable habitat. The zooplankton community in both lakes is comprised of several genera of ciliates, a calanoid copepod *Arctodiaptomus salinus*, rotifers *Brachionus plicatilis* and *Hexarthra* sp. and an amphipod *Gammarus lacustris*. This amphipod seems to have migrated from the lake's littoral-benthic regions to the pelagic region now. In Lake Shira, the zooplankton, dominated by calanoid copepod *A. salinus*, feeds mostly on phytoplankton, whereas the contribution of bacterial community in the chemocline to the grazer's food web is insignificant. Zooplankton in Lake Shunet feeds on both bacterial community and *Cryptomonas* sp. in the chemocline. Thus, the food-web structure of the two lakes differs due to the differences in their mean and maximum depths and depth and stability of the chemocline.

Laminated sediment of Lake Shira was studied to reconstruct the lake's mixing regime (Kalugin et al. 2013). Geochemical analyses show that Lake Shira has two contrasting mixing states: meromictic and holomictic. These two states were identified by the content of organic matter and the concentration of okenone, the main carotenoid pigment of the PSB inhabiting the lake as these markers differ in sediments depending on the lake mixing regime (see further for the differences). For holomictic conditions, sediments with low content of organic matter appeared in the periods BC 220–275, AD 480–565, AD 900–1045, AD 1437–1603 and AD 1853–1947 (Kalugin et al. 2013). Presumably, during holomictic conditions most organic matter produced in the waterbody was oxidised due to complete mixing of the water column. Simultaneously, the biomass of phototrophic sulphur bacteria seems to have decreased/disappeared due to the lack of H_2S. Meromictic periods are comprised of black, thin laminated mud coloured by organics and hydrotroilite (FeS_nH_20), which are predominant in the core.

Between 1920 and 1940, the water level of Lake Shira decreased by about 7 m but salinity level increased. The low content of okenone in the sediment of the lake at this time indicates that meromixis was quite weak or absent. A peak concentration of okenone during 1945–1970 indicates the lake to be meromictic. Concurrently, water level of the lake has increased, and the lake has become strongly meromictic. This regime shift was most probably caused by the inflow of fresh water with the surface runoff onto the saltier lake. Therefore, the transition between holomictic and meromictic regimes has been caused by changes in the lake's water level.

Lake Rogoznica (Chap. 6) Lake Rogoznica is a typical, euxinic marine lake, situated on the eastern Adriatic coast of Croatia. It is a karstic depression filled with sea water, surrounded by high karstic cliffs, which restrain the wind-induced mixing. The lake is stratified, with an oxic upper water layer and an anoxic deeper layer. Phytoplankton photosynthetic activity in upper layers is high, and decomposition process dominates in deeper layers, which are hypoxic on top and anoxic below, and produce H_2S microbially. The mixolimnion varies seasonally in depth and thickness due to influence of temperature, wind and rainfall.

The lake switches back and forth between meromictic and holomictic conditions, which affect the lake's ecology. During meromixis, water strata below 10 m usually become anoxic, and concentrations in the anoxic region of reduced sulphur species, nutrients and DOC are high. In the anoxic deep water, laminated sediment, trace metals, especially Fe and Mo, are rich. Extreme ecological conditions prevail in the lake. The species diversity of phytoplankton and zooplankton is low. Among phytoplankton, diatoms are the dominant phytoplankters, while among zooplankton, oligotrich ciliates, copepods and heterotrophic are important organisms in the lake's food web, especially in the post-holomictic anoxic period. Densities of PSB are high in the chemocline region. Ciliates graze upon PSB and thus transfer organic carbon to higher trophic levels. They themselves form an important food source for nauplii and copepodites of *Acartia italica*.

Shift to holomictic conditions cause anoxic conditions in the whole water column, with high sulphide and ammonium concentrations. They cause a mass mortality of phytoplankton and benthic organisms. The simulation with 1-D General Ocean Turbulence Model (GOTM) (Ciglenečki et al. 2015) shows that mixing between water layers in lake Rogoznica can be fast if the water-column stability is reduced by abrupt changes in the air temperature and strong wind. This mixing causes a rapid oxidation of the sulphide, depleting oxygen in the surface waters and leading to water-column anoxia.

Lake Cadagno (Chap. 7) Lake Cadagno is a small (26 ha), crenogenic meromictic lake located in the Swiss Alps at 1921 m a.s.l. The lake is maximally 21 m deep, with a redox transition zone at about 12 m depth, which is also the depth of chemocline. The water layers are stabilised by density differences of salt-rich water entering from sub-aquatic springs to the monimolimnion. In contrast, the mixolimnion contains electrolyte-poor surface water. The geology of the lake's catchment area is marked by both crystalline rocks and a dolomite gypsum-rich strip (hydrated calcium sulphate) making the lake a typical 'sulphuretum' dominated by the coupled carbon and sulphur cycles.

Steep sulphide and light gradients in the chemocline support the growth of a large bacterial plume (up to 10^7 cells mL^{-1}) dominated by the GSB of genus *Chlorobium* and PSB (*Chromatium okenii*); the latter represents only 0.3 % of the total cell number, but because of its big size, it comprises about 70 % of the total carbon uptake in the chemocline. Since the early Holocene (10.5–8 cal kyr BP), PSB and GSB have shown long-term changes in their abundance and relative dominance (Wirth et al. 2013).

Whereas *Cryptomonas phaseolus* dominates the eukaryotic primary producers in the anoxic zones, ciliates and choanoflagellates represent the eukaryotic heterotrophs in this zone. Zooplankton and fish are abundant in the mixolimnion of the lake and are linked via food web to microbial plume in the chemocline. In short, the Lake Cadagno has been studied for >30 years for its prokaryotic diversity, especially its microorganisms and the related sulphur cycle. The study also highlights the discovery of novel eukaryotic microbes and provides an overview of phytoplankton and zooplankton populations and fishes of this lake.

Lake La Cruz (Chap. 8) Laguna de la Cruz (Lake La Cruz) is a small and deep karstic meromictic waterbody located in central-eastern Spain. Meromixis in the lake has a biogenic origin supported by the high relative depth and wind protection by its steep walls. Calcium bicarbonate and magnesium bicarbonate are the main dissolved salts. The monimolimnion permanently spans from ca. 18–21 m to the lake bottom. The water strata above the monimolimnion are thermally stratified from April to October. The chemocline (16–18 m) splits mixolimnion from monimolimnion. The concentrations of both inorganic phosphorus and nitrogen are high in deep layers.

The development of phytoplankton is related to the physico-chemical gradient: carbon fixation rates are the highest in oxic-anoxic interface during the stratification period. Picocyanobacteria form a deep chlorophyll maximum below the metalimnion. Photosynthetic sulphur bacteria, both purple and green, thrive mainly in the anoxic yet sulphide-poor water of the hypolimnion. Dark inorganic carbon fixation by the chemolithoautotrophic bacteria contributes notably to the total carbon fixation in the lake, although oxygenic photosynthesis dominates. Most of the primary production sinks to the anoxic waters, unconsumed by metazooplankton (Camacho et al. 2001). Later, rotifers and, to a lesser extent, *Tropocyclops prasinus*, a copepod, dominate.

Some anaerobic ciliates partly consume the sinking organic matter. The lake sediment forms a dark layer that annually alternates with a white layer of calcium carbonate, resulting from the latter's massive precipitation in summer because of carbonate oversaturation and increase in pH due to phytoplankton photosynthesis. The lake sediment thus forms palaeoecological archives, and, together with the remains of organisms, photosynthetic pigments, lipids, pollen, etc. facilitate verifying the lake's past ecological conditions and how land use and climate could have affected it.

Open Pit Lakes (Chap. 9) Pit lakes are the product of open-pit, surface mining excavations that are deep and intersect the local water table. Many pit lakes are meromictic, having developed in the former rock quarries and in abandoned mines. The numbers of meromictic pit lakes are growing worldwide as the former pit-mining operations are being closed (and flooded) and new open pit mines are being quarried (Castendyk and Eary 2009).

Earlier, almost all pit lakes were considered to become meromictic because of their limited surface area and large relative depth. Studies now reveal that geochemical variability of water (i.e. total dissolved solids, chemical composition and pH) plays an important role in lake stratification. Hence, the meromixis in existing pit lakes is often due to crenogenic processes and occasionally also to biogenic causes. Meromixis may also result from the treatment of mine drainage within the initially holomictic pit lakes. Many countries now entail an ameliorated management of meromictic pit lakes.

The three meromictic pit lakes reviewed here are (1) Cueva de la Mora, Spain; (2) the disappearance and recovery of meromixis in Lake Goitzsche, Germany; and (3) the Island Copper Mine pit lake, Canada, where meromixis was created deliberately in order to use the monimolimnion as a disposal and treatment site for the harmful mine waste and acid rock drainage. The authors briefly discuss the management options and predictive modelling as essential tools.

Pit lake Cueva de la Mora (Spain) was intensively investigated from 2008 to 2014 for its physical limnology, water chemistry, sediment biogeochemistry and microbial ecology. The lake's connection to the former underground mine galleries and shafts is crucial for the lake's stratification. The lake has usual mixolimnion, a 2-m thick transitional chemocline and a 30-m thick monimolimnion, which is sharply stratified.

Use of stable isotopes allowed inferring the hydrological dynamics, residence times and evaporation rates of the different waters layers of the lake (Sánchez-España et al. 2014). The water in the lake is very acidic having high sulphate concentration due to oxidative dissolution of pyrite and other metal sulphides. The concentrations of base metals Zn, Mn, Co, Ni, Cd and metalloids (e.g. As) are quite high. Most of these can be toxic to aquatic microorganisms.

The deeper layers of the lake are rich in CO_2, which is related to both the dissolution of trace carbonates, e.g. calcite, dolomite and ankerite, and decomposition of dead phytoplankton in deeper layers. The continuous release of carbon dioxide in this acidic lake is potentially dangerous. The CO_2 concentrations in the deep waters of the lake are high, varying from 1600 to 2600 mg L^{-1} CO_2, which amounts to a total gas pressure of ca. 2.3 bar. To avoid any risk of limnic eruption, a degassing strategy may be needed.

Three essential elements (iron, sulphur and carbon) in the lake are recycled in the pelagic redoxcline, which is a thick turbid layer at 10–20 m depth stratum. This layer comprises bacterially induced metal sulphides, with a little of iron colloids. The primary production of microalgae provides organic compounds for the growth of bacterial consortia, including iron oxidisers, iron reducers, sulphur oxidisers and sulphate reducers. Phototrophic organisms in the mixolimnion of the lake comprise filamentous green algae, diatoms and *Chlamydomonas* sp. The acidic nature of the oxic mixolimnion explains the absence of higher organisms. Lake's food web is restricted to microorganisms.

Lake Goitzsche (Germany) is a lake where both the disappearance and reformation of meromixis have been studied. The lake was formed in a former lignite mine about 100 km southwest of Berlin. The initial formation of the monimolimnion was probably caused by (1) the inflow of groundwater having high concentrations of dissolved solids and (2) the local morphology of the lake basin. As a result, chemically distinct layers were formed and persisted in the local depressions.

A flood in 2002 caused an uncontrolled flooding and a 7-m rise of the lake level. Prior to the flood, the monimolimnion remained unaltered by the seasonal overturn of the upper layer. Due to a high load of suspended solids, the flood water entering the lake was denser than the monimolimnion water, and it partly replaced monimolimnion. Deep recirculation of the lake took place in the winter after the flood (2002/2003). Monimolimnion was, however, re-established during summer stratification next year (2003) from the groundwater flowing into the lake. The thermocline safeguarded the newly evolved chemocline from the turbulent disturbance caused by wind-induced currents. The resulting stability prevented recirculation during the winter of 2003/2004. During summer of 2004, the conductivity and density continued to rise in the monimolimnion due to the groundwater inflow. Microbial decay and dissolution of suspended minerals in the monimolimnion may also have contributed to its stability. The chemocline rapidly recovered to the previous level.

Island Copper Mine Pit Lake (Canada) was created during 1996–1998 after the mine closure by capping the sea water in the mine with a 7-m thick layer of brackish water. The waste rock surrounding the pit generated acid and required long-term treatment after the mine closure. The monimolimnion of the lake was used to deposit and treat harmful mine waste. For creating a stable meromictic lake, sea water was diverted into the open mine. Freshwater from the nearby river was then pumped on top of the sea water.

Management measures were planned to both stabilise the lake and cut down the long-term operational costs. The upward movement of the injected acid rock drainage led to lateral disparities in the thickness of the mixolimnion. The injection of acid rock drainage also increased volume of the upper monimolimnion. The risk arose that the mixolimnion could become too thin to survive and serve as an effective cap, to prevent metal-enriched water from the upper monimolimnion reaching the lake surface.

Fertilisation of the mixolimnion with liquid ammonium phosphate and urea ammonium nitrate (UAN) stimulates the growth of algae so that the metals could be removed from the water column by adsorption onto the algae, which then apparently sink on becoming senescent. The year-round fertilisation of algae and adsorption of metals onto algae caused a significant drop in metal concentrations in the mixolimnion. The remaining metal concentrations in the mixolimnion of ca. 0.02 mg L^{-1} zinc, 0.005 mg L^{-1} copper and 0.003 mg L^{-1} cadmium were all three well below their permitted values (Pelletier et al. 2009) for waters. Undoubtedly, the adsorption of metals onto the growing algae and then the sinking of dead or senescent algae worked well.

Management Options for Meromictic Pit Lakes. Because of generally poor quality of water in the pit lakes, environmental regulations and mining industry standards require their management, from both chemical and biological viewpoints. Meromixis may need maintaining for stabilising stratification for preventing the hazardous substances from the monimolimnion reaching the mixolimnion. For stable meromictic conditions, processes sustaining meromixis must be ascertained. Uninterrupted introduction of freshwater into the mixolimnion will preserve the density difference between the monimolimnion and mixolimnion. However, accumulation of dissolved gases such as CO_2 in monimolimnion may destabilise stratification and, in extreme instances, cause a limnic eruption similar to that in the naturally meromictic lakes Nyos and Monoun (Cameroon; Kling 1987). A controlled removal of CO_2 (degassing) from the monimolimnion should be considered to avoid such disasters. To maintain meromixis, degassed water can be redirected into the monimolimnion using devices similar to hypolimnetic aerators, used in lake restoration (Cook et al. 2005).

Purging the meromixis may become necessary if (1) it cannot be stabilised, (2) if its stabilisation by active management measures is too expensive, and if (3) the prevention of flow from monimolimnion into groundwater is not practicable. The lakes may be destratified by blowing compressed air at their bottom. In order to prevent contamination of the mixolimnion by substances enriched in the monimolimnion before destratification, the selective withdrawal of monimolimnetic water and its treatment are also an option. The addition of oxygen to the monimolimnion using hypolimnetic aerators or deep oxygen injection systems to remove substances by oxidation can prevent adverse effects in the mixolimnion. Moreover, if the monimolimnion contains ferrous iron, as in many mine pit lakes, the subsequent precipitation of ferric iron and associated sorption of trace metals and metalloids will benefit the water quality. Adding iron or aluminium flocculants can also remove substances from the monimolimnion without damaging the water quality of mixolimnion. If necessary, this can be combined with capping the sediment.

Modelling of Meromictic Pit Lakes. Hydrodynamic models coupled with geochemical and groundwater models are a good decision-making tool that allows the planners and stakeholders to assess the risks before starting a pit lake. The available instruments include simple 1-D models that can explore the significance of specific processes that cause meromixis in small lakes. Large pit lakes with intricate shoreline are more likely to be subjected to horizontal variability in heating, wind speed and water inputs. Such lakes are, therefore, better characterised by 2-D models. Use of 3-D models for the management of meromictic pit lakes with complex geometry is also considered as relevant (Vandenberg 2011).

Modellers must consider some basic questions relating to physical limnology before they develop and deploy geochemical models. These questions relate to:

- The vertical stability within the mixolimnion under the effect of regular storm and wind
- The magnitude of landslide required to induce whole-lake mixing and the prospects of such an event to happen
- The temperature, salinity or injected wastes that can generate buoyant plumes leading to circulation in the monimolimnion, destabilising the water column

If the physical conditions are well defined, geochemical models can provide answer to questions vis-à-vis the monimolimnion. These comprise the residence time, chemistry and toxicity of the predicted monimolimnion and changes herein over time. Also important here are the concentrations and partial pressures of O_2, CO_2, CH_4 and H_2S and sulphate reduction, sulphide mineral precipitation and methanogenesis in the monimolimnion. Finally, it is important to assess the conditions under which dissolved gases within the monimolimnion may pose a threat to the surface ecosystem and lake users in the immediate vicinity.

Conceptual model of a meromictic pit lake should include the processes creating and stabilising the density difference between monimolimnion and mixolimnion. Processes associated with nutrient loading and biological productivity are also important in pit lakes. Plankton influences the settling of organic matter, turbidity and light extinction and adsorbs trace metals onto its organic surfaces. Models allow mine managers to compare different lake-filling strategies and management options.

It is important to remember that alterations in boundary conditions can cause significant errors in model predictions. For example, climate change can directly affect the energy, heat and water balances of the lakes and would be important for limnological modelling of pit lakes that do not achieve steady-state lake levels for long periods. Landslide impacts on stratification need to be simulated since pit-wall breakdowns often happen in post-mining pits. The impact of potential extreme scenarios such as strong storm events needs to be examined in modelling studies. Information on changes in the water balance is needed to predict long-lasting droughts, floods, temperature fluctuations and evaporation and precipitation in response to climate change. Regional climate models need to predict changes in water balance and meteorological inputs and adjust accordingly.

Meromictic Lakes at Tropical Latitudes (Chap. 10) Perennially high temperatures and high irradiance levels in tropics create conditions that markedly diverge from those prevailing in the temperate regions. Meromixis in tropical waters does not require a salinity gradient to stabilise the water column against thermal convection. In other words, the deep tropical lakes are usually quite stable without such a salinity gradient. But similar to temperate meromictic lakes, tropical lakes also accumulate nutrients in their anoxic monimolimnia so that the primary production rates in these waters are generally higher than in temperate lakes. The factors imparting stability to stratification are rates of evaporation, vertical water exchange and wind patterns. However, the responses to climate change may vary. Physical, chemical, biological and microbiological studies in the large lakes reviewed are those from the East African lakes Tanganyika and Malawi and Lake Matano in Indonesia, with a brief mention to other tropical meromictic lakes.

Lake Tanganyika is the second deepest lake (1470 m) in the world. The lake is meromictic from about 100–200 m depth, depending on the basin and season. Below this depth, the lake is permanently anoxic and relatively rich in nutrients. The large size of the lake and its complex physical limnology are important for the functioning of its ecosystem. Seasonal upwelling is important for the recycling of nutrients and sustaining food webs. The lake fauna is characterised by high diversity and endemicity. Species richness in the littoral zones is especially high. The pelagic food web is a simple one with a few species and trophic levels.

The effects of climate warming may be very important for the deep tropical stratified lakes. In Lake Tanganyika, since 1913, the average deep-water temperatures (c. 1000 m) have increased by 0.2 °C, while the average water temperatures in upper 100 m increased by almost 1 °C. This differential heating has increased the density gradient, restricting vertical mixing and reducing nutrient fluxes to the photic zone, as well as decreasing primary production rates. Oxygen concentration and penetration depth into the water column have decreased in the lake. Concomitantly, H_2S accumulation in the monimolimnion has increased, as SO_4 reduction rates have increased. Since H_2S is now found closer to the surface than earlier in the century, the inhabitable depth range of deep-dwelling species has reduced (Verburg et al. 2003). Air temperature rise is likely to continue to increase the stability of the lake's water column. The lake annual productivity will be diminished if nutrient loading from deep waters and depth of oxygen penetration is reduced. Potential fish yields from Lake Tanganyika are also likely to decline.

Lake Malawi (Nyasa) is the fourth largest freshwater lake in the world by volume. Together with lakes Tanganyika and Baikal, Lake Malawi is one of the oldest lakes in existence. The lake is known as a hotspot of biodiversity, with almost 1000 endemic species of fish alone. It has both a permanently stratified temperature gradient and a weak salinity gradient. Meromixis in Lake Malawi has resulted in permanent bottom-water anoxia.

Despite its low concentrations of nutrients, organic carbon and chlorophyll, Lake Malawi's has relatively high estimated rates of primary productivity (mesotrophic to eutrophic, Bootsma and Hecky 2003). There appears to be a tighter coupling of primary production, secondary production and eventually fish production, i.e. 'trophic efficiency' in Lake Malawi is greater than in temperate lakes (Guildford et al. 2000). Hypotheses attempting to explain this higher trophic efficiency include better nutrient concentrations in Lake Malawi phytoplankton, which has fairly balanced N:P ratios relative to those in some temperate lakes (Guildford et al. 2000), and has better temperatures the year round for fish growth (Boostsma and Hecky 2003).

Lake Malawi has only four prevalent zooplankton species, a calanoid copepod (*Tropodiaptomus cunningtoni*), two cyclopoid copepods (*Mesocyclops aequatorialis aequatorialis* and *Thermocyclops neglectus*) and a cladoceran (*Diaphanosoma excisum*). The lake is characterised by great diversity of fish species as already mentioned (~1000 fish species, mostly endemic) (Snoeks 2000). The penetration of oxygen below the thermocline in Lake Malawi allows the subsistence of bottom-dwelling fauna, notably *Chironomidae*. Emergence of adult Chironomids results in dense clouds of midges, often carried hundreds of metres into the air.

Lake Matano on the Sulawesi island of Indonesia is a relatively small lake (161 km^2), but being 590 m deep, it is the seventh deepest lake on the planet. The lake is meromictic, with the persistent pycnocline, situated at ~100 m depth, and has an oxic surface layer separated from anoxic bottom waters. The lake has been meromictic for at least the past 130 years, and the renewal of the deep water is reckoned to occur every 200–1000 years (Katsev et al. 2010).

Lake Matano's deep and permanently anoxic waters contain high concentrations of dissolved ferrous (Fe^{2+}) iron, a feature exceptionally rare today but characteristic of the deep ocean waters. The lake waters are exceptionally transparent. Light penetration to 120 m depth allows photosynthesis by a low-light-adapted community of microorganisms that perform anoxygenic photosynthesis. A significant fraction of microbial primary production in the lake is associated with photoferrotrophic bacteria in the upper part of the anoxic water column. These bacteria produce Fe oxides instead of oxygen, a pathway that has been assumed to have dominated the productivity of the ancient Archean Ocean (Crowe et al. 2008). The pycnocline of Lake Matano is characterised by significant concentrations of bacterial Chl *e*, which underlies a layer enriched in Chl *a*. The redoxcline region of Lake Matano may also support several other rare biogeochemical pathways, such as anaerobic oxidation of methane with solid-phase ferric iron as an electron acceptor (Crowe et al. 2011).

Mono Lake (USA) (Chap. 11) Mono Lake (USA) is a hypersaline meromictic lake with sodium as major cation and carbonate, chloride and sulphate as major anions. The levels of soluble reactive phosphorus are high, and concentrations of inorganic nitrogen variable. The lake alternates between meromictic and monomictic states depending on volume of freshwater introduced during snowmelt. Three episodes of meromixis occurred in about three and half decades between 1979 and 2013.

The biological communities of the Mono Lake, as in many meromictic lakes, can be separately described for microbial part of the food web and classical grazer food web. Most of the 16S rRNA gene sequences of microbes from the lake are in five major lineages: α- and γ-proteobacteria, bacteriodetes, high G+C Gram positive (actinobacteria) and low G+C Gram positive bacteria (*Bacillus/Clostridium*) (Humayoun et al. 2003). The microbial communities in chemocline and monimolimnion are more diverse than epilimnetic microbial communities and have a broader phylogenetic representation. Biogeochemical processes driven by bacteria fluctuate both vertically and seasonally; they include aerobic and anaerobic methane oxidation and a coupling between anaerobic methane oxidation and dissimilatory sulphate reduction.

The phytoplankton abundance varies seasonally, with high densities from autumn to winter and maximum in early spring and low values in summer due to grazing by the brine shrimp *Artemia monica*, both a dominant and often the sole species of zooplankton. The shrimps hatch in late winter and spring from the cysts produced in previous years; they produce in summer a second generation ovoviviparously. In autumn, the *Artemia* population declines, because of predation by migratory grebes, decrease in water temperature and their senescence. In near-shore habitats, the alkali fly (*Ephydra hians*) is abundant seasonally. Both the brine shrimp and alkali fly larvae provide food for a large number of birds including breeding colonies of the California gull (*Larus californicus*), migratory eared grebes (*Podiceps nigricollis*) and some other bird species. The lake has no fish probably because of high salinity.

The lake mixes completely in winter, and subsequently chlorophyll levels increase in the spring period. However, the response of brine shrimp to these variations in algal biomass is weak. Thus, the bottom-up control of the food web due to mixing event and release of nutrients from the monimolimnion is not very high.

Mexican Meromictic Lakes (Chap. 12) The results of a survey of meromictic lakes in Mexico are summarised. The study demonstrates that meromictic lakes in Mexico display a mosaic of environmental and ecological features. There are reports of only a few meromictic lakes in Mexico. The lakes range widely in latitude (≈16–26 °N), longitude (≈87–102 °W) and altitude (≈71–700 m a.s.l.). The lakes (a) are located in the coastal region where the freshwater and sea water in karst zones interact; (b) lie in semiarid and arid regions where annual water budget is negative, i.e. evaporation far exceeds precipitation; and (c) are deep with small surface area due to which they do not completely circulate. The lakes vary widely in area and depth, with turbid or transparent mixolimnion and generally a turbid chemolimnion. Some meromictic lakes exhibit thermal inversion or subsurface maximum or double thermocline; there are one to three haloclines with moderate to quite large salinity ranges. The dissolved oxygen profiles are typically clinograde or with subsurface maxima. The pH in the mixolimnion is alkaline, but in the monimolimnion it varies from acidic to alkaline. Nutrients concentrations vary from low to high. Notably, the documentation of meromictic lakes in Mexico is not yet complete. Moreover, as we discussed earlier in Chap. 1, the identification of a lake as meromictic is not always correct. Lake can be identified as meromictic based on short sampling period, but longer study period could reveal irregular circulation patterns. Therefore, long-term sampling programmes are required to identify the mixing regime of lakes in a region and to reaffirm these phenomena over years.

13.4 Final Remarks

Meromixis is now viewed much more dynamic than it was some years ago. At the same time, we still can consider the definition of meromixis by Hutchinson (1957) 'A lake in which [a chemically different] water remains partly or wholly unmixed with the main water mass at the circulation periods is said to be **meromictic'** as valid definition. The major progress is not made to improve the definition of meromictic lakes but in the overall studies to understand the different processes in meromictic lakes. We understand that hydrophysics and factors that govern mixing are much more complex than we expect (Chap. 2). We know now about various biogeochemical reactions that can sustain meromixis (Chap. 3). We defined regions on the Earth that tend to host such lakes (Chap. 1). With the advancement of limnology and more frequent and extensive sampling of lakes in different geographical regions, the number of meromictic lakes is likely to increase. Also, the number of mining pit lakes and the associated meromictic lakes is likely to increase due to exploitation of the earth for metals and even disposal of mining waste products.

The mixing regime of many lakes is not consistent over the years on a long-term basis; there are also episodes of meromixis when anoxic bottom sediments remain well preserved. Paleolimnological studies of the lake sediment reveal that many important proxies of lake biogeochemistry, productivity and species composition are 'buried' in lake sediment and well preserved and intact for years. Therefore, the paleolimnological studies of lake sediments are very essential to both understand the dynamics of lakes in the past and to predict the future conditions of many lakes.

Meromixis can be considered as an undesirable mixing regime for the water quality. Indeed, if large volumes of the water become anoxic, benthic fauna and fish will die en masse or suffer high mortalities. Moreover, the monimolimnion may trap such dissolved gases as CO_2 and H_2S in very high concentrations, and gas pressure may increase to the point that there is a limnic eruption of these gases. Such catastrophic events as mentioned above have happened in the recent past in lakes Nyos and Monoun (in Cameroon, Africa: Kling 1987) catastrophic for the people inhabiting the lake surroundings. Moreover, such a calamitous overturn will impact most biota and water quality in the mixolimnion of lakes. In this case, the degassing strategy is needed to reduce the risk of limnic eruption. Thus, it is important to have tools to predict processes that may result in either the mixing of meromictic lake or switch lake from holomictic to meromictic regime. Such predictions are usually made with the numerical simulations. The development of mathematical models and use of computers significantly have greatly increased our ability to simulate the lake dynamics and predict the future conditions. Early modelling attempts were focused using on 1-D models and perhaps only rarely on 2-D models, but nowadays 3-D models for lake circulation are already in wide use and their further sophistication is going on.

Notably, many meromictic lakes are man-made, small lakes created for the treatment and storage of harmful mine wastes and waste water. Even though such lake systems may have quite peculiar physical, chemical or biological features, they still can be viewed and analysed as lakes. As many of these lakes are usually owned by commercial people or by environmental governmental agencies, they provide a great opportunity to study the processes that sustain meromixis stably. The prediction of stability of meromixis in such lakes is of great importance as the failure to sustain meromixis may lead to local and regional episodes of pollution and serious water-quality problems. Thus, the business and environmental interests towards such lakes as water management tools can also drive the progress in limnology of meromictic lakes.

The ecology of meromictic lakes is quite complex. Microbial communities and microbially mediated biogeochemical reactions play important role in cycling of nutrients in these lakes. Even though oxygenic photosynthesis in the oxic waters is the main process to fix inorganic carbon (DIC) into organic matter, processes such as anoxygenic photosynthesis, performed by phototrophic sulphur bacteria in anoxic layers, and chemolithoautotrophy (both aerobic and anoxic layers) are exceptionally important for fixation of inorganic carbon in meromictic lakes. Sulphur-oxidising bacteria can achieve high carbon fixation rates; the same is true for photoferrotrophy and iron-based chemolithotrophy. Other important process is anaerobic oxidation of ammonia by obligately anaerobic chemolithoautotrophs. These widely observed activities in meromictic lakes to fix carbon link the carbon cycle with the sulphur (chemolithotrophic sulphur oxidisers and anoxygenic photosynthetic bacteria and nitrogen (nitrifiers and anammox bacteria) and iron (photoferrotrophs and chemolithoautotrophs) cycles (Chap. 4).

The classical grazer food web in pelagics of meromictic lakes is often truncated, fish being absent. However, the diversity of chemical setting and environmental and climate conditions observed in meromictic lakes make it difficult to draw a uniform concept of food web in these lakes. Very often the drivers of food-web structure and interactions in a given lake are local chemical conditions or external climatic forcing. What is exceptionally important for the food-web structure is a shift from meromictic conditions to complete mixing (holomixis) of lakes. When a lake experiences such a shift, the food web is altered under stress.

Finally, in this book we have attempted to present state-of-the-art information on some of the general physical, chemical and biological features of meromictic lakes. Also, we demonstrate the variety of processes that promote meromixis in different lakes. We provide a broad picture of meromixis in lakes in both warm tropical and cold regions, in man-made pit lakes and euxinic marine lakes and in freshwater as well as hypersaline lakes. We can state that meromictic lakes are found in all continents and in different climate and geographic zones. Mostly physical and chemical properties but also microbial and biochemical reactions are important to support meromixis in lakes. At the same time, food-webs and biota respond to the permanently stratified, anoxic deep water with contrasting chemistry of layered water column. We encourage all interested in lake ecology to study meromictic lakes as they are also perfect natural laboratories to improve our understanding of the limnology of inland waters.

Acknowledgement EZ was partially supported by the Council on grants from the President of the Russian Federation for support of leading scientific schools (grant NSh-9249.2016.5).

References

Boehrer B, Schultze M (2008) Stratification of lakes. Rev Geophys 46:RG2005. doi:10.1029/2006RG000210

Bootsma HA, Hecky RE (2003) A comparative introduction to the biology and limnology of the African Great Lakes. J Great Lakes Res 29:3–18

Camacho A, Erez J, Chicote A et al (2001) Microbial microstratification, inorganic carbon photoassimilation and dark carbon fixation at the chemocline of the meromictic Lake Cadagno (Switzerland) and its relevance to the food web. Aquat Sci 63:91–106

Castendyk DN, Eary LE 2009 The nature and global distribution of pit lakes. In: Castendyk DN, Eary LE (eds) Mine pit lakes: characteristics, predictive modeling, and sustainability. Society for Mining, Metallurgy, and Exploration, Littleton, CO, pp 1–11

Ciglenečki I, Janeković I, Marguš M et al (2015) Impacts of extreme weather events on eutrophic seawater ecosystem (Rogoznica Lake, Adriatic coast). Cont Shelf Res 108:144–155

Cooke GD, Welch EB, Peterson SA, Nichols SA (2005) Restoration and management of lakes and reservoirs, 3rd edn. Taylor & Francis, Boca Raton, FL

Crowe SA, Jones C, Katsev S et al (2008) Photoferrotrophs thrive in an Archean Ocean analogue. Proc Natl Acad Sci 105:15938–15943

Crowe SA, Katsev S, Leslie K et al (2011) The methane cycle in ferruginous Lake Matano. Geobiology 9:61–78

Dietz S, Lessmann D, Boehrer B (2012) Contribution of solutes to density stratification in a meromictic lake (Waldsee/Germany). Mine Water Environ 31:129–137

Findenegg I (1935) Limnologische Untersuchungen im Kärntener Seengebiete. Ein Beitrag zu Kenntnis des Stoffhaushaltes in Alpenseen. Int Rev Gesamten Hydrobiol Hydrogr 32:369–423

Guildford SJ, Bootsma HA, Fee EJ et al (2000) Phytoplankton nutrient status and mean water column irradiance in Lakes Malawi and Superior. Aquat Ecosyst Health Manag Soc 3:35–45

Humayoun S, Bano N, Hollibaugh JT (2003) Phylogenetic composition of the bacterioplankton from an alkaline, hypersaline lake, Mono Lake, California. Appl Environ Microbiol 69:1030–1042

Hutchinson GE (1957) A treatise on limnology, Geography, physics and chemistry, vol 1. Wiley, New York, NY

Kalugin I, Darin A, Rogozin D, Tretyakov G (2013) Seasonal and centennial cyclesof carbonate mineralization during the past 2500 years from varved sediment in Lake Shira, South Siberia. Quat Int 290–291:245–252

Katsev S, Crowe SA (2015) Organic carbon burial efficiencies in sediments: The power law of mineralization revisited. Geology 43:607–610

Katsev S, Crowe SA, Mucci A et al (2010) Mixing and its effects on biogeochemistry in the persistently stratified, deep, tropical Lake Matano, Indonesia. Limnol Oceanogr 55:763–776

Kling GW (1987) Seasonal mixing and catastrophic degassing in tropical lakes, Cameroon, West Africa. Science 237:1022–1024

Overmann J, Sandmann G, Hall KG, Northcote T (1993) Fossil carotenoids and paleolimnology of meromictic Mahoney Lake, British Columbia, Canada. Aquat Sci 55:1015–1621

Pelletier CA, Wen ME, Poling GW (2009) Flooding pit lakes with surface water. In: Castendyk DN, Eary LE (eds) Mine pit lakes. Society for Mining Metallurgy and Exploration, Littleton, CO, pp 187–202

Sánchez-España J, Diez-Ercilla M, Pérez Cerdán F, Yusta I, Boyce AJ (2014) Hydrological investigation of a multi-stratified pit lake using radioactive and stable isotopes combined with hydrometric monitoring. J Hydrol 511:494–508
Snoeks J (2000) How well known is the ichthyodiversity of the large East African lakes? Adv Ecol Res 31:17–38
Vandenberg J (2011) Use of water quality models for design and evaluation of pit lakes. In: McCullough CD (ed) 2011 Mine Pit Lakes: closure and management. Australian Centre for Geomechanics, Perth, WA, pp 63–80
Verburg P, Hecky RE, Kling H (2003) Ecological consequences of a century of warming in Lake Tanganyika. Science 301:505–507
Vollmer MK, Weiss RF, Bootsma HA (2002) Ventilation of Lake Malawi/Nyasa. In: Odada EO, Olago DO (eds) The East African great lakes: limnology, paleolimnology and biodiversity. Kluwer, Dordrecht, pp 209–233
von Rohden C, Boehrer B, Ilmberger J (2010) Evidence for double diffusion in temperate meromictic lakes. Hydrol Earth Syst Sci 14:667–674
Wetzel RG (2001) Limnology: lake and river ecosystems. Academic Press
Wirth SB, Gilli A, Niemann H et al (2013) Combining sedimentological, trace metal (Mn, Mo) and molecular evidence for reconstructing past water-column redox conditions: the example of meromictic Lake Cadagno (Swiss Alps). Geochim Cosmochim Acta 120:220–238

Index

© Springer International Publishing AG 2017

R.D. Gulati et al. (eds.), *Ecology of Meromictic Lakes*, Ecological Studies 228,
DOI 10.1007/978-3-319-49143-1

MIX
Papier aus verantwortungsvollen Quellen
Paper from responsible sources
FSC® C105338

If you have any concerns about our products,
you can contact us on
ProductSafety@springernature.com

In case Publisher is established outside the EU,
the EU authorized representative is:
Springer Nature Customer Service Center GmbH
Europaplatz 3, 69115 Heidelberg, Germany

Printed by Libri Plureos GmbH
in Hamburg, Germany